I0047000

George Chrystal

# Introduction to Algebra

For the Use of Secondary Schools and Technical Colleges

George Chrystal

**Introduction to Algebra**
*For the Use of Secondary Schools and Technical Colleges*

ISBN/EAN: 9783744645805

Printed in Europe, USA, Canada, Australia, Japan

Cover: Foto ©berggeist007 / pixelio.de

More available books at **www.hansebooks.com**

# INTRODUCTION

TO

# ALGEBRA

FOR THE USE OF

SECONDARY SCHOOLS AND TECHNICAL COLLEGES

BY

G. CHRYSTAL, M.A., LL.D.

HONORARY FELLOW OF CORPUS CHRISTI COLLEGE, CAMBRIDGE;
PROFESSOR OF MATHEMATICS IN THE UNIVERSITY OF EDINBURGH

LONDON
ADAM AND CHARLES BLACK
MDCCCXCVIII

THIS BOOK IS DEDICATED

BY THE AUTHOR

TO

DAVID RENNET, LL.D.

IN MEMORY OF

HAPPY HOURS SPENT IN HIS CLASS-ROOM

IN DAYS OF OLD

# PREFACE

THERE are many signs that the departure of the old-fashioned English Elementary Algebra is at hand, not the least of them being the appearance of numerous competitors for the heirship. That I should have entered into a suit where there are already so many litigants is partly due to recent changes in the University system of Scotland, which cannot be discussed here, and partly to a long-standing promise to my publishers to provide an introduction to my larger text-book, which is suitable only for the highest classes in schools, and which contains too little of practical application, and not enough of graphical illustration for the purposes of a technical college.

It is somewhat surprising to me to find myself in the rôle of a reformer of the methods of elementary instruction—*non ita nutritus.* I began to teach in the old-fashioned way, and have been driven, simply by the stress of experience, until I find myself more or less at one in most of their positions with the reforming party of mathematical teachers, whether academic or (I suppose I must say) technical. The experience in question, extending now over more than twenty years, has been gained in laboratory work, in examining schoolboys and entrants to the universities, and (until 1893) in teaching the junior mathematical class in a Scottish University, which, like the pass work in an English University, was essentially the work of a schoolmaster. I have therefore had a better opportunity than most of learning exactly where the old methods were defective.

The English text-books of Algebra in vogue during the latter part of this century have tended to degenerate into a mere farrago of rules and artifices, directed to the solution of examination puzzles of a somewhat stereotyped character having little visible relation to one another and still less bearing on practice. If general principles appeared at all, they were usually huddled apologetically into a chapter of "Miscellaneous Theorems,"—an arrangement which we might parallel by building a man of muscle and tendons, etc., and putting all his bones into his coat-tail pocket. It has been often and loudly complained that Algebra thus taught will not bear the superstructure of a university course, and is totally useless in practice. My own experience has convinced me that both complaints are in great measure just.

The present attempt to remedy these evils is a compromise, destined, I hope, to be superseded presently by something better. Nothing but a compromise is at present practicable, because *natura non agit per saltum.*

In the first place, I have kept the fundamental principles of the subject well to the front from the very beginning; I may instance the treatment of the derivation of equations in Chapter VI., a subject usually dealt with as if it were a separate science. At the same time I have not forgotten, what every mathematical (and other) teacher should have perpetually in mind, that a general proposition is a property of no value to one that has not mastered the particulars. The utmost rigour of accurate logical deduction has therefore been less my aim than a gradual development of algebraic ideas. Sometimes I have contented myself with stating a general proposition, and then proceeded to apply it, referring elsewhere for the demonstration; quite as often I have led up to the general principle by means of a series of suggestive examples, and finally stated it with or without formal demonstration; but in all cases the clarifying principle has been insisted upon. A mathematical truth is not made part of the mental furniture of a pupil merely by furnishing him with an irrefragable demonstration; it is not until he has tried it in particular cases, and seen not only where it succeeds, but where it fails to apply, that it becomes a sword loose in the scabbard and ready for emergencies. The rigorous demonstration is but the last polish given to the blade. It is better now and then to lead a learner to feel the need of a weapon before we place it in his hands. Accordingly, it will be noticed that towards the end of some of the Exercises in this book little problems are given which more or less anticipate the succeeding book-work.

I have gone as far as I dared, in the face of existing examination programmes, in cutting out book-work which has nothing to do with elementary theory or with practice. In particular, I have excluded the treatment of subjects that depend on the theory of limits and convergency. The premature introduction of such subjects with loose and even misleading or false demonstrations has been one of the most glaring defects of our elementary mathematical text-books. In this respect it is scarcely too much to say that many of them are half a century behind the age. Not only is teaching of this kind a waste of time, but it is an absolute obstruction to further progress. How deplorable the results are is well known to every examiner and university teacher.

In arranging the Exercises I have acted on a similar principle of keeping out as far as possible questions that have no theoretical or practical interest. Many of the Exercises were constructed expressly to illustrate theoretical points; and there will be found a larger infusion than usual of problems that occur in practice, and of pieces of work that the student will meet with later on in the Applications of Algebra to Geometry and Physics. At the same time I have borne in mind that we are expected to

> wisely tell what hour o' th' day
> The clock doth strike by Algebra,

and to exhibit other little accomplishments of the kind at the call of the ever-present examiner. My chief object in this matter has been variety; if any one using the book finds defect in quantity, there is a superabundance of sources from which he can supplement.

A prominent feature of the present book is the constant use of graphical illustration; it is introduced in a simple form very early; and altogether about fifty pages are devoted to it exclusively. This proportion may startle some; but will not astonish those who are familiar with the tendency of the best modern teaching. The graphic method furnishes the most valuable antidote to the tendency of school algebra to degenerate into puzzle-solving and legerdemain. By the constant exercise of graph-tracing the beginner acquires through his fingers three fundamental mathematical notions, viz. the Idea of a Continuously Varying Function, the Conception of a Limit, and the Method of Successive Approximation. These notions he will find to be more valuable in the higher mathematics and in applications to practice than all the rest of his algebraic accomplishments put together. In order to get the full educative benefit of graph-tracing, it will not be sufficient merely to read the relative paragraphs. The teacher must trace the graphs before his pupils; and also cause them to work the curves out independently. To facilitate this kind of work, I strongly recommend that a blackboard, permanently ruled into small squares, like a sheet of plotting paper, should be part of the furniture of every mathematical class-room.

A word regarding the first steps in teaching Algebra. I hold, in common, I believe, with most teachers of Mathematics who have deeply considered their business, that the teaching of Algebra—that is, of the science of arithmetical operations—should commence with the teaching of Arithmetic itself. For example, the beginner should not be allowed simply to learn that 3 and 5 together make up 8, and to write mechanically the scheme

$$\begin{array}{rr} & 3 \\ & 5 \\ \hline \textit{Ans.} & 8 \end{array}$$

and the like; but ought as well to be made to write $3+5=8$, or even $+3+5=+8$. It should also be pointed out to him that $3+5=8=5+3$; that $3+(3-2)=3+3-2$; and so on. The laws herein involved need not be named to him at first by their long forbidding names; but they should be illustrated by means of concrete instances, and especially by geometric figures. After a course of this kind, extended over the earlier years of his arithmetical training, the learner should be made to state the solutions of the little problems which he works as concatenations of numerical operands and operating symbols. The next stage is to learn to generalise a problem by substituting letters or hypothetical operands for the actual numbers of earlier essays. Then, and not till then, should a book expressly dealing with Algebra be put into the pupil's hands. In brief, the best Beginner's Algebra is a good book on arithmetic in the hands of a good teacher.

It remains to acknowledge my obligations. As I have already hinted, my debt to the traditional text-book is greater than I could have wished. Of the books recently in extensive use the excellent little work by the Master of Sidney Sussex College, Cambridge, is almost the only one that treats the subject from what I consider to be the proper point of view. I have not consciously borrowed from that book, but I have profited from it as every one must that carefully studies a conscientious piece of work by another. For proof-reading and valuable criticism, I am indebted to my assistant, Mr. Charles Tweedie, and to Messrs. J. Alison, J. B. Clark, D. B. Mair, and J. A. Macdonald, friends and former pupils. To Mr. Archibald Milne, one of my students, who undertook the laborious work of checking the Answers to the Exercises, I am much indebted for the celerity and phenomenal accuracy with which he performed this task.

To the Oxford and Cambridge Schools Examination Board, and to the Cambridge Local Examinations Syndicate, I owe acknowledgment for liberty kindly accorded me to take exercises from their examination papers, of which I have taken considerable advantage.

Hints for improvement and corrections of errors, of which not a few must remain after all our care, will be gratefully received; and I take this opportunity of again thanking the many friends, some of whom I have never yet had the pleasure to meet face to face, that have given me assistance of this kind in improving my larger work.

G. CHRYSTAL.

*April* 1898.

# CONTENTS

# INDEX OF TECHNICAL WORDS AND PHRASES

*The Numbers refer to Pages*

# CHAPTER I

## GENERALISED ARITHMETIC

§ **1.** Newton, who was one of the greatest masters of Algebra, called his work on that subject *Arithmetica Universalis*, which may be freely translated *Generalised Arithmetic.* From the point of view of this chapter no better name for Algebra could have been found. In every arithmetical calculation there is a part which is special to the particular concrete subject to which it is applied. But there is also a general, or in Newton's phrase, universal part. In the first place, there are certain operations performed, viz. addition, subtraction, multiplication, or division, the laws of which are altogether independent of the particular case contemplated; and, again, it is possible to consider the particular concrete case as merely an example of a general concrete case in which the actual quantities are not definitely specified.

Consider, for example, the following simple problem. A grocer has five cases, each containing a dozen eggs, and a broken case containing only seven; how many eggs has he altogether. If a child were asked to arrange the calculation on his slate, he would probably put down something like this—

$$\begin{array}{r} 12 \\ 5 \\ \hline 60 \\ 7 \\ \hline \text{Answer } 67 \text{ eggs.} \end{array}$$

The first step towards an algebraical view of this simple matter is to distinguish the operations performed by appropriate signs (× and +), to introduce an abbreviation for the copula, or statement of equality, (=), and write

$$12 \times 5 + 7 = 67.$$

We thus recognise better than before that there are in the calculation certain **operations** with certain **operands** (12, 5, 7), the general nature of which is unaffected by the fact that we are counting eggs and not, say, chickens.

Again, we may generalise our little problem in another direction. Instead of considering five cases, each containing twelve eggs, we may consider any number of cases, say $a$, each containing any number, say $b$, of eggs, and the number in the broken case may be taken to be any number $c$; then, if $d$ be the whole number of eggs, we have, by precisely the same use of the operations as in the particular case—

$$b \times a + c = d.$$

We have now obtained a "general formula," or a so-called "algebraic formula," for solving all problems of the same concrete type as the one originally proposed. We pass back to the solution of any particular problem by substituting the appropriate values of $a$, $b$, and $c$ in $b \times a + c$, and carrying out the arithmetical operations indicated.

Ultimately the first of these generalisations, which consists in regarding Arithmetic as the science of a set of operations conducted under certain rules, is the more important in modern Algebra; but for a beginner the latter generalisation, which consists merely in replacing the operands, or numbers operated with, by letters which may stand for any number, is perhaps more important; because it is a useful mental exercise in itself, and because it leads by degrees inevitably to the other generalisation. It is, in fact, obvious that we cannot use generalised operands without using symbols of operation; and the construction of algebraic formulæ for the solution of special problems tends to bring the general nature of the arithmetical operations into prominence. Practice in this kind of exercise should be begun very early. Every arithmetical problem, almost every step in arithmetical theory, and every application of Geometry to the mensuration of figures will furnish an opportunity. A few examples are appended to the present chapter.

§ 2. The fundamental operations of Arithmetic, sometimes

called the "Four Species," are addition, subtraction, multiplication, and division, and the operands for the present may be taken to be the integral numbers 1, 2, 3, etc., together with any fractions, such as $\frac{123}{334}$, which can be formed by means of a finite number of digits.* Hereafter it will be seen that the operations and the operands of Algebra may be defined in a perfectly abstract manner; but in the meantime the learner is to attach to the operations the meanings to which he has been accustomed in arithmetic; and he may think of the operands as denoting concrete quantities of any kind with which he happens to be familiar, *e.g.* lengths of lines, sums of money, volumes or weights of matter, etc.

§ 3. The symbol for addition is always $+$; and for subtraction $-$.

For multiplication the symbol $\times$ is used. When no ambiguity is to be feared, the $\times$ is generally omitted, and the multiplicand and multiplier merely written in close succession; thus $ab$ means $a \times b$; and $3a$ means $3 \times a$. When both multiplicand and multiplier are numbers, this second notation is sometimes ambiguous: thus, 12 means $10 + 2$, and not $1 \times 2$; and $2\frac{3}{4}$ means $2 + \frac{3}{4}$, and not $2 \times \frac{3}{4}$. In such cases a dot placed as low as possible between the multiplicand and multiplier is used, thus, $1.2$ means $1 \times 2$; $2.\frac{3}{4}$ means $2 \times \frac{3}{4}$. In using the dot notation care must be taken to avoid confusion with the decimal point, which is placed higher: thus, $1.2$ means $1 \times 2$, but $1 \cdot 2$ means $1 + \frac{2}{10}$. As in arithmetic, we speak of $a \times b$ either as "$a$ multiplied by $b$" or as "the product of $a$ and $b$, or of $b$ into $a$." The multiplicand $a$ and the multiplier $b$ are often spoken of as the "factors" of the product.

Division is indicated by the symbol $\div$, viz. $a \div b$ means "$a$ divided by $b$," or "the quotient of $a$ by $b$," $a$ and $b$ being spoken of as the dividend and divisor exactly as in arithmetic. Alternative notations are $\frac{a}{b}$, $a/b$, $a : b$. The symbols / (solidus notation) and : (ratio notation) are equivalent to $\div$, with an excep-

* Any such arithmetical number is spoken of as Commensurable or Rational (in the arithmetical sense). An example of an incommensurable number is a non-terminating, non-repeating decimal, *e.g.* the ratio of the circumference of a circle to its diameter, usually denoted by $\pi$.

tion to be stated presently; and the former of these two, owing to the readiness with which it can be either written or printed, is now coming much into use. The fractional notation $\frac{a}{b}$ has certain advantages, but it is difficult to print and takes up more room than the others. We shall have occasion to remark later on that the fractional notation is not in all cases exactly equivalent to $\div$ or to $/$; it has in some cases the effect of a bracket, *e.g.* $\frac{a}{b+c}$ is not the same thing as $a \div b + c$ or $a/b + c$, but is really equivalent to $a \div (b + c)$ or $a/(b + c)$.

The fact that when the dividend and divisor are integral numbers, as in $\frac{3}{4}$, the fractional and divisional notation are not distinguishable is of no consequence, because in Algebra, whenever we regard $\frac{3}{4}$ not as a whole, but with respect to 3 and 4 separately, we regard it as a quotient, and, on the other hand, when $\frac{3}{4}$ is regarded as a whole, *i.e.* as an operand, it has the same abstract properties whether we consider it as "three-fourths" or as three divided by four: *e.g.* from either point of view $\frac{3}{4} \times 4 = 3$; and this may be regarded as the fundamental or defining property of $\frac{3}{4}$.

It may be mentioned here that $1 \div a$, or $1/a$, or $\frac{1}{a}$, where $a$ is any quantity whatever, is often spoken of as the **Reciprocal of** $a$.

§ 4. Whatever concrete or other meaning the learner may have hitherto attached to addition and subtraction, he will see that the two operations are mutually **Inverse** in the sense that, if we first add any quantity and then subtract the same, or first subtract any quantity and then add the same, the result is the same as if we had not operated at all; that is to say—

$$a + b - b = a, \quad a - b + b = a.$$

Multiplication and division are inverse to each other in exactly the same sense, viz. we have

$$a \times b \div b = a, \quad a \div b \times b = a.$$

Rightly considered, the above remark leads us to see that when addition and multiplication are fully defined by concrete interpretation or otherwise, the nature and laws of their inverses, subtraction and division, are determined (see A. Ch. I.).*

* In references A. signifies my larger work on Algebra.

§ **5.** In addition to the four species it is usual, even in arithmetic, to introduce another pair of mutually inverse operations, viz. **Involution** (Raising to a Power), and **Evolution** (Radication or Root Extraction). In the first instance, at least, these new operations are not independent of those already enumerated. Involution is in fact repeated multiplication: thus $3$, $3 \times 3$, $3 \times 3 \times 3$, $3 \times 3 \times 3 \times 3$, . . . are represented by $3^1$, $3^2$, $3^3$, $3^4$, . . . and are described as three to the first power, three to the second power or three square, three to the third power or three cube, three to the fourth power, . . . and in general $a \times a \times a \times$ . . . ($n$ factors), $n$ being of course an arithmetical integer, is contracted into $a^n$. This operation is called **Involution** or **Raising to the nth Power**; $a^n$ is called the $n$th power of $a$, $a$ to the $n$th power, or briefly $a$ to the $n$th. Also $a$ is called the **Base of the Power**; and $n$ the **Index** or **Exponent of the Power.** Hereafter we shall extend this notation to cases where $n$ is not integral, or indeed a mere arithmetical quantity at all; but it must be observed that, according to our present definition, $a^n$ has no meaning unless $n$ be an integer in the ordinary arithmetical sense.

The quantity whose $n$th power is $a$ is called the $n$th root of $a$, and is denoted by $\sqrt[n]{a}$, $a$ being called the **Radicand** and $n$ **the Order of the Root**; and the operation of deriving $\sqrt[n]{a}$ from $a$ is called **Evolution, Root Extraction,** or **Radication**; special cases are the second root or square root, written $\sqrt{a}$; the third root or cube root, written $\sqrt[3]{a}$. If $a = b^n$, $b$ being any ordinary arithmetical quantity, it is at once obvious that $b$ satisfies the definition of $\sqrt[n]{b^n}$. It also follows from the definition that the $n$th power of $\sqrt[n]{a}$ is $a$. From these remarks the mutual inverseness of Evolution and Involution, regarded as arithmetical operations, follows at once.

It is important to notice that, if we confine ourselves to mere arithmetical values of the radicand, and to mere arithmetical values of the root, there can only be one value of an $n$th root; and that, according to the nature of the radicand, there are two distinct senses in which the root can be said to exist. Consider, for simplicity, the case of the square root. If the radicand be the square of any commensurable number, say the square of $b$, the square root is of course $b$, and is commensurable; and if the square root be commensurable, the radicand must be the square of a commensurable number. If, on the other hand, the radicand be not the square of any commensurable number, then the square root must be incommensurable, and does not exist in the same simple sense as before. It will readily be seen, however, by considering a table of numbers and their squares, that we can determine a commensurable number whose square shall differ from any given number by as little as we please. We thus determine a **commensurable approximation** to the square root of any required degree of accuracy ; and it is clear, from the nature of the process, that successive approximations will differ less and less from each other, and, therefore, approach more and more nearly to one particular value. In this sense there exists always one and only one arithmetical value of the square root of every arithmetical number which is not itself the square of a commensurable number.

The same order of ideas applies exactly in the case of an $n$th root.

§ 6. In algebra, just as in arithmetic, it frequently happens that the result of a series of operations becomes itself an operand. It thus becomes necessary, especially in algebra, where the operations are symbolised, to have some means of indicating that the result of several operations is to be taken as a whole, and regarded as a single operand preceding or following a particular symbol of operation.

This is effected by inclosing the complex operand between a pair of **Brackets** or parenthesis-marks, ( ), { }, or [ ], or by drawing over it a line or "vinculum," $\overline{\qquad}$, $|\overline{\qquad}$, or $\overline{\qquad}|$. For example, if we have to represent the subtraction from 8 of the result of subtracting 2 from 4, we write $8-(4-2)$; and not $8-4-2$, which would mean quite a different thing, viz. the result of first subtracting 4 from 8, and from that result subtracting 2.

When there are complex operands within complex operands, we have to use brackets within brackets ; and in such cases it is usual to vary the forms of the pairs of parenthesis-marks for greater clearness ; thus, instead of writing $a-(b-(c-d))$, it is usual, although not absolutely necessary, to write $a-\{b-(c-d)\}$, $a-(b-\{c-d\})$, $a-(b-\overline{c-d})$, or $a-|\overline{b-|\overline{c-d}}$. It should be mentioned, however, that in mathematical work which is to be printed, the use of the vinculum should as far as possible be avoided, because it cannot be printed by means of a single type of the ordinary construction, and consequently is both troublesome to the printer and costly to the author and publisher.

In accordance with the usual convention of writing in all European languages, a chain of algebraical operations is read, as it is written, from left to right. There is a tacit understanding, *so long as the same symbol of operation or its inverse is repeated*, that all that *precedes* the operating symbol is to be regarded as a single (complex) operand. By this convention we are saved considerable complication in the use of brackets. Thus, for example, $a+b-c-d+f$ really means $[\{(a+b)-c\}-d]+f$, and $a\times b\div c\div d\times f$ means $[\{(a\times b)\div c\}\div d]\times f$. If, however, there is passage from addition and subtraction, on the one hand, to multiplication or division on the other, or *vice versa*, then the bracket cannot be omitted without affecting the meaning of the chain. Thus, for example, $a-b\times c$ means, when written with full symbolism, $a-(b\times c)$, and must not be confused with $(a-b)\times c$. For instance, $13-2\times 5=3$; but $(13-2)\times 5=55$.

When division is indicated by the use of the horizontal line, this line also performs the part of a vinculum or bracket; thus $\dfrac{a}{b+c}$ is equivalent to $\dfrac{a}{(b+c)}$, *i.e.* to $a\div(b+c)$ or $a/(b+c)$, the bracket being suppressible without ambiguity in the first notation, but evidently not in the other two.

A similar remark applies to the solidus in so far as multiplications and divisions, unbroken by intervening additions or subtractions, is concerned. Thus $a/b\times c\div d$ means $a\div(b\times c\div d)$, and not $a\div b\times c\div d$. *The solidus has, in short, the effect of a bracket upon an immediately following chain of multiplications and divisions.* On the other hand, $a/b+c$ means $a\div b+c$, and not $a\div(b+c)$.

It should be noticed that the radical symbol $\sqrt[n]{\ }$ does not act as a bracket, like $/$: thus $\sqrt{4}\times 2$ means $(\sqrt{4})2$, *i.e.* $2\times 2$, and not $\sqrt{(4\times 2)}$, *i.e.* $\sqrt{8}$; and, again, $\sqrt{4}/2$ means $2/2$, and not $\sqrt{(4/2)}$, *i.e.* $\sqrt{2}$.

The use of brackets is one of the fundamental parts of the algebraic art, and it should be carefully studied by means of gradually generalised examples taken from the arithmetical exercises of the beginner; it is also of primary theoretical importance, as we shall see, when we lay down the Law of Association. Some of the exercises on this chapter are framed for the purpose of testing the student's grasp of the bracket notation.

§ 7. For the purposes of Algebra the fundamental operations just enumerated are classified as follows:—Addition, Subtraction, and Multiplication, including, of course, Involution, which is simply repeated multiplication, are called the **Integral Operations.** Division is called the **Fractional Operation.**

Addition, Subtraction, Multiplication, and Division, taken together, are called the **Rational Operations,** and in contradistinction Radication or Evolution is called the **Irrational Operation.**

The operations of Addition, Subtraction, Multiplication, Division, and Radication, taken together, are called the **Algebraic Operations.**

It will be observed that in the above definitions no mention is made of the operands at all, and that the definitions are quite distinct from those of the corresponding terms in Arithmetic. Thus, for example, we call the extraction of the square root in Algebra an irrational operation, quite irrespective of the fact that the result may be arithmetically rational (*i.e.* commensurable), or arithmetically irrational (*i.e.* incommensurable), according to circumstances.

§ 8. Any concatenation of operands and operating symbols which has an intelligible meaning according to the fundamental definitions or interpretations of these operands and operating symbols, we call a **Function** of the operands in question, or of any number of them that may be selected for special notice. Thus, for example, $3 \times 2 + 6$ is said to be a function of 3, 2, and 6; we may also say that it is a function of 3 and 2, or a function of 6, etc.; again, $ab^2 + \sqrt{c}$ may be spoken of as a function of $a$, $b$, $c$; a function of $a$ and $b$; a function of $a$ and $c$; and so on, as may be convenient.

The operands which for the moment are selected for notice are commonly spoken of as the **Variables,** and any other operands involved in the function are called in contradistinction **Constants.**

The word **Expression** is often used in the same sense as the word function, and is at times convenient. "Function" enters more conveniently into composition, *e.g.* we can say "function of $a$," whereas if we use "expression," we must say "expression involving $a$." Moreover, "function" is the word generally used in all parts of mathematics.

It will be observed that we define a function at present synthetically, *i.e.* with reference to the operations or steps in its construction or synthesis; and it is understood that the number of such operations is finite. As the student proceeds, he will find that the notion of a function is gradually extended. Whenever it is necessary for clearness to do so, we may more fully describe the kind of function which can be constructed by means of a finite number of the algebraic operations as a Synthetic Algebraic Function. Synthetic indicates that the function is to be constructed directly by steps or operations; Algebraic means, of course, that the operations are to be merely one or more of the five algebraic operations above enumerated. In the meantime, we may call any function, which is not an algebraic function, a **Transcendental Function.**

§ **9.** Parallel to the classification of the five fundamental operations given in § 7, there is a classification of Synthetic Algebraic Functions.

An algebraic function which, so far as any selected set of operands is concerned, is constructed by means of a finite number of integral operations, is called an **Integral Algebraic Function,** or simply an **Integral Function** of the selected set of operands.

If an algebraic function involves division with respect to any one of a selected set of operands, it is said to be a **Fractional Function** of these operands.

Integral and fractional functions are classed together as **Rational Functions,** so that a function is rational with respect to any set of operands when it involves every one of these only by way of addition, subtraction, multiplication, or division. If, on the contrary, root extraction with respect to any of the selected operands is involved, the function is called an **Irrational Function** of these operands.

As examples of these distinctions we may give the following:— $3 \div 2 + 6 \div (5+4)$ is an integral function of 3 and 6; a fractional function of 2; a fractional function of 3, 5, and 4; a fractional function of 4; a rational function of any or of all of its operands. $(a+\sqrt{b})/(c+\sqrt{d})$ is a rational function of $a$ and $c$, or of $a$, $\sqrt{b}$, $\sqrt{d}$; an integral function of $a$, or of $\sqrt{b}$, or of both; a fractional function of $c$, or of $\sqrt{d}$, or of both; an irrational function of $b$, or of $d$, or of both; and so on. $\frac{1}{2}+\frac{3}{4}x+x^5$ is an integral function of $x$; an integral function of $\frac{1}{2}$ or of $\frac{3}{4}$; but a fractional function of 2 or of 4, etc.

With respect to the meanings of the terms **Integral, Fractional, Rational,** and **Irrational in Algebra,** there are two points that must be constantly borne in mind. First, that the distinctions are a mere question of form, *i.e.* of the occurrence or non-occurrence of certain operations, and have nothing to do with the values of operands or with resulting values; second, that the terms are relative to a set of chosen operands or "variables" expressed or understood. If this be borne in mind, there will be no danger of confusion with the essentially distinct use of these words in relation to arithmetical quantity. As an instance of the distinction between the two usages, $4/2$ is in the algebraic sense a fractional function of 4 and 2; but as to its value (viz. 2) it is arithmetically integral; again, $\sqrt{9}$ is an irrational function of 9, but its value is rational in the arithmetic sense of the word. This double use of the same terminology is one of the difficulties in the way of the beginner in Algebra; and he must be warned once for all to pay close attention to the definitions and usage of algebraic terminology if he desires to acquire any but the loosest notions of the fundamental principles of the subject, and anything beyond the feeblest power of applying them independently. The present instance of the danger of confusion is merely one among many.

**§ 10.** Besides the predicative symbol $=$ meaning "is equal to," its negative, $\neq$, meaning "is not equal to," is also used in Algebra. Also the symbols $>$, $<$, $\ngtr$, $\nless$, meaning "is greater than," "is less than," "is not greater than," "is not less than," respectively, frequently occur. The last four symbols may at first be taken to have the ordinary arithmetical sense; but we shall hereafter assign to them an extended "algebraic sense" in connection with the use of purely negative quantity.

## EXERCISES I.

**1.** Evaluate $6a-3b+2c-x$ when $a=22$, $b=3$, $c=5$, $x=1$.

**2.** Evaluate $6a-2[b-4(3c-2x)+3\{a-(4c+x)\}]$ when $a=22$, $b=63$, $c=5$, $x=1$.

**3.** Evaluate $6x^3-5x^2+10x-3$ when $x=1$; when $x=2$; and when $x=1/3$.

Evaluate the following functions when $a=4$, $b=3$, $c=12$, $d=5$, working to three places of decimals when the result is not commensurable.

4. $3a^2+\dfrac{1}{a^2-b^2}$.  5. $20a+\dfrac{1}{a^2}-b^2$.

6. $a+1/(a^2-b^2)$.  7. $(a+b)/[4a-a\{6b-b(c-\overline{c-d})\}]$.

8. $\{(a+b)(c+d)+(a-b)(c+d)\}\{(a-b)(c+d)+(a+b)(c+d)\}$.

9. $\left\{\dfrac{a+b}{a-b}+\dfrac{a-b}{c+d}\right\}\dfrac{c}{a}$.  10. $\dfrac{(a+b)+(a-b)}{(a-b)+(c+d)}\dfrac{c}{a}$.

11. $\dfrac{a}{b+\dfrac{c+d}{c-d}}$.  12. $a/\{b+c/(c-d)\}$.

13. $a^2b^2$.  14. $(a^2b)^2$.  15. $(a^2b^3)^2/(a^3b^2)^2$.

16. $a^{b^2}$.  17. $\left(\dfrac{a}{b}\right)^2$.  18. $(a^3/a^2b)^2$.

19. $\dfrac{\sqrt{(a+b)}}{c-d}$.  20. $100\sqrt{(a+b)}/c-d$.

21. $\sqrt{\left(\dfrac{a+b}{c-d}\right)}$.  22. $\sqrt{(a+b)}/(c-d)$.

23. $\sqrt{\{(a+b)/(c-d)\}}$.  24. $\sqrt{(2ab^3)}$.  25. $\sqrt{\{\sqrt[3]{a^6}\}}$.

## EXERCISES II.

**1.** Represent symbolically the operation of reducing £3 6s. 8d. to pence.

**2.** Represent symbolically the operation of reducing $a$ pounds, $b$ shillings, and $c$ pence to pence.

**3.** Write down an expression for the number whose units digit is $x$, whose tens digit is $y$, and whose hundreds digit is $z$.

**4.** Write down the $n$th odd integer after 7.

**5.** Write down five consecutive odd integers of which $2n+1$ ($n$ being a positive integer) is the middle one.

**6.** A collector calls at 5 houses in street A, at 6 in street B, and at 8 in street C. In each house there are three flats. In street A he gets £$a$, £$b$, £$c$ from the respective flats, and the corresponding sums for streets B and C are £$a'$, £$b'$, £$c'$, and £$a''$, £$b''$, £$c''$ respectively. Find an expression for the whole sum collected; and evaluate it when $a=3$, $b=2$, $c=1$; $a'=4$, $b'=3$, $c'=1$; $a''=5$, $b''=3$, $c''=2$.

**7.** Find an expression for the number of pence in $a$ half-crowns, $b$ florins, $c$ shillings, and $d$ sixpences.

**8.** In a till there are $a$ five pound notes, $b$ pound notes, $c$ five shilling pieces, $d$ half-crowns, $e$ shillings. Taking £1 as unit, write down an expression for the value of the contents of the till.

**9.** A man walked from home a distance of 6 miles at the rate of $a$ miles an hour; he rested 20 minutes and returned at the rate of 3 miles an hour. How long was he out?

**10.** Write down symbolical expressions for the simple interest, and for the amount at simple interest of a given principal sum at a given rate per cent of interest per annum for a given number of years.

**11.** Write down an expression for the present value (at a given rate per cent per annum of simple interest) of a given sum due a given number of months hence.

**12.** The price of the $p$ per cents is P, of the $q$ per cents Q. A man sells out £A of $p$ per cent stock and with the price buys $q$ per cent stock. Neglecting brokerage, find an expression for his gain or loss of income.

**13.** A grocer mixes $a$ lbs. of tea worth $\alpha$ shillings per lb. with $b$ lbs. at $\beta$ shillings per lb., and sells the mixture at $\frac{1}{2}(\alpha+\beta)$ shillings per lb. How much does he gain or lose by selling $c$ lbs. of the mixture?

**14.** A man held a certain office for 30 years. He began with a salary of £P, and every 10 years he got an increase of £Q. Find a formula for his average salary during the 30 years.

**15.** If I mix $a$ oz. of a $p$ per cent solution of a salt with $b$ oz. of a $q$ per cent solution, what percentage of salt does the mixture contain?

Work out the result when $a=3{\cdot}5$, $p=5$, $b=4{\cdot}3$, $q=11$.

**16.** A is a $p$ per cent solution of pyrogallol; B a $q$ per cent solution of sodium sulphite; C an $r$ per cent solution of sodium carbonate. In a mixture of $a$ oz. of A, $b$ oz. of B, and $c$ oz. of C, how many oz. are there of pyrogallol, sodium sulphite, and sodium carbonate respectively; and what percentage of each of the three does the mixture contain?

**17.** A grocer lays out a sum of money in buying $a$ lbs. of tea at $p$ shillings per lb., $b$ lbs. of tea at $q$ shillings per lb., and $c$ lbs. at $r$ shillings per lb. He mixes the three quantities and sells the mixture at $s$ shillings per lb. Find a formula for the gain or loss per cent on his original outlay.

**18.** A certain vessel, which can never be entirely emptied, $p$ per cent of its contents always remaining behind, is full of a $q$ per cent solution of salt in water. The vessel is emptied as far as possible, filled up with water, shaken, and then emptied as far as possible. This is done three times. Find what percentage of salt there is in the remaining liquid.

**19.** Find an expression for the amount of a given principal sum for a given number of years at a given rate per cent of compound interest payable annually.

**20.** If the radix of a scale of notation be $r$, and the digits of a number be $p_0, p_1, \ldots, p_n$ from the units onwards, find an expression for the number.

**21.** Express by a formula the decimal fraction whose digits are $p_1, p_2, p_3, p_4$.

**22.** If 36,754 represent an integer in the scale whose radix is 8, find the representation of the same in the scale of 10.

**23.** Reduce ·3234, which represents a radix fraction in the scale of 5, to a decimal fraction.

**24.** From a vessel filled with spirit and containing $a$ gallons, $b$ gallons are removed, and the vessel filled up with water. Find an expression for the amount of spirit left after this has been done $n$ times.

# CHAPTER II

## FUNDAMENTAL LAWS OF ALGEBRA—LAWS OF COMMUTATION AND ASSOCIATION FOR ADDITION AND SUBTRACTION

§ **11.** It will be advisable for our present purposes that the learner should attach some convenient concrete meanings to addition and subtraction. In the first instance, we shall use the notion of credit and debit; later we shall employ a more important but perhaps less familiar illustration.

We shall suppose that $+a$ means $a$ pounds to be paid by some debtor to a merchant A, and that $-b$ means $b$ pounds to be paid by A to some creditor of his. It will facilitate matters if we suppose that A collects his debts and pays his creditors through an agent B, who may be supposed to have a certain amount of spare cash of his own, in case it may happen on his rounds that he may either have more to pay out than to collect, or that he may, in the first instance, have to pay out some money before the money he has to collect for A has come in to cover his outlay.

The chain of additions and subtractions $+a+b-c+d-e-f$ will then represent £$a$ collected, £$b$ collected, £$c$ paid out, £$d$ collected, etc., in a particular order on a certain round. $+b+a+d-c-e-f$ will evidently, so far as A is concerned, represent the same final result; it might indeed represent simply a different way of arranging B's round of business calls. There is, in fact, from our present point of view, no reason why any or all of the creditors should not be visited first, B in the meantime paying out of his own cash, and then we should have for the symbolic representation of B's round $-c-e-f+a+b+d$.

We are thus led to see that *in a chain of additions and*

*subtractions the order of the operations is indifferent, provided each operand carry with it the operating symbol + or − originally attached to it.*

Such a chain of operations is often called an **Algebraic Sum**, and the law just stated is spoken of as the **Law of Commutation for Addition and Subtraction.**

It may be here explained that when an algebraical sum begins with an addition it is usual, for shortness, to omit the corresponding + ; thus, instead of $+1+4-3$, we write $1+4-3$.

It will be immediately perceived that, although in what precedes we have not gone beyond the limits of common sense, we have already transcended the boundaries of Arithmetic as ordinarily understood; for, although $+1+4-3$ is at every step a perfectly intelligible arithmetical sequence, $+1-3+4$, which from the point of view above explained is the same thing, directs that 3 shall be subtracted from 1, and $-3+1+4$, according to the ordinary arithmetical notions, has no meaning at all. To this point we shall return hereafter; all that we need note at present is that the notion of debit and credit has led us to a generalisation of the operation of subtraction.

§ **12.** Since two separate debts of £1 each, both supposed good, are from the merchant's point of view the same thing as a single debt of £2, we may *associate* $+1+1$ into $+(+1+1)$, the bracket indicating that the two separate debts are regarded as one, and the + before the bracket meaning "payable to A," as before. We have therefore

$$+1+1=+(+1+1)=+2\ ;$$

and in like manner

$$+1+1+1=+(+1+1+1)=+3\ ;$$

and so on.

These results might, in fact, be regarded as the definitions of $+2$ and $+3$ from the algebraic point of view. It is, however, more important to note that we have here the simplest case of the **Law of Association** for addition, viz. *a chain of additions associated into a single addition by means of a bracket may be dissociated into the component additions by merely removing the bracket; and conversely.*

The familiar process of adding two integral numbers is a case of the law of association, *e.g.* we have

$$+2+3=+(+1+1)+(+1+1+1),$$

by the definitions of $+2$ and $+3$;

$$=+1+1+1+1+1,$$
$$=+(+1+1+1+1+1),$$

by the law of association; $=+5$ by the definition of $+5$.

§ **13.** The process of association may be carried further. Let us suppose that A's agent B in a day's round collects £$a$, pays out £$b$, and also pays out £$c$. Associating the whole of the day's business together, the result from A's point of view is $+(a-b-c)$. If we look at it from the point of view of B's cash, he owes to A £$a$, and A owes to him £$b$ and £$c$—that is to say, from B's point of view A owes him $-a+b+c$, therefore, if we look at the whole result of the day's transaction again from A's point of view, the result is $-(-a+b+c)$, the $-$ before the bracket meaning "due by A," as before. Combining the two results just arrived at with the original way of looking at each debit and credit separately, we have the following equalities:—

$$+a-b-c=+(+a-b-c);$$
$$+a-b-c=-(-a+b+c).$$

The last two equations exhibit fully the **Law of Association for Addition and Subtraction** which we may state as follows:—

*An algebraic sum associated into a single operation by means of a bracket may be dissociated into component additions and subtractions by removing the bracket, leaving all the signs $+$ or $-$ unchanged if the bracket is preceded by $+$, reversing each sign if the bracket is preceded by $-$; and conversely.*

§ **14.** Since every operand may arise by association as an algebraic sum, we may have complication of the process of association for addition and subtraction to any extent. In this way we have to consider such functions as $a-[b+\{d-(e-f-g)\}]$, for example, where there occur brackets within brackets. In reducing such functions to a simple algebraic sum by dissociation, we may remove the brackets in any order that may be convenient, the most usual perhaps being to begin with the outermost and proceed in order to the innermost. Thus—

$$\begin{aligned} & a - [b + \{d - (e - \overline{f - g})\}] \\ & = a - b - \{d - (e - \overline{f - g})\} \\ & = a - b - d + (e - \overline{f - g}) \\ & = a - b - d + e - \overline{f - g} \\ & = a - b - d + e - f + g. \end{aligned}$$

But we might also proceed thus—

$$\begin{aligned} & a - [b + \{d - (e - \overline{f - g})\}] \\ & = a - [b + \{d - (e - f + g)\}] \\ & = a - [b + \{d - e + f - g\}] \\ & = a - [b + d - e + f - g] \\ & = a - b - d + e - f + g. \end{aligned}$$

Or, again, we might suppose all the brackets removed at once; and consider the effect on the sign of each operand. Thus, for example, $f$ stands within four brackets; three of these are preceded by $-$; hence the sign of $f$ originally $+$ is thrice reversed; and is therefore finally $-$: and so on.

§ **15.** By means of the laws of commutation and association, we can deduce an important **Rule** for **Evaluating** an **Algebraic** sum. Thus, for example—

$$\begin{aligned} & a - b + d + e - f + g \\ & = a + d + e + g - b - f, \\ & = + \{(a + d + e + g) - (b + f)\}, \\ & = - \{(b + f) - (a + d + e + g)\}. \end{aligned}$$

Hence we see that *the reduced value of an algebraic sum is obtained by adding all the addends and all the subtrahends separately, taking the arithmetrical difference of these two sums, and prefixing + or − according as the sum of the addends or the sum of the subtrahends is the greater.* We here suppose that the operands are mere arithmetical quantities.

Example

$$\begin{aligned} & 6 + 3 - 7 - 9 + 10 - 11, \\ & = + 19 - 27, \\ & = - (27 - 19), \\ & = - 8. \end{aligned}$$

This rule for reducing an algebraic sum might equally have been deduced from our debit and credit illustration. For at the end of his day the agent B adds together the sums collected for his principal A, and also the sums paid out on behalf of A, takes the difference, and credits or debits A therewith according as the whole sum collected was greater or less than the whole sum paid out.

§ **16.** If we consider any two quantities which differ by $x$, say $a$ and $a+x$, we have

$$+(a+x)-a=+x,$$
$$+a-(a+x)=-x.$$

If now we make $x$ smaller and smaller, $a+x$ becomes more and more nearly $a$; and therefore if, as is usual in arithmetic, we denote *a quantity which is smaller than any assignable quantity* by 0, we have

$$+a-a=+0;$$
$$+a-a=-0.$$

These two equations may be taken as the definition of 0 as an operand in Algebra (so far as it is admissible in that capacity). We see at once that 0 has the special property possessed by no other operand, that

$$+0=-0.$$

This agrees perfectly with arithmetical notions; for we have

$$b+0=b=b-0;$$

and this again is consistent with our algebraical notion of the mutually inverse character of addition and subtraction; for if $+0$ stand for $+a-a$, we have $b+0=b+a-a=b$ (see § 4).

§ **17.** If we reduce the number of operations in a bracket to one, and formally apply the Law of Association for addition and subtraction, we get the following special results:—

$$+(+a)=+a, \quad +(-a)=-a,$$
$$-(+a)=-a, \quad -(-a)=+a,$$

which have a twofold interest.

In the first place, they lead us to the idea of the *cumulation of operative symbols* with the law that *the concurrence of two like symbols*, i.e. $+(+a)$ or $-(-a)$, *gives the direct symbol*, viz. in each case $+a$, while *the concurrence of two unlike symbols*, i.e. $-(+a)$ or $+(-a)$, *gives the inverse symbol*, viz. in each case $-a$.

We are thus led to break up the process of dissociation into two parts, viz. the removal of the bracket and the determination of the sign. Thus we may first write

$$-(-a+b+c)=-(-a)-(+b)-(+c);$$

then apply the "law of signs," which gives

$$-(-a)=+a, \quad -(+b)=-b, \quad -(+c)=-c.$$

The cumulation of the signs + and − thus suggested may be carried to any extent by repeated applications of the four fundamental cases, $+(+a)=+a$, $+(-a)=-a$, $-(+a)=-a$, $-(-a)=+a$. Thus we have

$$+a=+(+a)=+(+(+a))=+(+(-(-a)));$$
$$-a=+(-a)=+(-(+a))=+(-(-(-a)));$$

and so on; the rule for reduction to a single operation being obviously that the reduced sign is +, if there be no − or an even number of − signs in the sequence; and −, if there be an odd number of − signs in the sequence.

§ **18**. The Law of Association for an algebraic sum, in particular the four special cases $+(+a)=+a$, $+(-a)=-a$, $-(+a)=-a$, $-(-a)=+a$, leads us to another important idea, viz. the notion of **Algebraic Quantity** as distinguished from what may be called mere **Arithmetical Quantity**.*

In the first instance, the operands $a$, $b$, . . . were mere numbers (*e.g.* numbers of pounds in our debit and credit illustration); but in the expressions $+(+a)$ and $+(-a)$ the operand as regards the first + is not $a$, but $+a$ in the one case and $-a$ in the other. Such an operand, consisting of an arithmetical quantity with either + or − attached, we call an **Algebraic Quantity**, positive or negative according as the sign is + or −. Returning to our concrete illustration, we see that this amounts to dealing by way of addition and subtraction not with simple sums of money, but with such sums labelled as debits or as credits, thus $+2$ means £2 due to A, $-3$ means £3 due by A. In this kind of addition the effect of − is to turn a debit into a credit, and *vice versa*.

§ **19**. Just as we derive the positive integers from $+1$ by associating $+1+1$ into $+2$, $+1+1+1$ into $+3$, etc., and complete the series of positive quantity by inserting any required number of positive fractions between 0 and $+1$, $+1$

* Sometimes called **Scalar Quantity** or **Absolute Quantity**. The scalar or absolute value of an algebraic quantity $a$ is often denoted by $|a|$; thus $|-3|$ means 3. The absolute value of the difference between two algebraic quantities $a$ and $b$ is often denoted by $a \sim b$, thus $|a-b|=a \sim b$.

and $+2$, $+2$ and $+3$, etc., so from $-1$ we derive an infinite series of negative integers, $-2$, $-3$, etc., and we can complete the series of negative quantity by inserting any required number of negative fractions between 0 and $-1$, $-1$ and $-2$, $-2$ and $-3$, etc.

Strictly speaking, positive and negative quantities are not comparable as to magnitude, seeing that they are heterogeneous. Thus, for example, we cannot in the ordinary sense of the words say that "£5 due to A" is either greater or less than "£3 owed by A."

It is usual, however, to establish a conventional test of inequality between positive and negative quantities by laying down that *the algebraic quantity* a *is greater or less than the algebraic quantity* b *according as the reduced value of* a − b *is positive or negative.*

Applied to positive algebraic quantities merely this agrees with the ordinary arithmetical notion of inequality.

If $a$ be any positive quantity, say $+\alpha$, $b$ any negative quantity, say $-\beta$ (here $\alpha$ and $\beta$ are absolute quantities), then $a-b=+\alpha-(-\beta)=+\alpha+\beta$, and obviously has a positive reduced value. Hence any positive algebraic quantity, however small absolutely, is greater than any negative algebraic quantity.

$$\text{Ex. } +2>-3, \text{ since } (+2)-(-3)=+2+3=+5.$$

If $a$ be any negative quantity, say $-\alpha$, $b$ any negative quantity, say $-\beta$, then $a-b=-\alpha+\beta$, the reduced value of which is positive if $\beta>\alpha$, negative if $\beta<\alpha$. Hence one negative quantity is greater or less than a second according as the first is absolutely less or greater than the second.

$$\text{Ex. } -2>-3, \text{ since } (-2)-(-3)=-2+3=+1.$$
$$-3<-1, \text{ since } (-3)-(-1)=-3+1=-2.$$

If, therefore, we use $\infty$ to mean *a quantity greater than any assignable quantity*, then we may symbolise the whole series of algebraic quantity by

$$-\infty \ldots -2 \ldots -1 \ldots \pm 0 \ldots +1 \ldots +2 \ldots +\infty,$$

the order of ascending magnitude being from left to right. Owing to its exceptional property, $-0=+0$, 0 may be regarded as belonging to both the negative and the positive parts of the series, being the only quantity they have in common.

## Steps on a Line, Co-ordinates

§ **20.** There is another concrete representation of algebraic quantity which, although less familiar than the notion of debit and credit, is very important, because it makes its appearance in almost every application of mathematics.

Let X′X be an unlimited straight line; and let us distinguish the directions X′ to X, *i.e.* left to right, and X to X′, *i.e.* right to left, by calling them *positive* and *negative* respectively. Any limited portion of the straight line, say AB, whose ends are named in the order in which the letters are written, is called a **Right** (or **Positive**) **Step** or a **Left** (or **Negative**) **Step**, according as B is right or left of A. Thus in Fig. 1 AB is a right step; and A′B′ is a left step. We agree that two steps, wherever situated on the line X′X, are to be regarded as equal when their

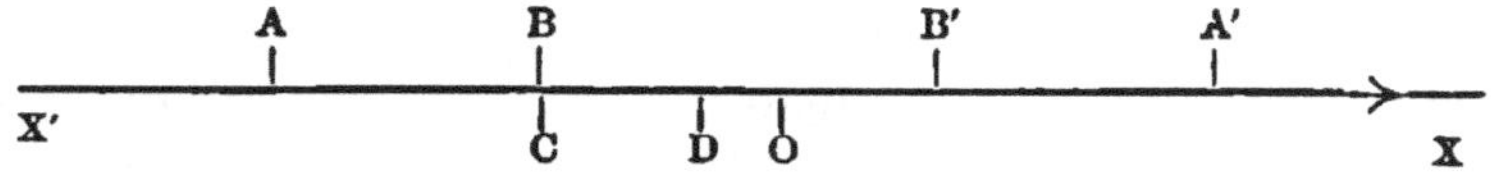


Fig. 1.

lengths and also their directions are the same. By the **composition of two steps**, AB and CD, is understood the operation of placing them without change of direction so that B coincides with C; the resulting step AD is called the **resultant** of AB and CD. By drawing the corresponding figures, or indeed intuitively, the learner will see at once the truth of the following statements :—

I. *The resultant of two, and therefore of any number of steps, is independent of the order in which they are compounded.* For example, the resultant of AB and CD is the same as the resultant of CD and AB.

II. *The resultant of two right steps is a right step whose length is the sum of the lengths of the components.*

III. *The resultant of a right step and a left step, or* vice versa, *is a step whose length is the difference of the lengths of the components, and which is a right step or a left step according as the right component is greater or less than the left component.*

IV. And, generally, *the resultant of any number of steps is a step whose length is the difference between the sum of the lengths of the right components and the sum of the lengths of the left com-*

*ponents, and which is right or left according as the former or the latter sum is the greater.*

We now see that the composition of steps on a line is exactly analogous to algebraic addition and subtraction. I. is the law of commutation; II., III., IV. are different cases of the law of association. In particular, IV. corresponds to the rule for reducing an algebraic sum. We may therefore specify a right step whose length is $a$ units by $+a$, a left step of the same length by $-a$. The statements I., II., III. then become

$$\left.\begin{aligned} +a+b &= +b+a, \\ +a-b &= -b+a, \\ &\text{etc.} \end{aligned}\right\} \text{I.};$$

$$+a+b = +(a+b), \quad \text{II};$$

$$\left.\begin{aligned} +a-b &= +(a-b), \\ &= -(b-a), \end{aligned}\right\} \text{III.};$$

And furthermore, if we agree that $+$ in general is to mean "set down a step" (whether right or left) in its proper direction; and $-$ to mean "set down the step reversed," then we see that we may operate with $+$ and $-$ on steps as well as on absolute lengths. Thus the four fundamental cases of the law of association

$$+(+a) = +a, \quad +(-a) = -a,$$
$$-(+a) = -a, \quad -(-a) = +a,$$

have their equivalents in the following statements regarding the right and left steps AB and BA :—

$$+\text{AB} = \text{AB}, \quad +\text{BA} = \text{BA},$$
$$-\text{AB} = \text{BA} \quad -\text{BA} = \text{AB}.$$

Corresponding to $+a-a=0$, we have $\text{AB}+\text{BA}=0$.

§ **21.** If on the unlimited line X′X we fix a reference-point O, usually called the **Origin,** then we may represent positive and negative algebraic quantities by steps from O: right for positive and left for negative quantities. In this representation to every algebraic quantity there will correspond one and only one point on the line X′OX, the point O corresponding to 0.

The series of algebraic (real) quantity

$$-\infty, \ldots, -2, \ldots, -1, \ldots, 0, \ldots, +1, \ldots, +2, \ldots, +\infty$$

will therefore be represented by an infinite succession of points arranged from left to right on the unlimited straight line X′OX. Here the notions of increase and decrease are associated with progress to the right and progress to the left respectively.

The right or left step from the origin to any point P, or the corresponding algebraic quantity, is often called the **Co-ordinate of the point P with respect to the origin O.**

Starting with this definition the learner will readily establish a series of propositions, such as the following:—If the co-ordinates of A and B be $x_1$ and $x_2$, then the step AB is in all cases represented by the algebraic quantity $x_2 - x_1$; the co-ordinate of the middle point of AB is $\frac{1}{2}(x_1 + x_2)$, and so on.

## EXERCISES III.

**1.** An agent has debts of £3 and £5 to collect, and a debt of £6 to pay for his employer. If he has £2 of his own cash with him, write down algebraic sums to represent all the different ways in which he can arrange his business round. All the payments to be made in cash.

Simplify the following by removing the brackets and removing mutually destructive operations:—

**2.** $(1+2-3)-(1-2+3)+(1-2-3)$.

**3.** $(3-5+8)-(3+5+8-9+3)-(6+7-8)$.

**4.** $\{9-(3-6)\}-\{9+(3-6)\}+(5-3)$.

**5.** $[9-\{8-(6+5)\}]-[9+\{8-(6-5)\}]$.

**6.** $1-[1-\{1-(1-\overline{1+1})\}]$.

**7.** $[x+\{x-(x-2)\}]-[x-\{x-(x+2)\}]$.

**8.** $[a+\{b+(c+d)\}]+[a+\{b+(c-d)\}]+[a+\{b-(c-d)\}]+[a-\{b-(c-d)\}]$.

**9.** $[\{(6+8)-(6-5)\}-\{(6+8)+(6-5)\}]+[\{(6-8)-(6+5)\}-\{(6-8)-(6-5)\}]$.

**10.** $\{(a+b+c)-(a-b-c)\}-\{(a-b+c)-(a+b+c)\}$.

**11.** $a-b-[a+b-\{a-b+(a+b-\overline{a-b})\}]$.

**12.** $+(-(+(-(+\ .\ .\ .\ 1))))$, $n$ pairs of brackets.

**13.** $a-(a-(a-(a-\ .\ .\ .\ )))$, $n$ pairs of brackets.

**14.** Find whether $3-(5-6)$ or $6-(5-2)$ is the greater.

**15.** Find whether $3-\{8-(10-9)\}$ or $6-\{8+(10-9)\}$ is the greater.

**16.** Represent by means of steps on a straight line the algebraic sum $+3-2+5-7$.

**17.** Illustrate graphically the algebraic identity $+2-3=-3+2$.

**18.** The co-ordinates of A and B are $(-1)$ and $(+3)$, and of C and D $(-3)$ and $(+5)$. Find the distance between the middle points of AB and CD.

# CHAPTER III

## FUNDAMENTAL LAWS OF ALGEBRA

### The Laws of Commutation and Association for Multiplication and Division—Laws of Indices

§ **22.** The reader who has been rationally taught the fundamental principles of Arithmetic is already aware that in a chain of multiplications and divisions, unbroken by additions or subtractions, the order in which these operations are performed is indifferent.

Thus, for example—

$$\times 3 \times 2 \times 6 = \times 3 \times 6 \times 2 = \times 6 \times 2 \times 3, \text{ etc.}$$

$$\times 16 \div 2 \times 8 = \times 16 \times 8 \div 2 = \times 8 \div 2 \times 16$$
$$= \div 2 \times 8 \times 16, \text{ etc.}$$

$$\times \tfrac{1}{2} \times \tfrac{1}{4} \div \tfrac{1}{3} \div \tfrac{1}{5} = \times \tfrac{1}{2} \div \tfrac{1}{3} \times \tfrac{1}{4} \div \tfrac{1}{5} = \text{etc.},$$

and, in general—

$$\times a \div b \times c \div d = \times a \times c \div b \div d,$$
$$= \times c \times a \div d \div b,$$
$$= \div b \times a \times c \div d, \text{ etc.}$$

It will be observed that we have here a law formally identical with the Law of Commutation already stated for an algebraic sum: it is called the **Law of Commutation for Multiplication and Division**, and may be verbally stated as follows:—

*In any chain of multiplications and divisions the order of the constituents is indifferent, provided the proper sign be attached to each operand and move with it.*

Just as in an algebraic sum, when the first operation is the direct one, in this case multiplication, the sign is usually omitted: thus we write $a \times b \div c$ instead of $\times a \times b \div c$.

§ **23.** Both from the principles and from the practice of

Arithmetic the learner is familiar with the following truths :— To multiply by 8, *i.e.* by 4 × 2, is the same as to multiply first by 4 and then multiply the product by 2, *i.e.* in symbols—

$$\times(\times 4 \times 2) = \times 4 \times 2 \quad . \quad . \quad (1).$$

Again to multiply by 4, *i.e.* by 8 ÷ 2, is the same as first to multiply by 8 and then divide by 2 ; that is—

$$\times(\times 8 \div 2) = \times 8 \div 2 \quad . \quad . \quad (2).$$

Also to divide by 8, *i.e.* by 4 × 2, is the same as to divide by 4 and then divide by 2 ; that is—

$$\div(\times 4 \times 2) = \div 4 \div 2 \quad . \quad . \quad (3).$$

Finally, to divide by 4, *i.e.* by 8 ÷ 2, is the same as to divide by 8 and then multiply by 2 ; that is—

$$\div(\times 8 \div 2) = \div 8 \times 2 \quad . \quad . \quad (4).$$

For simplicity we have chosen integers for operands ; but it will be recognised as an arithmetical truth that the same results hold when the operands are fractions.

The equivalences (1), (2), (3), (4) are examples of the Law of Association for multiplication and division, which may be stated thus :—

*A chain of multiplications and divisions associated into a single operand by means of a bracket may be dissociated into the constituent operations by removing the bracket, leaving all the signs unchanged if the bracket is preceded by ×, reversing each sign if the bracket is preceded by ÷.*

Or, in symbols—

$$\times(\times a \div b \div c \times d) = \times a \div b \div c \times d \quad . \quad (5);$$
$$\div(\times a \div b \div c \times d) = \div a \times b \times c \div d \quad . \quad (6).$$

§ **24.** As in the case of the Law of Association for addition and subtraction, we may resolve the process of dissociation for multiplication and division into two parts — the removal of the bracket and the determination of the signs. Thus (5) and (6) could be written

$$\times(\times a \div b \div c \times d) = \times(\times a) \times(\div b) \times(\div c) \times(\times d);$$
$$\div(\times a \div b \div c \times d) = \div(\times a) \div(\div b) \div(\div c) \div(\times d):$$

with the following **Law of Signs** :—

$$\times(\times a) = \times a, \quad \times(\div a) = \div a,$$
$$\div(\times a) = \div a, \quad \div(\div a) = \times a:$$

that is to say, the concurrence of like signs gives the direct sign ( $\times$ ), the concurrence of unlike signs the inverse sign ( $\div$ ).

This dissection of the Law of Association is of less importance in the present case; because there is in ordinary algebra, at least as yet, no important development of multiplicative and divisive quantity as there is of additive and subtractive quantity, which gave us the notion of so-called algebraic quantity.

§ **25.** In the association of multiplications and divisions we may have brackets within brackets, which may be resolved successively or simultaneously, exactly as in the case of additions and subtractions (see § 14). There is, however, for the beginner an element of perplexity in the variety of notations for multiplication and division.

Ex. 1.

$$\begin{aligned} &3\times[4\div\{3\div(6\times8)\}]\\ &\qquad=3\times4\div\{3\div(6\times8)\},\\ &\qquad=3\times4\div3\times(6\times8),\\ &\qquad=3\times4\div3\times6\times8. \end{aligned}$$

Ex. 2.

$$\begin{aligned} &3\div4/6\div8/3\times2\\ &\qquad=3\div4\div6\times8\times(3\times2),\\ &\qquad=3\div4\div6\times8\times3\times2. \end{aligned}$$

Ex. 3.

$$\begin{aligned} &\frac{\dfrac{3\div4}{6\div8}}{3\times2}=3\div4\div(6\div8)\times(3\times2),\\ &\qquad=3\div4\div6\times8\times3\times2. \end{aligned}$$

The bracketing effect of the "solidus" and of the "fraction line" should be observed in Examples 2 and 3. When it is desired to suspend this bracketing effect, a thicker fraction line is sometimes used. Thus while

$$\begin{aligned} &\frac{3\div4^{*}}{\dfrac{6\div8}{3\times2}}=3\div4\div\left(\frac{6\div8}{3\times2}\right),\\ &\qquad=3\div4\div6\times8\times3\times2. \end{aligned}$$

* $\dfrac{3\div4}{\dfrac{6\div8}{3\times2}}$ is sometimes written $\dfrac{3\div4}{\dfrac{6\div8}{3\times2}}$ (with a thicker fraction line) for the sake of emphasis; but this is unnecessary.

$$\frac{\frac{3 \div 4}{6 \div 8}}{3 \times 2} = 3 \div 4 \div (6 \div 8) \div (3 \times 2),$$
$$= 3 \div 4 \div 6 \times 8 \div 3 \div 2.$$

A similar convention might have, but so far as we know has not, been adopted for the solidus. In all cases where ambiguity is to be feared, it is better to insert the bracket.

§ **26.** If $a$ be any quantity differing from 0, we denote $\times a \div a$ by 1. This is perfectly in accordance with arithmetical notation ; and is analogous to our operational definition of 0 by the equation $+a - a = 0$, except that *we restrict the* a *in* a $\div$ a *to be different from* 0.

From the definition of 1 just given, and the laws of association and commutation for multiplication and division, and the mutual relation between multiplication and division, we have

$$b \times 1 = b \times (\times a \div a) = b \times a \div a = b.$$
$$b \div 1 = b \div (\times a \div a) = b \div a \times a = b.$$

Also

$$\times 1 = \times (\times a \div a) \quad = \times a \div a \quad = \div a \times a \quad = \div (\times a \div a);$$

that is to say, $\times 1 = \div 1$, which is analogous to $+0 = -0$.

§ **27.** It is interesting at this stage to notice that the laws for the transformation of fractions by multiplying or dividing numerator and denominator by the same quantity, and for multiplying and dividing fractions, are instances of the two laws which we have just been discussing. Thus

$$\frac{3 \times 5}{4 \times 5} = \frac{3}{4}$$

may be established as follows. We have

$$\frac{3 \times 5}{4 \times 5} = (3 \times 5) \div (4 \times 5),$$

since from the algebraical point of view the two sides of this equation are only different notations for the same function.

Next, by the law of association—

$$(3 \times 5) \div (4 \times 5) = 3 \times 5 \div 4 \div 5\ ;$$

and, by the law of commutation—

$$3 \times 5 \div 4 \div 5 = 3 \div 4 \times 5 \div 5.$$

Finally, from the definition of multiplication and division as mutually inverse operations—

$$3 \div 4 \times 5 \div 5 = 3 \div 4,$$

which establishes our result, since $3 \div 4$ and $\frac{3}{4}$ are the same thing.

In like manner, we could deduce in general the following results :—

$$\frac{a \times m}{b \times m} = \frac{a}{b}; \qquad (1)$$

$$\frac{a \div m}{b \div m} = \frac{a}{b}; \qquad (2)$$

$$\left(\frac{a}{b}\right) \times \left(\frac{c}{d}\right) = \frac{a \times c}{b \times d}; \qquad (3)$$

$$\left(\frac{a}{b}\right) \div \left(\frac{c}{d}\right) = \left(\frac{a}{b}\right) \times \left(\frac{d}{c}\right) = \frac{a \times d}{b \times c}. \qquad (4)$$

Take, for example, the second part of (4)—

$\left(\frac{a}{b}\right) \div \left(\frac{c}{d}\right) = (a \div b) \div (c \div d)$, by the meaning of the symbols ;
$= a \div b \div c \times d$, by the law of association ;
$= a \times d \div b \div c$, by the law of commutation ;
$= (a \times d) \div (b \times c)$, by the law of association ;
$= \frac{a \times d}{b \times c}$, by the meaning of the symbols.

Since the process we are engaged in is an analysis of the operations of arithmetic, whose laws we take for granted, into a few simple laws which are to be the laws of Algebra, it would not be logically consistent, from our present point of view, to regard the above deduction of certain rules of operation with arithmetical fractions from the laws of commutation and association as a *demonstration* of these rules. It is none the less interesting and important to see that these rules, in appearance so distinct, are really consequences of two very simple general principles. .

It is, however, very important to note that, since (1), (2), (3), and (4) are deducible from the fundamental laws of Algebra, they will hold, not merely for arithmetical operands, but for any algebraic operands whatsoever, simple or complex. Thus, for example—

$$\frac{a\times(-1)}{b\times(-1)}=\frac{a}{b},\quad i.e.\quad \frac{-a}{-b}=\frac{a}{b},$$

and

$$\frac{(x-1)(x+2)}{(x+3)(x+2)}=\frac{x-1}{x+3}$$

are cases of (1); and

$$\frac{x-1}{x+3}\div\frac{x+2}{x-2}=\frac{x-1}{x+3}\times\frac{x-2}{x+2}=\frac{(x-1)(x-2)}{(x+3)(x+2)}$$

is a case of (4).

## EXERCISES IV.

Simplify each of the following as much as you can :—

1. $3\times7\div3\times4\div7$.

2. $3\div\{5\div3\times(2\div3)\}$.

3. $\{6\div(2\times3)\}\div\{6\times(2\div3)\}$.

4. $(3a)\div(4b)\times b\div a$.

5. $2a\,.\,3a\,.\,4a\,.\,6a\div\{6a\,.\,12a\}$.

6. $\frac{1}{2}\times\frac{3}{4}/\frac{1}{2}\times\frac{1}{3}$.

7. $\left(\frac{a}{b}\right)\times\left(\frac{b}{c}\right)\times\left(\frac{c}{d}\right)\Big/\left(\frac{b}{a}\right)\times\left(\frac{c}{b}\right)\times\left(\frac{d}{c}\right)$.

8. $\dfrac{\frac{3}{4}\times\frac{2}{3}}{\dfrac{\frac{1}{2}\times\frac{5}{6}}{\frac{1}{3}\times\frac{1}{8}}}$.

9. $\dfrac{\dfrac{\frac{3}{4}\times\frac{2}{3}}{\frac{1}{2}\times\frac{5}{6}}}{\frac{1}{3}\times\frac{1}{6}}$.

10. $\dfrac{\frac{2}{3}\times\frac{3}{4}}{\frac{6}{8}\times\frac{4}{3}}\Big/\dfrac{\frac{1}{3}\times\frac{5}{4}}{\frac{5}{6}\times\frac{3}{4}}$.

11. $\dfrac{3a\,.\,3b\,.\,3c}{2d\,.\,2e\,.\,2f}\Big/\dfrac{2a\,.\,2b\,.\,2c}{3d\,.\,3e\,.\,3f}$.

12. $\dfrac{(x+1)(x-2)}{(x-1)(x+2)}\times\left(\dfrac{3(x+2)}{2(x+1)}\right)\div\left(\dfrac{6(x-2)}{4(x-1)}\right)$.

### MONOMIAL INTEGRAL FUNCTIONS—LAWS OF INDICES FOR INTEGRAL EXPONENTS

§ 28. **Technical Use of the Word Term.**—The word term is often used in Algebra in a technical sense, which it will be convenient here to define. *A function, or part of a function of any operands which involves only multiplication and division, and not addition and subtraction, is called a term.* Thus $3\times4$, $a\times b\div c$, and $3a^2$ are called terms. On the other hand, $a^2+b^2$, $ab-c^2/a$ are not in themselves terms; but $+a^2$ and $+b^2$ are the terms of $a^2+b^2$; and $+ab$ and $-c^2/a$ the terms of $ab-c^2/a$. A function which consists of a single term is called a **Monomial**; a function which is the algebraic sum of two terms a **Binomial**; and so on.

§ **29.** In dealing with rational terms (rational monomial functions), such as

$$(3a^9) \times (2a)^3 \div (abc)^3 \times (a^2)^3 \times \left(\frac{a}{b}\right)^4 \qquad (1),$$

it is found convenient to arrange so that all the multiplications or divisions by merely numerical operands shall be brought together and, usually, condensed into a single number; and, in like manner, all the multiplications and divisions by the same letter brought together and replaced by a multiplication or division by a single power of that letter.

Thus we shall presently show that the monomial (1) can be reduced to $24a^{19} \div b^7 \div c^3$ or $24a^{19}/b^7c^3$.

This reduction is greatly facilitated by the establishment of rules—

1. For expressing the product of any powers of one and the same base, or the quotient of two powers of one and the same base, by means of a single power of that base.

2. For expressing any power of a power of one base as a single power of that base.

3. For expressing a power of the product of any bases, or a power of the quotient of two bases, as a product or quotient of single powers of those bases.

These rules, commonly spoken of as the **Laws of Indices**, are as follows :—

I. ($\alpha$) $a^m \times a^n \times a^p \times \ldots = a^{m+n+p+\ldots}$

($\beta$) $a^m \div a^n = a^{m-n}$, if $m > n$;
$= 1 \div a^{n-m}$, if $m < n$.

II. $(a^m)^n = a^{mn}$.

III. ($\alpha$) $(a \times b \times c \times \ldots)^m = a^m \times b^m \times c^m \times \ldots$;

($\beta$) $(a \div b)^m = a^m \div b^m$;

or, in words—

I. ($\alpha$) *The product of any powers of one and the same base is a power of the base which is the sum of the given powers.*

($\beta$) *The quotient of two different powers of the same base is a power of the base which is the absolute difference of the two powers, or unity divided by the same, according as the index of the dividend is greater or less than the index of the divisor.*

II. *The* n*th power of the* m*th power of a base is the* mn*th power of that base.*

III. (α) *The* m*th power of a product of bases is the product of the* m*th powers of those bases.*

(β) *The* m*th power of the quotient of two bases is the quotient of the* m*th powers of those bases.**

The proof of these laws depends merely on the definition of an integral power, viz. that

$$a^m = a \times a \times a \times \ldots m \text{ factors},$$

and on the laws of association and commutation for multiplication and division.

To prove I. (α), let us consider first a special case, say $a^2 \times a^3 \times a^2$. By the definitions of $a^2$, $a^3$, $a^2$ we have

$$a^2 \times a^3 \times a^2 = (a \times a) \times (a \times a \times a) \times (a \times a),$$

by the law of association—

$$= a \times a \times a \times a \times a \times a \times a,$$

where there are $2+3+2$ factors; hence, finally, by the definition of a power—

$$= a^{2+3+2}.$$

The general proof may be stated thus—

$$a^m \times a^n \times a^p \times \ldots$$

by the definitions of $a^m$, etc.—

$$\begin{aligned} &= (a \times a \times \ldots m \text{ factors}) \\ &\times (a \times a \times \ldots n \text{ factors}) \\ &\times (a \times a \times \ldots p \text{ factors}) \\ &\times \ldots, \end{aligned}$$

by the law of association—

$$\left.\begin{aligned} &= a \times a \times \ldots \\ &\times a \times a \times \ldots \\ &\times a \times a \times \ldots \\ &\times \ldots \end{aligned}\right\} m+n+p+\ldots \text{ factors};$$

by the definition of a power—

$$= a^{m+n+p+\ldots}.$$

* The beginner may be cautioned against the frequent error of confusing these laws with one another; thus, for example, of putting $(a^3)^2 = a^{3+2}$, a confusion of I. (α) with II.

To prove I. ($\beta$), consider first a particular case, say $a^5 \div a^2$. We have

$$a^5 \div a^2.$$

by the definitions of $a^5$ and $a^2$—

$$= (a \times a \times a \times a \times a) \div (a \times a);$$

by the law of association—

$$= a \times a \times a \times a \times a \div a \div a;$$

by the law of commutation—

$$= a \times a \times a \times a \div a \times a \div a;$$

by the mutual inverseness of multiplication and division—

$$= a \times a \times a \times (5 - 2 \text{ factors});$$

by the definition of an index—

$$= a^{5-2}.$$

In general, by definition of a power—

$$a^m \div a^n = (a \times a \times \ldots \; m \text{ factors}) \div (a \times a \times \ldots \; n \text{ factors});$$

by law of association—

$$= a \times a \times \ldots \; m \text{ factors} \div a \div a \div \ldots \; n \text{ divisions}.$$

If now $m > n$, we have, by law of commutation—

$$a^m \div a^n = a \times a \times \ldots \; \overline{m-n} \text{ factors} \times a \div a \times a \div a \times \ldots \; n \text{ pairs};$$

by the mutual inverseness of multiplication and division—

$$= a \times a \times \ldots \; \overline{m-n} \text{ factors};$$

by the definition of a power—

$$= a^{m-n}.$$

If $m < n$, there are more divisions than multiplications, and we have

$$a^m \div a^n = a \div a \times a \div a \times \ldots \; m \text{ pairs} \div a \div a \; \ldots \; \overline{n-m} \text{ divisions}$$
$$= 1 \div a \div a \; \ldots \; \overline{n-m} \text{ divisions};$$

by the law of association—

$$= 1 \div (a \times a \times \; \ldots \; \overline{n-m} \text{ factors});$$

by the definition of a power—

$$= 1 \div a^{n-m}.$$

To prove II., consider the special case $(a^3)^2$. Since $(a^3)^2$ by the definition of a power means $a^3 \times a^3$, we have

$$\begin{aligned}(a^3)^2 &= a^3 \times a^3,\\ &= (a \times a \times a) \times (a \times a \times a);\end{aligned}$$

by law of association—

$$= a \times a \times a \ . \ . \ . \ 3 \times 2 \text{ factors};$$

by definition of a power—

$$= a^{3 \times 2}.$$

In general—

$$(a^m)^n;$$

by definition of a power—

$$= a^m \times a^m \times \ . \ . \ . \ n \text{ factors};$$

by definition of a power—

$$\left.\begin{aligned}&= (a \times a \times \ . \ . \ . \ m \text{ factors})\\ &\times (a \times a \times \ . \ . \ . \ m \text{ factors})\\ &\times \ . \ . \ .\end{aligned}\right\} n \text{ rows};$$

by law of association—

$$= a \times a \times \ . \ . \ . \ mn \text{ factors};$$

by definition of a power—

$$= a^{mn}.$$

As the reader has probably now grasped the simple principles involved, we give the general proof of III. ($\alpha$) and III. ($\beta$) at once.

We have, by the definition of a power—

$$(a \times b \times c \ . \ . \ .)^m = \left.\begin{aligned}&(a \times b \times c \times \ . \ . \ .)\\ &\times (a \times b \times c \times \ . \ . \ .)\\ &\times \ . \ . \ .\end{aligned}\right\} m \text{ rows};$$

by the laws of association and commutation—

$$\begin{aligned}&= (a \times a \times \ . \ . \ . \ m \text{ factors})\\ &\times (b \times b \times \ . \ . \ . \ m \text{ factors})\\ &\times (c \times c \times \ . \ . \ . \ m \text{ factors})\\ &\times \ . \ . \ . ,\end{aligned}$$

where we have in effect turned the columns of the first scheme into rows in the second. Hence, finally, by the definition of a power—

$$(a \times b \times c \times \ldots)^m = a^m \times b^m \times c^m \times \ldots$$

Again, by the definition of a power—

$$(a \div b)^m = (a \div b) \times (a \div b) \times \ldots \ m \text{ pairs};$$

by the law of association—

$$= a \div b \times a \div b \times \ldots \ m \text{ pairs};$$

by the law of commutation—

$$= a \times a \times \ldots \ m \text{ factors}$$
$$\div b \div b \div \ldots \ m \text{ divisions};$$

by the law of association—

$$= (a \times a \times \ldots \ m \text{ factors}) \div (b \times b \times \ldots \ m \text{ factors});$$

by the definition of a power—

$$= a^m \div b^m.$$

**§ 30.** Let us now return to the monomial—

$$(3a^9) \times (2a)^3 \div (abc)^3 \times (a^2)^3 \times \left(\frac{a}{b}\right)^4.$$

Using the laws of indices, we have

by III. (α), $(2a)^3 = 2^3a^3 = 8a^3$;

by III. (α), $(abc)^3 = a^3b^3c^3$;

by II., $(a^2)^3 = a^6$;

by III. (β), $\left(\frac{a}{b}\right)^4 = (a \div b)^4 = a^4 \div b^4$.

Hence

$$(3a^9) \times (2a)^3 \div (abc)^3 \times (a^2)^3 \times \left(\frac{a}{b}\right)^4$$
$$= (3 \times a^9) \times (8 \times a^3) \div (a^3 \times b^3 \times c^3) \times a^6 \times (a^4 \div b^4);$$

by the law of association—

$$= 3 \times a^9 \times 8 \times a^3 \div a^3 \div b^3 \div c^3 \times a^6 \times a^4 \div b^4;$$

by the law of commutation—

$$= 3 \times 8 \times a^9 \times a^3 \times a^6 \times a^4 \div a^3 \div b^3 \div b^4 \div c^3;$$

by the law of association—

$$= 24 \times (a^9 \times a^3 \times a^6 \times a^4) \div a^3 \div (b^3 \times b^4) \div c^3;$$

by the laws of indices (I. ($\alpha$))—

$$= 24 \times a^{22} \div a^3 \div b^7 \div c^3 ;$$

by the law of association—

$$= 24 \times (a^{22} \div a^3) \div (b^7 \times c^3) ;$$

and finally, since, by I. ($\beta$), $a^{22} \div a^3 = a^{22-3} = a^{19}$,

$$= 24a^{19}/b^7c^3.$$

The above calculation has been written at considerable length in order to show clearly how it depends on fundamental principles. This model should be followed by the beginner in every piece of work that is new to him and whenever he has fallen into doubt or perplexity. After the fundamental principles have become perfectly familiar, much of the work can be safely carried out mentally and the written calculation much shortened; but the learner should never fall into the bad habit of quoting formulæ or rules whose connection with fundamental principles he does not perfectly understand; by so doing he will strain his memory and retard his ultimate progress.

## Notion and Laws of Degree

§ **31.** It follows from the last paragraph that every integral monomial function of the operands $a, b, c, \ldots$ can be reduced to the standard form

$$Aa^\alpha b^\beta c^\gamma \ldots ,$$

where A is a factor not depending on $a, b, c, \ldots$, and $\alpha, \beta, \gamma, \ldots$ are positive integers.

The index of any operand or letter when an integral monomial is reduced to its standard form is called the **Degree of the Monomial with respect to that Letter.**

By the **degree of an integral monomial with respect to any set of named letters is meant the sum of the indices of those letters when it is reduced to the standard form.**

In other words, the degree in any set of named letters is the number of times that the whole of these letters occur as factors.

If no letters are specially mentioned, it is understood that all the letters appearing are to be taken into account in reckoning the degree.

Thus, for example, the degree of $3a^3b^2c^5$ in $a$ is 3; in $b$, 2; in $c$, 5; in $a$ and $b$, $3+2=5$, in $a$, $b$, and $c$, $3+2+5=10$.

The letters which for the time being are taken into account

in reckoning the degree are often called the **Variable Letters** or the **Variables.** In contradistinction, the remaining letters, including the numerical factor, are spoken of as the **Constant Letters** or the **Constants.** These definitions seem at first sight to involve a strain on the meaning of the words "variable" and "constant," but the convenience of the phraseology will be appreciated later on.

When it is necessary throughout a calculation of any length to distinguish the variables from the constants, the former are usually denoted by letters near the end of the alphabet—$x$, $y$, $z$, $u$, $v$, $w$, etc.; and the latter by letters near the beginning—$a$, $b$, $c$, $d$, $e$, etc.

When a monomial is thus for any purpose looked at as a function of constants on the one hand and variables on the other, the product of all the variables is spoken of as the **variable part** and the rest is called the **coefficient.** Thus, $a$, $b$, $c$ being constants, and $x$, $y$, $z$ variables in the monomial $3ab^2cx^2y^3z^4$, we call $x^2y^3z^4$ the variable part, and $3ab^2c$ the coefficient.

The beginner must not forget that the above distinctions, like other matters of algebraic form, depend on an arbitrary classification of the operands which may be made in one way for one purpose and in another way for another.

§ **32.** Since the multiplication of two powers of the same base is effected by adding the indices, and division of one power by another by subtracting the indices, we have immediately the two following **Laws of Degree.**

*The degree of the product of two integral monomials in any given set of letters is the sum of the degrees of the two factors in those letters.*

*If the quotient of one integral monomial by another be integral, the degree of the quotient in any given set of letters is the difference between the degrees of the divisor and dividend in those letters.*

Ex. The degree of $2abx^3y^4z^2$ in $x$, $y$, $z$ is 9, and the degree of $3ab^2x^2y^3z$ in $x$, $y$, $z$ is 6; the degree of the product, viz. $6a^2b^3x^5y^7z^3$ in $x$, $y$, $z$ is $15=9+6$; the quotient, viz. $\left(\dfrac{2}{3b}\right)xyz$, is integral so far as $x$, $y$, $z$ is concerned, and its degree in these letters is $3=9-6$.

The theory of degree is of great importance in Algebra; in fact, degree will be found in many respects to play the same part in Algebra as absolute magnitude does in Arithmetic.

## EXERCISES V.

Simplify the following monomials by reducing them to the standard form :—

$$Aa^{\alpha}b^{\beta} \ldots / k^{\kappa}l^{\lambda} \ldots,$$

and finally to a single number in cases where the operands are all numerical.

1. $5^8 \times 15^4 \times (2^2 \times 3^{15} \div 5^2)^3/(5 \times 60 \times 3^8)^5$.

2. $\dfrac{5^7 \times 7^8/35^3}{7^5 \times 5^8/70^3}$.

3. $(3^{2^2})^{2^2}/\{(3^2)^2\}^{2^3}$.

4. $[5^3 \times (2^3)^2 \times 6^2]^3/[5^7 \times 3^2 \times 2^{12}]^2$.

5. $(ab^2c^3)(bc^2a^3)^2(ca^2b^2)^3$.

6. $(3ayz)^2(6bczx)^3/(6axy)^3(3bcyz)^3$.

7. $\left(\dfrac{35^3a^4b^5c^6d^8}{1225a^3b^4c^3d^4}\right)\left(\dfrac{a^2b^4c^3d^2}{(abc)^2(bcd)^2}\right)$.

8. $(a)^2(a^2)^3(a^3)^4(a^4)^5$.

9. $(\frac{2}{3}a^2)^3(\frac{3}{2}a^3)^2$.

10. $[\{(a)^2\}^3]^5$.

11. $(\ldots(((a)^1)^2)^3 \ldots)^n$.

12. $(\frac{5}{6}a^2by^2z^2) \times (\frac{3}{5}b^2cz^2x^2) \times (2c^2ax^2y^2)$.

13. $\left(\dfrac{a^2bc}{b^2cayz}\right)^2\left(\dfrac{b^2ca}{c^2abzx}\right)^2\left(\dfrac{c^2ab}{a^2bcxy}\right)^2$.

14. $\dfrac{(x^2yz)^l(y^2zx)^m(z^2xy)^n}{\left(\dfrac{yz}{x^2}\right)^l\left(\dfrac{zx}{y^2}\right)^m\left(\dfrac{xy}{z^2}\right)^n}$

15. Show that $\left(\dfrac{y^m}{z^n}\right)^l\left(\dfrac{z^n}{x^l}\right)^m\left(\dfrac{x^l}{y^m}\right)^n$, where $l$, $m$, $n$ are unequal positive integers, cannot be an integral function of $x$, $y$, $z$. What is its value when $x=y=z$.

16. Exhibit $360 \times 21 \times 150$ as a product of powers of primes.

17. Reduce $(252)^2(365)^5(121)^3/(144)^3(108)^2(110)^3$ to a product and quotient formula in which the operands are powers of primes.

18. Show that every integer which is a perfect square must be the product of factors, each of which is an even power of a prime.

19. Show that every composite number must have a factor which does not exceed its square root.

20. In how many ways can $a^{\alpha}b^{\beta}c^{\gamma}$, where $a$, $b$, $c$ are primes and $\alpha$, $\beta$, $\gamma$ positive integers, be decomposed into a pair of factors; and in how many ways into a pair of factors which are prime to each other?

21. Find the largest number, not exceeding 731, which has only 2 or 5 or both as prime factors.

22. If A be a coefficient independent of $x$, $y$, $z$, find the multiplier of lowest degree in $x$, $y$, $z$ which will render $Ax^2y^3z^5$ a complete square as regards $x$, $y$, $z$. Is the problem determinate?

23. Find the monomial of highest degree in $x$, $y$, $z$, such that the quotients formed by dividing $3x^2y^3z^5$, $6xy^3z^4$, and $8x^2y^2z^2$ by it are all integral. Is the problem determinate?

**24.** If $\left(\frac{yz}{x}\right)^l\left(\frac{zx}{y}\right)^m\left(\frac{xy}{z}\right)^n=\left(\frac{x^2}{yz}\right)^l\left(\frac{y^2}{zx}\right)^m\left(\frac{z^2}{xy}\right)^n$, show that $(x^2y^2z^2)^{l+m+n}=(x^ly^mz^n)^5$.

**25.** If $x$, $y$, $z$ be all positive integers, and $x=y^z$, $y=z^x$, $z=x^y$, show that $x=y=z=1$.

**26.** What are the degrees of **12** and **14** above in $x$, $y$, and $z$ separately, and in $x$, $y$, and $z$?

**27.** Find the monomial of lowest degree in $x$, $y$, and $z$ whose quotients by $3x^2y^3z^5$, $6xy^3z^4$, and $8x^2y^2z^2$ are all integral. Is this problem determinate?

**28.** An integral monomial function of $x$, $y$, and $z$ is of the 3rd, 2nd, and 1st degrees in these variables respectively; and the numerical value of the function is 4 when $x=1$, $y=2$, $z=3$. Find the monomial.

**29.** The product of two integral monomial functions of $x$ and $y$ is $8x^4y^6$, and their quotient is $2x^2y^2$. Find them.

# CHAPTER IV

## FUNDAMENTAL LAWS OF ALGEBRA

### The Law of Distribution

§ **33.** The primitive meaning of multiplication is repeated addition. Thus $8 \times 3$ is a contraction for $8 + 8 + 8$.

In our discussion of the laws of commutation and association for multiplication and division we considered only the case where the operands are absolute, *i.e.* merely arithmetical quantities. The further points that arise when the operands are algebraical quantities—that is to say, absolute quantities with the signs + or − attached—are most conveniently considered in connection with the Law of Distribution, which is the last of the three fundamental laws of algebraical operation.

Reverting to $8 \times 3$, let us write the product more fully as $(+8) \times (+3)$, and notice that we may also write $8 + 8 + 8$ more fully in the form $+8 + 8 + 8$, or if we choose $+8 \times 1 + 8 \times 1 + 8 \times 1$. Remembering that $+3$ is a contraction for $+1 + 1 + 1$, we may therefore write the equation $8 \times 3 = 8 + 8 + 8$ in the forms—

$$(+8) \times (+3) = +(+8) + (+8) + (+8)$$
$$= +8 + 8 + 8 = +24 \qquad (1);$$

or $$(+8) \times (+1 + 1 + 1) = +8 \times 1 + 8 \times 1 + 8 \times 1 \qquad (2).$$

We thus look upon multiplication by a positive multiplier as a contraction for repeated addition. In like manner, it is natural to regard multiplication by a negative multiplier as a contraction for repeated subtraction. Taking this view, we have—

$$(+8) \times (-3) = -(+8) - (+8) - (+8),$$
$$= -8 - 8 - 8 = -24 \qquad (3);$$

or $$(+8)(-1 - 1 - 1) = -8 \times 1 - 8 \times 1 - 8 \times 1 \qquad (4).$$

Also $(-8)\times(+3) = +(-8)+(-8)+(-8),$

$= -8-8-8 = -24 \qquad (5);$

or $(-8)(+1+1+1) = -8\times1-8\times1-8\times1 \qquad (6).$

$(-8)\times(-3) = -(-8)-(-8)-(-8),$

$= +8+8+8 = +24 \qquad (7);$

or $(-8)(-1-1-1) = +8\times1+8\times1+8\times1 \qquad (8).$

§ **34.** Consider now the case where the multiplier is an algebraic sum, say $+8-5$. To multiply $+8$ by $+8-5$ may, according to our present view, be taken to mean: add $+8$ eight times and subtract $+8$ five times—that is to say, using multiplication by positive and negative multipliers as before to denote repeated additions and subtractions respectively, we have—

$$(+8)\times(+8-5) = +8\times8-8\times5 \qquad (9).$$

In like manner

$$(-8)\times(+8-5) = -8\times8+8\times5 \qquad (10).$$

The equations (1), (3), (5), and (7) suggest the laws for the sign of the product of two algebraic quantities, viz. *the product is positive if the factors have the same signs, negative if they have opposite signs.*

The equations (2), (4), (6), (8), (9), and (10) suggest the following rule for multiplying any algebraic quantity by an algebraic sum. *Write down all the partial products formed by multiplying the multiplicand by each term of the multiplier, and determine the sign of each partial product by the law of signs just given.* Or, in general symbols—

$$(+A)(+a-b-c+d) = +Aa-Ab-Ac+Ad;$$
$$(-A)(+a-b-c+d) = -Aa+Ab+Ac-Ad \qquad (11).$$

This process we call **Distributing the Multiplier.**

The order of ideas which we are now following suggests that the multiplicand may also be distributed. For we have, by the meanings attached to positive and negative multipliers—

$$(+7-5)\times(+3) = +(+7-5)+(+7-5)+(+7-5),$$
$$= +7-5+7-5+7-5,$$
$$= +7+7+7-5-5-5,$$

by the laws of association and commutation for algebraic sums. Hence, finally, by the primitive signification of multiplication—

$$(+7-5)\times(+3) = +7\times3-5\times3 \qquad (12).$$

In like manner—

$$(+7-5)\times(-3)=-7\times3+5\times3 \qquad (13).$$

Hence *to multiply an algebraic sum by an algebraic quantity write down all the partial products obtained by multiplying each term of the multiplicand by the multiplier, and determine the sign of each partial product by the law for the sign of the product of two algebraic quantities.* Or, in general symbols—

$$(+A-B+C-D)(+a)=+Aa-Ba+Ca-Da\,;$$
$$(+A-B+C-D)(-a)=-Aa+Ba-Ca+Da \qquad (14).$$

§ **35.** Consider finally the product of two algebraic sums, say $(+A-B+C)\times(+a-b)$. We may, in the first instance, consider $+A-B+C$ as associated into a single operand $+(+A-B+C)$. If we distribute the multiplier, we have

$$\{+(+A-B+C)\}(+a-b)=+(+A-B+C)a$$
$$-(+A-B+C)b.$$

Since we may also distribute the multiplicand, we have, reading $(+A-B+C)a$ as $(+A-B+C)(+a)$,

$$(+A-B+C)a=+Aa-Ba+Ca.$$

Also $(+A-B+C)b=+Ab-Bb+Cb\,;$

and $-(+A-B+C)b=-Ab+Bb-Cb.$

Hence, finally—

$$(+A-B+C)(+a-b)=+Aa-Ba+Ca-Ab+Bb-Cb \qquad (15).$$

This last result suggests the **Law of Distribution** in its full form, viz. *to multiply one algebraic sum by another write down the algebraic sum of all the partial products obtained by multiplying each term of the multiplicand by each term of the multiplier, determining the sign of each partial product by the law that the product of two terms having the same sign is to have the sign* $+$, *and the product of two terms having opposite signs the sign* $-$.

The law of distribution might also be stated in symbols for the particular case (15) thus—

$$(+A-B+C)(+a-b)=(+A)(+a)+(-B)(+a)$$
$$+(+C)(+a)+(+A)(-b)+(-B)(-b)+(+C)(-b),$$

with the following laws of sign :—

$$(+A)(+a)=+Aa,\quad (-A)(-a)=+Aa,$$
$$(+A)(-a)=-Aa,\quad (-A)(+a)=-Aa.$$

This form of statement has the advantage of separating the

operation of "distribution" properly so called from the "law of signs."

It will be observed that in the product of an algebraic sum of $m$ terms into an algebraic sum of $n$ terms there are $mn$ partial products, if they are all written down directly in accordance with the law, without any collection of like or suppression of mutually destructive terms. This rule will sometimes enable the beginner to correct a mistake in his calculations.

§ **36.** It will be seen that the law of distribution for multiplication has been suggested to us by the consideration of arithmetical operations with integral numbers. A little thought will convince the learner that the law holds whether the operands be integral or fractional. Thus, for example, the arithmetical truth of the following equations—

$$\left(\tfrac{5}{4}-\tfrac{3}{5}\right)\times\left(\tfrac{5}{8}-\tfrac{1}{2}\right)=\left(\tfrac{5}{4}-\tfrac{3}{5}\right)\times\tfrac{5}{8}-\left(\tfrac{5}{4}-\tfrac{3}{5}\right)\times\tfrac{1}{2},$$
$$=\tfrac{5}{4}\times\tfrac{5}{8}-\tfrac{3}{5}\times\tfrac{5}{8}-\tfrac{5}{4}\times\tfrac{1}{2}+\tfrac{3}{5}\times\tfrac{1}{2},$$

as arithmetical statements will be readily seen; and they are simply particular applications of the law of distribution. We therefore lay down this law as one of the fundamental principles of Algebra in the assurance that it agrees with the fundamental principles of ordinary arithmetic; beyond this, all we have to consider is merely the mutual consistency or non-contradiction of the various laws we adopt and of the consequences that follow therefrom. On this latter point the learner will gradually acquire conviction as he proceeds with the study of the subject and of its applications.

In the meantime we remark that, as in the case of the other laws, and in Algebra generally, we shall not confine the operands to be arithmetical or even algebraic quantities in a reduced form, the operands may be complex functions of other quantities. Thus—

$$\left(\frac{x-1}{x+2}-\frac{x}{x-1}\right)\times\left(\frac{x+1}{x^2+1}\right)$$
$$=\left(\frac{x-1}{x+2}\right)\times\left(\frac{x+1}{x^2+1}\right)-\left(\frac{x}{x-1}\right)\times\left(\frac{x+1}{x^2+1}\right),$$

where $x$ is not specifically assigned, is a particular case of the law of distribution.

§ **37. Law of Distribution for Division.**—The law of distribution has a limited application to division.

In the first place, we may point out that the laws of signs for the division of algebraic quantities follow from the corresponding laws for multiplication. For example, since

$+(a\div b)\times(-b) = -\{(a\div b)\times b\}$, by law of signs for multiplication,

$= -a$, by the mutual inverseness of multiplication and division,

it follows, if $b$ be any finite quantity differing from 0, that

$$+(a\div b)\times(-b)\div(-b)=(-a)\div(-b).$$

Hence, suppressing the mutually destructive operations $\times(-b)\div(-b)$ on the left, and interchanging the two sides of the equation, we get

$$(-a)\div(-b) = +(a\div b).$$

In like manner we establish all the four cases—

$$(+a)\div(+b) = +(a\div b),\quad (+a)\div(-b) = -(a\div b),$$
$$(-a)\div(+b) = -(a\div b),\quad (-a)\div(-b) = +(a\div b) \qquad (1).$$

Again, by the law of distribution for multiplication, if A be any finite quantity or operand differing from 0, we have

$$\begin{aligned}(a\div \mathrm{A} - b\div \mathrm{A} + c\div \mathrm{A} - d\div \mathrm{A})\times(+\mathrm{A})\\ = +\{(a\div \mathrm{A})\times \mathrm{A}\} - \{(b\div \mathrm{A})\times \mathrm{A}\} + \{(c\div \mathrm{A})\times \mathrm{A}\}\\ - \{(d\div \mathrm{A})\times \mathrm{A}\}\\ = +a-b+c-d. \qquad (2).\end{aligned}$$

If now we divide both sides of (2) by $+\mathrm{A}$, and suppress the mutually destructive operations $\times(+\mathrm{A})\div(+\mathrm{A})$ on the left, we get, after interchanging the two sides of the equation—

$$\begin{aligned}(+a-b+c-d)\div(+\mathrm{A})\\ = +(a\div \mathrm{A}) - (b\div \mathrm{A}) + (c\div \mathrm{A}) - (d\div \mathrm{A}) \qquad (3).\end{aligned}$$

And we could in the same way deduce that

$$\begin{aligned}(+a-b+c-d)\div(-\mathrm{A})\\ = -(a\div \mathrm{A}) + (b\div \mathrm{A}) - (c\div \mathrm{A}) + (d\div \mathrm{A}) \qquad (4).\end{aligned}$$

Equations (3) and (4) are evidently particular cases of the following general law.

*To divide an algebraic sum by any algebraic quantity write down all the partial quotients obtained by dividing each term of the dividend by the divisor, attaching the sign + if the term and the divisor have like signs, the sign − if they have opposite signs.*

We may express this result briefly by saying that the dividend may be distributed. The same is not true of the divisor, at least not as a general rule of algebraic operation; and it is only with such rules that we are now concerned. To establish this it is sufficient to advance a single arithmetical exception. If the divisor could be distributed, then we should have $3 \div (2+1) = 3 \div 2 + 3 \div 1$; in other words, $1 = 4\frac{1}{2}$, which is false.

§ 38. **Distributive Properties of 0.**—If $a$ be any finite quantity, and $b$ any finite quantity or 0, then we have, by the laws already established—

$$(+a-a) \times b = +(ab) - (ab);$$

that is to say—

$$0 \times b = 0 \qquad (1);$$

and, in particular—

$$0 \times 0 = 0 \qquad (2).$$

Again, if $b$ be any finite quantity, *excluding* 0, we also have

$$(+a-a) \div b = +(a \div b) - (a \div b);$$

in other words—

$$0 \div b = 0 \qquad (3).$$

The equations (1), (2), and (3) may be called the distributive properties of zero.

§ 39. **Excepted Operands.**—In laying down the laws of Algebra we have assumed throughout that all operands are finite definite quantities. Otherwise the operands so far may have any finite value in the series of real quantity, except only that 0 may not be a divisor. It may be of interest to satisfy the reader that the admission of division by 0 as a general algebraic operation would lead to contradiction. Let us suppose that $a$ is any finite quantity whatever; then, if division by 0 is to be admissible as an algebraic operation, $a \div 0$ must have some finite value, $b$ say. We should then have

$$a \div 0 = b \qquad (1).$$

Again, we should deduce from (1) that

$$a \div 0 \times 0 = b \times 0 \qquad (2).$$

If 0 is to be admitted as an ordinary operand, $a \div 0 \times 0$ is, by the mutual relation of multiplication and division, simply $a$. On the other hand, $b \times 0$, since $b$ is finite, is 0. Hence (2) asserts that $a=0$, which contradicts our hypothesis that $a$ is *any* finite quantity whatever.

It is a very common beginner's mistake to suppose that $0 \div 0$ is an admissible algebraic operation, and that its value is 1. No such operation can be admitted as we have seen; but it is perhaps well that it should be seen that, even were it to be admitted, it could not be asserted that the result is 1 in all cases. This will be seen from the consideration of the three quotients, $x^2 \div x$, $2x \div x$, $x \div x^2$. If it were the case that $0 \div 0$ is admissible and equivalent to 1 in all cases, it would follow that as $x$ is made smaller and smaller each of these quotients should approach more and more nearly to 1; whereas it is obvious that the first becomes more and more nearly 0; the second is 2; and the third becomes greater and greater, and can be made to exceed any given quantity whatever.

§ **40.** As we have now discussed all the fundamental laws of Algebra, it will be convenient to give a synoptic table of them for convenience of reference.

For the sake of brevity we have condensed the statements by the use of double signs. Thus, instead of writing all the different cases $+(+a+b)=+(+a)+(+b)$, $+(+a-b)=+(+a)+(-b)$, etc., we have written $\pm(\pm a \pm b)=\pm(\pm a)\pm(\pm b)$, the understanding being that the signs are to be taken from corresponding places on both sides.

## SYNOPTIC TABLE OF THE LAWS OF ALGEBRA

### Definitions connecting the Direct and Inverse Operations

| Addition and subtraction— | Multiplication and division— |
|---|---|
| $+a-b+b=+a,$ | $\times a \div b \times b = \times a,$ |
| $+a+b-b=+a.$ | $\times a \times b \div b = \times a.$ |

### Law of Association

| For addition and subtraction— | For multiplication and division— |
|---|---|
| $\pm(\pm a \pm b)=+(\pm a)\pm(\pm b),$ | $\overset{\times}{\div}(\overset{\times}{\div}a\overset{\times}{\div}b)=\overset{\times}{\div}(\overset{\times}{\div}a)\overset{\times}{\div}(\overset{\times}{\div}b),$ |

with the following law of signs:—

The concurrence of like signs gives the direct sign;

The concurrence of unlike signs the inverse sign.

Thus—

$$+(+a) = +a, \quad +(-a) = -a; \qquad \times(\times a) = \times a, \quad \times(\div a) = \div a,$$
$$-(-a) = +a, \quad -(+a) = -a. \qquad \div(\div a) = \times a, \quad \div(\times a) = \div a.$$

## Law of Commutation

For addition and subtraction—

$$\pm a \pm b = \pm b \pm a,$$

For multiplication and division—

$$\substack{\times \\ \div} a \substack{\times \\ \div} b = \substack{\times \\ \div} b \substack{\times \\ \div} a,$$

the operand always carrying its own sign of operation with it.

### *Properties of* 0 *and* 1

$$0 = +a - a,$$
$$\pm b + 0 = \pm b - 0 = \pm b,$$
$$+0 = -0.$$

$$1 = \times a \div a,$$
$$\substack{\times \\ \div} b \times 1 = \substack{\times \\ \div} b \div 1 = \substack{\times \\ \div} b,$$
$$\times 1 = \div 1.$$

## Law of Distribution

For multiplication—

$$(\pm a \pm b) \times (\pm c \pm d) = +(\pm a) \times (\pm c) + (\pm a) \times (\pm d)$$
$$+ (\pm b) \times (\pm c) + (\pm b) \times (\pm d),$$

with the following law of signs :—

If a partial product has constituents with like signs, it must have the sign $+$;

If the constituents have unlike signs, it must have the sign $-$.

Thus—

$$+(+a) \times (+c) = +a \times c, \quad +(+a) \times (-c) = -a \times c,$$
$$+(-a) \times (-c) = +a \times c, \quad +(-a) \times (+c) = -a \times c.$$

### *Property of* 0

$$0 \times b = b \times 0 = 0.$$

For division—

$$(\pm a \pm b) \div (\pm c) = +(\pm a) \div (\pm c) + (\pm b) \div (\pm c),$$

with the following law of signs :—

If the dividend and divisor of a partial quotient have like signs, the partial quotient must have the sign $+$;

If they have unlike signs, the partial quotient must have the sign $-$.

Thus—

$$+(+a)\div(+c)=+a\div c, \quad +(+a)\div(-c)=-a\div c,$$
$$+(-a)\div(-c)=+a\div c, \quad +(-a)\div(+c)=-a\div c.$$

*N.B.*—The divisor cannot be distributed.

*Property of* 0

$$0\div b=0.$$

RESTRICTIONS ON THE OPERANDS

In applying the general laws every algebraic operand is supposed to be finite, and division by 0 is excluded.

Although the laws of integral indices are derived and not fundamental principles, it will be convenient to repeat them here for reference.

LAWS OF INDICES

I.

$$(\alpha)\quad a^m a^n a^p \ldots = a^{m+n+p+\ldots}$$
$$(\beta)\quad a^m/a^n = a^{m-n}, \text{ if } m>n;$$
$$= 1/a^{n-m}, \text{ if } m<n.$$

II.

$$(a^m)^n = a^{mn}.$$

III.

$$(\alpha)\quad (abc\ldots)^m = a^m b^m c^m \ldots$$
$$(\beta)\quad (a/b)^m = a^m/b^m.$$

§ 41. **Meaning of the Sign =, Identical and Conditional Equations.**—The reader should here mark the exact signification of the sign = as hitherto used. It means "is transformable into by applying the laws of Algebra and the definitions of the symbols or functions involved, without any assumption regarding the operands involved."

Any "equation" which is true in this sense is called an "Identical Equation," or an "Identity"; and must, in the first instance at least, be carefully distinguished from an equation the one side of which can be transformed into the other by means of the laws of Algebra *only when the operands involved have particular values or satisfy some particular condition.*

For example, $(x+1)(x-1)=x^2-1$, and $3\times2+2\times2=5\times2$ are identities; but $2x-3=x+1$, and $x+y=x^2+y^2$ are conditional equations.

Some writers constantly use the sign $\equiv$ for the former kind of equation, and the sign $=$ for the latter. There is much to be said for this practice, and teachers will find it useful with beginners. We have, however, for a variety of reasons,* adhered, in general, to the old usage; and have only introduced the sign $\equiv$ occasionally in order to emphasise the distinction in cases where confusion might be feared.

## Elementary Applications of the Laws of Algebra, more particularly of the Law of Distribution

§ **42.** It will tend to clearness in what follows if we dwell for a little on certain ways of classifying monomial rational functions more briefly called terms. As we have already mentioned, the operands in any algebraical calculation are usually divided into two classes—constants (including almost always any numerical operands) and variables. When nothing is said to the contrary, all the operands which are indicated by letters are to be taken as variables, unless some are indicated by initial letters of the alphabet, and others by final letters, in which case the latter are to be taken as variables.

Two terms that have the same variable part are spoken of as **like terms,** *e.g.* $3a^2b$ and $6a^2b$ are like terms; so also are $7ab/c$ and $\frac{1}{2}ab/c$; and again, $ab^2xy/z$ and $a^2bxy/z$, $x$, $y$, $z$ being variables according to the understanding mentioned above.

§ **43.** When the variable part of one term can be derived from the variable part of another by interchanging one or more pairs of the variables the terms are said to be of the same **type.** For example, $3yz/x$, $6zx/y$, $7xy/z$ are all of the same type; for the variable part of the second can be derived from the first by interchanging $x$ and $y$; and the variable part of the third from

* Chief among them is the view which will be found to pervade this book, that all algebraic equality (which is not approximate) is, at bottom, identity. The same is true of arithmetic equality. Algebra is, in short, "The Calculus of Identity."

the variable part of the first by interchanging $x$ and $z$. When any one term of a particular type is given, the variable parts of all the others of the same type can at once be written down. The following are examples of terms of given types (for shortness we have written only the variable parts, omitting the coefficients):—

I. VARIABLES $x$, $y$.

| | |
|---|---|
| Type $x^2$ | $x^2$, $y^2$ |
| „ $x^2y$ | $x^2y$, $xy^2$ |
| „ $x^2/y$ | $x^2/y$, $y^2/x$. |

II. VARIABLES $x$, $y$, $z$.

| | |
|---|---|
| Type $x^2$ | $x^2$, $y^2$, $z^2$ |
| „ $xy$ | $yz$, $zx$, $xy$ |
| „ $xyz$ | $xyz$ |
| „ $x^2yz$ | $x^2yz$, $xy^2z$, $xyz^2$ |
| „ $x^2/y^2z^2$ | $x^2/y^2z^2$, $y^2/z^2x^2$, $z^2/x^2y^2$. |

III. VARIABLES $x$, $y$, $z$, $u$.

| | |
|---|---|
| Type $x^2$ | $x^2$, $y^2$, $z^2$, $u^2$ |
| „ $xy$ | $xy$, $xz$, $xu$, $yz$, $yu$, $zu$ |
| „ $xyz$ | $yzu$, $xzu$, $xyu$, $xyz$. |

Terms which are integral are also classified according to degrees, thus $3x^2y$, $4xy^2$, $6x^3$ are all of the third degree in $x$ and $y$. It should be noticed that, although integral terms of the same type are necessarily of the same degree, the converse is not true, as the example just quoted shows.

§ **44.** The following definitions are also of great importance:—

By the **degree of an integral function** with respect to any assigned variables is meant *the degree of the term of highest degree in the function.*

An integral function is said to be **homogeneous** with respect to any set of variables *when each of its terms is of the same degree with respect to those variables.*

Ex. **1.** The degree of $3x^3-2x+1$ in $x$ is the 3rd.

Ex. **2.** The degree of $abx^2y+b^2cxyz+acx^3y$ in $x$, $y$, $z$ is the 4th; in $a$, $b$, $c$ the 3rd; in $a$, $b$, $c$, $x$, $y$, $z$ the 6th.

Ex. **3.** $3x^2+2y^2-xy$ is homogeneous in $x$ and $y$.

We speak of all the products of the 1st, 2nd, 3rd . . . degrees that can be formed with a given set of variables, say $x$, $y$, $z$, as the **unary, binary, ternary, . . . products** of these variables. For example, all the ternary products of $x$, $y$, $z$ are $x^3$, $y^3$, $z^3$ ; $y^2z$, $yz^2$, $z^2x$, $zx^2$, $x^2y$, $xy^2$ ; $xyz$. It will be noticed that they fall into three distinct types, viz. type $x^3$, type $y^2z$, type $xyz$.

When we have enumerated all the different ternary products of $x$, $y$, $z$, we can write down all the different integral terms of the 3rd degree which can be constructed by means of these variables, viz. using letters to denote the constant parts or coefficients, these terms are the following ten :—$ax^3$, $by^3$, $cz^3$ ; $dy^2z$, $eyz^2$, $fz^2x$, $gzx^2$, $hx^2y$, $ixy^2$ ; $jxyz$ ; and so in general.

§ **45.** By using the law of distribution read backwards, *e.g.* $ab-ac=a(b-c)$, which may be called the process of **Collection,** we can condense into one term all the terms in an algebraic sum which have the same variable part.

Ex. 1. $3x-(6x-2y)+2x+3y-z$

$$\begin{aligned}&=3x-6x+2y+2x+3y-z, &&\text{(assoc. law)},\\ &=3x-6x+2x+2y+3y-z, &&\text{(comm. law)},\\ &=(3-6+2)x+(2+3)y-z, &&\text{(dist. law)},\\ &=(-1)x+5y-z, &&\text{(assoc. law)},\\ &=-x+5y-z, &&\text{(dist. law)}.\end{aligned}$$

Ex. 2. $(x^2-3x+1)(x+1)$

$$\begin{aligned}&=x^2x-3xx+x\\ &\quad+\ x^2-3x+1, &&\text{(dist. law and prop. of 1)},\\ &=x^3-3x^2+x\\ &\quad+\ x^2-3x+1, &&\text{(laws of indices)},\\ &=x^3+(-3+1)x^2+(1-3)x+1, &&\text{(dist. law)},\\ &=x^3-2x^2-2x+1, &&\text{(assoc. law)}.\end{aligned}$$

In transforming and reducing integral functions and functions which reduce to an algebraic sum of rational terms, it is convenient and usual to arrange side by side terms which are of the same type, and also in the case of integral terms to group together those of the same degree. The functions so arranged may be said to be in **Standard Form.**

Ex. 3. $(a^2+b^2+c^2)(a+b+c)$

$$\begin{aligned}&=a^3+ab^2+ac^2+a^2b+b^3+bc^2+ca^2+b^2c+c^3,\\ &\qquad\qquad\text{(dist. law, comm. law, and ind. law)},\\ &=a^3+b^3+c^3+b^2c+bc^2+c^2a+ca^2+a^2b+ab^2,\\ &\qquad\qquad\qquad\qquad\text{(comm. law)}.\end{aligned}$$

Ex. 4. $(x+y+1)(x+y)$

$$\begin{aligned}&=x^2+yx+x+xy+y^2+y, &&\text{(dist. law)},\\ &=x^2+2xy+y^2+x+y, &&\text{(comm. law, dist. law)}.\end{aligned}$$

**Sigma-notation.**—In cases, such as Example 3, *where terms of the same type having the same coefficient occur*, it is usual to employ a contraction for the sum of the terms of a particular type, thus we write $\Sigma a^3$ in place of $a^3+b^3+c^3$; and $\Sigma b^2c$ or $\Sigma a^2b$ or $\Sigma ab^2$ (any term of the type may follow the $\Sigma$) in place of $b^2c+bc^2+c^2a+ca^2+a^2b+ab^2$. Since all the terms of a particular type are known, this can lead to no confusion, provided the variables concerned are clearly understood. If there be any doubt about this last point, the variables should be subscribed to the $\Sigma$; thus $\Sigma_{abc} a^3=a^3+b^3+c^3$, but $\Sigma_{ab} a^3=a^3+b^3$; usually, however, it is clear from the context what the variables are.

With this **Sigma-notation** we should write the result of Example 3—

$$(a^2+b^2+c^2)(a+b+c)=\Sigma a^3+\Sigma a^2b\,;$$

or

$$\Sigma_{abc} a^2\Sigma_{abc} a=\Sigma_{abc} a^3+\Sigma_{abc} a^2b.$$

And the result of 4

$$(x+y+1)(x+y)=\Sigma x^2+2xy+\Sigma x\,;$$

or

$$(\Sigma_{xy} x+1)\Sigma_{xy} x=\Sigma_{xy} x^2+2xy+\Sigma_{xy} x.$$

Ex. 5. $\{3(x+y-1)x-(3x+3y-3)y\}/(x-y)$.

Since $3x+3y-3=3(x+y-1)$, (dist. law),

$$\begin{aligned}&\{3(x+y-1)x-(3x+3y-3)y\}/(x-y)\\ &=\{3(x+y-1)x-3(x+y-1)y\}/(x-y),\\ &=3(x+y-1)(x-y)/(x-y), &&\text{(dist. law)},\\ &=3(x+y-1), &&\text{(def. of}\times\text{and}\div),\\ &=3x+3y-3, &&\text{(dist. law)}.\end{aligned}$$

Ex. 6. Reduce

$$\frac{\dfrac{x-1}{(x+1)(x+2)}-\dfrac{x+1}{(x+2)(x+3)}}{\dfrac{x-2}{(x-1)(x+1)}+\dfrac{x+2}{(x+1)(x+3)}}$$

to the quotient of two integral functions of $x$.

Since

$$\begin{aligned}&\frac{(x-1)}{(x+1)(x+2)}\\ &=(x-1)\div\{(x+1)(x+2)\}\times(x+3)\div(x+3),\\ &\qquad\text{(def. of }\times\text{and}\div),\\ &=\{(x-1)(x+3)\}\div\{(x+1)(x+2)(x+3)\},\\ &\qquad\text{(comm. law and assoc. law)},\\ &=\frac{(x-1)(x+3)}{(x+1)(x+2)(x+3)}\,;\end{aligned}$$

and, in like manner—

$$\frac{x+1}{(x+2)(x+3)}=\frac{(x+1)(x+1)}{(x+1)(x+2)(x+3)},$$

we have

$$\frac{x-1}{(x+1)(x+2)}-\frac{x+1}{(x+2)(x+3)}=\frac{(x-1)(x+3)}{(x+1)(x+2)(x+3)}-\frac{(x+1)(x+1)}{(x+1)(x+2)(x+3)}$$

$$=\frac{(x-1)(x+3)-(x+1)(x+1)}{(x+1)(x+2)(x+3)},$$

(dist. law for $\div$),

$$=\frac{(x^2-x+3x-3)-(x^2+x+x+1)}{(x+1)(x+2)(x+3)},$$

(dist. law),

$$=\frac{-4}{(x+1)(x+2)(x+3)},$$

(assoc. comm. dist. laws).

In like manner—

$$\frac{x-2}{(x-1)(x+1)}+\frac{x+2}{(x+1)(x+3)}=\frac{(x-2)(x+3)+(x-1)(x+2)}{(x-1)(x+1)(x+3)},$$

$$=\frac{(x^2-2x+3x-6)+(x^2-x+2x-2)}{(x-1)(x+1)(x+3)}$$

$$=\frac{2x^2+2x-8}{(x-1)(x+1)(x+3)}.$$

Using F to denote the given function, we therefore have

$$F=\{-4/(x+1)(x+2)(x+3)\}/\{(2x^2+2x-8)/(x-1)(x+1)(x+3)\}.$$

If we now apply the law of association for multiplication and division we get

$$F=-4\div(x+1)\div(x+2)\div(x+3)\div(2x^2+2x-8) \times(x-1)\times(x+1)\times(x+3).$$

Commutating and removing the mutually destructive operations $\div(x+1)$ and $\times(x+1)$, etc., we deduce

$$F=-4(x-1)/(2x^2+2x-8)(x+2).$$

Since

$$(2x^2+2x-8)(x+2)$$
$$=2x^3+2x^2-8x$$
$$+4x^2+4x-16,$$
$$=2x^3+6x^2-4x-16, \qquad \text{(dist. law)},$$

and $-4(x-1)=-4x+4$, (dist. law), we have finally

$$F=\frac{-4x+4}{2x^3+6x^2-4x-16}.$$

**§ 46.** We now give three sets of exercises, the object of which is to accustom the beginner to the direct application of the laws of Algebra. In view of their purpose he must in working them avoid all use of derived principles, even if he happen to be partially acquainted with such ; he must not speak of "cancelling" this or that, of "cross multiplying," or the like ; he must not use any of the rules for the multiplication or addition of fractions, or any mechanical rule or process whatsoever. For every step a reference to a definition, or to one of the fundamental laws as given in § 41, or to one of the laws of indices must be assigned.

When, in Exercises VI., VII., VIII., IX. a function is set down to be *transformed* by the laws of Algebra, the understanding is, that if an integral result be attainable, it is to be arranged in standard form so that terms of like degree and of like type are placed together. When initial and final letters of the alphabet both occur in the same function, it is to be understood that the latter are the variables, and the former, together with mere numbers, are constants, and the coefficients are to be constructed accordingly, brackets being used for this purpose where necessary. If the final result is not integral, it should be arranged as a sum of fractional terms grouped according to type; or else as a rational fraction of the simplest possible kind—that is to say, as the quotient of two integral functions, each of which is of the lowest possible degree. Both these operations are usually indicated by the word "simplify," which, however, scarcely conveys a definite meaning to the mind of a novice, and is unfortunately used in many different and partly contradictory senses.

Neatness, facility, and above all, accuracy in using the various signs should be carefully studied in writing out the working of each exercise. For example, care must be taken not to write $a \div b + c$ when $a \div (b + c)$ is meant. One of the most important educational uses of elementary Algebra, next to the logical training it ought to give, is that it cultivates neatness and forethought in arranging details—in short, the power of organisation on a small scale. Like many other sciences, Algebra consists largely in skilfully fitting together a number of very simple considerations about very simple matters; and the difficulty that untrained minds find in it arises simply from deficiency in the capacity for taking pains.

The beginner should also be careful to give clear explanations in good English wherever such are required. The ideal of a piece of good algebraic work is not a page of symbols without a word of the Queen's English anywhere, but a piece of *consecutive* reasoning, partly in symbolic shorthand no doubt, but still so written that it could, if need were, be wholly translated into non-symbolic language.

After the beginner feels that he has thoroughly mastered the application, both direct and inverse, of the algebraic laws, and has attained sufficient neatness and accuracy of working, he should not waste his time by the vain repetition of familiar exercises, but go on and try some that are more difficult or less familiar, or proceed with the book work.

## EXERCISES VI.

Transform the following into standard forms :—

1. $(-a)^5$.
2. $(-a)^{4n+1}$.
3. $(-a)^2 \times (-a^2)^3 \times (-a^3)^4 \times (-a^4)^5$.
4. $(-\frac{2}{3}a^2)^3 \times (-\frac{3}{2}a^3)^3$.
5. $(-\frac{5}{6}a^2by^2z^2) \times (-\frac{3}{5}b^2cz^2x^2) \times (2c^2ax^2y^2)$.
6. $[-\{-(-a)^2\}^3]^5$.
7. $[(-5)^3\{-(-2)^3\}^2(-6)^4]^3$.
8. $(- \, . \, . \, . \, (-(-(-a)^1)^2)^3 \, . \, . \, .)^{4n}$.
9. $3a+2b-3a+2c-a-4b$.
10. $\frac{1}{2}x-\frac{1}{3}x^2+\frac{4}{5}x-\frac{2}{3}x^2$.
11. $(a+b)+(a-b)$.
12. $(a+b)-(a-b)$.
13. $(b-c)+(c-a)+(a-b)$.
14. $3(a-b)-\{2(a-b)-(a+b)\}$.
15. $(a+b+c)-(a+b+c)-(a-b+c)$.
16. $(x+2y-3z)-(3x+4y)-(y+z-2x)$.
17. $\{x-(2x-3y)\}+\{x-(2x+3y)\}-\{x-(3y-2x)\}$.
18. $(x^2-xy+y^2)+(y^2-yz+z^2)+(z^2-zx+x^2)$.
19. $(x^3-x+1)+(x^3-x^2+1)+\{x^3+2(x^2+x)+1\}$.
20. $a-[b-\{a-(b-x)\}]$.
21. $a-[2a-\{3a-(4a-x)\}]$.
22. $\frac{1}{2}\{a-\frac{1}{2}(a-\frac{1}{2}[a-2b])\}$.
23. $(a_1-a_2)+(a_2-a_3)+(a_3-a_4)+ \, . \, . \, . \, +(a_{n-1}-a_n)$.
24. $1(3-2)+2(3-1)+3(1-2)$.
25. $a(b-c)+b(c-d)+c(d-e)+d(e-a)$.
26. $6\left(\frac{x}{2}+\frac{x}{3}\right)-6\left(\frac{x}{2}-\frac{x}{3}\right)$.
27. $\{(a+b)x+(a-b)y\}-\{(a-b)x-(a+b)y\}$.
28. $a[1-b\{1-c(1-d\overline{1-e})\}]$.
29. $(a_1-a_2)+2(a_2-a_3)+3(a_3-a_4)+ \, . \, . \, . \, +(n-1)(a_{n-1}-a_n)$.
30. $2[x-y-2\{x-3(y-2x)-y\}]$.
31. $a(3b-2b)+b(3a-2a)+(2a-3a)(2b-3b)$.
32. $a(a+b-c)+b(-a+b+c)+c(a-b+c)$.
33. $x^2(x^2-x+1)+x(x^3-x^2+x+1)-(2x^4-3x^3+4x^2-x+1)$.
34. $(yz+zx-xy)x+(zx+xy-yz)y+(xy+yz-zx)z$.
35. $(x-6)(x+8)$.
36. $(-x-6)(x+8)$.
37. $(x-\frac{1}{2})(x+\frac{1}{3})$.
38. $(\frac{1}{2}x+1)(x-\frac{1}{2})$.
39. $(\frac{1}{3}x^3+\frac{1}{4}x^2+\frac{1}{2}x+1)(x-\frac{1}{2})$.
40. $(ax+1)(bx+1)$.
41. $(ax-b)(cx-d)$.
42. $(x/a+1)(y/b+1)$.
43. $\frac{1}{2}(x+1)(x-1)-\frac{1}{3}(3x+3)(x-1)-\frac{1}{4}(x^2-4x+4)$.
44. $(x-y+2z-xy)-(2xy-y+2x)-2(x-1)(y-1)$.

45. Show that, if $x=1$ or $x=5$, then $x^2-6x+5\equiv 0$.
46. Show that, if $x=-a$ or $x=-a^2$, then $x^2+(a+a^2)x+a^3\equiv 0$.

Find by inspection values of $x$ that will make the following equations identities :—

47. $(x-1)(x-2)=0$.
48. $(3x-3)(x+4)=0$.
49. $(2x+4)(x+2)=0$.
50. $x(x-1)+2x(x-1)=0$.

**51.** Find by inspection values of $x$ and $y$ which will render the following pair of equations identities :—

$$x/a^2+y/b^2=2, \quad x/a+y/b=a+b.$$

**52.** The variables being $x$, $y$, $z$, write down all the terms of the following types :—1°, $x^3y$ ; 2°, $x^2y^2z$ ; 3°, $xy/z$.

**53.** The variables being $x$, $y$, $z$, $u$, write down all the terms of the following types :—1°, $x^2y$ ; 2°, $x^3y$ ; 3°, $xy/z$ ; 4°, $x^2y/z$ ; 5°, $x^2y^2/z^2$.

**54.** Write down all the ternary products of the four letters $x$, $y$, $z$, $u$.

**55.** Write down general forms (indicating coefficients by distinct letters) for integral functions of $x$, $y$, $z$, $u$, 1° which contain all the terms of the second degree in the variables, 2° which contain all terms of the type $x^2y$.

**56.** Write down an integral function of $x$ and $y$ which contains all possible terms whose degree does not exceed the third.

**57.** Find a general form for an integral function of $x$, $y$, $z$ which contains all terms of degree not exceeding the second, and which does not alter in value when any pair of the variables are interchanged.

**58.** The variable part of a rational monomial function is $x^2y/z^2$ ; if its value be $-3$ when $x=-1$, $y=-2$, $z=-3$, find its value when $x=1$, $y=3$, $z=-5$.

**59.** A rational function of $x$, $y$, $z$ is the sum of all terms of the type $\mathrm{A}yz/x$. If its value be $-1$ when $x=y=z=1$, find its value when $x=y=z=2$.

**60.** An integral function of $x$, $y$, $z$ contains only the terms of the type $xy$, and is unaltered in value by the interchange of any pair of the variables. If the value of the function be 10 when $x=1$, $y=2$, $z=3$, find the function.

**61.** If the value of $\mathrm{A}yz/x+\mathrm{B}zx/y+\mathrm{C}xy/z$, where A, B, C are coefficients independent of $x$, $y$, $z$, be unaltered in value when $y$ and $z$ are interchanged, show that $\mathrm{B}=\mathrm{C}$.

## EXERCISES VII.

Transform the following into standard forms :—

**1.** $(-\frac{2}{3}x^2-\frac{4}{3}x+1)(-\frac{1}{2}x^2)$. **2.** $(3x^2-3x+1)(x-1)$.

**3.** $(3x^2-3x+1)(x+1)$. **4.** $(\frac{1}{2}x^3-\frac{5}{6}x^2+\frac{3}{2}x-1)(2x-\frac{6}{5})$.

**5.** $\frac{1}{2}(2x^2+6x+4)(2x+\frac{1}{2})$. **6.** $(x-\frac{1}{2})(x-\frac{1}{5})(x+\frac{1}{6})$.

**7.** $(x^3+x^2y+xy^2+y^3)(x-y)$. **8.** $\{(a-b)x^2+(b-c)x+(c-d)\}(x+1)$.

**9.** $\left(\frac{a}{b}x+\frac{b}{a}y\right)\left(\frac{b}{a}x-\frac{a}{b}y\right)$.

**10.** $(x+a)(x+b)+(x+a)(x-b)+(x-a)(x+b)+(x-a)(x-b)$.

**11.** $(x^2-3xy+y^2)(x^2+xy+y^2)$. **12.** $(x^2+y^2)(x^2-y^2)$.

**13.** $\{(a+b)x^2-(a+b)y^2\}(x^2+y^2)$. **14.** $(3x^2-y^2)(3x^2+y^2)$.

**15.** $(x^n+a^n)(x^n-a^n)$. **16.** $(x^n-1)^2$. **17.** $(x^n-1/x^n)^2$.

**18.** $(2^nx^2+2^{n-1}x+1)(x+2)$. **19.** $(3x^n+2x^{n-1}+1)(x-1)$.

**20.** $(x^{2n}-x^n+1)(x^n+1)$. **21.** $(x^{p+q}+x^p+1)(x^{p-q}-x^p+1)$.

**22.** $(x^{p-a}+x^{p-b}+x^{p-c})(x^{p+a}+x^{p+b}+x^{p+c})$.

**23.** $(x+\sqrt{(xy)}+y)(x-\sqrt{(xy)}+y)$.

**24.** $(x+\sqrt{3}-1)(x-\sqrt{3}-1)=0$.

**25.** Express $\{(x-1)^3+3(x-1)^2+2(x-1)+1\}(3x+2)$ in the form $A(x-1)^4+B(x-1)^3+C(x-1)^2+D(x-1)+E$, where A, B, C, D, E are numerical coefficients.

**26.** Show that $c$ can be determined so that

$$x^2+3x+c\equiv(x+1)(x+2).$$

**27.** Find A so that $(x+1)(x^2+Ax+1)\equiv x^3+1$.

**28.** Determine the numerical coefficients A, B, C, so that

$$(x^2+x+2)(x^2-2x+1)-(Ax^2+Bx+C)\equiv x^4-x^3+x+1.$$

**29.** Show that $(1/\sqrt{x}+3/\sqrt{x}-1)(1/\sqrt{x}+3/\sqrt{x}+1)$ is a rational function of $x$.

**30.** Show that a numerical value of $a$ can be found such that $(x+\sqrt{x}+1)(x+a\sqrt{x}+1)$ is a rational integral function of $x$.

## EXERCISES VIII.

Transform the following into standard forms :—

**1.** $\{3-\frac{5}{4}(\frac{1}{2}-\frac{1}{3})\}/\{\frac{3}{2}-\frac{5}{6}(\frac{1}{2}-\frac{1}{3})\}$.

**2.** $\left(\dfrac{a_1}{a_2}\right)\left(\dfrac{a_2a_3}{a_3a_4}\right)\left(\dfrac{a_3a_4a_5}{a_4a_5a_6}\right)\left(\dfrac{a_4a_5a_6a_7}{a_5a_6a_7a_8}\right)$.

**3.** $\dfrac{a^2b-a^2c}{a}+\dfrac{b^2c-b^2a}{b}+\dfrac{c^2a-c^2b}{c}$.

**4.** $\left\{\dfrac{x^2+1}{x-1}-x-1\right\}(x+1)$.

**5.** $\dfrac{1}{a+b}(a-c+b)=1-\dfrac{c}{a+b}$.

**6.** $xyzu(1/x+1/y+1/z-1/xyz)-u\{(1+y)(1+z)+(1+z)(1+x)$
$+(1+x)(1+y)\}+2u(x+y+z+4)$.

**7.** $\{a(x+y)-b(x+y)+3x+3y\}/(x+y)$.

**8.** $(2x+2)^2(x-1)^3/(2x-2)$.

**9.** $(x-1)/(x-2)=(x^2+x-2)/(x^2-4)$.

**10.** $\dfrac{1+1/x^2}{1+1/x^3}+x\dfrac{x^2+1}{x^3+1}$.

**11.** $\left(1-\dfrac{y}{z}\right)\left(1+\dfrac{z}{x}\right)+\left(1-\dfrac{z}{x}\right)\left(1+\dfrac{x}{y}\right)+\left(1-\dfrac{x}{y}\right)\left(1+\dfrac{y}{z}\right)$.

**12.** $\left(\dfrac{x}{y}+1\right)\left(\dfrac{y}{x}+1\right)+\left(\dfrac{x}{y}-1\right)\left(\dfrac{y}{x}-1\right)$.

**13.** If $x=(d-b)/(a-c)$, then $ax+b\equiv cx+d$.

**14.** $\dfrac{x+xy}{x-xy}+\dfrac{y+y^2}{y-y^2}$.

**15.** $2/(1/x+1/y)+(x^2+y^2)/(x+y)$.

**16.** $\left(1+\dfrac{y}{x-y}\right)\left(1-\dfrac{2y}{x+y}\right)\left(1+\dfrac{y}{x}\right)$.

**17.** $(a-b)/(a+d)+(b-c)/(a+d)+(c+d)/(a+d)$.

**18.** $\left(\dfrac{x}{a+b}+\dfrac{1}{a-b}\right)\left(\dfrac{x}{a-b}+\dfrac{1}{a+b}\right)$.

**19.** $(a/b+b/a)^2(a/b-b/a)^2-a^4/b^4-b^4/a^4+2$.

**20.** $1+1/(x-1)-x^2/(x^2-x)+(3x+3)/(x+1)$.

**21.** $\dfrac{x+1}{x-1}\left[\dfrac{x-1}{x+1}-\dfrac{x+1}{x-1}\left\{\dfrac{(x-1)^2}{(x+1)^2}-\dfrac{x+1}{x-1}\left(\dfrac{(x-1)^3}{(x+1)^3}-1\right)\right\}\right]$.

22. Express $x/(y-z)+y/(z-x)+z/(x-y)$ as a single fraction whose denominator is $(y-z)(z-x)(x-y)$.

23. $x-\cfrac{x}{x-\cfrac{x}{x-\cfrac{x}{x-1}}}.$ 24. $x-\cfrac{y}{x-\cfrac{y}{x-\cfrac{y}{x-y}}}.$

§ 47. **Separate Determination of the Coefficient of a Term of given Variable Part.**—As soon as the learner has thoroughly mastered the direct application of the laws, he should begin to practise dissecting the distribution of a product into the determination of the coefficients of all the separate terms that can possibly occur in the distributed product.

Consider for example $(x^2-x+1)(x+2)$. Here we can have a term which does not contain the variable (so-called **absolute term**), and terms whose variable parts are $x$, $x^2$, $x^3$. Let us fix our attention on $x^2$. Remembering that each term in the distributed product arises by multiplying a term from one of the brackets by a term from the other, we see that the terms in the distributed product which have $x^2$ for variable part are $x^2\times 2$ and $-x\times x$, that is, $2x^2$ and $-x^2$, the sum of which is $x^2$; hence the coefficient of $x^2$ is $+1$. Again, the terms whose variable part is $x$ are obviously $+1\times x$ and $-x\times 2$, the sum of which is $-x$; hence the coefficient of $x$ is $-1$; and so on. In this way the work of distributing a product can, after a little practice, be done in most cases mentally; and the method has many other advantages which will be appreciated more and more as we go on.

## Use of Standard Identities

§ 48. Ease and rapidity of algebraic calculation are much increased by the judicious use of **Standard Identities.** We proceed to establish a few of these by the direct application of the fundamental laws, reserving a more careful study of the methods of construction for a later chapter.

By the law of distribution combined with the first law of indices we have

$$(a+b)(a-b)=a^2+ba-ab-b^2.$$

Hence, since, by the law of commutation, $ba=ab$, we have

$$(a+b)(a-b)=a^2-b^2 \qquad (1).$$

Since $(a+b)^2$ stands for $(a+b)(a+b)$, we have, by distributing—

$$(a+b)^2 = a^2 + ba + ab + b^2.$$

Hence, collecting like terms—

$$(a+b)^2 = a^2 + 2ab + b^2 \qquad (2).$$

In like manner—

$$(a-b)^2 = a^2 - 2ab + b^2 \qquad (3).$$

The identity (2) may readily be generalised. Consider, say, four variables, $a$, $b$, $c$, $d$, then

$$\begin{aligned}(a+b+c+d)^2 &= (a+b+c+d)(a+b+c+d),\\ &= (a+b+c+d)a\\ &\quad + (a+b+c+d)b\\ &\quad + (a+b+c+d)c\\ &\quad + (a+b+c+d)d \qquad (\alpha),\end{aligned}$$

where for clearness in what follows the multiplier alone has been distributed. Since each partial product in the final result of the distribution consists of two factors (viz. one from each bracket), the only possible variable parts are $a^2$, $b^2$, $c^2$, $d^2$, *i.e.* the squares of all the variables; and $ab$, $ac$, $ad$, $bc$, $bd$, $cd$, *i.e.* all the products of pairs of the variables. Looking at the column of products ($\alpha$), we see at once that $a^2$ can only occur once, viz. from the distribution of the first line; and the same is true of $b^2$, $c^2$, $d^2$. Again, each of the products will occur twice and no more, *e.g.* $ab$ will come once from the line in which $a$ is multiplier, and once from the line in which $b$ is multiplier.

Hence we have

$$\begin{aligned}(a+b+c+d)^2 &= a^2 + b^2 + c^2 + d^2 + 2ab + 2ac\\ &\quad + 2ad + 2bc + 2bd + 2cd,\\ \text{or} \quad &= \Sigma a^2 + 2\Sigma ab \qquad (4).\end{aligned}$$

This result obviously holds for any number of letters, provided the sigmas be properly interpreted. We may state it in words as follows:—*The square of an algebraic sum is the algebraic sum of the squares of its terms and of twice all the products of its terms taken in pairs.*

By direct distribution we have

$$\begin{aligned}&(a^2 + ab + b^2)(a - b)\\ &\quad = a^3 + a^2b + ab^2\\ &\qquad\quad - a^2b - ab^2 - b^3.\end{aligned}$$

Hence, collecting—

$$(a^2+ab+b^2)(a-b)=a^3-b^3 \qquad (5).$$

In the same way—

$$(a^2-ab+b^2)(a+b)=a^3+b^3 \qquad (6).$$

Since, by the laws of indices, $(a+b)^3=(a+b)^2(a+b)$, we have, using (2)—

$$\begin{aligned}(a+b)^3 &= (a^2+2ab+b^2)(a+b),\\ &= a^3+2a^2b+ab^2\\ &\quad +a^2b+2ab^2+b^3.\end{aligned}$$

Hence, collecting—

$$(a+b)^3=a^3+3a^2b+3ab^2+b^3 \qquad (7).$$

Again—

$$\begin{aligned}(a-b)^3 &= (a-b)^2(a-b),\\ &= (a^2-2ab+b^2)(a-b),\\ &= a^3-2a^2b+ab^2\\ &\quad -a^2b+2ab^2-b^3.\end{aligned}$$

Hence, collecting—

$$(a-b)^3=a^3-3a^2b+3ab^2-b^3 \qquad (8).$$

We leave the direct deduction of the following, all of them important identities, as exercises for the learner:—

$$(a\pm b)^2\mp 4ab=(a\mp b)^2 \qquad (9),$$

$$(a^2+ab+b^2)(a^2-ab+b^2)=a^4+a^2b^2+b^4 \qquad (10),$$

$$(a^2+\sqrt{2}ab+b^2)(a^2-\sqrt{2}ab+b^2)=a^4+b^4 \qquad (11),$$

$$(a+b+c)^3=\Sigma a^3+3\Sigma a^2b+6abc \qquad (12),$$

$$\begin{aligned}&(a+b+c)(a^2+b^2+c^2-bc-ca-ab)\\ &\quad=\Sigma a^3-3abc \qquad (13),\end{aligned}$$

$$\begin{aligned}&(a+b+c)(-a+b+c)(a-b+c)(a+b-c)\\ &\quad=2\Sigma a^2b^2-\Sigma a^4 \qquad (14),\end{aligned}$$

$$\begin{aligned}(b-c)(c-a)(a-b) &= -a^2(b-c)-b^2(c-a)-c^2(a-b),\\ &= a(b^2-c^2)+b(c^2-a^2)+c(a^2-b^2),\\ &= -bc(b-c)-ca(c-a)-ab(a-b),\\ &= bc^2-b^2c+ca^2-c^2a+ab^2-a^2b \qquad (15),\end{aligned}$$

$$(x+a)(x+b)=x^2+(a+b)x+ab \qquad (16),$$

$$(x+a)(x+b)(x+c)=x^3+\Sigma ax^2+\Sigma abx+abc \qquad (17).$$

Several of these identities will be discussed from different points of view later on; and an extensive reference table of **Standard Identities** will be found in Chap. XI.

§ 49. **Principle of Substitution.**—The use of the law of association enables us to deduce from any of the above standard identities an infinity of others.

Consider for example $(a-b)^2$. By the law of association this may be written $\{a+(-b)\}^2$. Since there is no restriction upon the operands in an algebraic identity (except that they must be finite and in some cases not 0), we may replace the $b$ in identity (2) by $-b$ throughout; it then becomes

$$\{a+(-b)\}^2=a^2+2a(-b)+(-b)^2.$$

Since, by the law of distribution, $2a(-b)=-2ab$, and $(-b)^2=+b^2$, we have, replacing $+(-b)$ by $-b$ on the left—

$$(a-b)^2=a^2-2ab+b^2.$$

It thus appears that the identity (3) is really a particular case of (2).

Again $(3a-2b)^2$ may be written $\{(3a)+(-2b)\}^2$. Hence, from (2) we have $(3a-2b)^2=(3a)^2+2(3a)(-2b)+(-2b)^2$; whence, since, by the laws of indices and by the law of distribution $(3a)^2=3^2a^2=9a^2$, $2(3a)(-2b)=-2\times3\times2ab=-12ab$, $(-2b)^2=2^2b^2=4b^2$, we have

$$(3a-2b)^2=9a^2-12ab+4b^2 \qquad (\beta).$$

It will be observed that in the identity ($\beta$) we have lost all trace of the original identity (2). We can deduce ($\beta$) from (2); but not (or at least not so readily) (2) from ($\beta$). Here we see clearly the advantage of having a simple standard formula, and deducing others from it.

Since the method now explained consists in *substituting* for the operands in a standard formula other more or less complex operands, it is sometimes called the **Principle of Substitution.** The principle, however, is nothing but an assertion of the perfect generality of algebraic operands; in practice it confers upon algebraic formulæ the quality of a Chinese box, within which we find another box, within that another, and so on. The derived formulæ do not, however, necessarily become less capacious like the successive boxes; they may, in fact, become more capacious, *i.e.* more general, as the first of the following examples will show:—

Ex. 1.

$$\begin{aligned}(a+b+c)^3&=\{a+(b+c)\}^3,\\ &=a^3+3a^2(b+c)+3a(b+c)^2+(b+c)^3, \quad \text{by § 48 (7)};\\ &=a^3+3a^2(b+c)+3a(b^2+2bc+c^2)+(b^3+3b^2c+3bc^2+c^3),\end{aligned}$$

by § 48 (2) and (7).

Hence, distributing, dissociating, and collecting together terms of like type, we have finally

$$(a+b+c)^3=a^3+b^3+c^3+3b^2c+3bc^2+3c^2a+3ca^2+3a^2b+3ab^2+6abc \quad (\gamma),$$

which is identity (12) of § 48.

Ex. 2. If we substitute $-c$ for $c$ in the result of last example—

$$(a+b-c)^3=a^3+b^3-c^3-3b^2c+3bc^2+3c^2a-3ca^2+3a^2b+3ab^2-6abc \quad (\delta).$$

Ex. 3. By the law of association—

$$\begin{aligned}
&(a+b+c)(-a+b+c)(a-b+c)(a+b-c)\\
&=[\{(b+c)+a\}\{(b+c)-a\}][\{a-(b-c)\}\{a+(b-c)\}],\\
&=[(b+c)^2-a^2][a^2-(b-c)^2], \quad \text{by § 48 (1)},\\
&=[(b^2+2bc+c^2)-a^2][a^2-(b^2-2bc+c^2)], \text{ by § 48 (2) and (3)},\\
&=[2bc+(-a^2+b^2+c^2)][2bc-(-a^2+b^2+c^2)],\\
&=(2bc)^2-(-a^2+b^2+c^2)^2, \quad \text{by § 48 (1)},\\
&=4b^2c^2-(a^4+b^4+c^4-2a^2b^2-2a^2c^2+2b^2c^2),\\
&\qquad \text{by laws of indices and § 48 (4)}.
\end{aligned}$$

Finally, dissociating and collecting terms of the same type—

$$=2\Sigma b^2c^2-\Sigma a^4.$$

This identity is of great importance in connection with the mensuration of triangles.

**§ 50. Factorisation by Means of Standard Identities.**— The reader should observe that every algebraical identity can be read either forwards or backwards; it states, in short, what logicians call a convertible proposition. Thus $(a+b)(a-b)=a^2-b^2$ may equally be read $a^2-b^2=(a+b)(a-b)$. In the latter form the identity enables us to express $a^2-b^2$ as the product of two integral functions of the variables $a$ and $b$, or, as it is expressed, *to resolve $a^2-b^2$ into integral factors*, or more briefly, to *factorise* $a^2-b^2$. Every one of the seventeen identities of § 48, therefore, contains a factorisation theorem, and from these many others may be obtained by the principle of substitution. We shall return to the systematic discussion of this subject in a later chapter; meantime, a few examples may be given.

Ex. 1.
$$\begin{aligned}
a^2-b^2-2b-1&=a^2-(b^2+2b+1),\\
&=a^2-(b+1)^2,\\
&=\{a+(b+1)\}\{a-(b+1)\}, \quad \text{by § 48 (1)},\\
&=(a+b+1)(a-b-1).
\end{aligned}$$

Ex. 2.
$$\begin{aligned}
a^4-b^4&=(a^2)^2-(b^2)^2, \quad \text{by laws of indices},\\
&=(a^2+b^2)(a^2-b^2), \quad \text{by § 48 (1)},\\
&=(a^2+b^2)(a+b)(a-b), \quad \text{by § 48 (1)}.
\end{aligned}$$

Ex. 3.
$$\begin{aligned}
a^6+b^6&=(a^2)^3+(b^2)^3, \quad \text{by laws of indices},\\
&=\{(a^2)+(b^2)\}\{(a^2)^2-(a^2)(b^2)+(b^2)^2\}, \quad \text{by § 48 (6)},\\
&=(a^2+b^2)(a^4-a^2b^2+b^4).
\end{aligned}$$

Arithmetical calculations can often be shortened by means of algebraical transformations, such as factorisation, etc.

Ex. 4. To calculate

$$\{(37655)^2-(37649)^2\}\div 24.$$

We have

$$\begin{aligned}&\{37655^2-37649^2\}\div 24,\\ &=(37655+37649)(37655-37649)\div 24,\\ &=75304\times 6\div 24,\\ &=75304\div 4,\\ &=18826.\end{aligned}$$

## EXERCISES IX.

1. Show that, if $x=a+b$, then $ax+b^2=bx+a^2$.
2. Show that, if $x=a-b$, or $x=a+b$, then $x^2-2ax+a^2-b^2=0$.
3. Show that, if $x=1+\sqrt{2}$, or $x=1-\sqrt{2}$, then $x^2-2x-1=0$.
4. Show that $(x^2+y^2)(z^2+u^2)=(xz-yu)^2+(xu+yz)^2$.

Distribute and arrange the following :—

5. $(xy-1)(x^2y^2+xy+1)+(xy+1)(x^2y^2-xy+1)$.
6. $(ab+cd)(ad+bc)(ac+bd)$.
7. $(a+b)(a-b)+(b+c)(b-c)+(c+d)(c-d)$.
8. $(x-y+z-u)(x-y-z+u)$.
9. Show by using brackets that $(2a+2b-c)(2a+2b+c)=4(a+b)^2-c^2$.
10. $\{(b+c)^2-a^2\}+\{(c+a)^2-b^2\}+\{(a+b)^2-c^2\}=(a+b+c)^2$.
11. Distribute $(x-y)(x+y)(x^2+y^2)$, and deduce the distribution of $(x-y-z)(x+y+z)(x^2+\overline{y+z}|^2)$.
12. Prove that $a^3-b^3=(a-b)(a^2+ab+b^2)$; and hence show that $(a-b)^3-(a^3-b^3)=-3ab(a-b)$.
13. If $x+y+z=0$, show that $x^3+y^3+z^3-3xyz=0$.

Find independently the coefficient of—

14. $x^2$ in $\{x^3+(x-1)^2\}(x-1)$.
15. $x^3$ in $(x^4+x^3-x^2+x-1)(x^2+1)$.
16. $x^2$ in $(x^2-x+1)(x-1)(x+2)$.
17. $yz$ in $\{(b-c)x+(c-a)y+(a-b)z\}\{ax+by+cz\}$.
18. $yz/x$ in $(ax+by+cz)(x/a+y/b+z/c)(1/x+1/y+1/z)$.

Distribute and arrange the following in standard forms by means of the standard identities of § 48 :—

19. $(x^2+y^2-z^2)(x^2-y^2+z^2)$. 20. $(x+2y-z)^3$.
21. $(x+y-z)^4$. 22. $(x+y-z)^2(x+y+z)$.
23. $(ax+by+cz)(a^2x^2+b^2y^2+c^2z^2-bcyz-cazx-abxy)$.
24. $(\sqrt{a}+\sqrt{b}+\sqrt{c})(-\sqrt{a}+\sqrt{b}+\sqrt{c})(\sqrt{a}-\sqrt{b}+\sqrt{c})(\sqrt{a}+\sqrt{b}-\sqrt{c})$.
25. $(b+c)(c+a)(a+b)(b-c)(c-a)(a-b)$.
26. $(x^2-y)(x^2+z)(x^2-u)$. 27. $(x-1)(x-2)(x-3)(x-4)$.
28. $(x-1)(x-2)(x-3)(x+1)(x+2)(x+3)$.

**29.** $(b-c)\{a-(b+c)\}+(c-a)\{b-(c+a)\}+(a-b)\{c-(a+b)\}$.

**30.** $\{a^2(b-c)+b^2(c-a)+c^2(a-b)\}(a+b+c)$.

**31.** $\{a(b^2-c^2)+b(c^2-a^2)+c(a^2-b^2)\}(a^2+b^2+c^2)$.

Factorise—

**32.** $x^2-a^2+2ab-b^2$. **33.** $x^4+2a^2x^2+a^4-b^4$.

**34.** $x^6-2x^5+x^4$. **35.** $x^4+16$.

**36.** $x^4+y^4+1-2x^2y^2-2x^2-2y^2$.

**37.** $bc^2+b^2c-ca^2-c^2a+ab^2-a^2b$. **38.** $x^4-(a-b)x^2-ab$.

**39.** $x^6-(a^2+b^2+c^2)x^4+(b^2c^2+c^2a^2+a^2b^2)x^2-a^2b^2c^2$.

**40.** Show, by properly arranging and bracketing the terms in pairs, that $a^2(b-c)+b^2(c-a)+c^2(a-b)$ contains $b-c$ as a factor.

# CHAPTER V

## GRAPHICAL REPRESENTATION OF FUNCTIONS OF A SINGLE VARIABLE

§ 51. Hitherto we have dwelt chiefly on the formal construction of functions by means of the operations $+$, $-$, $\times$, $\div$, $\sqrt[n]{}$; in practice, however, we are very often mainly concerned with the numerical values which a function assumes when different numerical values are given to its variable or variables. Indeed, the values of a function may be accessible to us only through experiment or observation, and may not be wholly, or may be only roughly, constructible by means of assigned algebraic operations. It is therefore essential to have means for representing all the different values that a function can assume when different values are given to its variables ; in other words, means for studying its **variation.** We shall confine ourselves to the simplest and most important case, viz. **Functions of a Single Variable.**

One way of studying the variation of a function of a single variable is to form a numerical table of the function, by giving to its variable a succession of different (usually equi-different) values and calculating and tabulating opposite to these the corresponding values of the function. Barlow's table of squares, cubes, square roots, cube roots, and reciprocals, for example, gives the values of the functions $x^2$, $x^3$, $\sqrt{x}$, $\sqrt[3]{x}$, $1/x$ for $x=1, 2, 3, \ldots$, up to 10,000. Such tables are bulky (Barlow's table occupies 200 pages) ; they do not suggest at a glance the leading features of the variation of a function ; and, moreover, have only been constructed for a few of the more important functions. It is important to have a method of representing a function which will suggest the leading characteristics of its variation at a glance, and which can be made less or more accurate according to the requirements of practice. The Graphic Method, which we now proceed to explain and illustrate, fulfils these conditions perfectly; it has of late years come so extensively into all kinds of practical uses that it may be said to have popularised the notion of a function.

§ 52. **Specification of Points in a Plane.**—Let X′OX, Y′OY (called the *axes of x and y*) be two straight lines fixed for reference in a plane (see Fig. 2, p. 67); and let their positive directions be fixed as indicated by the arrows. We shall invariably suppose the angle between them to be a right angle, but this is not essential, and for special purposes other angles are occasionally chosen. The intersection O we call the *origin;* and the four regions XOY, YOX′, X′OY′, Y′OX are spoken of as the first, second, third, fourth quadrants. Consider any point P in the plane, and draw through P parallels to Y′OY and X′OX, meeting X′OX and Y′OY in M and N respectively. We call M and N the *projections of P on the axes of x and y respectively.* It is obvious that when P is given, M and N are uniquely determined; and when M and N are given, P is uniquely determined. Now, if we take O as origin of co-ordinates on X′OX and Y′OY, we have seen that we can uniquely specify the positions of M and N by means of two algebraic quantities, $x$ and $y$ representing right or left steps from O on X′OX, and upward or downward steps from O on Y′OY, according to the latent signs of $x$ and $y$ respectively. Since $x$ and $y$ together specify the projections M and N, they also specify P uniquely; we therefore speak of $x$ and $y$ or $(x, y)$ as the co-ordinates of P.* $x$ is commonly called the **abscissa**, and $y$ the **ordinate** of P.

It will be observed that the signs of the co-ordinates determine the quadrant in which P lies. For example $(+1, +2)$ is a point in the first quadrant, whose distances from the axes of $y$ and $x$ are 1 and 2 units respectively; $(-1, -2)$ is a point in the third quadrant, whose distances from the axes are the same as before.

Unless the contrary is implied, the scale units for $x$ and $y$ are taken to be the same; for many graphical purposes, however,

* Since there are other ways of specifying a point by means of co-ordinates, the above is sometimes, for distinction, called the Cartesian System, after Descartes, the founder of Analytical Geometry.

it is convenient to take the scale unit for $y$ larger or smaller than the scale unit for $x$. This, it will readily be seen, amounts to uniformly stretching or contracting the paper on which the diagram is drawn in a direction parallel to the axis of $y$.

The operation of marking in the diagram just described any point whose co-ordinates are given is called **plotting the point.**

§ **53.** Graphical representation is much facilitated by the use of paper ruled twice over by two sets of equidistant parallel straight lines, the one set being perpendicular to the other. Every tenth line is ruled a little thicker, and every fiftieth or every hundredth a little thicker still. The axes of co-ordinates are taken parallel to the ruled lines, and by means of them distances of any given number of units, tenths, and hundredths can be set off at once by merely counting the divisions on the paper itself. Such paper is sold at most shops that furnish drawing materials; but if it is not immediately available, it can be replaced by ordinary paper conjoined with the use of a scale, set square, and dividing compass.

### EXERCISES X.

1. Plot the points whose co-ordinates are $(+1\frac{1}{2}, -3)$, $(+1, 0)$, $(-1, 0)$, $(0, -\frac{3}{2})$, $(0, 0)$, $(-2, +3)$.
2. Plot the locus of all the points which have the abscissa $-3$.
3. Plot the locus of all the points which have the ordinate $(+5)$.

§ **54. The Graph of a Function.**—Let us now consider any function of $x$, say $x^3$, and denote the value of the function, corresponding to the value $x$ of the variable, by $y$, so that $y=x^3$. If we assign any particular algebraic value to $x$, say $x= +\cdot 8$, and calculate (or find in a table ready calculated for us) the value of $y$, we get $y= +\cdot 512$. We now plot in our co-ordinate diagram (Fig. 2, p. 67) a point P whose co-ordinates are $(+\cdot 8, +\cdot 512)$. To every point, M, which we like to choose on the $x$-axis there is a corresponding positive or negative value of $x$, and for every such value of $x$ a corresponding algebraic value of $y$. In the present case $y$ is positive when $x$ is positive, and negative when $x$ is negative. To every point on the $x$-axis there corresponds a point P. The points P will therefore constitute a continuous curve, which we call the *Graph of the Function* $x^3$.

In practice we cannot of course plot all the infinity of points on the graph. In constructing the actual curve, marked $x^3$ in Fig. 2, before the reader, the following table of values was used :—

$x = +\cdot100, \ +\cdot200, \ +\cdot300$, etc. . . . $-\cdot100, \ -\cdot200, \ -\cdot300$, etc.
$y = +\cdot001, \ +\cdot008, \ +\cdot027$, etc. . . . $-\cdot001, \ -\cdot008, \ -\cdot027$, etc.

The corresponding points were plotted, and then a curve was drawn through them with a free hand. In this way we obtain of course merely an approximation to the graph, quite sufficient, however, to give the leading features of the variation of the function. If a more accurate representation be required, all we have to do is to choose the values of $x$ closer together ; the points on the graph then lie closer, and the free hand curve will deviate less from the actual graph.

§ **55.** Looking at the completed graph of $y = x^3$ we see at once that as $x$ varies continuously from $-\infty$ through 0 to $+\infty$ $y$ varies continuously from $-\infty$ through 0 to $+\infty$ ; that $y$ continually increases as $x$ increases, very rapidly for large values of $x$, very slowly for very small values of $x$. No value of $y$ occurs twice ; and the function becomes 0 and changes sign as $x$ passes through 0.

For the comparison and contrast, we have also drawn in figure 2 the graphs of $y = x$ and $y = x^2$, the former of which is a straight line.

In most respects, except that the variation is uniform, the function $y = x$ resembles $y = x^3$.

On the other hand, $y = x^2$ presents distinct features. As $x$ varies from $-\infty$ to 0, $y$ varies from $+\infty$ to 0, continually decreasing, rapidly for numerically large values of $x$, slowly for small values; as $x$ varies from 0 to $+\infty$, $y$ varies from 0 to $+\infty$, repeating the same values as before in the opposite order. In the present case, therefore, each value of the function occurs twice. The function vanishes, *i.e.* passes through the value 0, without changing its sign; it has a minimum value, viz. 0 ; and it has no negative values.

The reader should extend the diagram of figure 2 by adding the graphs of $y = x^4$, $y = x^5$, etc. He will then be able to compare the characters of the powers of $x$ regarded as continuously varying functions. The graphs of the even powers all resemble the graph of $y = x^2$ ; and the graphs of the odd powers the graph of $y = x^3$ In particular, he will notice that the larger $n$ the

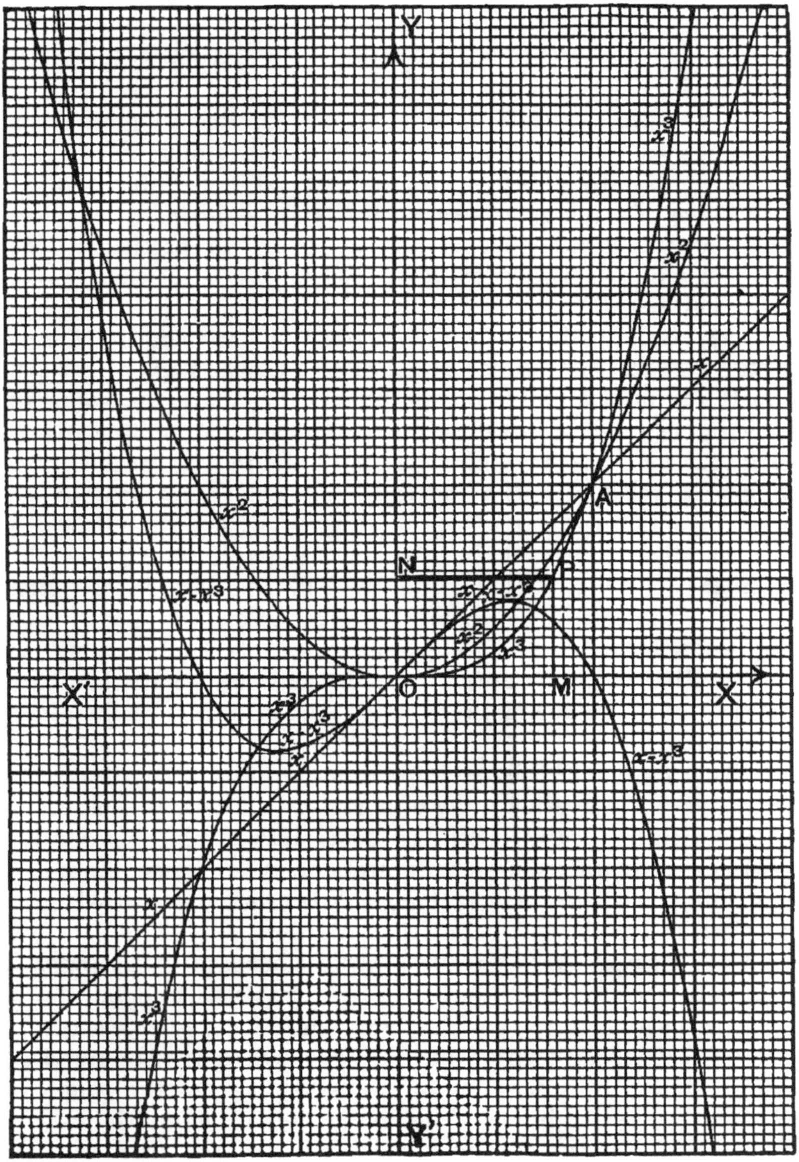


Fig. 2.

more slowly does $x^n$ increase between $x=0$ and $x=+1$, but the more rapidly between $x=+1$ and $x=+\infty$.

As another illustration, we have also given in figure 2 the graph of $y=x-x^3$. This function presents several new features; it has a minimum value $-\cdot385$, and a maximum value $+\cdot385$; each value of $y$ lying between $-\cdot385$ and $+\cdot385$ occurs three times over as $x$ varies from $-\infty$ to $+\infty$; in particular, the function vanishes three times, viz. when $x=-1$, when $x=0$, and when $x=+1$.

§ 56. **Calculation of the Value of the Function by Means of the Graph.**—When the graph of a function is given, we can find the numerical value of the function for any given value of its variable. All we have to do is to take OM on the $x$-axis such that the number of abscissa-scale-units in OM is the given numerical value of $x$, M being right or left of O according as the given algebraic value of $x$ is positive or negative; draw through M a parallel to the $y$-axis to meet the graph in P; find the number of ordinate-scale-units in MP, and attach the sign + or − according as P is above or below the $x$-axis, and the result is the required algebraic value of the function. If the parallel meets the graph in more than one point, there are more values than one of the function corresponding to the given value of $x$.

The accuracy of this determination depends of course on the magnitudes of the scale units of $x$ and $y$, and on the accuracy with which the graph itself has been drawn. If considerable accuracy is required, large scale units must be chosen for $x$ and $y$, and the points by which the graph is originally plotted must be close together.

§ 57. **Inverse Use of the Graph—Inverse Functions.**—Not only does the graph, say of $y=x^3$, enable us to determine $y$ when $x$ is given, it also enables us to determine $x$ when $y$ is given, viz. we lay down a point N on the $y$-axis whose co-ordinate is $y$; draw through N a parallel to the $x$-axis to meet the graph in P, then the $x$-co-ordinate of the projection of P on the $x$-axis, *i.e.* $\pm$ NP, is the algebraic value of $x$. *We thus solve the problem of determining the value of the variable of a function when the value of the function is given.*

The result just obtained may be looked at in another way. If we confine ourselves to the real value of the cube root of $y$, which we shall here denote by $\sqrt[3]{y}$, the equation

$$y = x^3 \quad (1).$$

is equivalent in meaning to the equation

$$x = \sqrt[3]{y} \quad (2).$$

It follows, therefore, that the problem we have just been solving in this paragraph is to determine the value of the function $\sqrt[3]{y}$ when the algebraic value of its variable $y$ is given.

It thus appears that the graph originally drawn for $y = x^3$ may be regarded as the graph of two functions, viz. of $x^3$ when $x$ is regarded as variable, and of $\sqrt[3]{y}$ when $y$ is regarded as variable.* Two functions connected in this way by the same graph are said to be *inverse to each other.*

It should be noticed that the graph of figure 2 is the graph of $x = \sqrt[3]{y}$. If it be desired to obtain the graph of $y = \sqrt[3]{x}$, we must turn the diagram round and over, so that X′OX comes into the position formerly occupied by Y′OY, and *vice versa;* and then rename the axes in the usual way.

§ **58.** The reader will readily understand that the graphs of $y = ax$, $y = ax^2$, $y = ax^3$ can be obtained from the graphs of $y = x$, $y = x^2$, $y = x^3$ by simply increasing (or decreasing) the ordinates of the latter in the proportion $a : 1$. For example, the graph of $y = 3x$ is obtained from that of $y = x$ simply by trebling every ordinate of the latter. In particular, it results from this consideration, by the simplest properties of similar figures, that the graph of $y = ax$, where $a$ is an algebraic constant, is always a straight line passing through the origin. Let now $b$ be any constant algebraic quantity, and add $b$ to each of the ordinates of the graph of $y = ax$, the result on the figure is the same as if we moved the straight line which constitutes the graph parallel to itself through a distance parallel to the $y$-axis of $|b|$ ordinate-scale-units, upwards or downwards according as $b$ is positive or negative; the result as regards the equation is that $y$ is now determined by

$$y = ax + b \quad (3).$$

We have thus established the important theorem that *the graph of an integral function of the first degree is a straight line.*

* For distinction's sake, when values are given to $x$ and values calculated or deduced for $y$, $x$ is called the "*independent variable or argument*" and $y$ the *dependent variable.* Thus above in the one case we take *$x$ as independent variable,* in the other *$y$ as independent variable.*

On account of the property just mentioned the function $ax+b$ is usually called a **linear integral function**; or, when there is no danger of confusion with the linear fractional function $(ax+b)/(cx+d)$, a **linear function** simply.

## EXERCISES XI.

Trace the graphs of the following functions:—

**1.** $y=-x$. **2.** $y=-3x$. **3.** $y=0$.
**4.** $y=3$. **5.** $y=-4$. **6.** $y=x+1$.
**7.** $y=-x+1$. **8.** $y=x-1$. **9.** $y=-x-1$.
**10.** $y=3x-2$. **11.** $y=-2x+3$. **12.** $y=-2x+2$.
**13.** $y=x+x^3$. **14.** $y=x^2-2x+1$. **15.** $y=(x-1)(x-2)$.
**16.** $y=-x^2-x-1$. **17.** $y=1-x^3$. **18.** $y=1/x$.
**19.** $y=1/x^2$. **20.** $y=1/x^3$. **21.** $y=1/(x-1)$.
**22.** $y=1/(x+1)$. **23.** $y=x/(x-2)$. **24.** $y=(x-1)/(x-2)$.
**25.** $y=(x-1)/(x-2)^2$.

**26.** Find graphically approximations to the values of $x$ for which $x^2-4x+2$ has the value zero.

**27.** Find graphically approximations to the values of $x$ for which the functions $x-x^3$ and $x^2$ have equal values.

# CHAPTER VI

## SIMPLE EQUATIONS

§ **59.** Attention has already been called to the difference between an **identity**, *i.e.* an equation the one side of which can be transformed into the other by merely applying the laws of Algebra and the definitions of the symbols and functions involved, such, for example, as $x^2-(x+1)(x-1)=1$, $x^2\times x=x^3$, and a **conditional equation,** where such transformation is possible only when the variable or variables involved have certain values.

We now propose to study the latter class of equations more closely. The following are examples of conditional equations:—

$$3x+6=\tfrac{1}{2}x+8\,;$$
$$x^2+3x+2=0\,;$$
$$x+y^2=x^2+y\,;$$
$$x/(x-1)+x/(x-2)=1\,;$$
$$2^x=4.$$

For the present, we confine ourselves to the case where there is only **one variable**, or, as it is often called in the present connection, **unknown quantity**, $x$, say. Further, we shall limit our discussion in this chapter to **integral equations**, *i.e.* equations in which both sides are integral functions of $x$. By the **degree of an integral equation** in one variable $x$ is meant the index of the highest power of $x$ that occurs on either side of the equation. Ultimately in this chapter we shall narrow our view to the case of **integral equations of the first degree**, commonly called **Simple Equations.***

* Integral equations of the first degree in one or more variables are also often called *linear equations*.

§ **60.** It will contribute to clearness if we study the genesis and solution of a conditional equation in a simple concrete instance. Consider the following problem:—A man had a certain number of shillings in his purse. He found that, if he were to give each of six beggars a certain sum, he would have five shillings left, but that if he were to give each of three beggars the same sum he would have fourteen shillings left. What sum did he propose to give to each beggar?

Let $x$ denote the number of shillings which he proposed to give to each beggar; then on the first supposition the number of shillings in his purse is $6x+5$; on the second supposition the number is $3x+14$; we must therefore have

$$6x+5=3x+14 \qquad (1).$$

By the equation (1) taken in the abstract we mean that if $x$ be replaced by a certain number (in this case the number of shillings proposed to be given to each beggar), then the left-hand side of (1) can be transformed into the right by means of the laws of Arithmetic, which are of course included in the laws of Algebra. In short, we mean that, if $x$ have a certain numerical value, (1) will be an identity. Our object, therefore, is *to find the value of* x *which will render* (1) *an identity;* this we call **solving the equation.** The particular value of $x$ thus found we call the **solution of the equation** (or, in the case of equations in a single variable, the **root of the equation**).

Although we do not as yet know the value (or values) of $x$ which makes (1) an identity, we may proceed to transform the equation on the hypothesis that $x$ has such a value. To emphasise the fact that we make this hypothesis, we use in what follows the distinctive sign of identity, viz. $\equiv$.

Since

$$6x+5\equiv 3x+14 \qquad (2),$$

we have

$$(6x+5)-(3x+5)\equiv(3x+14)-(3x+5) \qquad (3).$$

Hence, by the laws of Algebra and the definition of the numerical symbols, we have

$$3x\equiv 9 \qquad (4).$$

From (4) we have

$$3x\div 3\equiv 9\div 3 \qquad (5);$$

whence, by the laws of Algebra and the definition of the symbol 9, we get

$$x\equiv 3 \qquad (6).$$

It follows that any value of $x$ that makes (2) an identity must be 3, or some number reducible to 3. In other words, if (2) have any solution, the solution is $x = 3$. But the steps taken are all reversible—that is to say, (5) follows from (6), (4) from (5), (3) from (4), and (2) from (3). Hence $x = 3$ actually makes (2) an identity. This might also be seen by substituting 3 for $x$ in (2), and showing that $6 \times 3 + 5 \equiv 3 \times 3 + 14$, which is sometimes described as **verifying the solution.** We conclude, therefore, that the simple equation (2) has one and only one root, viz. $x = 3$ ; and, as regards the problem considered, that the sum which the man proposed to give to each beggar was three shillings and no other.

§ **61.** A study of the example of last paragraph leads us to the following conclusions regarding a conditional equation in general :—

(1) *Every conditional equation is a hypothetical identity ; and the process of solving it consists in finding a set of values for the variable or variables, in terms of given or constant quantities,* such as will make the equation an actual identity.*

This value or set of values is said to *satisfy the equation.*

(2) *In every transformation of the equation we suppose the variable or variables to have values such that the equation is an identity. Every such transformation is therefore merely an application of one or more of the laws of Algebra.*

(3) *In each step of the process of solution we deduce from a previous equation* (A) *another* (B), *which has all the solution or solutions of* (A).

An equation (B) which has all the solutions that (A) has is said to be a **derivative of (A).**

If all the solutions of (B) are also solutions of (A), the derivation is said to be **reversible ;** and the two equations, which then have exactly the same solutions, are said to be **equivalent.**

All the derivations in § **60** are reversible, and each of the equations (2), (3), (4), (5), (6) is equivalent to every other. Derivations, however, are not necessarily reversible. For example, from

$$x - 1 = 0 \qquad (7)$$

* A very common beginner's mistake is to assign as a solution a value of the variable which depends on the variable itself or on other variables, *e.g.* if the value $(3x+9)/6$ be substituted for $x$ on the left-hand side of equation (1) above, the equation will become an identity, but $x=(3x+9)/6$ is not "a solution of the conditional equation."

we derive $$(x-2)(x-1)=0 \qquad (8);$$

for (8) is satisfied by every value of $x$ that satisfies (7). The converse, however, is not true, for $x=2$ is a solution of (8), but not of (7). Hence (7) and (8) are not equivalent.

This may be seen from another point of view. In attempting to pass from (8) to (7) we should naturally divide both sides of (8) by $x-2$. It must, however, be remembered that in operating on (8), we suppose $x$ to have one or other of the values that render (8) an actual identity, *i.e.* $x$ must be either 1 or 2. When $x=1$, $x-2 \neq 0$; and there is no difficulty; but when $x=2$, $x-2=0$; and we should have to perform the inadmissible operation of dividing by 0 (see § 39). It thus appears that the irreversibility of the derivation is closely connected with the illegitimacy of division by 0.

§ 62. The following **theorem regarding the vanishing of a product** is so useful in the theory of conditional equations that it deserves explicit statement. *If* P *and* Q *be functions of* x, *and for any value of* x, $P \times Q \equiv 0$, *and for this value of* x, P *be finite both ways,* i.e. *finite and not zero, then for this value of* x *must* $Q \equiv 0$; *and* vice versa, *if* Q *be finite both ways, then must* $P \equiv 0$.

The proof is immediate; for since we have $P \times Q \equiv 0$ and $P \neq 0$, it follows, by the laws of Algebra, that $P \times Q/P \equiv 0/P$, that is, $Q \equiv 0$.

*From this it follows that the solutions of the equation* $PQ=0$, *if we exclude solutions, if any, for which either* P *or* Q *is not finite,* are values of* x *which satisfy either* $P=0$ *or* $Q=0$, *or both* $P=0$ *and* $Q=0$.

By means of this principle the solution of many equations can be obtained by inspection, which is the best method of solution when it is applicable. For example, the equation $x^2-3x=0$ may be written $x(x-3)=0$, hence its solutions are the solutions of $x=0$ and $x-3=0$, *i.e.* $x=0$ and $x=3$. Again, $x^2-3x+2=0$ can be written $(x-2)(x-1)=0$, from which it is obvious that it has the solutions $x=2$ and $x=1$.

§ 63. We now give two theorems of fundamental importance in the derivation of equations.

I. *If* P *and* Q *be functions of* x *and* R *a constant, or any function of* x *which remains finite for all values of* x *in question, the equations*

$$P=Q \qquad (9),$$

* The reason for this exception will be better understood by studying the cases where $P=1/(x-2)$, $Q=(x-1)(x-2)^2$, and $P=1/(x-2)$, $Q=(x-1)(x-2)$, both with reference to $x=2$.

*and*

$$P \pm R = Q \pm R \qquad (10)$$

*are equivalent.*

For, if for any value of $x$ $P \equiv Q$, then by the laws of Algebra $P \pm R \equiv Q \pm R$—that is to say, any value of $x$ which makes (9) an identity makes (10) an identity. Conversely, if for any value of $x$ $P \pm R \equiv Q \pm R$, then for that value of $x$ $P \pm R \mp R \equiv Q \pm R \mp R$, that is, $P \equiv Q$. Hence every value of $x$ that makes (10) an identity also makes (9) an identity.

COR. 1. *Any term or part on one side of an equation preceded by the signs + or − may, without affecting the equivalence, be transferred to the other side, provided the sign be reversed.*

For example, from $x^2 - 3x = 2x + 1$ we deduce, by subtracting $2x$ from both sides, the equivalent equation $x^2 - 3x - 2x = 1$. Again, from $2/x + 1/(x-1) = 3/x - 2/(x-1)$, by adding $2/(x-1)$ to both sides, we deduce $2/x + 1/(x-1) + 2/(x-1) = 3/x$.

COR. 2. *From any equation we can deduce an equivalent equation by reversing the sign of every term on both sides of the equation.*

For this is tantamount to subtracting the sum of the right and left hand sides of the equation from both sides, and then interchanging the two sides of the equation.

COR. 3. *Every equation can be reduced to an equivalent* **standard form** $P = 0$. To effect this we have merely to subtract from both sides the right-hand side of the given equation.

For example, from $3x^3 - 2x = x - 1$ we deduce, by subtracting $x - 1$ from both sides, $3x^3 - 3x + 1 = 0$.

II. *If* P *and* Q *be functions of* x, *then all the solutions of*

$$P = Q \qquad (11)$$

*satisfy*

$$RP = RQ \qquad (12)$$

*where* R *is a constant or any function of* x *which remains finite for all values of* x *which are solutions of* $P = Q$; *but the two equations are not in general equivalent unless* R *be a constant* (*i.e.* independent of x) *differing from zero.*

PROOF. If, for any value of $x$, $P \equiv Q$, then for that value of $x$, R being finite, we have $RP \equiv RQ$—that is, every solution of (11) is a solution of (12).

Next consider the equation $RP = RQ$. This is equivalent by our first theorem to $RP - RQ = 0$—that is, to $R(P - Q) = 0$.

If R be a constant differing from 0, any value of $x$ which gives $R(P-Q)\equiv 0$ gives $P-Q\equiv 0/R\equiv 0$, whence $P\equiv Q$—that is, every solution of (12) is a solution of (11).

If, however, R be a function of $x$, then, in general, by § 62, the solutions of $R(P-Q)=0$ will be the solutions of $R=0$ and of $P-Q=0$—that is to say, of $R=0$ and $P=Q$. In other words, (12) has, in addition to the solutions of (11), the solutions of the equation $R=0$.

As an example of an irreversible derivation we may give $x-3=1$, from which we derive $(x-2)(x-3)=(x-2)$, the latter is satisfied by the solution of the former, viz. $x=4$, but it has in addition the solution $x=2$.

Again, if we were to multiply both sides of $(x-2)(x-3)=(x-2)$, which has the solutions $x=2$ and $x=4$, by $1/(x-2)$ we should deduce the equation $x-3=1$, which has only the solution $x=4$. This is therefore not an admissible derivation at all, the reason being that the multiplier $1/(x-2)$ does not remain finite when $x=2$, which is one of the solutions of the original equation.*

COR. *If the coefficients of any integral equation are fractional, either in the arithmetical or in the algebraic sense, we can derive an equivalent equation in which the coefficients are all integral in the arithmetical or in the algebraic sense.*

To this end we have merely to multiply both sides of the equation by the product of the denominators of the fractional coefficients, or by some simpler number or integral function of the coefficients which is exactly divisible by all the denominators. This is a reversible derivation, since these coefficients are constant.

Ex. 1. From $\frac{1}{6}x^2+\frac{1}{4}x+\frac{1}{3}=0$ we derive, by multiplying by $6\times 4\times 3$, the equivalent equation $12x^2+18x+24=0$. Since 12 is the L.C.M. of 6, 4, 3, we could attain the same end more simply by multiplying by 12. The resulting equivalent equation would then be $2x^2+3x+4=0$.

Ex. 2. From

$$\frac{x^2}{p-q}+\frac{x}{p+q}+\frac{1}{p^2-q^2}=0$$

we deduce, by multiplying by $p^2-q^2$—

$$(p+q)x^2+(p-q)x+1=0.$$

§ 64. Returning to the point from which we started we may now give the **general theory of the simple equation**. Since both sides are integral functions of $x$ and no higher power of $x$

* One of the commonest of beginners' blunders is to lose solutions of an equation by rejecting factors which contain the variables.

than the first is admissible, we can by proper transformations reduce every such equation to the form

$$ax+b=cx+d \qquad (13),$$

where $a$, $b$, $c$, $d$ are constants.

From (13) we derive the equivalent equation

$$(a-c)x+(b-d)=0,$$

or, say,

$$Ax+B=0 \qquad (14),$$

where A and B are constants.

We may therefore regard (14) as the *standard form to which every simple equation can be brought.*

For the present we shall suppose $A \neq 0$, merely remarking that in general the equation $0 \times x + B = 0$ evidently does not admit of a finite solution.

From (14) we derive the equivalent equation

$$Ax = -B \qquad (15);$$

and from this, since $A \neq 0$, the equivalent

$$x = -B/A \qquad (16).$$

It follows that *the solution of* (14) *is* $x = -B/A$, *and that there is no other solution.*

*Two simple equations*

$$Ax+B=0 \qquad (17),$$
$$A'x+B'=0 \qquad (18),$$

*have not in general a common solution. The condition that they have such a solution is* $AB'-A'B=0$.

For the only solution of (17) is $x = -B/A$, and the only solution of (18) is $x = -B'/A'$. The necessary and sufficient condition that the two have a common solution is therefore $-B/A = -B'/A'$, which, since $A \neq 0$, $A' \neq 0$, is obviously equivalent to $AB'-A'B=0$.

§ **65. Graphical Solution.**—Since, as we have seen, the graph of the linear function $Ax+B$ is a straight line, to effect the solution of the equation $Ax+B=0$, *i.e.* to find the value of $x$ for which $y=0$, we have merely to measure the abscissa of the point where the graph of $y=Ax+B$ meets the $x$-axis, and attach the proper sign to the resulting number of abscissa-scale-units.

If we take the simple equation in the form $ax+b=cx+d$,

then we must trace the graphs of $y_1 = ax + b$ and $y_2 = cx + d$, and find the value of $x$ for which $y_1$ and $y_2$ are equal ; in other words, find the abscissa of the point of intersection of the two graphs.

§ 66. As regards the **Practical Algebraic Solution** of simple equations, all that need be said is that the equation should be reduced as rapidly as possible to the standard form $\mathrm{A}x + \mathrm{B} = 0$, or, if preferred, to the form $\mathrm{A}x = \mathrm{B}$, the solutions of which are evident. In the course of the reduction every advantage afforded by the special form of the equation should be taken to remove redundant parts before distributing or otherwise reducing the same : useless and confusing work will thus be avoided. It may also be pointed out that many equations which at first sight seem to be of higher degree than the first prove, after reduction, to be merely simple equations. Also, an equation which professes to be a conditional equation may turn out on reduction to be an identity ; it is then satisfied by any finite value of $x$. Again it may happen that the $x$ disappears entirely, and the equation reduces to the form $c = 0$, where $c$ is some constant ; in this case the inference is that the equation cannot be satisfied by any finite value of $x$. The following examples will illustrate these remarks :—

Ex. 1. $(x-6)/6 + 2(x-3)/3 = x - 1/3$.

By the law of distribution we may arrange the equation thus—

$$x/6 - 1 + 2x/3 - 2 = x - 1/3.$$

We derive an equivalent equation by adding to both sides $-x+1+2$ ;

viz.— $$x/6 + 2x/3 - x = 1 + 2 - 1/3 ;$$

that is to say— $$(\tfrac{1}{6} + \tfrac{2}{3} - 1)x = \tfrac{8}{3} ;$$

that is— $$-\tfrac{1}{6}x = \tfrac{8}{3} ;$$

whence, multiplying both sides by $-6$, we get finally

$$x = -16.$$

We may verify that this is the correct solution by substituting in the original equation. The result is

$$(-16-6)/6 + 2(-16-3)/3 = -16 - 1/3 ;$$

that is— $$-11/3 - 38/3 = -49/3,$$

which is a numerical identity as it ought to be.

In solving this equation we might have begun by clearing the coefficients of fractions. This is obviously effected by multiplying both sides by 6, which will not affect the equivalence. We thus get

$$x - 6 + 4(x-3) = 6x - 2,$$

that is— $$x - 6 + 4x - 12 = 6x - 2,$$

or $$5x - 18 = 6x - 2.$$

If we now add $-6x+18$ to both sides, we get

$$-x=16,$$

whence, multiplying both sides by $-1$, we get $x=-16$.

Ex. 2.

$$3[(x-5)-2\{(x-3)-6(x-1)\}]=2[x-6-3\{x-3-6(x-1)\}].$$

Here, by proceeding cautiously, we can effect some economy of calculation. Removing the square bracket on both sides, we see that the equation may be written

$$3(x-5)-6\{x-3-6(x-1)\}=2(x-6)-6\{x-3-6(x-1)\}.$$

If we add to both sides $6\{x-3-6(x-1)\}$, which is a reversible operation, neither adding to nor subtracting from the solutions of the given equation, we get

$$3(x-5)=2(x-6);$$

that is—

$$3x-15=2x-12,$$

whence, adding $-2x+15$ to both sides, we have

$$x=3.$$

Ex. 3. $(x-1)^3+(x-2)^3-x^3=x(x-5)(x-4)$.

Distributing the powers and products we see that the equation may be written

$$x^3-3x^2+3x-1+x^3-6x^2+12x-8-x^3=x^3-9x^2+20x,$$

that is—

$$x^3-9x^2+15x-9=x^3-9x^2+20x;$$

whence, adding $-x^3+9x^2-20x+9$ to both sides—

$$-5x=9.$$

Therefore

$$x=-9/5.$$

Ex. 4. $(x-1)^3+(x-2)^3+(3-2x)^3=0$.

We observe that $(x-1)+(x-2)+(3-2x)\equiv 0$.

Now it follows from the identity $\Sigma X^3-3XYZ\equiv\Sigma X(\Sigma X^2-\Sigma XY)$; that, if $\Sigma X\equiv 0$, then $\Sigma X^3-3XYZ\equiv 0$; in other words, $\Sigma X^3\equiv 3XYZ$. In the present case we have, therefore—

$$(x-1)^3+(x-2)^3+(3-2x)^3\equiv 3(x-1)(x-2)(3-2x).$$

Hence the given equation is neither more nor less than

$$3(x-1)(x-2)(3-2x)=0.$$

Now this last becomes an identity when, and only when, $x-1=0$, or $x-2=0$, or $3-2x=0$—that is to say, when $x=1$, or $x=2$, or $x=3/2$; these, therefore, are the required solutions.

§ 67. If an integral equation of higher degree than the first contains only a single power of $x$, say $x^n$, it can be brought by the methods above explained and illustrated into the equivalent form $x^n=A$. The solution then depends merely on the extraction of the $n$th root of A. In the particular case where the equation is quadratic, that is, only $x^2$ occurs, the equation can be reduced to the equivalent form $x^2+a^2=0$, or $x^2-a^2=0$, where

$a$ is a real quantity. It is obvious, since the square of every real number (whether positive or negative) is positive, that $x^2 + a^2 = 0$ has no real solution. On the other hand, $x^2 - a^2 = 0$ is equivalent to $(x-a)(x+a) = 0$, and has obviously the two solutions $x = a$ and $x = -a$, and no other finite solution.

Ex. $(x-1)^2 + (x+1)^2 = 6$ is equivalent to

$$2x^2 + 2 = 6,$$

that is, to $$2x^2 - 4 = 0,$$

that is, to $$x^2 - 2 = 0,$$

which may be written $$x^2 - (\sqrt{2})^2 = 0,$$

or $$(x - \sqrt{2})(x + \sqrt{2}) = 0.$$

Hence the solutions are $$x = \sqrt{2},\ x = -\sqrt{2}.$$

## EXERCISES XII.

Solve the following equations :—

**1.** $4x - 4 = 0.$

**2.** $6x - 18 = 0.$

**3.** $3x + 3 = 0.$

**4.** $3x + 7 = 0.$

**5.** $x + 5 = 2x + 6.$

**6.** $3x + 5 = 2x + 6x.$

**7.** $3x - 3 + 6(x-1) - 5(1-x) = 0.$

**8.** $6(4 - 3x) + 6(3x - 4) = 0.$

**9.** $4(1-x) - 6(x+1) + 3(x-3) = 0.$

**10.** $3[4 - 5\{6 - 7(8-x)\}] = 0.$

**11.** $3 - [x + \{x - 3(x-4)\}] = 5 + [x - \{x - 3(x-4)\}].$

**12.** $x/6 + x/3 = 1.$

**13.** $x + \frac{3}{4}x = 1.$

**14.** $\frac{1}{5}x + \frac{1}{15}x + \frac{1}{20}x = \frac{19}{60}.$

**15.** $(x-2)/2 + (x-3)/3 + (x+2)/2 + (x+3)/3 = 0.$

**16.** $(5x-1)/7 + (9x-5)/11 = (9x-7)/5.$

**17.** $\frac{1}{3}(x+1) + \frac{1}{5}(x+4) + \frac{1}{6}(2x+5) = x - 1.$

**18.** $3(x - \frac{1}{3}) - 5(x - \frac{1}{5}) = 8(x - \frac{1}{8}) - 6(x - \frac{1}{6}).$

**19.** $\frac{3}{4}(2x-3) + \frac{4}{5}(3x-4) = \frac{1}{2}(3x-2) + \frac{1}{4}(4x-3).$

**20.** $(x+1)^2 + (x+4)^2 = (x+2)^2 + (x+5)^2.$

**21.** $(x+1)(x+5) = (x+2)^2.$

**22.** $(6-x)(1+2x) + 3x(5+x) = (x+1)^2 - x.$

**23.** $(2x-2)(1+2x) + \frac{1}{2}(x-1)(x+2) + \frac{3}{4}(x-1)(x+2)$
$= \frac{5}{4}\{(x^2 + x - 2) - x + 1\}.$

**24.** $(x+1)(x-1) + 2(x+2)(x+3) = 3(x+1)^2.$

**25.** $(x - \frac{1}{2})(x - \frac{1}{3}) = (x + \frac{1}{2})(x + \frac{1}{3}).$

**26.** $x(x + \frac{1}{2})(x + \frac{2}{3}) + x = x(x + \frac{1}{3})(x + \frac{5}{6}).$

**27.** $(x-1)\left\{x + 3 + \dfrac{1}{x-1}\right\} = (x+1)\left\{(x+1) + \dfrac{2}{x+1}\right\}.$

**28.** $\cdot 384x + \cdot 5(x - 1\cdot 03) = 3\cdot 68.$

**29.** $\cdot 016(2x-1) + \cdot 032(2x-3) = 2x - 2.$

**30.** $(x - 3\cdot 5)/2\cdot 5 + (x - 2\cdot 3)/3\cdot 5 + (x - 1\cdot 3)/1\cdot 5 = 0.$

Find $x$ to three decimal places from—

**31.** $3\cdot 68(x - \cdot 03) + 1\cdot 34(10x - 1\cdot 52) = 13\cdot 314.$

**32.** $(x - 1\cdot 23)(x + 2\cdot 31)(x + 3\cdot 21) = x(x + 2\cdot 145)^2.$

Find $x$ to six decimal places from—

**33.** $6\cdot 3127x - 1\cdot 56832x = (x-1)/\cdot 06834.$

**34.** $3\cdot 689543(x - 6\cdot 31489) = 2\cdot 15632x - 10\cdot 687321.$

Solve the following by inspection :—

**35.** $3x^2-1=0$.

**36.** $4(x-1)^2-x^2=0$.

**37.** $(x-3)^2+5(x-3)=0$.

**38.** $(x-1)^2=(x-1)(2x+3)$.

**39.** $(x^2-2x+1)^2-(x-1)^2(x-3)^2=0$.

**40.** $(2x-1)^2=(3x-1)^2$.

**41.** $x^2(x-1)^2=4(x-1)^2$.

**42.** $x(x^2-1)=3x^2-3$.

**43.** $32(x+1)^2(x+2)^2-8(x-3)^2(x+1)^2=0$.

**44.** $x^2(x-1)=3x-3$.

**45.** $2x^2(x-1)=5(x-1)$.

**46.** $(x-1)(2x-5)+(x-1)(x-3)=(1-x)(2x-4)$.

**47.** Point out the fallacy in the following paradox :—Let $x=1$ ; then $x^2=x$, whence $x^2-1=x-1$. If we divide both sides of this last equation by $x-1$, we get $x+1=1$—that is, since $x=1$, $2=1$.

Remark on the solution of the following equations :—

**48.** $(x-1)(x-3)=(x-2)^2$.

**49.** $3(x-\frac{2}{3})-5(x-\frac{2}{5})=6(x-\frac{1}{3})-8(x-\frac{1}{4})$.

**50.** $(x+1)^2+(x+4)^2=(x+2)^2+(x+3)^2$.

Solve the following :—

**51.** $(x-1)(x-3)+(x+1)(x+3)=15$.

**52.** $(x+1)(x-3)(x+5)+(x-1)(x-3)(x+5)=4x^2$.

**53.** $(x-1)(x-2)(x-7)=(x-1)(5x+14)$.

**54.** Construct an integral equation whose roots are $+3$ and $-5$.

**55.** Construct an integral equation whose coefficients are integral numbers which has the roots $-\frac{1}{2}$, $\frac{3}{2}$, $-\frac{1}{4}$.

**56.** Construct an equation whose roots are $3+\sqrt{5}$ and $3-\sqrt{5}$.

**57.** Construct an equation whose roots are $-2$, $1+\sqrt{2}$, $1-\sqrt{2}$, $3+\sqrt{2}$, $3-\sqrt{2}$.

§ **68.** The solution of equations in which the coefficients are not actual numbers but letters standing for such numbers, or for quantities in general supposed to be given, involves no new principle. The calculations required in reducing the equations to the standard form and in expressing the solution obtained in its simplest form demand, however, a firm grasp of the laws of Algebra, and some familiarity with the elementary standard formulæ of § 48. This kind of work, although somewhat difficult for the beginner, is an invaluable exercise at the present stage. In all cases he should verify his solution by showing that it renders the given equation an identity. Every equation thus furnishes two exercises. The verification is not infrequently more troublesome than the process of solution : at other times it is immediate ; and in such cases the solution can usually be guessed without going through the process of solution. To guess a solution and append the verification is sufficient, provided we can be sure that there is only one solution—as, for example, is always the case when the equation is linear.

Ex. 1. $(x-a)/p+(x-a)/q=1/p^2-1/q^2$.
Noticing that $1/p^2-1/q^2\equiv(1/p)^2-(1/q)^2\equiv(1/p+1/q)(1/p-1/q)$, we see, by the law of distribution, that our equation may be written

$$(1/p+1/q)(x-a)=(1/p+1/q)(1/p-1/q).$$

If we suppose that $1/p+1/q\neq 0$,* then we may divide both sides by $1/p+1/q$ without altering the equivalence of the equation. The result is

$$x-a=1/p-1/q,$$

whence, if we add $a$ to both sides, we get $x=a+1/p-1/q$, which is the solution.

Ex. 2. $(x+a)/(a+b)^2-(x-a)/(a-b)^2=2a/(a^2-b^2)$.

We can integralise the coefficients of this equation, without affecting its equivalence, by multiplying both sides by the factor $(a+b)^2(a-b)^2$—that is, $(a^2-b^2)^2$. We thus get

$$(a-b)^2(x+a)-(a+b)^2(x-a)=2a(a^2-b^2).$$

Distributing $(x+a)$ and $(x-a)$ and collecting, we get

$$\{(a-b)^2-(a+b)^2\}x+a\{(a-b)^2+(a+b)^2\}=2a(a^2-b^2).$$

Now

$$(a-b)^2-(a+b)^2\equiv a^2-2ab+b^2-a^2-2ab-b^2\equiv -4ab\,;$$
$$(a-b)^2+(a+b)^2\equiv a^2-2ab+b^2+a^2+2ab+b^2\equiv 2a^2+2b^2.$$

The equation to be solved is therefore equivalent to

$$-4abx+a\{2a^2+2b^2\}=2a(a^2-b^2),$$

that is to say—

$$-4abx+2a^3+2ab^2=2a^3-2ab^2,$$

whence, adding $-2a^3-2ab^2$ to both sides—

$$-4abx=-4ab^2.$$

Finally— $$x=(-4ab^2)/(-4ab)=b,$$

the correctness of which may be readily verified.

Ex. 3. $\sum_{abc}(x-a)^2-\sum_{abc}(x-a)(x-b)=0$.

Writing the equation in full and expanding the squares and products, we get

$$x^2-2ax+a^2+x^2-2bx+b^2+x^2-2cx+c^2$$
$$-x^2+(b+c)x-bc-x^2+(c+a)x-ca-x^2+(a+b)x-ab=0,$$

that is— $$a^2+b^2+c^2-bc-ca-ab=0.$$

The equation has therefore, in general, no finite solution. If $a$, $b$, $c$ be such that $a^2+b^2+c^2-bc-ca-ab\equiv 0$, then the original equation is an identity and may be regarded as being satisfied by any finite value of $x$ whatever.

Ex. 4. $(b-c)(x-a)+(c-a)(x-b)+(a-b)(x-c)=0$.

It will be found that this equation is an identity and is therefore satisfied by any finite value of $x$ whatever.

---

* If $1/p+1/q=0$ then the equation is an identity.

## EXERCISES XIII.

1. $\frac{1}{2}(x-2a)+\frac{1}{3}(x+3a)+\frac{1}{6}(x-6a)=0.$
2. $\frac{1}{2}\left(x-\frac{a}{2}\right)+\frac{1}{3}\left(x-\frac{a}{3}\right)+\frac{1}{4}\left(x-\frac{a}{4}\right)=0.$
3. $ax+b=bx+a.$
4. $b(x+a)=a(x+b).$
5. $a^2(x-p)^2-b^2(x+q)^2=0.$
6. $ax+a=x/a+1/a.$
7. $b(x+a)+a(x+b)=a(a+2b)+b(a+b).$
8. $(x-a)(x-b)=(x-a-b)^2.$
9. $a^2(x-a)-b^2(x-b)=(a-b)^3.$
10. $(a+x)(a-b)+(a-x)(a+b)=2a^2-2b^2.$
11. $(x-a)(x+a)^2=(x+a)(x-a)^2.$
12. $(x-a)/(p-a)+(x-b)/(p-b)=2.$
13. $a(x-b)-b(x-a)+c(x-a-b)=(a-b)(a+b+c)+c^2.$
14. $x/a-\{1-bx/(a-b)\}/a=x/(a+b).$
15. $(px+q)/(p+q)=(mx+n)/(m+n)+(mx-n)/(m-n).$
16. $(x+a)/(2a+b)+(x+b)/(a+2b)=2.$
17. $\frac{1}{2}(ax+b^2)/(a^2+b^2)+\frac{1}{2}(ax-b^2)/(a^2-b^2)=1.$
18. $(x-a)\{(x-b)+(x+b)\}=(x+b)\{(x-a)+(x+a)\}.$
19. $\{(a^2-b^2)x-1\}^2+\{2abx-1\}^2=\{(a^2+b^2)x+1\}^2.$
20. $(x+b)(x+c)+(x+c)(x+a)+(x+a)(x+b)$
    $=(x-b)(x-c)+(x-c)(x-a)+(x-a)(x-b)+(a+b+c)^2.$
21. $(x+b+c)(x+b-c)+(x+c+a)(x+c-a)$
    $+(x+a+b)(x+a-b)=0.$
22. $(b-c)(x-a)^3+(c-a)(x-b)^3+(a-b)(x-c)^3=0.$
23. $\sum_{abc}(x-a)\{(x-a)^2-(x-b)(x-c)\}=0.$

24. Determine $h$ so that the two equations $hx+a=bx+h$, $ax+h=hx+b$ may be consistent. What peculiarity arises when $a=b$?

25. If $a$, $b$, $c$ be all finite both ways, and if the three equations $ax+b-c=0$, $bx+c-a=0$, $cx+a-b=0$ be all consistent, then must $a=b=c$.

26. $(x-a)(x+a)+(x+2a)(x-2a)-(x+3a)(x-3a)=40a^2.$
27. $(x-a)(x-b)+(x+a)(x+b)=a^2+b^2+x^2.$
28. $(x^2-b^2)/(a^2-b^2)-(x^2+b^2)/(a^2+b^2)=2b^4/(a^4-b^4).$
29. $(b^2-c^2)(x-a)^3+(c^2-a^2)(x-b)^3+(a^2-b^2)(x-c)^3=0.$

# CHAPTER VII

## SOLUTION OF PROBLEMS BY MEANS OF SIMPLE EQUATIONS

§ **69.** An example of a problem leading to a simple equation has already been given. The solution of such problems is an excellent training for the beginner in mathematics, because it cultivates that faculty of dissecting things into their essential elements, which is one of the most valuable parts of a mathematician's power, and one of the most important practical results of a proper mathematical training.

In every problem which can be treated mathematically there are certain quantities that are directly given, or are supposed to be directly given, which we call the **constants** of the problem; and others that are not directly given, but are to be determined which we call the **unknown quantities** or **variables.** The premises by means of which we determine the variables are so many conditions; and the number of independent conditions must in general be equal to the number of variables if the problem is to be determinate — that is to say, have a limited number of solutions. In particular, if there be but one variable to determine, one condition will in general be sufficient.

In any problem before him the beginner should, as a first step, make perfectly clear to himself what are the given quantities on which the solution depends, and next choose an appropriate variable (or variables). The variable may be a magnitude whose determination is asked for; but very often it is more convenient to choose some other; and, when that other is determined, to calculate therefrom the magnitude or magnitudes required. The next step is to select the determining condition or conditions, and to state them as a conditional equation (or as conditional equations) connecting the constants with the variable or variables. Next, the analytical equation or equations must be solved; and finally, the solutions obtained must be examined, and, if necessary, tested in order to ascertain whether it or they satisfy the conditions of the concrete problem. It must be remembered that the abstract problem represented by the conditional equations is usually more general than the concrete problem from which it originated. It may happen, therefore, that only some, or that none of the analytical solutions are admissible in the concrete problem.

Ex. 1. An integer of two digits, the sum of which is 10, exceeds the integer represented by the same digits in reversed order by 18. Find the integer.

Let the tens digit be $x$; then the other digit will be $10-x$; and the integer in question may be represented by $10x+(10-x)$. The integer which has the same digits reversed may be represented by $10(10-x)+x$. The condition of the problem is therefore represented by the equation

$$\{10x+(10-x)\}-\{10(10-x)+x\}=18.$$

If the problem has any solution it will be furnished by this equation; but unless the value of $x$ given by the equation be a positive integer less than 10, the problem will have no solution.

We get $$10x+10-x-10(10-x)-x=18;$$
whence $$8x+10-100+10x=18;$$
whence $$18x=108,$$
which gives $x=6$, a suitable solution. Since $10-6=4$, the integer required is 64. In fact, we have $64-46=18$.

If the given difference in the present problem had been 19 instead of 18, the resulting equation would have been

$$\{10x+(10-x)\}-\{10(10-x)+x\}=19,$$

the solution of which is $x=6\frac{1}{18}$, which being fractional is unsuitable. We conclude that no integer can be found satisfying the altered conditions.

Ex. 2. At what time between four and five o'clock are the hour and minute hands of a clock oppositely directed in the same straight line?

Let $x$ be the number of minutes past four. When the minute hand is at XII. the hour hand is at IIII.; and, since the minute hand travels 60 minute divisions while the hour hand travels 5, *i.e.* twelve times as fast, it follows that during the time that the minute hand has travelled over $x$ minute divisions from XII., the hour hand has travelled over $x/12$ minute divisions from IIII. The distances of the minute and hour hand from XII. at the instant supposed are $x$ and $20+x/12$ minute divisions respectively. We have therefore, by the conditions of the problem—

$$x-(20+x/12)=30.$$

This gives $$11x/12=50,$$
whence $$x=600/11=54\tfrac{6}{11}\text{ min.},$$
obviously a suitable solution.

Ex. 3. A man leaves £569,000 to his wife, children, and a distant relation; the wife to receive a certain sum, his children half as much, and the distant relation the residue. The wife paid no legacy duty; the children paid 1 per cent, and the distant relation 10 per cent. It was then found that the distant relation had as much as the wife and children together. How much was left to the wife, children, and distant relation respectively?

Let $x$ be the legacy to the wife, then the legacy to the children may be represented by $\frac{1}{2}x$, and that to the relation by $569,000-x-\frac{1}{2}x \equiv 569,000-\frac{3}{2}x$, the unit, be it observed, being £1 throughout.

After paying 1 per cent of legacy duty the children had 99/100 of their legacy left, and after paying 10 per cent the relation had only 9/10 of his. By the condition of the problem we have therefore

$$x+\frac{99}{100}(\tfrac{1}{2}x)=\frac{9}{10}(569,000-\tfrac{3}{2}x);$$

whence

$$\left(1+\frac{99}{200}+\frac{27}{20}\right)x=\frac{9}{10}\times 569,000;$$

that is—

$$\frac{569}{200}x=\frac{9}{10}\times 569,000;$$

whence

$$x=\frac{180}{569}\times 569,000,$$
$$=180,000.$$

This is evidently an admissible solution, and we see that the respective legacies were £180,000, £90,000, and £299,000.

Ex. 4. Two couriers, A and B, who travel at the rate of $a$ and $b$ miles per hour respectively, start along the same road. If B have a start of $h$ hours, at what distance from the starting-point will A overtake B?

Instead of taking the distance from the starting-point as variable, it will be convenient to take the number of hours that A travels before overtaking B. Let us denote this by $x$; then A has travelled a distance of $ax$ miles, and B, who has been travelling $h$ hours longer, has travelled $b(x+h)$ miles. By our conditions these distances are equal. Hence

$$ax=b(x+h)$$

From this equation we deduce immediately $x=hb/(a-b)$. This is an admissible solution, provided $a>b$; the distance from the starting-point is $xa$, that is, if we substitute the value of $x$ just found, $ahb/(a-b)$.

Ex. 5. A grocer buys $p$ lbs. of tea at $r$ shillings per lb., and $p'$ lbs. at $r'$ shillings per lb. He first sells the same quantity of each at cost price, and then mixes what remains and sells the mixture at the average of the two cost prices, viz. $\frac{1}{2}(r+r')$ shillings per lb. If the whole amount of money drawn from selling separately be the same as that drawn by selling the mixture, find how much he gained or lost by the transaction.

Let $x$ be the number of lbs. of each tea sold before mixing; the corresponding amount of money drawn is $xr+xr'$ shillings.

The number of lbs. of the mixture is $p-x+p'-x \equiv p+p'-2x$; the amount drawn for this is $(p+p'-2x)(r+r')/2$ shillings. The condition of the problem is therefore expressed by

$$xr+xr'=(p+p'-2x)(r+r')/2,$$

that is—

$$x(r+r')=(p+p'-2x)(r+r')/2.$$

If we multiply both sides of this equation by $2/(r+r')$, we derive the equivalent equation

$$2x=p+p'-2x;$$

whence

$$4x=p+p';$$

hence $x=\frac{1}{4}(p+p')$, obviously an admissible solution, provided $\frac{1}{4}(p+p')$ be less than the smaller of $p$ and $p'$.

The whole amount of money drawn from sales is $2(xr+xr') \equiv 2(r+r')x=\frac{1}{2}(p+p')(r+r')$ shillings, since $x=\frac{1}{4}(p+p')$. Hence the algebraic excess of the sale over the cost price in shillings is

$$\begin{aligned}\tfrac{1}{2}(p+p')(r+r')-pr-p'r' \\ &=\tfrac{1}{2}pr+\tfrac{1}{2}pr'+\tfrac{1}{2}p'r+\tfrac{1}{2}p'r'-pr-p'r', \\ &=-\tfrac{1}{2}pr+\tfrac{1}{2}pr'+\tfrac{1}{2}p'r-\tfrac{1}{2}p'r', \\ &=\tfrac{1}{2}p(r'-r)-\tfrac{1}{2}p'(r'-r), \\ &=\tfrac{1}{2}(p-p')(r'-r);\end{aligned}$$

this will be gain if $p>p'$ and $r'>r$, or if $p<p'$ and $r'<r$, otherwise loss—that is to say, the grocer will lose unless he bought less of the higher-priced tea.

## EXERCISES XIV.

1. Find a number which, when multiplied by 5, exceeds 18 by as much as the number itself falls short of 18.

2. Divide 64 into two parts, such that the half of one part shall exceed the other part by 8.

3. I subtract 10 from a certain number, halve the result, and add twice the original number. If the final result be 85, what was the original number?

4. Find a number such that the sum of its third and fourth parts exceeds its sixth part by 150.

5. Find two consecutive integers such that the fourth and eleventh parts of the less together exceed by 1 the sum of the fifth and ninth parts of the greater.

6. In 10 years A will be twice as old as B was 10 years ago. A is 9 years older than B. Find their ages now.

7. The united ages of a man and his wife at the present day amount

to 120 years. Thirty years ago when they were married, the man's age was double the wife's. What are their ages now?

**8.** A man said to his son, "Seven years ago I was seven times as old as you were, and three years hence I shall be three times as old as you." Find their ages.

**9.** A sum of £500 is to be divided among three men, so that the share of the first shall be double that of the second, and the share of the third equal to the sum of the shares of the first and second. Find the share of the third.

**10.** A man left half his estate to his wife, one-ninth to each of his four children, and the residue, amounting to £360, to a charitable institution. What was the value of his estate?

**11.** A sum of £8 was paid in half-crowns, shillings, and sixpences. There were 140 coins altogether, and twice as many half-crowns as shillings. How many coins of each kind were used?

**12.** A collection, amounting to £3 : 17 : 6, is made up of shillings, florins, and half-crowns; the number of florins is one more than half the number of shillings, and the number of half-crowns is two less than the number of florins. Find how many coins there are of each kind.

**13.** A man finds that during four weeks his average expenditure on a week-day is six shillings and threepence more than his average expenditure on a Sunday. If he spends altogether £12 : 15s. in the four weeks, find his average expenditure on a week-day.

**14.** On a certain day 186 single and 216 return tickets are taken out between two stations on a railway, the total receipts from these amounting to £37 : 12s. If the return ticket cost 1s. more than the single ticket, find the cost of the single ticket.

**15.** A factory hand is paid 4s. 6d. per day, but is fined 1s. for every working day that he is absent from work. After 30 working days he received £5 : 2s. How many days was he absent from work?

**16.** A tradesman buys a certain number of yards of cloth at 7d. per yard, and sells 8 yards of it at 9d. more per yard, and the rest at 1s. more per yard than he originally paid for it. With the money thus obtained he buys 40 yards at the same cost price as before. How many yards did he originally buy?

**17.** The receipts of a railway company are apportioned as follows:— 49 per cent for working expenses, 10 per cent for the reserved fund, a guaranteed dividend of 5 per cent on one-fifth of the capital, and the remainder, £40,000, for division amongst the holders of the rest of the stock, being a dividend at the rate of 4 per cent per annum. Find the capital and the receipts.

**18.** If the cost of 7 lbs. of coffee be equal to that of 5 lbs. of tea, and if 6 lbs. more of coffee than of tea can be bought for £1 : 15s., what is the price per lb. of tea?

**19.** A bag contains sixpences, shillings, and half-crowns—102 coins altogether. If the sums of money represented by the different coins be the same, find the number of each.

**20.** The length of a rectangle is thrice its breadth. If each side be increased by 1 foot, its area is increased by 65 square feet. Find its dimensions.

**21.** The length of a rectangle is $a$ times its breadth. If the length be increased by $b$ feet, and the breadth diminished by $c$ feet, the area is unchanged. Find the breadth.

**22.** A man bought a certain number of pounds of apples at $a$ pence per lb., and 3 pounds more of pears at $b$ pence per lb. He expended altogether 3s. 6d. How many pounds of apples did he buy?

**23.** A market-woman first sells half her eggs and 7 eggs more, then half of the remainder and 7 eggs more, and finally half of the remainder and 7 eggs more; all her eggs were then sold. How many had she to begin with?

**24.** A and B begin to play with £5 each. A loses the first stake; he also loses the second, which is half of what he has left, and then he gains 4s. He has now a quarter of what B has. What was the amount of his first stake?

**25.** A gentleman who met a beggar every morning promised to give him every time they met as much money as he possessed, on condition that he should beg from no one else during the day. The beggar observed the condition of the agreement, but every day spent 1s. 4d. The arrangement began on Monday morning, and on Friday morning the beggar had nothing left. What sum had he to start with, and how much did the gentleman give him altogether?

**26.** A and B played three games. They started with certain sums of money, B having £12 more than A. In each of the three games each player staked half the money in his possession at the beginning of that game. A won in the first two games, and lost in the third. Finally, B was left with £21. How much had each at first?

**27.** If a man's net income, after paying 6d. per pound of income tax, be £500, what will it be when income tax is $6\frac{3}{4}$d., and what would it have been had there been no income tax?

**28.** A merchant has two qualities of tea—one at 2s. per lb., the other at 3s. per lb.; in what proportion must he mix them so as to sell at 2s. 6d. per lb. and gain a profit of 10 per cent?

**29.** A man wishes to invest £250,000 partly at 4 per cent and partly at $3\frac{1}{2}$ per cent, so that the resulting income shall be £9000. How much was invested at each rate?

**30.** A man desires to invest £1000 partly at 5 per cent, partly at 3 per cent, so that the return on his whole capital shall be at the average rate of $4\frac{1}{2}$ per cent. What sum was invested at each rate?

**31.** At what rate per cent above cost must a tradesman mark his goods so that he may make a profit of 5 per cent after allowing a reduction of 10 per cent on the marked price?

**32.** In an examination 11 out of every 13 candidates who entered passed; but of the failures 18 were due to candidates leaving without giving in a paper. Had 9 of these stayed and succeeded in passing, then 10 out of every 11 candidates entered would have passed. How many candidates were there?

**33.** A debtor is just able to pay his creditors 5s. in the pound; but if his assets had been five times as great and his debts two-thirds of what they really were, he would have had in his favour a balance of £140. How much did he owe?

**34.** A merchant buys $a$ articles for £1 each and sells $b$ of them at a gain of 5 per cent. At what price must he sell the rest of them so that he may gain 7 per cent on the whole transaction?

**35.** The expenses of a tram-car company are fixed, and when it only sells threepenny tickets for the whole journey it loses 10 per cent. It then divides the route into two parts, selling twopenny tickets for each part, thereby gaining 4 per cent and selling 3300 more tickets every week. How many persons used the cars weekly under the old system?

**36.** A owes B £100 due at the beginning of the year; B owes A £110 : 4s. due at the end of the year. Find at what time or times during the year A and B may be held to be quits, allowing interest and true discount at 10 per cent simple interest.

**37.** A cistern can be filled by one pipe in two hours, emptied by a second in one hour, and by another in an hour and a half. If all these pipes are open together and the cistern starts full, how long will it be before it is empty?

**38.** A miner contracted to hew a certain quantity of coal, intending to do it all himself. After working for 14 days he found that two-thirds of the work remained to be done. He then took a partner, and together they finished the job in 10 days. How long would it have taken each of them to do it alone?

**39.** Two men started at a distance of 30 miles from each other to walk towards each other on a road; the one walks at the rate of $3\frac{1}{2}$ and the other at the rate of $3\frac{3}{4}$ miles per hour. Find where, and how long after the start they meet.

**40.** A and B start $a$ yards apart. The length of A's stride is to the length of B's as $b$ is to $c$; and A takes $d$ steps while B takes $e$; if they run towards each other, where will they meet?

**41.** A clock which gains 2 minutes per day was set right at 9 A.M. on Sunday morning. Find the true time when this clock indicates 12 noon on Saturday following.

**42.** One traveller walking 5 miles an hour starts from A to walk to B, another walking 3 miles an hour starts simultaneously to walk from B to A. When the first traveller has got half-way the two are still 5 miles apart. How far is it from A to B?

**43.** A boat's crew row up stream in $h$ hours; and down in $k$ hours. If their speed in dead water be $l$ miles per hour, find the velocity of the stream.

**44.** Windsor is 32 miles from London by river. By rowing half-way and walking half-way I can go from Windsor to London in 6 hours, from London to Windsor in 8 hours. Find the speed of the Thames, given that my rate of walking is 4 miles an hour.

**45.** Two cyclists run round a closed course whose length is 500 yards. The one goes $m$ miles an hour, and the other $m'$ miles an hour ($m>m'$); if the latter have a start of 10 yards, find how long it will be before the first is seen to pass the second for the first time, and how long before he is seen to pass for the second time.

**46.** At what hour between 2 and 3 o'clock are the hands of a clock perpendicular?

**47.** A man could walk from A to B and back in a certain time at the rate of $3\frac{1}{2}$ miles per hour; if he walk there at 3 miles per hour and back at 4 miles per hour he would take 5 minutes longer; find the distance from A to B.

**48.** It was observed that the interval between two successive coincidences of the hour and minute hand of a clock was 67 minutes true time; find the error of the clock.

**49.** The co-ordinates of two points A and B on a line are $x_1$ and $x_2$. C and D divide A and B internally and externally in the same ratio, and the co-ordinate of the middle point between C and D is $a$; find the co-ordinates of C and D.

**50.** From a regiment of soldiers a solid square is formed leaving $c^2+pc$ men over, where $c$ and $p$ are positive integers. It is found that in order to form a square with $c$ more men in the side would require $pc$ more men than there are in the regiment. Find the number of men in the regiment.

**51.** When a solid body is weighed in water its real weight is diminished by the weight of the water which it displaces, *i.e.* by the weight of a quantity of water equal in volume to itself. A piece of an alloy of gold and silver weighs W grammes *in vacuo* and $w$ grammes in water. Assuming that the two metals alloy without change of volume, calculate the proportion (by volume) of gold and silver in the alloy, given that the ratio of the weight of a piece of gold to the weight of the same volume of water is $\rho$, and that the corresponding ratio for silver is $\sigma$.

**52.** A man's business pays every year $p$ per cent on the capital invested in it. He invests a certain sum, and at the end of each year withdraws £$a$ to meet his private expenses. At the end of $n$ years he finds that his capital has increased $q$-fold; find the sum originally invested.

Show that if he invests £100 $a/p$, his capital will remain stationary, and that, if he cannot invest so much as this, he will sooner or later become bankrupt. [It may be found necessary to use the identity $X^n-1=(X-1)(X^{n-1}+X^{n-2}+\ .\ .\ .\ +X+1)$.]

# CHAPTER VIII

## LINEAR EQUATIONS WITH TWO OR MORE VARIABLES

§ **70.** In many respects the theory of equations with more than one variable is the same as the theory of equations with only one variable; there are, however, some important new points which must be carefully studied. We begin as before with a special problem :—To find two algebraic quantities such that three times the one together with twice the other is 5. If we denote the two quantities by $x$ and $y$, the single condition of the problem is represented by the equation

$$3x + 2y = 5 \qquad (1).$$

If now $x$ and $y$ were actually replaced by a pair of quantities satisfying the condition of the problem, the equation (1) would become a numerical identity. Assuming $x$ and $y$ to have such values, and writing for emphasis

$$3x + 2y \equiv 5 \qquad (2),$$

we have, by the laws of Algebra—

$$2y \equiv 5 - 3x \qquad (3);$$

and

$$y \equiv (5 - 3x)/2 \qquad (4).$$

It appears from (4), since the steps are all reversible, that, if we give to $x$ any algebraical value whatever, then a corresponding value of $y$ is determined, which together with the assumed value of $x$ will render (2) an identity ; we have, in fact, $a$ being any assumed value of $x$ whatever—

$$3a + 2(5 - 3a)/2 \equiv 5 \qquad (5).$$

The equation (1) has therefore an infinity of solutions; for $a$

may be any quantity in the series $-\infty, \ldots, -1, \ldots 0, \ldots, +1, \ldots +\infty$.* In the following table we give a few of these solutions chosen at random :—

$$\begin{array}{lllllll} x=-1, & -0{\cdot}20, & -0{\cdot}10, & 0{\cdot}0, & +0{\cdot}010, & +1, & +2{\cdot}0 \\ y=+4, & +2{\cdot}80, & +2{\cdot}65, & +2{\cdot}5, & +2{\cdot}485, & +1, & -0{\cdot}5 \end{array}$$

We are thus led to define *the solution of a conditional equation involving two variables* x *and* y *as a pair of corresponding values of* x *and* y *in terms of the constants or given quantities such that when these values are substituted for* x *and* y *respectively in the conditional equation it becomes an actual identity;* and a corresponding definition is naturally suggested for an equation involving any number of variables.

The peculiarity just established for the equation (1) evidently belongs to any linear equation involving two variables ; for directly we attribute to $x$ any constant value the equation becomes a linear equation to determine $y$, and such an equation has, as we have seen, one and only one solution.

*Every linear equation involving both* x *and* y *has therefore a onefold infinity of solutions.*

§ **71.** If we bear in mind that two variables $x$ and $y$ are now in question, and that "a solution" means a value of $x$ and a corresponding value of $y$, we see that the two theorems regarding the derivation of equations already established (§ 63) for one variable will apply to equations involving two variables (and, indeed, to equations involving any number of variables). The demonstrations will be exactly the same, provided we remember that the letters P, Q, R now stand for functions of $x$ and $y$, instead of functions of $x$ merely as before ; and all the corollaries will hold in like manner.

Ex. 1. The equations

$$\begin{aligned} 3x+2y&=6x+1, \qquad (6)\\ 2y&=3x+1,\\ -3x+2y-1&=0,\\ -12x+8y-4&=0, \end{aligned}$$

are all equivalent—that is to say, a solution that satisfies any one of them will satisfy any other ; and this is true of every one of the onefold infinity of solutions which each possesses.

---

* It is usual to say in such a case that $a$ may have any one of a *one-fold* infinity of values.

Ex. 2. The equation

$$(x-y)(3x+2y)=(x-y)(6x+1) \qquad (7)$$

is a derivative of (6) above, but is not equivalent to (6). The multiplier, by means of which we have derived it from (6), viz. $x-y$, is not constant; and as a matter of fact (7) possesses in addition to the one-fold infinity of solutions of (6) also the one-fold infinity of solutions belonging to $x-y=0$.

§ 72. Since no term of higher degree in $x$ and $y$ than the first can occur on either side, the most general conceivable form of an integral equation of the first degree in $x$ and $y$ is

$$Ax+By+C=A'x+B'y+C' \qquad (8),$$

where A, B, C, A′, B′, C′ stand for constants, *i.e.* quantities independent of $x$ and $y$.

By subtracting $A'x+B'y+C'$ from both sides, we can always reduce (8) to the form

$$(A-A')x+(B-B')y+(C-C')=0,$$

or say—

$$ax+by+c=0 \qquad (9),$$

where $a$, $b$, $c$ are constants.

*We may therefore regard* (9) *as the general standard form for an integral equation of the first degree in* x *and* y, *or say, for shortness, a linear equation in* x *and* y.

If $y$ actually occur in the equation—that is to say, if $b\neq 0$, we may, by subtracting $ax+c$ from both sides and dividing by $b$, reduce (9) to the equivalent form

$$y=\left(-\frac{a}{b}\right)x+\left(-\frac{c}{b}\right),$$

or say—

$$y=mx+n \qquad (10),$$

where $m$ and $n$ are constants.

This last result is important among other reasons, because it shows that *the linear equation in two variables involves in reality only two independent constants.*

§ 73. As the conception of the one-fold infinity of integral solutions belonging to a given linear equation may present some difficulty to a beginner, it will be well at this stage to look at this matter from the graphical point of view.

Confining ourselves for the present to the case where $y$ is not absent from the equation, we have seen that it can be reduced to the form (10). It has been shown that the graph of any linear function $mx+n$ is a straight line. That is to say, if we assign to $x$ any particular algebraic value, $\alpha$, and calculate the corresponding algebraic value of $y$, say $\beta$, and plot the point whose co-ordinates are $(\alpha, \beta)$, then all such points lie on a certain fixed straight line, whose position depends on the constants $m$ and $n$. In other words, *to every one of the one-fold infinity of solutions of the linear equation there corresponds a definite point on a certain fixed straight line, the co-ordinates of this point being the* x *and the* y *respectively of that solution.*

We have excluded the case of the general equation (9) in which $b=0$. In this case the equation reduces to $ax+c=0$. Supposing $a \neq 0$, this gives $x=-c/a$; in other words, $x$ has a definite value, and $y$ may have any value whatsoever. The corresponding graph is obviously a straight line parallel to the $y$-axis. Combining this with our previous result, we now see that *any linear equation in which one at least of the two variables* x *and* y *occurs may be regarded as represented graphically by a straight line, in the sense that every point whose co-ordinates constitute a solution of the equation is a point on the straight line.* We have thus incidentally established one of the fundamental theorems of co-ordinate geometry; and we may now speak of the graph of a linear equation, just as we speak of the graph of a function.

It is very easy to plot in a co-ordinate diagram the graph of any given linear equation, say (1). We have merely to find any two solutions whatever, plot the corresponding points, and join them with a ruler. It is usually most convenient to select the two solutions for which $y=0$, and $x=0$ respectively. These give the points where the line meets the axes of $x$ and $y$ respectively. For example, in the case of (1), when $y=0$, we get $3x=5$, *i.e.* $x=5/3$; and, in like manner, when $x=0$, $y=5/2$; so that $(+5/3, 0)$, $(0, +5/2)$ are points on the line. We lay off $+5/3$ on the $x$-axis right of O, $+5/2$ on the $y$-axis upwards from O, and the straight line joining the two resulting points is the graph of $3x+2y=5$.

§ 74. **Diophantine Solutions.**—It is interesting to notice that the indeterminateness of the solution of a linear equation may often be removed by imposing a **restriction** * upon the nature of

* The reader should notice the difference between the uses of the words *condition* and *restriction* here adopted. To require a point to lie on the circumference of a given circle is a condition on the point; to require it to lie within the given circle is merely a *restriction.*

the solution. The most common case arises when we require the values of $x$ and $y$ belonging to the solution to be both arithmetically integral, in special cases integral and positive; such solutions may be called Diophantine Solutions.* For example, if we take the equation $3x+2y=5$, and require that the $x$ and $y$ of the solution be both positive integers, it is easy to see that there is only one such solution, viz. $x=1$, $y=1$; for, if we write the equation in the equivalent form $2y=5-3x$, we see instantly that no integral value of $x$ greater than 1 will render $y$ positive; also $x=0$ gives $y=5/2$, which is not integral. Hence $x=1$, which gives $y=1$, is the only solution of the required species which exists.

Ex. 1. I have in my purse only florins and half-crowns; in how many ways can I pay a bill amounting to £1 : 9s.?

Let $x$ be the number of florins and $y$ the number of half-crowns used; then by the condition of the problem $2x+\frac{5}{2}y=29$, or

$$4x+5y=58 \qquad (11).$$

Now (11) may be written in the equivalent form

$$4x=58-5y \qquad (12).$$

By hypothesis $x$ is integral; therefore the left of (12) is exactly divisible by 2; so also must the right be, which is hypothetically identical with the left. But 58 is exactly divisible by 2; hence, since 5 is prime to 2, $y$ must be an even integer; also by hypothesis it is positive. Now try $y=0$, $y=2$, $y=4$, $y=6$, $y=8$, $y=10$ ($y=12$, etc., are obviously inadmissible, since $x$ must be positive), and we find at once that $x=12$, $y=2$; $x=7$, $y=6$; and $x=2$, $y=10$ are the only possible solutions. The debt can therefore be paid in three ways only.

Ex. 2. Find general formulæ for the Diophantine solutions of the equation (11) of last example. If we put the equation in the form (12), and divide by 4, we get

$$x=58/4-5y/4.$$

Separating out the greatest possible integral multiples of 4, *i.e.* putting $58=14\times4+2$ and $5y=(4+1)y$, we may write

$$x=14-y+(2-y)/4 \qquad (13).$$

Since $x$ must be an integer positive or negative, $(2-y)/4$ must be some integer, $t$ say. Hence $2-y=4t$. It follows that $y=2-4t$; and on substituting this value of $y$ in (13) we get $x=14-(2-4t)+t=12+5t$. Hence the required general formulæ for $x$ and $y$ are $x=12+5t$, $y=2-4t$. Here $t$ may have any integral value positive or negative; the corresponding values of $x$ and $y$ are all integral solutions of (11),

* After the Alexandrine mathematician Diophantos, in whose works many problems of the kind are solved.

but not all *positive* integral solutions. If we desire to obtain solutions of the latter kind, it is clear that $t$ cannot be greater than 0 or less than $-2$. The admissible values are, in fact, $t=0$, $t=-1$, $t=-2$, and these give the positive integral solutions (12, 2), (7, 6), (2, 10), already obtained in Example 1 by simpler considerations.

## EXERCISES XV.

**1.** Find that solution of $6x-3y=2$ for which $y=-3$.

**2.** Find the solutions of $(x-2y)(3x+y-1)=0$ for which $x=0$, 1, $-1$, respectively, and plot them.

**3.** Find that solution of $x+6y-1=0$ for which $y=2x$.

**4.** Find that solution of $3x+2y=3x+2$ for which $y=-x$.

**5.** Write down a general form for a linear equation in $x$ and $y$ which has the solution $x=-2$, $y=+3$.

**6.** What is the general form for all linear equations in $x$ and $y$ which have the solution $x=0$, $y=0$?

**7.** If the two linear equations $ax+by+c=0$, $a'x+b'y+c'=0$ be equivalent, show that $a'=Pa$, $b'=Pb$, $c'=Pc$, where P is some constant.

Draw the graphs of the following equations:—

**8.** $2x-5y=10$. **9.** $3x+6y=12$. **10.** $3x+y=0$.

**11.** $3x-y=0$. **12.** $x-1=0$. **13.** $y-3=0$.

**14.** $x^2-1=0$. **15.** $x^2-xy=0$. **16.** $x^2-y^2+3(x-y)=0$.

Find all the positive integral solutions of—

**17.** $3x+2y=24$. **18.** $6x+7y=123$.

**19.** $7x+13y=111$. **20.** $13x+17y=141$.

**21.** In how many ways can two fractions be found whose denominators are 3 and 5, and whose sum is $2\frac{1}{15}$?

**22.** The agent of a charity went out to pay allowances to its pensioners and to collect subscriptions. Each allowance was 8s.; and each subscription 6s. At the end of the day he was left with 10s. belonging to the charity, but had lost his note of the number of subscriptions collected and the number of allowances paid. Show that he can recover these numbers provided the former did not exceed 6.

**23.** Find the general integral solution of the equation $6x-8y=10$, and determine how many positive integral solutions it has for which the values of $y$ lie between 10 and 20.

**24.** A and B have in their purses 10 florins and 10 half-crowns respectively; find in how many ways they can liquidate a debt of 11s. which A owes B.

**25.** Find general formulæ to represent all the integral solutions of the equation $4x-5y=22$.

**26.** Find the positive integral solution of $13x+15y=189$ for which the value of $y$ is largest.

**27.** Show that $2x+6y=13$ has no Diophantine solutions.

## Systems of two or more Equations

**§ 75.** Consider the problem to find two algebraic quantities such that the sum of thrice the first and twice the second is 5, and the excess of twice the first over thrice the second is 1.

We have now two conditions; and we should expect a determinate solution in accordance with the general principle that $n$ conditions are required to determine $n$ variables. If $x$ and $y$ denote the two quantities required, the two conditions of the problem are expressed by the system of conditional equations

$$\left.\begin{aligned}3x+2y=5,\\ 2x-3y=1\end{aligned}\right\} \qquad (14).$$

The algebraic problem is to determine a pair (or it may be pairs) of numerical values of $x$ and $y$ which render (14) actual numerical identities. Supposing $x$ and $y$ to have such values, we see by principles already familiar that (14) are equivalent to

$$\left.\begin{aligned}3x+2y\equiv 5,\\ y\equiv(2x-1)/3\end{aligned}\right\} \qquad (15).$$

Now, if $y\equiv(2x-1)/3$, as supposed, then $3x+2y\equiv 3x+2(2x-1)/3$; and it follows from (15) that

$$\left.\begin{aligned}3x+2(2x-1)/3\equiv 5,\\ y\equiv(2x-1)/3\end{aligned}\right\} \qquad (16).$$

If we multiply both sides of the first of these equations by the non-evanescent constant 3, and subtract 15 from both sides, we derive

$$\left.\begin{aligned}13x-17\equiv 0,\\ y\equiv(2x-1)/3\end{aligned}\right\} \qquad (17);$$

whence

$$\left.\begin{aligned}x\equiv 17/13,\\ y\equiv(2x-1)/3\end{aligned}\right\} \qquad (18).$$

If we now substitute the value of $x$ assigned by the first equation of (18) in the second, we derive

$$x\equiv 17/13, \quad y\equiv 7/13 \qquad (19).$$

If, therefore, the system (14) have any solution, it must be the solution $x=17/13$, $y=7/13$, and no other. But the steps are all reversible, so that (14) can be derived from (19); hence $x=17/13$, $y=7/13$ actually makes (14) a pair of identities. In other words, $x=17/13$ and $y=7/13$ is the unique solution of the system (14). In verification, we find

$$3\times 17/13+2\times 7/13\equiv(51+14)/13\equiv 65/13\equiv 5,$$
$$2\times 17/13-3\times 7/13\equiv(34-21)/13=13/13\equiv 1.$$

§ **76.** Careful consideration of the foregoing special example leads us to two new ideas.

In the first place, we have the notion of *a solution of a system of equations—that is to say, a set of values of the variables in terms of constants or given quantities which render each of the equations of the system an identity.*

Again, we have used a principle for the transformation of systems which deserves formal statement, viz. *In working with a system of conditional equations we may, without affecting the equivalence of the system* (i.e. *without adding to or subtracting from the totality of its finite solutions*), *transform any one of its equations by using any other as if it were an identity.*

This is, of course, an immediate consequence of our notion of the nature of a solution of the system and the fact that we work with the system on the supposition that the variables have a set of values which constitute a solution, so that the equations are treated as if they were actual identities. This principle, which we shall call **Inter-equational Transformation,** can be used in various ways to simplify systems of equations with a view to solution.

Ex. 1. Solve the system

$$x-y+\tfrac{3}{2}(2x+y)=3 \qquad (20),$$
$$2x+y=2 \qquad (21).$$

By (21) we have $2x+y\equiv 2$; hence (20) may be written $x-y+\frac{3}{2}\times 2=3$, whence $x-y=0$ or $y=x$. The system is therefore equivalent to

$$y=x \qquad (22),$$
$$2x+y=2 \qquad (23).$$

By means of (22) we reduce (23) to $3x=2$. It now appears that the solution is $x=2/3$, $y=2/3$.

Ex. 2.

$$x^2-y^2+3x+y=10 \qquad (24),$$
$$x+y=3 \qquad (25).$$

Since by (25) $x+y\equiv 3$ it follows that $x^2-y^2+3x+y\equiv(x+y)(x-y)+3x+y\equiv 3(x-y)+3x+y$. Hence (24) may be written

$$6x-2y=10 \qquad (26).$$

It follows that, so far as finite solutions are concerned, the system (24), (25) is equivalent to the system (25), (26), which is a linear system and can be readily solved.

§ **77.** The method of solution employed in § 75 may be described as the **elimination** of $y$ from one of the equations of the system **by substitution** from the other equation; it is applicable to the discussion of the general linear system in two variables, as we shall now show.

The most general conceivable linear system may be reduced to the form

$$\left.\begin{aligned} ax+by+c=0,\\ a'x+b'y+c'=0 \end{aligned}\right\} \quad (27).$$

We shall suppose that $y$ occurs in at least one of the equations, say in the second, so that $b' \neq 0$. (If $y$ occurred in neither equation, $y$ would be subject to no condition; and the solution would be indeterminate, at least in so far as $y$ is concerned.)

Since $b' \neq 0$, the system is equivalent to

$$\left.\begin{aligned} ax+by+c=0,\\ y=-(a'x+c')/b' \end{aligned}\right\} \quad (28).$$

Using the second equation to modify the first, we derive the equivalent system

$$\left.\begin{aligned} ax-b(a'x+c')/b'+c=0,\\ y=-(a'x+c')/b' \end{aligned}\right\} \quad (29).$$

Since $b' \neq 0$, (29) is equivalent to

$$\left.\begin{aligned} (ab'-a'b)x+b'c-bc'=0,\\ y=-(a'x+c')/b' \end{aligned}\right\} \quad (30).$$

If now $ab'-a'b \neq 0$, **(30)** is equivalent to

$$\left.\begin{aligned} x&=(bc'-b'c)/(ab'-a'b),\\ y&=-(a'x+c')/b' \end{aligned}\right\} \quad (31).$$

Since

$$\begin{aligned} &a'(bc'-b'c)/(ab'-a'b)+c'\\ &\quad=(a'bc'-a'b'c+ab'c'-a'bc')/(ab'-a'b),\\ &\quad=-b'(ca'-c'a)/(ab'-a'b), \end{aligned}$$

we get from **(31)**, by substituting in the second equation the value of $x$ given by the first—

$$\left.\begin{aligned} x&=(bc'-b'c)/(ab'-a'b),\\ y&=(ca'-c'a)/(ab'-a'b) \end{aligned}\right\} \quad (32).$$

As the steps are all reversible, we conclude that *if* $ab'-a'b \neq 0$, *then the general linear system* **(27)** *has one and only one solution, viz.* **(32)**.

If we represent the function $ab' - a'b$ by the scheme

$$a \diagdown \quad b$$
$$\diagup\diagdown$$
$$a' \diagup \quad b',$$

where the line sloping downwards from left to right is a direction to form a product and attach the sign $+$, and a line sloping upwards from left to right a direction to form a product and attach the sign $-$, then we may represent the common denominators and the numerators of $x$ and $y$ respectively in the solution (32) by the following scheme:—

$$a \diagdown\ b \diagdown\ c \diagdown\ a$$
$$a' \diagup\ b' \diagup\ c \diagup\ a',$$

the formation of which from the coefficients of the system (27) is obvious. This scheme is often used as a *memoria technica* for writing down the solution of a given linear system at sight. For instance, taking (14), we have

$$+3 \diagdown\ +2 \diagdown\ -5 \diagdown\ +3$$
$$+2 \diagup\ -3 \diagup\ -1 \diagup\ +2.$$

Hence the common denominator is $(+3)(-3) - (+2)(+2) = -13$; the numerator of $x$ is $(+2)(-1) - (-5)(-3) = -17$; the numerator of $y$ is $(-5)(+2) - (+3)(-1) = -7$. Therefore $x = -17/-13 = 17/13$, $y = -7/-13 = 7/13$.

§ 78. The solution of a linear system may be arranged in an elegant manner by employing the following general principle of equivalence, which is often useful in dealing with systems of equations :—

*If* l, m, l′, m′ *be four finite constants, such that* $lm' - l'm \neq 0$ *and* P *and* Q *be functions of any variables* x, y, *etc., then the system*

$$P = 0, \quad Q = 0 \tag{33},$$

*is equivalent to the system*

$$lP + mQ = 0, \quad l'P + m'Q = 0 \tag{34}.$$

To prove this we remark, in the first place, that it is obvious, since *l, m, l′, m′* are all finite, that any values of $x$, $y$, etc., which render (33) identities will also render (34) identities.

Again, since

$$m'(lP + mQ) - m(l'P + m'Q) \equiv (lm' - l'm)P,$$

and

$$l(l'P + m'Q) - l'(lP + mQ) \equiv (lm' - l'm)Q,$$

it follows that any values of $x$, $y$, etc., which render (34) identities also render

$$(lm' - l'm)\mathrm{P} = 0, \quad (lm' - l'm)\mathrm{Q} = 0 \qquad (35)$$

identities. But, if $lm' - l'm \neq 0$, these necessitate that

$$\mathrm{P} \equiv 0, \quad \mathrm{Q} \equiv 0.$$

Hence every solution of (33) is a solution of (34), and every solution of (34) is a solution of (33); in other words, the two systems are equivalent.

In practice, the four constants are chosen so that the derived system (34) shall in some respect be simpler than (33).* Thus in the case of the general linear system (27), we choose $l = +b'$, $m = -b$, $l' = -a'$, $m' = +a$; the object being to derive two new equations, of which one does not contain $y$, and the other does not contain $x$. In fact, the new system equivalent to (27) is, provided $ab' - a'b \neq 0$—

$$\left.\begin{array}{r} b'(ax + by + c) - b(a'x + b'y + c') = 0, \\ -a'(ax + by + c) + a(a'x + b'y + c') = 0, \end{array}\right\}$$

that is to say—

$$\left.\begin{array}{l} (ab' - a'b)x + b'c - bc' = 0, \\ (ab' - a'b)y + c'a - ca' = 0 \end{array}\right\} \qquad (36),$$

from which, provided always $ab' - a'b \neq 0$, we derive the same result and the same general conclusions as before.

This process is sometimes described as solution by **Cross Multiplication**; it is usually preferable when the coefficients in the equation are literal, and especially where the equations are symmetrical, as in Example 2 below. This leads us to remark that the expression for $y$ given in (32) is derivable from the expression for $x$, by interchanging the letters $a$ and $b$, and at the same time $a'$ and $b'$.

Ex. 1. If we treat the system (14) by cross multiplication, we derive

$$\left.\begin{array}{l} 3(3x+2y)+2(2x-3y)=3\times5+2\times1, \\ 2(3x+2y)-3(2x-3y)=2\times5-3\times1; \end{array}\right\}$$

that is—

$$13x=17,$$
$$13y=7;$$

whence $x=17/13$, $y=7/13$, as before.

---

* It should also be noticed that one of the four may be zero, *e.g.* if we put $l'=1$, $m'=0$, we get the theorem that $\mathrm{P}=0$ and $\mathrm{Q}=0$ is equivalent to $l\mathrm{P}+m\mathrm{Q}=0$, $\mathrm{P}=0$ provided $m \neq 0$.

Ex. 2. $$x/(a+\lambda)+y/(b+\lambda)=1,$$
$$x/(a+\mu)+y/(b+\mu)=1.$$

If we multiply first by $1/(b+\mu)$, and by $1/(b+\lambda)$, and subtract; and then by $1/(a+\mu)$, and by $1/(a+\lambda)$, and subtract, we get the equivalent system

$$\{1/(a+\lambda)(b+\mu)-1/(a+\mu)(b+\lambda)\}x=1/(b+\mu)-1/(b+\lambda),$$
$$\{1/(a+\lambda)(b+\mu)-1/(a+\mu)(b+\lambda)\}y=1/(a+\lambda)-1/(a+\mu).$$

Since

$$\frac{1}{(a+\lambda)(b+\mu)}-\frac{1}{(a+\mu)(b+\lambda)}\equiv\frac{(a+\mu)(b+\lambda)-(a+\lambda)(b+\mu)}{(a+\lambda)(a+\mu)(b+\lambda)(b+\mu)},$$
$$\equiv\frac{a\lambda+b\mu-b\lambda-a\mu}{\text{etc.}},$$
$$\equiv\frac{(a-b)\lambda-(a-b)\mu}{\text{etc.}},$$
$$\equiv\frac{(a-b)(\lambda-\mu)}{(a+\lambda)(a+\mu)(b+\lambda)(b+\mu)};$$

and

$$\frac{1}{b+\mu}-\frac{1}{b+\lambda}\equiv\frac{(b+\lambda)-(b+\mu)}{(b+\lambda)(b+\mu)},$$
$$\equiv\frac{\lambda-\mu}{(b+\lambda)(b+\mu)};$$
$$\frac{1}{a+\lambda}-\frac{1}{a+\mu}\equiv\frac{\mu-\lambda}{(a+\lambda)(a+\mu)};$$

the last pair of equations (provided none of the quantities $a-b$, $\lambda-\mu$, $a+\lambda$, $a+\mu$, $b+\lambda$, or $b+\mu$ vanish) are equivalent to

$$x=(a+\lambda)(a+\mu)/(a-b),$$
$$y=(b+\lambda)(b+\mu)/(b-a),$$

which is the solution of the given system.

It should be noticed here that the value of $y$ is obtainable from the value of $x$ by interchanging simultaneously $a$ and $b$ and $\lambda$ and $\mu$. That this must be so is evident *à priori* from the fact that this interchange, along with the interchange of $x$ and $y$, does not alter the given system, the result being in fact to produce the system

$$y/(b+\lambda)+x/(a+\lambda)=1,$$
$$y/(b+\mu)+x/(a+\mu)=1,$$

in which the terms in $x$ and $y$ are merely commutated. Hence any consequence derived regarding $x$, $y$, $a$, $b$, $\lambda$, $\mu$, will also hold regarding $y$, $x$, $b$, $a$, $\mu$, $\lambda$.

§ **79**. If a system of linear equations in two variables $x$ and $y$ contain more than two equations, it is obvious that in general there will be no solution which satisfies all the equations of the system or, as it is commonly put, *a system of more than two equations in two variables* x *and* y *is in general inconsistent.* This is merely a particular case of the general principle that we cannot in general determine two variables so as to satisfy more than two independent conditions. In the particular case of a linear system, say—

$$\left.\begin{array}{c} ax+by+c=0, \\ a'x+b'y+c'=0, \\ a''x+b''y+c''=0, \\ a'''x+b'''y+c'''=0, \\ \text{etc.} \end{array}\right\} \qquad (37),$$

a definite proof may be given as follows. Take any two of the equations of the system, say the two first. We may suppose that $ab'-a'b \neq 0$, since we are at present only considering systems in general. Then these two equations have a definite common solution, viz. $x=(bc'-b'c)/(ab'-a'b)$, $y=(ca'-c'a)/(ab'-a'b)$; but this solution will not in general satisfy the third or any other equation of the system, for the simple reason that a particular solution will not satisfy any equation taken at random.

In order that the third equation of the above written system may be consistent with the two first, we must have

$$a''(bc'-b'c)/(ab'-a'b)+b''(ca'-c'a)/(ab'-a'b)+c''=0,$$

which, since $ab'-a'b \neq 0$, is equivalent to

$$a''(bc'-b'c)+b''(ca'-c'a)+c''(ab'-a'b)=0 \qquad (38).$$

*This is the necessary and sufficient condition in general that a system of three linear equations be consistent. If there are more than three equations in the system, then there is a corresponding condition for each additional equation; so that, if there are $n$ equations, there are $n-2$ independent conditions of consistency.*

The left-hand side of the equation (38), which often occurs in the applications of Algebra, may be recovered by means of the *memoria technica*—

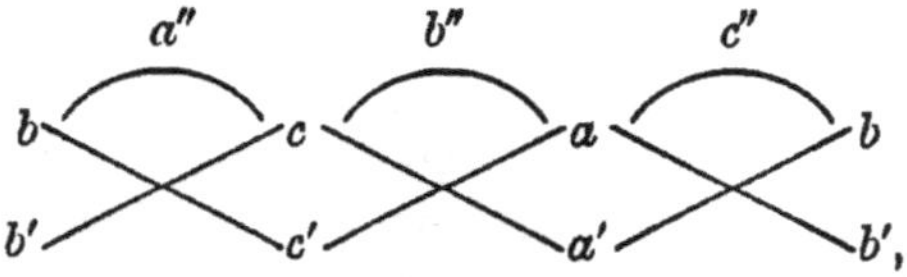


the construction and interpretation of which will be understood from § 77.

Ex. 1. Are the three equations $3x+2y=5$, $2x-3y=1$, $x+y-2=0$ consistent?

The common solution of the first two is $x=17/13$, $y=7/13$. Now, when $x$ and $y$ have these values, $x+y-2=17/13+7/13-2=24/13-2 = -2/13 \neq 0$. The only common solution of the first two equations does not satisfy the third, so that there is no common solution.

Ex. 2. Determine the value of the constant $c$ in order that $3x+2y=5$, $2x-3y=1$, $(1+c)x+(1-c)y=1$ may be a consistent system.

The common solution of the two first equations is $x=17/13$, $y=7/13$; and this must satisfy the third. Hence $c$ must be such that

$$(1+c)17/13+(1-c)7/13=1.$$

This last is a simple equation to determine $c$. Solving it we get $c=-11/10$.

Or we may quote the condition (38), which gives

$$(1+c)\{(+2)(-1)-(-5)(-3)\}+(1-c)\{(-5)(+2)-(+3)(-1)\} \\ +(-1)\{(+3)(-3)-(+2)(+2)\}=0,$$

that is—

$$-17(1+c)-7(1-c)+13=0,$$

from which we get $c=-11/10$, as before.

§ 80. **Graphical Solution of a Linear System.**—Since to every linear equation in $x$ and $y$ corresponds a straight line such that the values of $x$ and $y$ belonging to any solution of the equation are the co-ordinates of a point on the line, it follows that the common solution of two given linear equations in $x$ and $y$ is simply the values of $x$ and $y$, which are the co-ordinates of the common point of the two straight lines which are the graphs of the two equations. To obtain the solution of a system graphically all that is necessary is to plot the straight lines represented by the two equations; measure the $x$ and $y$ co-ordinates of their intersection and attach the proper algebraic signs.

Since two straight lines in a plane have in all cases a single finite intersection, unless they are coincident or parallel, we see that the case, excepted in the analytical theory of § 77, where $ab'-a'b=0$, must correspond to the cases now excepted, where the two graphs are coincident or parallel. In the former of these cases there are an infinity of points common, every point on either graph being also a point on the other. Algebraically this means that every solution of either equation is a solution of the other—that is to say, the two equations of the system are not independent but equivalent. In the latter of the two excepted cases there is no finite point of intersection. Algebraically this means that for finite values of $x$ and $y$ the two equations are inconsistent.

Ex. 1. The system $x-2y+1=0$, $3x-6y+3=0$ has a one-fold infinity of solutions, the two equations being equivalent.

Ex. 2. The system $x-2y+1=0$, $3x-6y+1=0$ has no finite solution, its two equations being inconsistent for finite values of $x$ and $y$, as may be seen more clearly by writing the system in the equivalent form $x-2y=-1$, $x-2y=-1/3$, the two equations of which are obviously contradictory for finite values of $x$ and $y$.

## EXERCISES XVI.

Solve the following systems :—

**1.** $3x-2y=9$, $2x-3y=1$. **2.** $6x+y=23$, $2x+3y=21$.

**3.** $13x-14y+1=0$, $16x+y-17=0$.

**4.** $13x+15y-280=0$, $12x-6y=60$.

**5.** $101x+100y=1607$, $x+y=16$.

**6.** $7x+101y=777$, $67x+47y=999$.

**7.** $4x-y-4=5x-2y=1$.

**8.** $\frac{1}{19}(11x+13y)=\frac{1}{71}(12x+7y)=1$.

**9.** $x/2+y/3=2$, $x/3+y/2=2\frac{1}{6}$. **10.** $x/2-y=\frac{1}{2}$, $x-y/2=\frac{5}{2}$.

**11.** $x/15-y/12=\frac{1}{6}$, $x/5-y/10=\frac{4}{5}$. **12.** $3x-2y=0$, $y=3x$.

**13.** $\frac{1}{2}(x-1)=\frac{1}{3}(y-1)$, $\frac{1}{3}(x-1)+\frac{1}{2}(y-1)=\frac{13}{6}$.

**14.** $(3x+y)/32=(11x-5y)/22$, $5y+1=8x$.

**15.** $\frac{1}{2}x-\frac{1}{3}(y-2)-\frac{1}{4}(x-3)=0$, $x-\frac{1}{2}(y-1)-\frac{1}{5}(x-2)=0$.

**16.** $(6x+3y)/4-(3y-2x)/6=7\frac{1}{4}-(x+y)/3$, $3x+11y=61$.

**17.** $(x+y+1)/3+(2x+3y+1)/10=x$,
$3x+(2y+1)/11+(x-y+1)/3=y$.

**18.** $(x+y)/4+(3x+y)/10=x$, $(3x-y)/11+(3y-x)/10=5$.

**19.** $(5x+7y)/(9x+33)=13/21$, $(11x+27)/(7x+5y)=19/11$.

**20.** $(x+y-2)/(x-y)=17$, $(x+y-2)/(x+y)=17/19$.

**21.** $\cdot 1x+\cdot 21y+\cdot 32=0$, $\cdot 01x+\cdot 01y+3=0$.

**22.** $\cdot 3x+1\cdot 2y=2$, $\cdot 6x-3\cdot 2y=3\cdot 5$, work out the values of $x$ and $y$ to three places of decimals.

**23.** $\cdot 30061x+1\cdot 6034y-3\cdot 6008=0$, $-\cdot 0136x+2\cdot 6165y+2\cdot 5361=0$, work to five places of decimals.

**24.** $(x-3)(y-4)=xy-84$, $5x-6y+27=0$.

**25.** $(x-10)(y-12)=(x-5)(y-6)$,
$(x-10)^2+(y-12)^2=(x-5)^2+(y-6)^2$.

## EXERCISES XVII.

**1.** $3x/a-2y/b=9$, $2x/a-3y/b=1$.

**2.** $6x/a^2+y/b^2=23$, $2x/a^2+3y/b^2=21$.

**3.** $x/(a+b)+y/(a-b)=1$, $x/(a+b)-y/(a-b)=1$.

**4.** $ax+by=ax-by=0$. **5.** $x/a+y/b=c$, $x/b+y/a=d$.

**6.** $ax+by=bx-ay=a^2+b^2$. **7.** $ax+by=ab$, $x+y=a+b$.

**8.** $(x+y)/(a+b)+(x-y)/(a-b)=2$, $ax+by=a^2+b^2$.

**9.** $a^3x+b^3y=a^3b^3$, $a^4x+b^4y=a^4b^4$.

10. $a^n x + b^n y = 2a^n b^n$, $b^n x + a^n y = a^{2n} + b^{2n}$.
11. $ax = by + \frac{1}{2}(a^2 + b^2)$, $(a - b)x = (a + b)y$.
12. $2(x - a)/a = 2(y - b)/b = (x + y)/(a + b)$.
13. $ax + by = c^2$, $a/(b + y) - b/(a + x) = 0$.
14. $x/(a + b) + y/(a - b) = 2$, $b(1/x + 1/y) = a(1/y - 1/x)$.
15. $(l + m)x + (l - m)y = 2(l^2 + m^2)$, $(l - m)x + (l + m)y = 2(l^2 - m^2)$.
16. $x/(p - a) + y/(p - b) = x/(q - a) + y/(q - b) = 1$.
17. $x/a^2 + y/b^2 = 2(x + y)/(a^2 + b^2) = 2(a + b)$.
18. $x/(c + a - b) + y/(c + b - a) = x/(c - a) + y/(c - b) = 1$.
19. $ax/(a^2 - \lambda) + by/(b^2 - \lambda) = 1/a + 1/b$, $x/a + y/b = 1$.
20. $(p - q)x/(p + q) + (p + q)y/(p - q) = 2(p^2 + q^2)/(p^2 - q^2)$, $x + y = 2$.
21. $(a + b)(x/a + y/b) + (a - b)(x/a - y/b) = a + b$,
$(a - b)(x/a + y/b) - (a + b)(x/a - y/b) = a - b$.
22. $ax/(a^2 - c^2) + by/(b^2 - c^2) = 1 = a^2x/(a^2 - c^2) + b^2y/(b^2 - c^2)$.

## EXERCISES XVIII.

Discuss graphically the solution of the following systems pointing out any peculiarities :—

1. $x + y = 1$, $x - y = 1$.
2. $x + y = 1$, $-x - y = 1$.
3. $2x + 3y = 6$, $3x - 2y = 12$.
4. $x^2 - y^2 = 0$, $3x + 2y = 5$.
5. $x^2 - y^2 = 0$, $(x - y)(2x - 3y) + 12(x - y) = 0$.
6. $3x + 4y - 1 = 0$, $18x + 24y - 6 = 0$.

7. Construct a linear equation in $x$ and $y$ which has the solutions $x = 0$, $y = 0$, and $x = -1$, $y = 3$.

8. Construct a linear equation in $x$ and $y$ which has the solutions $x = 2$, $y = 5$, and $x = 9$, $y = -3$.

9. Given that one of the equations of a system is $6x - 4y + 8 = 0$, and that its solution is $x = 2$, $y = 5$, find the other equation. Is this problem determinate?

10. Show that the system $ax + by + c = 0$, $bx - ay + d = 0$ always has a finite determinate solution, unless $a = 0$ and $b = 0$.

11. For what value of $\lambda$ is the solution of the system $(3 + \lambda)x + (2 + \lambda)y + 4 = 0$, $(5 - \lambda)x + (4 - \lambda)y + 6 = 0$ not finite and determinate? Discuss the exceptional case graphically.

12. If $l$, $m$, as well as $a$, $b$, and $c$, be constants, show that the system $ax + by + c = 0$, $a'x + b'y + c' = 0$, $l(ax + by + c) + m(a'x + b'y + c') = 0$ is consistent.

13. Is the system $2x - 4y + 5 = 0$, $3x + 4y - 6 = 0$, $x - y + 3 = 0$ consistent or inconsistent?

14. Determine $k$ so that the system $3x - y = 5$, $x + 2y = 7$, $x - 3y = k$ may be consistent.

15. Find the value of $p$ in order that $4x - 6y + 1 = 0$, $3x + 4y - 1 = 0$, $px - 5y + 2 = 0$ may have a common solution.

16. Find the condition that $ax + by + c = 0$, $a^2x + b^2y + c^2 = 0$, $a^3x + b^3y + c^3 = 0$ may be consistent. What are the common solutions?

17. Find the value of $\lambda$ so that the system $x/(a + \lambda) + y/(b + \lambda) = x/(a + 2\lambda) + y/(b + 2\lambda) = x/(a + 3\lambda) + y/(b + 3\lambda) = 1$ may be consistent.

18. Show that the system $(ax + by)/(a^2 + b^2) = (bx + cy)/(b^2 + c^2) = (cx + ay)/(c^2 + a^2) = 1$ will not be consistent unless $\Sigma a^3 b = 2\Sigma b^2 c^2$.

19. Discuss the solution of the system $ax + by = 0$, $a'x + b'y = 0$—

first, when $ab' - a'b \neq 0$; second, when $ab' - a'b = 0$, and illustrate graphically.

## LINEAR SYSTEMS WITH MORE THAN TWO VARIABLES

§ **81.** We have seen that, by using appropriate theorems regarding the equivalence of systems, the theory of systems of linear equations involving two variables can be made to depend on the theory of linear equations involving only one variable. In like manner, the theory of linear systems with three variables $(x, y, z)$ can be made to depend on the corresponding theory for systems of two variables, and on the theory of equations with only one variable, and so on. It will be sufficient to sketch the theory for three variables.

We remark, in the first place, that *a single equation of this nature, say*

$$ax + by + cz + d = 0 \qquad (39),$$

*has a twofold infinity of solutions.* We may give to $y$ and $z$ any definite values we please, and then (39) becomes a linear equation for $x$ which has only one solution. But $y$ may have any one of a one-fold infinity of values, and $z$ any one of a one-fold infinity of values, and any value of $y$ may be coupled with any value of $z$. Thus there arise, as it were, infinity times infinity different pairs of values of $y$ and $z$, and with each of these goes a definite value of $x$ to make up a solution of (39).

Next suppose we have a system of two linear equations in $x, y, z$, say (39) along with—

$$a'x + b'y + c'z + d' = 0 \qquad (40).$$

We have now to consider the sets of values of $x, y, z$ that render both (39) and (40) identities. If we assign to $z$ any definite value, (39) and (40) will, by the principles already established, in general furnish a definite pair of values for $x$ and $y$, and these with the assumed definite value of $z$ make up a solution of the system of two equations. *Such a system has therefore, in general, a one-fold infinity of solutions.*

Next consider a system of three linear equations in $x, y, z$, say (39), (40), and (41)—

$$a''x + b''y + c''z + d'' = 0 \qquad (41).$$

By cross multiplication, or by substitution for $z$, it is easy to derive from the given system an equivalent system of the form*

$$\left.\begin{aligned} ax+by+cz+d&=0,\\ px+qy+r&=0,\\ p'x+q'y+r'&=0 \end{aligned}\right\} \qquad (42),$$

two of whose equations do not contain $z$. If $pq'-p'q\neq0$, the two last of (42) will give unique definite values for $x$ and $y$; and when these are substituted in the first, it gives a corresponding definite value of $z$. *We thus, save in certain special cases, arrive at a unique solution for any system of three linear equations in three variables.* In practice we may use any theorem of equivalence to simplify the system before proceeding to its final solution. The following examples will help to give definiteness to the foregoing remarks:—

Ex. 1. Solve the system

$$\begin{aligned} 3x-2y+z&=1,\\ x+y-2z&=2,\\ 2x+3y+4z&=3. \end{aligned}$$

If we multiply both sides of the first equation by 2, and both sides of the second by 1, and add; and again multiply both sides of the second by 2, and both sides of the last by 1, and add, we derive the equivalent system

$$\begin{aligned} z&=1-3x+2y,\\ 7x-3y&=4,\\ 4x+5y&=7. \end{aligned}$$

If we solve the last two of these three as a system in $x$ and $y$, we get

$$x=41/47, \quad y=33/47.$$

Finally, we substitute these values of $x$ and $y$ in the first of the three equations and obtain $z=-10/47$. The solution is therefore

$$x=41/47, \quad y=33/47, \quad z=-10/47.$$

Ex. 2. Solve the system

$$lx+a(x+y+z)=p, \quad ly+b(x+y+z)=q, \quad lz+c(x+y+z)=r.$$

By addition from all three equations, we get

$$l\Sigma x+\Sigma a\,.\,\Sigma x=\Sigma p.$$

Hence, assuming that $l+\Sigma a\neq0$, $\Sigma x=\Sigma p/(l+\Sigma a)$. Using this result in the first of the three equations, we get

$$lx+a\Sigma p/(l+\Sigma a)=p,$$

* The following simple theorem of equivalence, which the reader may easily prove, is here involved. If P, Q, R be functions of $x$, $y$, $z$, the system $P=0, Q=0, R=0$ is equivalent to $P=0, lP+mQ=0, l'P+m'R=0$, where $l$, $m$, $l'$, $m'$ are finite constants and $m\neq0$, $m'\neq0$.

whence

$$x=p/l-a\Sigma p/l(l+\Sigma a),$$

we get the corresponding values of $y$ and $z$ by interchanging $\begin{pmatrix} x & y & z \\ a & b & c \\ p & q & r \end{pmatrix}$ simultaneously, viz.—

$$y=q/l-b\Sigma p/l(l+\Sigma a),$$
$$z=r/l-c\Sigma p/l(l+\Sigma a).$$

Ex. 3. Show that the one-fold infinity of solutions of the system

$$ax+by+cz=0,$$
$$a'x+b'y+c'z=0,$$

is represented by

$$x=(bc'-b'c)\rho, \quad y=(ca'-c'a)\rho, \quad z=(ab'-a'b)\rho,$$

where $\rho$ is a new variable.*

We shall suppose one at least of the three $bc'-b'c$, $ca'-c'a$, $ab'-a'b$ to be different from zero. This will certainly be the case if the two equations are independent. Let say $bc'-b'c\neq 0$, then the given system is equivalent to

$$-c'(ax+by+cz)+c(a'x+b'y+c'z)=0,$$
$$b'(ax+by+cz)-b(a'x+b'y+c'z)=0\,;$$

that is, to

$$(ca'-c'a)x-(bc'-b'c)y=0,$$
$$(ab'-a'b)x-(bc'-b'c)z=0.$$

Since $bc'-b'c\neq 0$, we can always choose a new variable $\rho$ so that $x=(bc'-b'c)\rho$. This assumption reduces the last two equations to

$$(ca'-c'a)(bc'-b'c)\rho-(bc'-b'c)y=0,$$
$$(ab'-a'b)(bc'-b'c)\rho-(bc'-b'c)z=0.$$

Since $bc'-b'c\neq 0$, those are equivalent to

$$(ca'-c'a)\rho-y=0,$$
$$(ab'-a'b)\rho-z=0\,;$$

that is to say, we have

$$x=(bc'-b'c)\rho, \quad y=(ca'-c'a)\rho, \quad z=(ab'-a'b)\rho.$$

It is easy to verify *a posteriori* that these values do, in fact, satisfy the two equations; and they are sufficiently general, so long as the two equations are independent, since $\rho$ is susceptible of a one-fold infinity of values, and the formulæ for $x$, $y$, $z$ therefore give a one-fold infinity of solutions. These last two remarks afford indeed a proof of the theorem, although they give no clue to its discovery.

## EXERCISES XIX.

1. $3x+y-2z=9$, $4x+y+z=5$, $x+2y+z=6$.
2. $3x-4y+6z=-12$, $x+2y-3z=1$, $2x-3y+8z=-5$.

---

* This result is often useful in practice.

3. $y+z=3,\ z+x=2,\ x+y=1.$
4. $4x+3y-z=2x+4y+z=x+y+z=1.$
5. $3x-6y+4z=2x+5y-6z=4x+5y+6z=1.$
6. $3x-5y+3z=4,\ 4(x-3)=3(y-2)=2(z-5).$
7. $x+\frac{1}{2}(y+z)=17,\ y+\frac{1}{3}(z+x)=11,\ z+\frac{1}{4}(x+y)=11.$
8. $\frac{1}{7}(x+y-2z)=\frac{1}{4}(2x+3y+z)=\frac{1}{5}(3x-y+2z)=1.$
9. $\frac{1}{3}x-\frac{1}{4}y+\frac{1}{5}z=\frac{11}{20},\ \frac{1}{2}x-\frac{1}{3}y+\frac{1}{4}z=\frac{1}{4}x+\frac{1}{3}y+\frac{1}{6}z=\frac{11}{12}.$
10. $(x-3)/4=(y-2)/-6=(z-1)/-8,\ x+3y-2z=1.$
11. Discuss the system $x+9y+6z=16,\ 2x+3y+2z=7,$
$3x+6y+4z=13.$
12. Find the positive integral solutions of the system $3x-5y+7z=14$, $2x-7y+6z=6$; also the general formulæ which give all the integral solutions of the system positive or negative.
13. $4x-6y=4y-6z=4z-6y=a.$
14. $2x+15y+3z=0,\ x+y+z=0,\ ax+by=c.$
15. $-x+y+z=x-y+z=x+y-z=a.$
16. $\sum_{xyz} x=\sum_{abc} a,\ \Sigma(x-a)/a+3=0,\ \Sigma(x-a)/a^2=1.$
17. $lx+my+nz=\Sigma mn,\ \Sigma x=\Sigma l,\ \Sigma(m-n)x=0.$
18. $x/a=y/b=z/c,\ \Sigma b^4c^4x^4=3a^4b^4c^4.$
19. $\dfrac{y+z-x}{b+c}=\dfrac{z+x-y}{c+a}=\dfrac{x+y-z}{a+b}=\Sigma a.$
20. Find the condition that $bx+ay=cz,\ cy+bz=ax,\ az+cx=by$ have a solution different from $x=0,\ y=0,\ z=0.$
21. Show that the equations $(x-a)/l=(y-b)/m=(z-c)/n,\ (x-a')/l'=(y-b')/m'=(z-c')/n'$ will be consistent provided $\Sigma(a-a')(mn'-m'n)=0.$

## Systems of Higher Order whose Solution depends on Linear Equations

§ 82. The solution of systems of integral equations of higher degree than the first can often be made to depend on the solution of linear systems by taking certain functions of the original variables as variables instead of the original variables themselves. This will be sufficiently understood from the following special examples :—

Ex. 1. $4x^2+y^2=61,\ 7x^2-2y^2=13.$

If we treat $x^2$ and $y^2$ as the variables, multiply the first equation by 2 and add the second, we derive the equivalent system $15x^2=135,\ y^2=61-4x^2$. These give $x^2=9,\ y^2=25$, or $(x-3)(x+3)=0$, $(y-5)(y+5)=0$; from which we infer the four solutions :—

$$x=+3,\quad y=+5;\quad x=+3,\quad y=-5;$$
$$x=-3,\quad y=+5;\quad x=-3,\quad y=-5.$$

Ex. 2.
$$a/(a+x)+b/(b+y)=1$$
$$a^2/(a+x)+b^2/(b+y)=a+b.$$

Here we shall treat $1/(a+x)$ and $1/(b+y)$ as variables. If we first multiply the first equation by $b$ and subtract, and then multiply the first equation by $a$ and subtract, we get the equivalent system

$$a(a-b)/(a+x)=a, \quad b(b-a)/(b+y)=b.$$

If we suppose $a \neq 0$, $b \neq 0$, and disregard any solutions for which $a+x=0$, or $b+y=0$, the last system is equivalent to

$$a-b=a+x, \quad b-a=b+y,$$

from which we get $\quad x=-b, \quad y=-a.$

Ex. 3.
$$2x-5y+3z=0,$$
$$x+3y+2z=0,$$
$$x^2+y^2+z^2=1.$$

Introducing the auxiliary variable $\rho$, we have, by the method of § 81, Ex. 3.

$$x=-19\rho, \quad y=-\rho, \quad z=11\rho.$$

If we substitute these values in the third equation, we get

$$483\rho^2=1.$$

Hence $\rho=\pm\sqrt{(1/483)}$; and we get two solutions of the system, viz.—

$$x=-19\sqrt{(1/483)}, \quad y=-\sqrt{(1/483)}, \quad z=11\sqrt{(1/483)};$$
$$x=19\sqrt{(1/483)}, \quad y=\sqrt{(1/483)}, \quad z=-11\sqrt{(1/483)}.$$

§ **83**. If an integral system of, say, two equations in two variables can be thrown into the form

$$P.Q.R. \ldots =0, \quad P'.Q'.R' \ldots =0,$$

where P, Q, R, . . . , P′, Q′, R′, . . . are linear functions of $x$ and $y$, then it can be completely solved by means of linear systems. For it is obvious, by the principle of § 62, that its solutions are the solutions of the systems

$$P=0, \ P'=0; \quad P=0, \ Q'=0; \quad P=0, \ R'=0; \ \ldots$$
$$Q=0, \ P'=0; \quad Q=0, \ Q'=0; \quad Q=0, \ R'=0; \ \ldots$$
$$\ldots \quad \ldots \quad \ldots \quad \ldots$$

all of which are linear.

Ex. 1. $(x-y)(x+y-1)=0, \ (x-2y)(x+3y-2)=0.$

The given system is equivalent to the assemblage of the following systems :—

$$x-y=0, \quad x-2y=0;$$
$$x-y=0, \quad x+3y-2=0;$$
$$x+y-1=0, \quad x-2y=0;$$
$$x+y-1=0, \quad x+3y-2=0.$$

The system has therefore the four solutions

$$(0, 0), \quad (\tfrac{1}{2}, \tfrac{1}{2}), \quad (\tfrac{2}{3}, \tfrac{1}{3}), \quad (\tfrac{1}{2}, \tfrac{1}{2}).$$

Ex. 2. $(x-y)(x+y-1)=0,\ (x-y)(x+3y-2)=0.$

This system is equivalent to

$$\begin{aligned} x-y&=0, & x-y&=0\,;\\ x-y&=0, & x+3y-2&=0\,;\\ x+y-1&=0, & x-y&=0\,;\\ x+y-1&=0, & x+3y-2&=0. \end{aligned}$$

The first of these systems is indeterminate, viz. it admits of all the one-fold infinity of solutions belonging to $x-y=0$. The other three are determinate, and give the solutions $(\frac{1}{2}, \frac{1}{2})$, $(\frac{1}{2}, \frac{1}{2})$, $(\frac{1}{2}, \frac{1}{2})$ respectively.

There is a noteworthy peculiarity in this case, viz. *that the system is in part determinate and in part indeterminate*—a possibility which the beginner should bear in mind, as he may meet with cases of the kind in practice.

## EXERCISES XX.

1. $2x-3/y=3,\ 8x+15/y+6=0.$
2. $3/x-1/y=4,\ 1/x-2/y+2=0.$
3. $(x+y)/(a-b)+(x-y)/(a+b)=1,\ (x-y)/(a-b)$
   $+(x+y)/(a+b)=1.$
4. $l/x+m/y=1,\ (l+3m)/x+(l-m)/y=2.$
5. $(2x+3y+z)(x+y+z)=0,\ 2x+3y=0,\ 4x+y-5z=1.$
6. $(3x-4y+5)(x+7y-2)=0,\ (2x+y-3)(9x-y)=0.$
7. $(3x-4y+2)(x-5y)=0,\ (x-6y+1)(2x-3y+1)=0.$
8. $3x^2+4y^2=111,\ 5x^2-6y^2=71.$
9. $26x^2-y^2=95,\ 18x^2+3y^2=219.$
10. $(x+y)^2-4(x-y)^2=0,\ 3x+2y-1=0.$
11. $a/(x+a)+b/(y+b)=b/(x+a)+a/(y+b)=1.$
12. $(a^2+\lambda^2)/x^2-(b^2+\lambda^2)/y^2=1,\ (a^2-\lambda^2)/x^2-(b^2-\lambda^2)/y^2=1.$
13. Tabulate all the solutions of $x^2+y^2+z^2=3,\ 3x^2+5y^2+z^2=9,$
    $8x^2+y^2+7z^2=16.$

# CHAPTER IX

## PROBLEMS DEPENDING ON SYSTEMS OF LINEAR EQUATIONS

§ **84.** After what has been said in Chapter VII., there is nothing new in principle in the treatment of problems which depend on more than one variable. Such problems, being more complicated, require closer thought to enable the calculator to fix clearly in his mind an appropriate set of variables, and to dissect the independent conditions out of the statement of the problem. The number of these conditions must in general be the same as the number of variables to be determined; but it happens in special cases that a problem which is under-conditioned may become determinate, owing to certain *restrictions* on the nature of the variables, such, for example, as that they must be positive integers. Of this class of problem one or two examples have already occurred, and others will be given below. It is important to take care that the conditions selected are really *independent*. It will not do, for example, first to select two conditions and then another which is a logical consequence of the two first. A logical error of this kind would be reflected in the resulting conditional equations, which would be interdependent in the sense that *any* solution *whatever* common to the two first would also be a solution of the third.

Regarding the choice of variables, it may be remarked that it may be necessary to introduce into the solution of a problem variables the values of which are not required in the statement of the result aimed at; and, as in the case of one variable, the simplicity of the algebraic work may often be greatly increased by an adroit choice of variables. It may at the same time be added, in order to counteract a fetish worship of mere symbols not uncommon with beginners, that choice of variables, or indeed any device symbolical or other, can by no means change the essential character of a problem—cannot, for example, make a problem determinate which is indeterminate, or cause a problem which has two solutions to have one, or effect any other logical revolution. The symbolical statement of the solution of a problem can no more alter the nature of a problem than language can alter an idea which it professes to express.

Ex. 1. The rents of three farms were £275, £775, and £1325 respectively. The whole rent of each was raised by the same amount, and then it was found that the rent per acre was the same for all three. Given that the acreages of the first and second were 150 and 350 respectively, find the rise of rent and the acreage of the third.

Let the rise of rent be £$x$, and the acreage of the third farm be $y$; then, by the conditions of the problem, we must have

$$(275+x)/150=(775+x)/350=(1325+x)/y.$$

From these equations we have in the first place

$$350(275+x)=150(775+x),$$

whence $$200x=150\times 775-350\times 275=20,000;$$

hence $x=100$. If we now use this value of $x$, we get from the original system

$$375/150=1425/y.$$

Since $y=0$ is not in question, this last is equivalent to $y=1425\times 150/375=570$.

Hence the rise of rent was £100, and the third farm contained 570 acres.

Ex. 2. The numerator and denominator of a certain proper fraction exceed the numerator and denominator of another respectively by unity. The difference between the two fractions is 1/36 and their sum is 55/36; find the numerators and denominators of these fractions.

Let the numerator and denominator of the second fraction be $x$ and $y$, then the two fractions are $(x+1)/(y+1)$ and $x/y$, where $x<y$. We have $(x+1)/(y+1)-x/y\equiv y(x+1)/y(y+1)-x(y+1)/y(y+1)$ $\equiv\{y(x+1)-x(y+1)\}/y(y+1)\equiv(y-x)/y(y+1)$, which is positive, since $y>x$. We may therefore state the conditions of the problem thus:—

$$(x+1)/(y+1)-x/y=1/36,$$
$$(x+1)/(y+1)+x/y=55/36.$$

Instead of solving this system as it stands, we replace it by the following equivalent system obtained by addition and subtraction:—

$$2(x+1)/(y+1)=56/36, \quad 2x/y=54/36.$$

Since solutions involving $y+1=0$ or $y=0$ are obviously out of the question, we may replace the last pair by

$$9(x+1)=7(y+1), \quad 4x=3y;$$

that is— $$9x-7y+2=0, \quad 4x-3y=0,$$

the unique solution of which is readily found to be $x=6$, $y=8$.

The two fractions required are therefore 7/9 and 6/8.

Ex. **3**. Construct a homogeneous integral function of $x$ and $y$ of the second degree which shall vanish when $x=y$; have the value 1 when $x=4$, $y=3$; and the value 2 when $x=3$, $y=4$.

The kind of integral function meant is one in which each term is of the second degree in $x$ and $y$; it will therefore be of the form $Ax^2+Bxy+Cy^2$, where A, B, C are constant, and must be determined so that the function may satisfy the conditions imposed. A, B, C are therefore the variables or unknown quantities of the present problem.

The first condition gives $Ay^2+By^2+Cy^2=0$, *i.e.* $(A+B+C)y^2=0$, whatever $y$ may be. This necessitates that

$$A+B+C=0.$$

The other two conditions give

$$16A+12B+9C=1,$$
$$9A+12B+16C=2.$$

From the last two, by subtraction, we derive

$$7A-7C=-1;$$

multiplying both sides of the first equation by 12 and subtracting from the second, we get

$$4A-3C=1.$$

From the last two equations, multiplying by 3 and by 7 and subtracting, and again multiplying by 4 and by 7 and subtracting, we get $A=10/7$, $C=11/7$, and the first equation immediately gives $B=-21/7$.

The function required is therefore $\frac{1}{7}(10x^2-21xy+11y^2)$.

Ex. **4**. A testator leaves to his eldest son £$a$ and the $m$th part of the residue of his estate, and to his second son £$b$ and the $n$th part of what remains after fulfilling the previous provisions of his will. It is found, when the estate is divided, that the two sons have equal shares. What did each get, and what was the value of the whole estate?

Let the value of the whole estate be £$x$, and the share of each son £$y$.

After £$a$ has been set apart for the eldest son, the residue is £$(x-a)$; hence, by the first condition—

$$y=a+(x-a)/m.$$

After setting apart £$y$ for the eldest son, and the definite legacy of £$b$ for the youngest, the residue is $x-y-b$; hence by the second condition—

$$y=b+(x-y-b)/n.$$

These two equations are equivalent to

$$x-my=(1-m)a, \quad x-(n+1)y=(1-n)b.$$

From these equations, by subtraction—

$$(n-m+1)y=(n-1)b-(m-1)a.$$

If we multiply the first by $n+1$, and the second by $m$ and subtract, we get

$$(n-m+1)x = m(n-1)b - (n+1)(m-1)a.$$

Hence the share of each son was $\{(n-1)b-(m-1)a\}/(n-m+1)$; and the whole value of the estate was $\{m(n-1)b-(n+1)(m-1)a\}/(n-m+1)$.

Ex. 5. The numerator and denominator of a certain proper fraction each consist of the same two digits written in different order. If the value of the fraction be 4/7, find the numerator and denominator.

So far as the conditions are concerned the problem is obviously indeterminate, the only condition properly so called being that the value of the fraction shall be 4/7.

Let the digits be $x$ and $y$, then we may represent the numerator and denominator by $10x+y$ and $10y+x$. The single condition is then represented by the equation

$$(10x+y)/(10y+x) = 4/7.$$

Since $10y+x \neq 0$, this is equivalent to

$$7(10x+y) = 4(10y+x),$$

which leads to

$$66x = 33y,$$

or

$$y = 2x.$$

Now, by the nature of our problem, $x$ and $y$ are both restricted to be positive and integral and $\ngtr 9$. Hence the only admissible values of $x$ are 1, 2, 3, 4, and the corresponding values of $y$ are 2, 4, 6, 8.

There are therefore four fractions which satisfy the conditions of the problem, viz. 12/21, 24/42, 36/63, 48/84.

## EXERCISES XXI.

1. A bill of £7 : 15s. was paid with florins and half-crowns. There were 70 coins altogether. How many were there of each?

2. A said to B, "Give me £100 and I shall have as much money as you;" B replied, "Give me £100 and I shall have double as much as you." How much had each?

3. A and B owe £1200 and £2550 respectively. A said to B, "If you would lend me the eighth part of your money, I could pay my debts;" and B replied, "If you would lend me the sixth part of your money, I could pay mine." How much money had each?

4. Find an integer of two digits which is seven times the sum of its digits and twenty-one times their difference, the tens digit being the greater.

5. A certain fraction is equal to 2/3, and its denominator exceeds its numerator by 21. Find the numerator and denominator.

6. An integer of two digits is multiplied by 4, and the product is less by 3 than the number formed by inverting its digits; if it be multiplied by 5, the tens digit in the product is greater by 1, and the units digit less by 2 than the units digit in the original number; find the number.

7. A man can walk a certain distance in four hours; if he were to

increase his rate by one-fifteenth he could walk one mile more in that time. What is his rate?

**8.** A has twice as many pennies as shillings; B, who has 8d. more than A, has twice as many shillings as pennies; together they have one penny more than they have shillings. How much has each?

**9.** If the numerator of a certain fraction be doubled and its denominator increased by 7, it becomes $\frac{1}{2}$; if the denominator be doubled and the numerator increased by 7, it becomes unity. Find the numerator and denominator of this fraction.

**10.** A certain number of two digits exceeds the number obtained by reversing the digits by 9; also the sum of the two numbers is 77. Find them.

**11.** If 3 be added to both numerator and denominator of a certain fraction, it is increased to 8/7 of its original value; but if 3 be subtracted from both numerator and denominator, it is reduced to 16/21 of its original value. Required the numerator and denominator of the fraction.

**12.** A certain number of three digits exceeds the sum of its digits by 180. If reversed, it exceeds the same sum by 378. But if divided by the sum of its digits, the quotient is 14 and the remainder 11. Find the number.

**13.** Divide 100 into three parts such that, if the second be divided by the first the quotient is 2, and the remainder 1, and, if the third be divided by the second, the quotient is 2, and the remainder 6.

**14.** Divide the number 123 into four parts such that, if the first be increased by 7, the second diminished by 6, the third multiplied by 5, and the fourth divided by 4, the results may be all equal.

**15.** If the joint fortunes of three heiresses, A, B, and C, taken in pairs be given, say of B and C £$a$, of C and A £$b$, of A and B £$c$, find the fortune of each. What conditions must the numbers $a$, $b$, $c$ satisfy in order that the concrete problem may be possible?

**16.** A said to B, "I am now twice as old as you were when I was your age; and if you and I both live till you are my present age, I shall be 100." Find the ages of A and B.

**17.** If 3 cows and 8 horses cost £245, and 5 cows and 7 horses cost £250, how much do 2 cows and 3 horses cost?

**18.** If the length and breadth of a rectangle be increased and diminished respectively by 5 feet, its area is diminished by 45 square feet; and if the length and breadth be each increased by 19 feet, the length is then $\frac{14}{9}$ of the breadth. Find the dimensions of the rectangle.

**19.** Two persons, A and B, agree to pay a bill of £10, each to contribute half his money and A to pay what is left. It is found that A is left with £2 to pay; and he ends by having as much money as B had originally. How much had each?

**20.** A cyclist after riding a certain distance has to stop for half an hour to repair his machine, after which he completes the whole journey of 30 miles (at a slower pace) in 5 hours. If the breakdown had occurred 10 miles farther on, he would have done the journey in 4 hours; find where the breakdown occurred and his original speed.

**21.** Two vessels A and B contain mixtures of spirit and water. A mixture of one part from A and three parts from B is found to contain 30% of spirit; and a mixture of two parts from A and three parts from B 27% of spirit. Required the percentages of spirit in A and B respectively.

**22.** A man had two creditors, his debt to the one being double his debt to the other. After paying his larger creditor 4 shillings in the pound and the other creditor in full, he had £10 left. If he had divided all his estate fairly between them, each would have got 10 shillings in the pound. What was the value of his estate, and how much did he owe each of his creditors?

**23.** In a mile race A can beat B by 50 yards and C by 80; by how much can B beat C?

**24.** A sum of money amounted in a certain number of years to £$a$ at $i$% simple interest. Lent for $m$ years longer at $j$%, it amounted to £$b$. What was the sum?

**25.** Find a linear integral function of $x$ which shall have the values 3 and 10, when $x$ has the values 4 and 5 respectively.

**26.** Find a linear integral function of $x$ whose value is doubled when $x$ is doubled, and which has the value 2 when $x=2$.

**27.** Find a linear integral function of $x$ and $y$ which has the values 3, 9, 11, corresponding to the values (1, 5), (1, $-4$), and ($-1$, $-3$) of $x$ and $y$ respectively.

**28.** Construct a homogeneous symmetric function of $x$ and $y$ of the second degree, which shall vanish when $x=2$, $y=2$, and have the value 1, when $x=4$, $y=2$.

**29.** Construct a quadratic integral function of $x$, whose values shall be 3, 4, 5, when the values of $x$ are 1, 2, 3 respectively.

**30.** Construct a quadratic integral function of $x$ which has the values 0, 1, 2 when $x$ is equal to 1, 2, 4 respectively.

**31.** If $y$ be an integral function of $x$ of the second degree, and its values be 3, 5, 7, when $x=1, 2, 3$ respectively, find its value when $x=4$.

**32.** Show that, when two solutions of a linear equation in $x$ and $y$ are given, all its solutions are known.

**33.** Construct a linear equation in $x$ and $y$ which has the solutions $x=2$, $y=-3$; $x=3$, $y=5$.

**34.** Show that, if $(x_1, y_1)(x_2, y_2)(x_3, y_3)$ be three solutions of the same linear equation in $x$ and $y$, then $x_1(y_2-y_3)+x_2(y_3-y_1)+x_3(y_1-y_2)=0$.

**35.** A man invested his money partly in the 3 per cents at 80, and partly in the 4 per cents at 90, and his income was £85. If the 3 per cents had been at 90, and the 4 per cents at par, his income would have been £76. Find the whole sum invested.

**36.** A cistern is supplied by two pipes A and B, and emptied by a pipe C. If the cistern be empty, and all the pipes open, the cistern will be filled in 10 minutes; if A and C only be opened, in 30 minutes; and, if B and C only be opened, in 45 minutes. A supplies 10 gallons more per minute than B. How many gallons does the cistern hold?

**37.** A and B start to walk 2 miles. A gives B a start of a mile, overtakes him in 20 minutes, and finishes the whole distance in 10 minutes more. Find the speed of A and B.

**38.** Two passengers have together 600 lbs. of luggage, and are charged

3s. 4d. and 11s. 8d. respectively for excess above the weight allowed. If the luggage had all belonged to one of them, he would have been charged £1. How much free luggage is allowed to each passenger?

**39.** A and B run two mile-races. In the first A gives B a start of 20 yards and beats him by $45\frac{5}{11}$ seconds; in the second he gives B a start of 30 seconds and beats him by 352 yards. Required the number of seconds in which A and B can each run a mile.

**40.** A German tourist said, "I have travelled in Germany, in France, and in England, and spent 8325 thalers, viz. 1520 thalers in Germany, 7540 francs in France, and 820 pounds in England." Having been asked the value of the pound and of the franc in German money, he answered, "£5 is equivalent to 4 th. more than 108 fr." How much at this rate are the franc and the pound worth in thalers?

**41.** 37 lbs. of tin loses 5 lbs. when weighed in water. In like manner, 23 lbs. of lead loses 2 lbs. in water. An alloy of lead and tin weighing 120 lbs. loses 14 lbs in water. Required the quantities of lead and of tin in the alloy.

**42.** Find all the pairs of unequal positive integers which are such that the difference of their squares is equal to three times the difference of the numbers themselves.

**43.** Show that it is impossible to find two unequal positive integral numbers, neither of which is zero, such that the difference of their squares is equal to the difference of the numbers themselves.

**44.** Find two consecutive even integers twice the product of which shall exceed the sum of the squares of two consecutive odd integers by 166.

**45.** Show that it is impossible to find two consecutive odd integers the sum of whose squares is the sum of the squares of two consecutive even integers.

**46.** A landlord had three farms, originally rented at £$p$, £$q$, £$r$ respectively, the number of acres in the first two were $a$ and $b$ respectively; and it was discovered after adding the same sum to the rent of each of the three farms that the rent per acre was the same for each farm. Find the rise of rent and the number of acres in the third farm.

**47.** How many days will it take three workmen, A, B, C, to finish a job which B and C together could do in $a$ days, C and A together in $b$ days, A and B together in $c$ days?

**48.** Three couriers start for a certain destination; the second rides $a$ miles an hour faster than the first, and starts $h$ hours later; the third rides $b$ miles an hour faster than the first, and starts $k$ hours later. They all arrive at the same time. Find the distance and the speed of the first courier.

**49.** Three couriers start for a certain destination, the second $h$ hours after the first, the third $k$ hours after the first. The second and third ride $a$ and $b$ miles an hour respectively; and all three arrive together. Required the distance and the speed of the first courier.

**50.** It is known that the distances, $x$ and $y$, of the object and image for a certain optical system are connected by the equation $Axy + Bx + Cy + D = 0$, where A, B, C, D are constants. If when $x = 1$, 3 5 inches, $y$ is 2, 4, 6 inches respectively, calculate $y$ when $x = 7$.

**51.** AB and CD are two straight lines of length $a$ and $b$ perpendicular to AC; DB and CA meet in O, and AD in BC in E. Find the distance between A and C in order that OE may be equal to $2ab/(a+b)$.

**52.** ABCD is a rectangle in which $AB=a$, $BC=b$. O is a point in BA produced such that $OA=C$. OPQ meets AD in P, and BC in Q. If OPQ bisect the area of the rectangle, calculate AP and BQ.

**53.** The figure being as before, find AP so that area OAP = area QCR, R being the point where OPQ meets CD.

**54.** In the same figure as before, find AP so that $OP=RQ$.

**55.** Two circles of radii $x$ and $y$ touch each other and are each inscribed in a semicircle of radius $r$. Show that $x$ and $y$ are connected by the equation $4x^2y^2+4r(x+y)xy+r^2\{(x+y)^2-8xy\}=0$. Hence determine $x$ and $y$: (1) when $y=mx$, $m$ being given; (2) when $x+y=a$ is given.

# CHAPTER X

## MULTIPLICATION OF STEPS—COMPOSITION OF CIRCUITS—MENSURATION

§ **85.** There is a geometrical interpretation of the multiplication of algebraic quantities which is interesting theoretically, and which contains as a particular case the fundamental principles of the application of Algebra to the mensuration of plane figures.

Let X′OX, Y′OY (Fig. 3) be two fixed lines at right angles to each other in a given plane; and, as in § 52, fix their positive directions as X′ to X, and Y′ to Y respectively. We shall represent the multiplicand and multiplier in any algebraic product as steps parallel to X′OX and Y′OY respectively. The small letters used will denote the absolute lengths of the steps; and the signs + or − attached will indicate the directions of the steps as in § 20. Consider now a magnetic pole P, fixed at any considerable distance from O, in the plane XOY say (but that is immaterial). Consider also any plane electric circuit in the plane XOY, whose linear dimensions and whose distance from O are very small compared with the distance of P from O. Then it is a well-known fact, which can be illustrated by simple experiments, that the action of the circuit on P depends (so long as P is fixed) merely on the area of the circuit, the strength of the current and the direction, counter- or cum-clock, in which the current circulates round the area. For simplicity, we shall suppose that the strength of the current is always the same, say unity; and, to suit our present purposes, that the circuit is always a rectangle whose sides are parallel to X′OX and Y′OY respectively. A circuit with a unit current flowing round it counter-clockwise we call a positive circuit: a circuit with a unit current flowing round it cum-clockwise a negative circuit. It is immediately obvious that, as regards their action on P, such circuits follow the laws of algebraic addition and subtraction, *e.g.* the order in which we set them down is a matter of indifference (Law of Commutation); we may replace any number of positive and negative circuits by

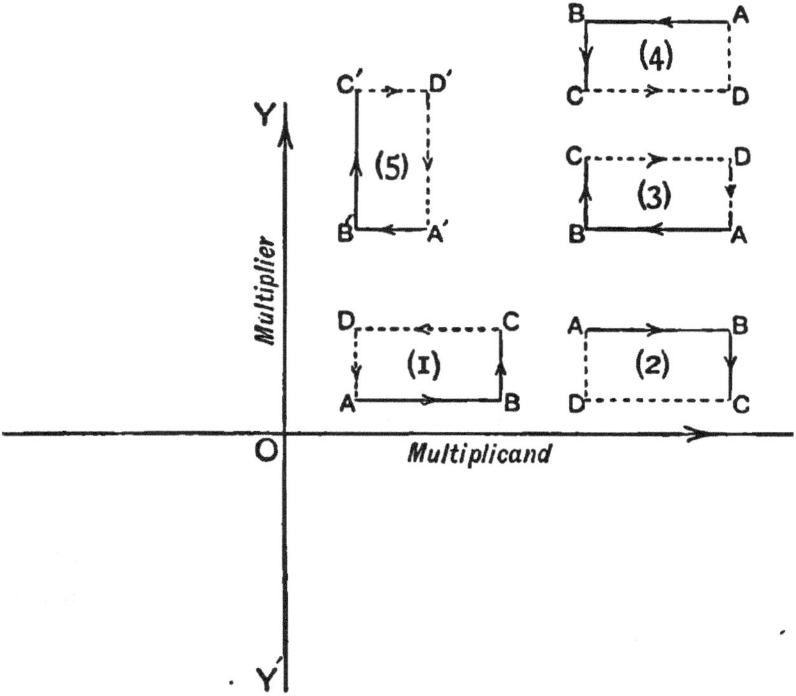


FIG. 3.

a single circuit whose area and sign are determined by the law for reducing an algebraic sum (Law of Association). A positive and a negative circuit of equal area annul each other's action $(+a-a=0)$, and so on.

Further, let us agree that the absolute product of two absolute lengths, $a$ and $b$, is to mean the rectangle contained by these two lengths, and $+ab$ and $-ab$ a positive and a negative circuit respectively, whose areas are each the rectangle contained by $a$ and $b$.

Finally, let us interpret the algebraic product $(+a)\times(+b)$

as a direction to construct a rectangular circuit as follows:—Draw a positive step AB parallel to X′OX; through B the end of this step a positive step BC parallel to Y′OY; complete the circuit (1) by drawing the remaining two sides of the rectangle, the direction of the unit current being determined by the order of the points ABCD. It will be seen that we have constructed a positive circuit, the area of which is the area of the rectangle contained by the lines $a$ and $b$, which may therefore be denoted by $+ab$. It will now be evident that the algebraical equation

$$(+a) \times (+b) = +ab$$

is merely the symbolic statement of the result of the above geometrical construction, provided we read $=$ as meaning "is magnetically equivalent to."

In the same way $(+a) \times (-b)$ directs us to construct the circuit (2); and the equation

$$(+a) \times (-b) = -ab$$

formally states the obvious fact that (2) is a negative circuit whose area is the rectangle contained by the lines $a$ and $b$.

The equations $(-a) \times (+b) = -ab$, $(-a) \times (-b) = +ab$ are interpreted in the same way by means of the circuits (3) and (4).

The Law of Commutation in the particular case

$$(+a) \times (-b) = (-b) \times (+a)$$

corresponds to the fact that the circuits (2) and (5), in which $A'B' = BC$, and $B'C' = AB$, are magnetically equivalent; which is obvious, since both are negative circuits and their areas are equal.

§ 86. It is easy to satisfy oneself that the Law of Distribution has its counterpart in the above geometrical interpretation. Consider, for example, the particular case

$$(+a-b) \times (-c) = -ac + bc \qquad (1),$$

and the corresponding diagram (Fig. 4), in which $AB = a$, $BC = b$, $BD = AE = CF = c$.

There are two cases according as $a$ is absolutely greater or less than $b$. In the first case (Fig. 4 (1)), the left-hand side is a direction to construct the negative circuit ACFE, by first stepping $+a$ parallel to OX and then $-b$; so that the result is the same as if we had taken the positive step AC, then to step $-c$, that is CF, parallel to OY. The equation (1) asserts that the magnetic effect of the negative circuit ACFE (Fig. 4 (I)) is the sum of the effects of the negative and positive circuits ABDE and BCFD, which is obvious.

In the second case (Fig. 4 (II)) the order of ideas is the same, only the positive circuit BCFD preponderates, and the resulting circuit ACFE is positive.

The other cases of the law of distribution may be similarly interpreted.

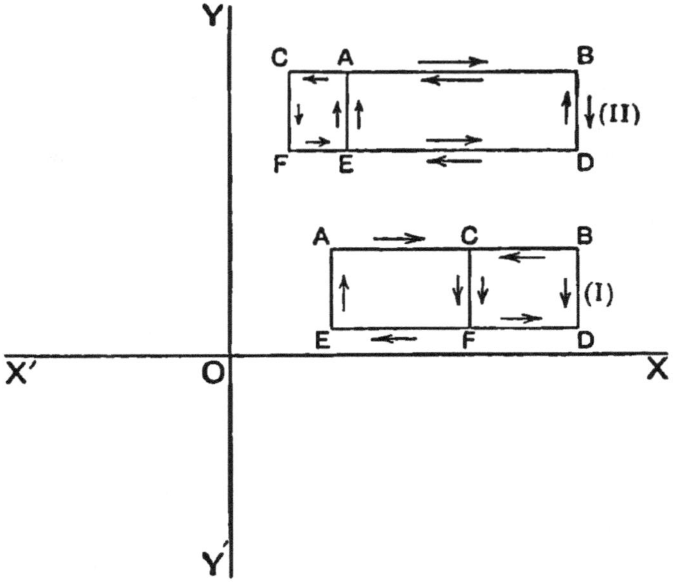


FIG. 4.

§ **87.** The operation of algebraic division also finds its interpretation in the theory of circuits. Let $c$ denote any area taken absolutely; then, since $(c \div b) \times b = c$, we see that $c \div b$ denotes the line which along with $b$ contains a rectangle whose area is $c$.

Again, since $\{(+c) \div (-b)\} \times (-b) = +c$, we see that $(+c) \div (-b)$ denotes the step parallel to X′OX which, followed by the step $-b$ parallel to Y′OY, leads to the construction of the positive circuit $+c$. By drawing the figure we see at once that the required step is a negative one, viz. $-c \div b$.

The other cases, $(+c) \div (+b)$, etc., can be dealt with in the same way, and the law of distribution for division may also be interpreted without difficulty.

§ **88.** We now see that the fundamental laws of Algebra, in so far as they apply to algebraic sums of single algebraic quantities and to products or quotients of pairs of such quantities, have their counterpart in the laws of the composition of steps and of the construction and composition of circuits. Hence every algebraic identity involving operations of the kind described can be interpreted as a theorem regarding the composition of steps or the equivalence of circuits.

§ **89.** If we avoid the occurrence of finally negative steps and of finally negative areas, we may neglect all considerations regarding the geometrical meaning of the sign of an area, and then all the circuits may be taken as rectangles merely. The application of the law of distribution then gives us simply propositions 1, 2, 3, 4, 5, 6, 7, 8, 9, 10 of the Second Book of Euclid, or their more succinct modern equivalents. For example—

$$A(a+b+c+\ \ldots) = Aa + Ab + Ac + \ \ldots \qquad (1)$$

$$(a+b)^2 = (a+b)a + (a+b)b \qquad (2)$$

$$(a+b)^2 = a^2 + 2ab + b^2 \qquad (3)$$

$$(a+b)(a-b) = a^2 - b^2 \qquad (4)$$

will be seen to be the equivalents of propositions 1, 2, 4, and 5, if it be remembered that $a^2$ means $a \times a$, which we interpret as the rectangle contained by two lines, each of which is $a$, *i.e.* the square on the line $a$.

It is easy by the interpretation of algebraic identities to obtain theorems regarding the rectangles contained by the segments of a straight line, the proof of which by ordinary geometrical methods would be complicated; and it is useful to note that, conversely, every such geometrical theorem may be established by verifying a particular algebraic identity.

Ex. Let A, B, C, D be four points in any order on a straight line. Take A as origin; and let $x$, $y$, $z$ be the co-ordinates of B, C, D with reference to A, see § 21. From the obvious identity

$$x(y-z)+y(z-x)+z(x-y)=0$$

we deduce

$$\overline{AB}\cdot\overline{DC}+\overline{AC}\cdot\overline{BD}+\overline{AD}\cdot\overline{CB}=0\,;$$

that is to say, the algebraic sum of the three rectangles contained by the steps $\overline{AB}$, $\overline{DC}$; $\overline{AC}$, $\overline{BD}$ and $\overline{AD}$, $\overline{CB}$ is zero, the sign of each rectangle to be taken as + or − according as the steps which contain it are like or oppositely directed. Thus, for example, if A, B, C, D stand in order from left to right, we have

$$-AB.CD+AC.BD-AD.BC=0;$$

or

$$AC.BD=AB.CD+AD.BC,$$

wherein AC, etc., now denote absolute lengths, and AC.BD, etc., absolute areas.

§ **90**. The fundamental theorem in the mensuration of plane areas is the proposition that the area of any parallelogram is equal to the area of a rectangle contained by its base and its altitude ; or the proposition, immediately derivable therefrom, that the area of a triangle is equal to half the area of the rectangle contained by its base and its altitude. If, therefore, $a$ denote the base and $h$ the altitude of a triangle, the symbolic expression for its area is $\frac{1}{2}ah$.

From this last result it is easy to deduce that a symbolic expression for the area of a trapezium is $\frac{1}{2}(a+b)h$, where $a$ and $b$ denote its parallel sides, and $h$ denotes the distance between these sides. This formula is often used in the approximate mensuration of plane figures.

Ex.—Show that the area of a regular hexagon is three times the rectangle contained by the radii of the inscribed and circumscribed circles.

Consider any two parallel sides and the parallel diagonal. This diagonal obviously divides the hexagon into two equal trapezia whose parallel sides are R and 2R, R being the radius of the circumscribed circle. Also the distance between the parallel sides of the trapezium is the radius of the inscribed circle. Hence the area of the hexagon is $2\times\frac{1}{2}(R+2R)r=3Rr$, which proves the theorem.

§ **91**. The fundamental theorem regarding the relations connecting the sides of plane rectilinear figures is the Pythagorean proposition regarding the squares on the sides of a right-angled triangle. If, following the general usage, we denote the sides of the triangle ABC opposite the angles A, B, C by $a$, $b$, $c$ respectively, then *the symbolical expression of the Pythagorean relation for a triangle which is right angled at* C *is*

$$c^2=a^2+b^2 \qquad (1).$$

*If the angle* C *be not a right angle, then we have*

$$c^2=a^2+b^2\pm 2bx \qquad (2),$$

*where* x *denotes the projection of the side* a *on the side* b, *and the upper or lower sign is to be taken according as the angle C is obtuse or acute.*

It is interesting to show that (2) is an algebraic consequence of the Pythagorean proposition. Take, for example, the case where C is acute. Let D be the projection of B on AC, so that $CD = x$, $AD = b - x$, and let $p$ denote the perpendicular BD. By the Pythagorean proposition, $a^2 = p^2 + x^2$, from which it follows that $p^2 = a^2 - x^2$. Hence, again, using the Pythagorean proposition, we have

$$\begin{aligned} c^2 &= p^2 + (b - x)^2, \\ &= a^2 - x^2 + b^2 + x^2 - 2bx, \\ &= a^2 + b^2 - 2bx. \end{aligned}$$

It is easy to deduce from the theorem (2) the following, which contains as particular cases Euclid II. 9 and 10, the theorem of Apollonius regarding the squares on the sides and on the median of a triangle, and other well-known propositions. *Let* A O B *be three points in a straight line such that* m . OA = n . OB ; *and let* a, b *denote the distances of* O *from* A *and* B *respectively* ; *x, y, z the distances of any point* P *from* A, B, *and* O *respectively* ; *then*

$$mx^2 \pm ny^2 = ma^2 \pm nb^2 + (m \pm n)z^2,$$

*the upper or the lower sign to be taken according as* O *does or does not lie between* A *and* B.

To prove this, let us take the case where O is between A and B ; let D be the projection of P on AB, which we shall suppose to lie between O and B. Let $u$ denote OD. Then we have, by the theorem (2) above—

$$\begin{aligned} x^2 &= a^2 + z^2 + 2au, \\ y^2 &= b^2 + z^2 - 2bu. \end{aligned}$$

If we multiply both sides of the first equation by $m$, and both sides of the second by $n$, and add, we deduce

$$mx^2 + ny^2 = ma^2 + nb^2 + (m + n)z^2 + 2mau - 2nbu.$$

Since $ma = nb$, we thus get

$$mx^2 + ny^2 = ma^2 + nb^2 + (m + n)z^2.$$

When D lies between A and O the proof is the same, only that the sign of $u$ is changed throughout.

§ **92.** It will be observed that in the immediately preceding paragraphs we have used the letters $a$, $b$, $x$, $y$, etc., merely as names for linear segments. We have not said, for example, "let $a$ be the number of units of length in the line BC." The introduction of the notion of measuring lines or areas by means of units is no necessary part of the application of Algebra to the composition of rectangular areas; the essential point is that the fundamental principles of this part of geometry are identical with certain cases of the fundamental laws of Algebra.*

We may, however, introduce the notion of units if we choose. Take, for example, the expression for the area of a triangle. Let the base be $a$ times the unit of length, which we shall denote by $l$, and the altitude $h$ times the unit of length. The expression for the area is now $\frac{1}{2}(al)(hl)$; this may be transformed into $\frac{1}{2}ahl^2$, which may be interpreted to mean that the area of the triangle is $\frac{1}{2}ah$ times the square on the unit of length. If now we take the square on the unit of length as our unit of area, and $\Delta$ be the number of units of area in the area of the triangle, we have $\Delta l^2 = \frac{1}{2}ahl^2$, whence $\Delta = \frac{1}{2}ah$—a result outwardly the same as before where $\Delta$, $a$, $h$ are now abstract numbers, and the formula has the meaning usually attached to it in rules for mensuration. It must not be forgotten, however, that this formula is not correct in this abstract sense unless all the lengths are measured in terms of a common unit, and the unit of area be taken to be the square on the unit of length. The conditions just mentioned are fulfilled in any rational system of space units; but unfortunately our British system is as yet partly irrational; hence certain difficulties in calculation with which beginners are sufficiently familiar.

## EXERCISES XXII.

1. If $a$ denotes the number of feet in the base of a triangle and $h$ the number of inches in its altitude, in terms of what unit of area does $\frac{1}{2}ah$ express the area of the triangle?

2. The lengths of the parallel sides of a trapezium are 6 feet and 8 feet respectively, and the distance between the parallel sides is 4 feet; find its area.

3. Find an expression for the area of a triangle the co-ordinates of whose vertices are $(x_1y_1)$, $(x_2y_2)$, $(x_3y_3)$.

4. If the square on one side of a right-angled triangle is $\frac{1}{10}$ of the square on the hypotenuse, that side is $\frac{1}{3}$ of the other side.

5. Deduce from the theorem $c^2 = a^2 + b^2 \pm 2bx$ that the sum of any two sides of a triangle is greater than the third, and their difference less.

---

* See Henrici, Art. "Geometry," *Encyclopædia Britannica*, 9th ed. vol. x. p. 270.

**6.** If $a$, $b$, $c$ denote the sides of a right-angled triangle, $c$ being the hypotenuse, $p$ the perpendicular from the right angle on the hypotenuse, $x$ and $y$ the projections of $a$ and $b$ respectively on $c$, and $\Delta$ the area, express each of $a$, $b$, $c$, $p$, $x$, $y$, $\Delta$ in terms of any other two.

**7.** Express the area of a triangle as a function of its sides.

**8.** ABC is a triangle right angled at C; and ABDE the square described externally on AB. If $CD=x$, $CE=y$, show that $x^2-y^2=a^2-b^2$; $x^2=a^2+(a+b)^2$; $y^2=b^2+(a+b)^2$.

**9.** If in a right-angled triangle one of the sides containing the right angle is 28, and the radius of the inscribed circle is 10, calculate the other sides.

**10.** Show that the triangle whose sides are $\sqrt{6}$, $\sqrt{2}+1$, $\sqrt{2}-1$ is right angled, and find its area and the length of the perpendicular from the right angle on the hypotenuse.

**11.** Find an expression for the area of a regular octagon whose side is $a$.

**12.** The co-ordinates of A and B are $(x_1)$, $(x_2)$; and C and D divide AB externally and internally in same ratio. If the distance $CD=a$, find the co-ordinates of C and D.

**13.** The co-ordinates of two points P and Q are $(-2)$, $(-3)$ respectively. Find the distance between the two points which divide PQ externally and internally in the ratio $2:3$.

**14.** The step between the points which divide the line joining $(x_1)$, $(x_2)$ externally and internally in the ratio $b:a$ is $\{2ab/(a^2-b^2)\}(x_2-x_1)$.

**15.** Find a point P in AB such that $AP^2-BP^2=c^2$.

**16.** The distance between two points A and B on a straight line is 10. If the position ratios of two points with respect to A and B be 3/5 and $-4/3$, calculate the distance between them.*

**17.** The hexagon ABCDEF is symmetrical about the diagonal AD, and $AB=3\sqrt{5}$, $BC=\sqrt{17}$, $CD=\sqrt{20}$, $BF=12$, $CE=4$. Calculate the area of the hexagon.

**18.** If C and $C_1$ be the internal and external points of "medial section" of the line AB, so that $AB.BC=AC^2$, and $AB.BC_1=AC_1^2$, show that $CC_1^2=5AB^2$.

**19.** CD is a diameter of a circle; AB and EF chords parallel to CD and each equal to half CD. Calculate the area ABDFEC.

**20.** C is a point in AB, so that $AB.BC=AC^2$. Show that $AB^2+BC^2=3AC^2$.

**21.** P and Q are two points in the line AB (both internal); if O be the middle point of AB and R the middle point of PQ, show that $AP^2 \backsim BQ^2=2AB.OR-2PQ.OR$.

**22.** If P and Q divide AB internally and externally respectively so that $AQ:BQ=AP^2:BP^2$, and if $AB=a$, $AP=x$, show that $PQ=x(a-x)/(2x-a)$.

**23.** If a point P divide the distance between A and B so that $AP^2-BP^2=c^2$, find the co-ordinate of P in terms of $c$ and the co-ordinates $x_1$ and $x_2$ of A and B respectively.

---

* By the *position ratio* of a point P on a straight line L with respect to two fixed points, AB on L, is meant the algebraic value of the ratio of the co-ordinates of P (§ 21) with respect to A and B respectively.

**24.** If A, B, C be any three points on a line and P any finite point whatever, then

$$\overline{AB}.\overline{CP}^2+\overline{BC}.\overline{AP}^2+\overline{CA}.\overline{BP}^2=-\overline{AB}.\overline{BC}.\overline{CA}.$$

**25.** ABCDE are five points in order on a line forming four segments such that any intermediate segment is half the sum of the two adjacent. Show

(i) $AB^2-CD^2=2BC.AB-2BC.CD$;

(ii) $9BC.CD-AB.DE=8PQ^2$, P and Q being the middle points of BC and CD.

**26.** A, B, $A_1$, $B_1$, P are points in order on a straight line. If $x^2=AP^2-BP^2$, $x_1^2=A_1P^2-B_1P^2$, show that $x^2/AB-x_1^2/A_1B_1=AA_1+BB_1$.

**27.** The radii of two circles are 205 and 85 inches respectively, and the distance between their centres is 200 inches; find the length of their common chord.

**28.** A, B, C, D are four points on a line, C being the middle point of BD. If $B_1$, $C_1$, $D_1$ are the middle points of AB, AC, AD respectively, show that $AD_1^2=AB_1^2+2AC_1.BC$.

**29.** If OACBD be five points in order on a straight line and C, D, be harmonically conjugate with respect to A, B, show that

$$2OA.OB+2OC.OD=OA.OD+OB.OC+OB.OD+OC.OA.$$

**30.** ABC is an equilateral triangle; ACDE a square described externally on AC. BD meets the perpendicular from A on BC in L. Calculate the distance AL, BC being 5 inches.

**31.** In a triangle whose sides are 17 and 18 and whose base is 19, find the altitude and median corresponding to the base.

**32.** ABC is a triangle, BE is perpendicular to AC and D bisects BC. Show that $4AD^2=4AE.AC+BC^2$.

**33.** ABCD is a square whose vertices are the middle points of the alternate sides of a regular octagon; if $a$ be the side of the octagon and $b$ the side of the square, show that $a^2=2(b-a)^2$.

**34.** Show that the area of the triangle whose sides are $3a$, $\sqrt{(2a^2+2b^2-c^2)}$, $\sqrt{(2a^2+2c^2-b^2)}$ is three times the area of the triangle whose sides are $a$, $b$, $c$.

**35.** If $a$ and $b$ be the lengths of the parallel sides of a trapezium, and each of its non-parallel sides be of length $c$, show that its area is $\frac{1}{4}(a+b)\sqrt{\{(2c+a-b)(2c-a+b)\}}$.

**36.** If $a$, $b$, $c$, $d$ be the lengths of the diagonals and of the two parallel sides of a trapezium, show that its area is

$$\tfrac{1}{4}\sqrt{\{(a+b+c+d)(-a+b+c+d)(a-b+c+d)(a+b-c-d)\}}.$$

**37.** ABC is a triangle, D the foot of the perpendicular from A on BC, O the middle point of BC, and P a point in CB produced such that $2AP^2=AB^2+AC^2$, show that $OP^2+OB^2=CP.CD-BP.BD$.

**38.** If $\alpha$, $\beta$, $\gamma$ be the lengths of the medians of a triangle, show that its area is $\frac{1}{3}\sqrt{(2\Sigma\alpha^2\beta^2-\Sigma\alpha^4)}$.

**39.** E and F are the feet of the perpendiculars from B and C on the opposite sides of the triangle ABC, and L and M the projections of E and F on BC; show that $LM=\{2a^2(b^4+c^4)+(b^2c^2-a^4)(b^2+c^2)-b^6-c^6\}/4ab^2c^2$.

**40.** P, Q, R are taken in the sides of a triangle (internally) so that $BP/PC=CQ/QA=AR/RB=\rho$; find the area of PQR in terms of the area of ABC.

**41.** The area of the triangle whose vertices are the feet of the internal bisectors of the angles of the triangle ABC (of area $\Delta$) is $2abc\Delta/\Pi(a+b)$.

**42.** ABC is a triangle whose base $BC=a$; $D_1$ is the middle point of AB, $D_2$ the middle point of $D_1B$, $D_3$ the middle point of $D_2B$, and so on; similarly, $E_1$ is the middle point of AC, $E_2$ of $E_1C$, and so on. Show that $D_nE_n=(2^n-1)a/2^n$; and that the area of $D_nE_nCB$ is $(2^{n+1}-1)/2^{2n}$ times the area of ABC.

**43.** If $A_1, A_2, \ldots, A_n$ be a series of points on a straight line, and O such that $\Sigma\overline{A_1O}=0$, then, if P be any other point, $\Sigma\overline{A_1P}=n\overline{OP}$.

If O be such that $\Sigma a_1\overline{A_1O}=0$, and O′ such that $\Sigma a_1\overline{A'_1O'}=0$, then $\Sigma a_1\overline{OO'}=\Sigma a_1\overline{A_1A'_1}$.

**44.** If $A_1, A_2, \ldots, A_n$ are $n$ points on a line, and O be such that $\Sigma\overline{AO}=0$, then, if P be any other point on the line, $\Sigma\overline{AP}=n\overline{OP}$; and $\Sigma AP^2=nOP^2+\Sigma AO^2$.

**45.** If the position ratios of P and Q be $\pm\rho$, and that of R be $\sigma$, find the position ratio of R with respect to P and Q.

**46.** If P and Q be points whose position ratios are $\pm\rho$, and C the point whose position ratio is $-1$, then $CP.CQ=CA^2$.

**47.** If $AB=a$, and the algebraic values of the position ratios of P and Q with respect to A and B be $\rho$ and $\sigma$, show that the distance between P and Q is $(\sigma-\rho)a/(\rho-1)(\sigma-1)$; and apply this result to calculate the distance between the points where the bisectors of the internal and external vertical angles of a triangle meet the base.

**48.** Taking the radius of the earth to be 3956 miles and $\pi=3{\cdot}1416$, calculate the plane area enclosed by the 60th parallel of latitude.

**49.** If $P_n$ and $Q_n$ be the areas of the regular inscribed and circumscribed n-gons for any given circle, then

$$P_{2n}^2=P_nQ_n, \quad 2/Q_{2n}=1/P_{2n}+1/Q_n.$$

**50.** A circle of 9 feet radius slides so as always to meet two perpendicular non-intersecting straight lines; if the distance between the two extreme points which it can reach on one of the lines be 10 feet, find the least distance betwen the lines.

**51.** Calculate the volume of the regular octohedron whose vertices are the centres of the faces of a unit cube.

# CHAPTER XI

## ELEMENTARY THEORY OF INTEGRAL FUNCTIONS

§ **93.** We return in the present chapter to the theory of algebraic identities; and we propose to discuss more especially those methods for constructing such identities which depend on the consideration of **Algebraic Form.** The fundamental notions on which such considerations depend have already been explained in previous chapters; and the beginner should revise paragraphs **31, 32, 42-50** before proceeding with what follows.

§ **94.** For our present purposes it is important to state the law of distribution in a generalised form, which is directly applicable to the product of any number of algebraic sums.

Consider the product $(a-b)(c-d)(e+f-g)$. If we apply the law of distribution in the form already established for the product of two factors, we have in succession

$$\begin{aligned}(a-b)(c-d)(e+f-g) &= (ac-ad-bc+bd)(e+f-g),\\ &= ace-ade-bce+bde\\ &\quad +acf-adf-bcf+bdf\\ &\quad -acg+adg+bcg-bdg \qquad (1).\end{aligned}$$

It will be seen that we first form all the partial products that can be obtained by taking one and only one term from each of the two first brackets. Into each of these partial products we multiply successively each term of the third bracket; so that we have finally the algebraic sum of all possible partial products that can be formed by taking in every possible way the product of three terms, one of which and no more is taken from each bracket, and determining the sign according to the laws already established in Chap. IV., for the multiplication of algebraic quantities. It will be observed that we thus get $2 \times 2 \times 3$ terms altogether, for each of the two terms of the first bracket may be combined with each of the two terms of the second; and each of the $2 \times 2$ resulting pairs with each of the three terms of the third bracket, giving $2 \times 2 \times 3$ partial products altogether. In this enumeration we suppose that each partial product is set down as it arises; and that there has been no collection of like terms or removal of mutually destructive terms.

The process detailed in the above instance is obviously applicable to any number of brackets, and leads us to the following rule:—

*To construct the product of any number of factors, each of which is an algebraic sum, form all possible partial products by taking one term, and only one, from each factor; determine the sign by the law of signs (that is, if there be an odd number of negative terms in the partial product, take the sign* $-$*; if an even number of such or none, take the sign* $+$*); set down the algebraic sum of all the partial products thus formed.*

COR. *If the number of terms in the factors of the product be* l, m, n, . . . *respectively, then the number of partial products in the distributed product as formed by the above rule will be* $l \times m \times n \times \ldots$

§ **95.** As an example of the use of the generalised form of the law of distribution, let us consider $(a+b)^3$, that is, $(a+b)(a+b)(a+b)$. There are here three factors; and each term in these consists of a single letter, $a$ or $b$; each partial product in the final distribution will contain three letters, each of which must be either $a$ or $b$. The only possible variable parts for the terms of the final distribution are therefore $a^3$, $a^2b$, $ab^2$, or $b^3$. Now there are $2 \times 2 \times 2 = 8$ partial products in all; hence some of the forms $a^3$, $a^2b$, $ab^2$, $b^3$ must be repeated more than once. It remains to find how often each is repeated. To get $a^3$ we must take $a$ from each bracket; and this can be done in one way only; and the same is obviously true for $b^3$. On the other hand, $a^2b$ can be obtained by taking $b$ from the first factor, and $a$ from each of the two others, or by taking $b$ from the second, and $a$ from each of the two others, or by taking $b$ from the third, and $a$ from each of the two others, altogether in three different ways; and the same holds for $b^2a$ (or $ab^2$), if we put $b$ for $a$ and $a$ for $b$. Hence, since the signs of all the partial products are $+$, we must have

$$(a+b)^3 = a^3 + 3a^2b + 3ab^2 + b^3 \qquad (2).$$

Next consider $(b+c)(c+a)(a+b)$. Here, as in last case, the variable part of every term in the distributed product will be of the third degree in $a$, $b$, $c$. The *à priori* possible forms are therefore $a^3$, $b^3$, $c^3$; $b^2c$, $bc^2$, $c^2a$, $ca^2$, $a^2b$, $ab^2$; $abc$. Of these, however, $a^3$ evidently cannot occur, because we cannot take $a$ from each of the three brackets simultaneously; and the same applies to $b^3$ and $c^3$, the other two terms of the same type. To get $b^2c$ we must take $b$ from two brackets and $c$ from the other, which can be done in one way only; the like obviously is true of all the other five terms of the same type. We can get $abc$ in two ways, viz. as $bca$ and as $cab$, the order of the letters indicating the brackets from which the terms in the partial product are taken. Hence, finally, we must have, since all the terms have the same sign—

$$(b+c)(c+a)(a+b)=b^2c+bc^2+c^2a+ca^2+a^2b+ab^2+2abc \quad (3).$$

As a contrast with last example, let us consider $(b-c)(c-a)(a-b)$. The forms of the terms that may occur are the same as before, viz. $b^2c$, $bc^2$, $c^2a$, $ca^2$, $a^2b$, $ab^2$; $abc$. Of the first six we can say as before that each can only occur once. There is, however, a difference. The term which has the form $b^2c$ occurs as $bc(-b)=-b^2c$; whereas the term which has the form $bc^2$ occurs as $(-c)c(-b)=bc^2$; the term in $c^2a$ has the sign $-$, in $ca^2$ $+$; in $a^2b$ $-$, in $ab^2$ $+$. Finally, the term whose variable part is $abc$ occurs twice, viz. as $bca$ and as $(-c)(-a)(-b)=-abc$; and these two partial products destroy each other. Hence

$$(b-c)(c-a)(a-b)=-b^2c+bc^2-c^2a+ca^2-a^2b+ab^2 \quad (4).$$

## EXERCISES XXIII.

Work out by direct application of the generalised Law of Distribution the distributed products of the following; and condense the results where possible by collecting together terms which have the same variable part:—

1. $(a+b+c)(d-e-f)(g-h)$.
2. $(a-b+c)(d+e-f)(g+h)$.
3. $(a+2b-c)(a-2b+c)$.
4. $(a+2b-c)^2$.
5. $(a+2b+2c+d)^2$.
6. $(a+2b-2c-d)^2$.
7. $(1+x+x^2+x^3)^2$.
8. $(1-x+x^2-x^3)^2$.
9. $(1+2x+3x^2)^2$.
10. $(a+2b+3c+4d)(a-2b+3c-4d)$.
11. How many terms are there in the distributed product of

$$(a_1+a_2+a_3)(b_1+b_2+b_3+b_4)(c_1+c_2+c_3+c_4+c_5);$$

and what is its value when each of the 12 variables, $a_1$, etc., has the value 2?

§ **96.** Since an integral function of any given set of variables, say $x$, $y$ $z$, contains no division with respect to any of these variables, it must be simply the algebraic sum of a number of terms each of which is integral with respect to these variables, and therefore of the form $Ax^l y^m z^n$, where $l$, $m$, $n$ are positive integers, and A is constant, *i.e.* independent of the variables $x$, $y$, $z$. The integral function may of course also contain a term which is a constant simply; such a term if present is spoken of as the **Absolute Term.** By the degree of the integral function, as has already been explained, is meant the degree of its term or terms of highest degree.

If an integral function does not contain any particular variable at all, it is said to be of degree 0 with respect to that variable.

The following are examples of integral functions:—

Ex. 1. $3+2x+3y+x^2+xy+2y^2$ is an integral function of $x$ and $y$ of the second degree; the most general function of the same description is $a+bx+cy+dx^2+exy+fy^2$, where $a$, $b$, $c$, $d$, $e$, $f$ are symbols denoting constant coefficients.

Ex. 2. The most general integral function of $x$, $y$, $z$ of the second degree is

$$a+bx+cy+dz+ex^2+fy^2+gz^2+hyz+izx+jxy \qquad (5).$$

§ **97.** Since the product of two or more integral terms is (see § 32) an integral term whose degree is the sum of the degrees of the factors, it follows immediately from the generalised law of distribution that

1. *The product of any number of integral functions is an integral function.*

2. *The highest terms in the distributed product of a number of integral functions are the product of the highest terms of the factors; and the absolute term the product of the absolute terms.*

3. *The degree of the product of a number of integral functions is the sum of the degrees of the factors.*

Ex. In the distributed product of $(2+x+y)(3+2x+3y+x^2+y^2)(1+xy)$ the absolute term is $2\times3\times1=6$, and the terms of highest degree are $(x+y)(x^2+y^2)xy=x^4y+x^3y^2+x^2y^3+xy^4$. The degree is $1+2+2=5$.

An important point to be noticed is that when an integral function of, say, $x$, $y$, $z$ is arranged in the standard form

$$a+bx+cy+dz+ex^2+fy^2+gz^2+hyz+\ldots,$$

the terms are algebraically independent of one another, *e.g.* the term $cy$ cannot be transformed (by the laws of Algebra merely) either wholly or partly into the term $a$, or into the term $bx$, or into any other term of the function, so long as we suppose $x$, $y$, $z$ unconnected by any relation. The standard form for a given integral function is therefore unique. It follows that *if two integral functions are equal in the identical sense* (i.e. *transformable into one another by the laws of Algebra when the variables are unrestricted*), *then their standard forms must be identical term by term.* In other words, *if*

$$a + bx + cy + dz + ex^2 + fy^2 + gz^2 + hyz + \ldots$$
$$\equiv a' + b'x + c'y + d'z + e'x^2 + f'y^2 + g'z^2 + h'yz + \ldots,$$

*there being of course a finite number of terms in each function, then*

$$a = a',\ b = b',\ c = c',\ e = e',\ f = f',\ g = g',\ h = h',\ \ldots$$

**§ 98. Even and Odd Functions.**—*When the change of sign of any variable in a function produces no alteration in the value of the function, it is said to be an even function of that variable; if the change of sign of the variable merely changes the sign of the value of the function, it is said to be an odd function of that variable.*

For example, since $2+(-x)^2+y^2+(-x)^4y \equiv 2+x^2+y^2+x^4y$, $2+x^2+y^2+x^4y$ is an even function of $x$. Again, since $(-x)y+(-x)^3y^2-(-x)^5 \equiv -(xy+x^3y^2-x^5)$, $xy+x^3y^2-x^5$ is an odd function of $x$. On the other hand, $2+x^2+xy+y^2$ is neither an even nor an odd function of $x$.

*If an integral function be even as regards any variable, it can only contain even powers * of that variable, if odd only odd powers.*

For, let the integral function be arranged according to powers of $x$, thus

$$A + Bx + Cx^2 + Dx^3 + \ \ldots,$$

where A, B, C, D depend on constants and, it may be, the other variables, *but not upon x.* If the function be even, we must have

$$A - Bx + Cx^2 - Dx^3 + \ \ldots \ \equiv A + Bx + Cx^2 + Dx^3 + \ \ldots$$

Since the arrangement of the function is unique as regards $x$, *i.e.* no one term can be algebraically transformed into any other so long as $x$ is unrestricted, it follows by last paragraph that we must have $-B \equiv B$, $-D \equiv D$, etc.; that is, $2B \equiv 0$, $2D \equiv 0$,

* A term that does not contain $x$ at all is reckoned as containing an even power of $x$.

etc., whence $B \equiv 0$, $D \equiv 0$, etc.; in other words, all the terms which contain odd powers of $x$ must be identically zero.

The reader will readily supply the corresponding proof for an odd function.

§ **99.** As already explained, an integral function is said to be homogeneous with respect to any set of variables when the degree of every term of the function with respect to those variables is the same. It follows immediately from this definition and from the generalised form of the law of distribution that—

*The product of a number of homogeneous functions of degrees* l, m, n, . . . *respectively is a homogeneous function of degree* l + m + n + . . .

This may be called the **Law of Homogeneity**, it is exemplified in the identities (1), (2), (3), and (4) of the present chapter; the identity $(x+y)(x^2+y^2)xy \equiv x^4y + x^3y^2 + x^2y^3 + xy^4$ is another instance.

The reader who has mastered what has been said regarding the construction of integral terms of given degree will have no difficulty in constructing homogeneous integral functions of a given degree the most general of their kind; all that is necessary is to write down all possible terms of the given degree, each multiplied by a letter to represent a possible coefficient.

The following are examples of general homogeneous integral functions:—

| Degree. | Variables. | Function. |
|---|---|---|
| 1 | $x$ | $ax$ |
| 2 | $x, y$ | $ax^2+bxy+cy^2$ |
| 2 | $x, y, z$ | $ax^2+by^2+cz^2+dyz+ezx+fxy$ |
| 4 | $x, y$ | $ax^4+bx^3y+cx^2y^2+dxy^3+ey^4$ |
| 1 | $x, y, z, u$ | $ax+by+cz+du$ |

§ **100. Symmetry.**—Another peculiarity which is often possessed by integral and other functions must already have struck the reader, viz. symmetry. It is seen, for example, in identities (2) and (3) of this chapter, in which each of the operands, $a$, $b$ in the one case, and $a$, $b$, and $c$ in the other, are involved in exactly the same way. We may give a precise definition as follows :—

*A function is said to be (absolutely) symmetric with respect to any set of variables when the interchange of any pair of the set of variables (which are supposed to be unconnected by any relation) does not alter the value of the function.*

Ex. 1. $2x+y+z$ is symmetric with respect to $y$ and $z$.

Ex. 2. $yz+zx+xy+x+y+z$ is symmetric with respect to $x$, $y$, $z$.

Ex. 3. $a/bc+b/ca+c/ab$ and $(a+b+c)/(b^2c^2+c^2a^2+a^2b^2)$ are symmetric with respect to $a$, $b$, $c$.

If we confine ourselves to integral functions, and bear in mind the fact that so long as the variables are unconditioned, no one term in the standard form of an integral function can be algebraically transformed into any other, we see at once that—

*The necessary and sufficient condition that an integral function be symmetric is that all the terms of any one type shall have the same coefficient.*

For the interchange of pairs of the variables simply changes any term of a particular type into another of the same type (§ 43). For example, if we interchange $x$ and $y$, $ax^2+bxy+cy^2$ passes into $cx^2+bxy+ay^2$ ; and, in order that $ax^2+bxy+cy^2 \equiv cx^2+bxy+ay^2$, we must have $a=c$.

The following are some examples of symmetric integral functions of given degree the most general of their kind :—

| Degree. | Variables. | Function. |
|---|---|---|
| 1 | $x$, $y$ | $a+bx+by$ |
| 2 | $x$, $y$, $z$ | $a+bx+by+bz+cx^2+cy^2+cz^2+dyz+dzx+dxy$ |
| 3 | $x$, $y$ | $a+bx+by+cx^2+dxy+cy^2+ex^3+fx^2y+fxy^2+ey^3$ |

The following are examples of homogeneous symmetric integral functions of given degree the most general of their kind :—

| Degree. | Variables. | Function. |
|---|---|---|
| 1 | $x$, $y$, $z$ | $ax+ay+az$ |
| 2 | $x$, $y$, $z$ | $ax^2+ay^2+az^2+byz+bzx+bxy$ |
| 4 | $x$, $y$ | $ax^4+bx^3y+cx^2y^2+bxy^3+ay^4$ |

Often there occurs what may be called **Collateral Symmetry** with respect to two or more sets of variables. Thus the function $(b+c)x+(c+a)y+(a+b)z$ is said to be symmetric with respect to the sets $\begin{pmatrix} x, & y, & z \\ a, & b, & c \end{pmatrix}$; because, if we interchange any two of $x, y, z$, and at the same time the corresponding two of $a, b, c$ as determined by the above array, the value of the function is unaltered, *i.e.* the new value can be transformed into the old without supposing $a, b, c$; $x, y, z$ to be connected by any relation whatever. In the same sense $ax^2+by^2+cz^2+dyz+ezx$ $+fxy$ is symmetrical with respect to $\begin{pmatrix} x, & y, & z \\ a, & b, & c \\ d, & e, & f \end{pmatrix}$.

Although transformation of a function by the laws of Algebra may bring it into a form in which the symmetry is not immediately obvious to the eye [*e.g.* $(bc+ca+ab)/(a+b+)c \equiv a+(bc-a^2)/(a+b+c)$], it is obvious from the nature of the definitions we have given that no such transformation can in reality either confer or remove the quality of symmetry. Hence we have the following important theorem, of which the identities (2) and (3) above are examples:—

*The sum or product of any number of symmetric functions, or the quotient of two such, is a symmetric function.*

§ 101. **Cyclic Symmetry.**—Another kind of symmetry is often to be observed in algebraic functions, which is connected with the notion of the **Cyclic Permutation** of a given set of variables. Consider three variables $x, y, z$, attending to the order in which they are written. If we replace $x$ by the connective one $y$, $y$ by $z$, and $z$, the last, by the initial one, $x$, we get $y, z, x$, which we call a *cyclic permutation* of $x, y, z$. In like manner, we derive $z, x, y$ from $y, z, x$. We thus have $x, y, z$; $y, z, x$; $z, x, y$ which we call the cyclic permutations of $x, y, z$. No more can be found, for the same process derives from $z, x, y$ the *original arrangement* $x, y, z$. Perhaps the simplest way to conceive of cyclic permutations is to write the variables in order at equal intervals round the circumference of a circle. If there be $n$ of them, we get the cyclic permutations, evidently $n$ in number, by turning the circle successively through the $1/n^{th}$, $2/n^{th}$, . . . parts of four right angles. Thus in the case of three variables we have—

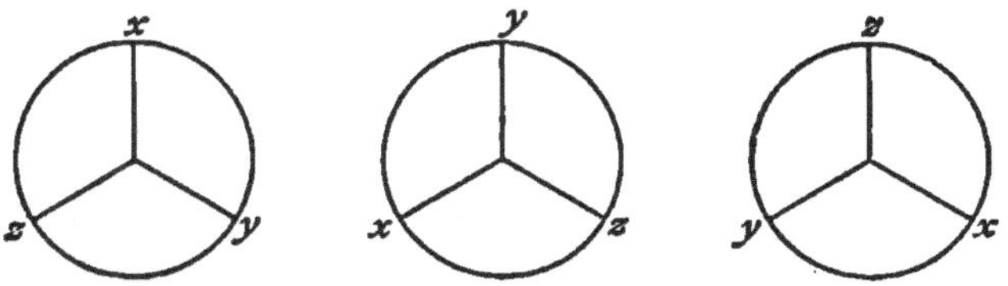


*When the terms of a function can be arranged in groups such that each of these groups can be derived from any one of the others by cyclic permutation of the variables, the function is said to have cyclic symmetry.*

The selected group from which the others are derived may be called the *typical group;* and the function is obviously determined when a typical group of its terms is given. The typical group may, of course, consist of a single term only.

The following are examples of cyclically symmetric functions:—

Ex. 1. $2y^2z+2z^2x+2x^2y$ : typical group $2y^2z$.
Ex. 2. $y^2/z+z^2/x+x^2/y$ : typical group $y^2/z$.
Ex. 3. $2x(y^2+z^2)+2y(z^2+x^2)+2z(x^2+y^2)$ : typical group $2x(y^2+z^2)$.
Ex. 4. $x^2(y-z)+y^2(z-x)+z^2(x-y)$ : typical group $x^2(y-z)$.

It will be readily seen, by considering Example 3, that a function which has cyclic symmetry may also have absolute symmetry; indeed, it is obvious that every absolutely symmetric function must be also cyclically symmetric. But the converse is not true, as will be seen at once by comparing Examples 1 and 3.

As in the case of absolutely symmetric functions, it is customary and convenient to abbreviate by writing only the typical group of a cyclically symmetric function preceded by a symbol of summation, say $\Sigma$.* In cases such as Example 2, where confusion might arise between absolutely and cyclically symmetric functions, a distinct symbol, say S, should be used. Thus, while $S2y^2z$ means $2y^2z+2z^2x+2x^2y$, $\Sigma 2y^2z$ means $2y^2z+2yz^2+2z^2x+2zx^2+2x^2y+2xy.^2$

It may be assumed, whenever the typical group consists of more than one term, that cyclic symmetry is meant. Thus, for example, when we write $\Sigma x^2(y-z)$ we mean $x^2(y-z)+y^2$

* The $\Pi$ notation may also be used to abbreviate a product whose factors are derivable from each other by cyclic permutations, *e.g.* $\Pi(b-c)$ means $(b-c)(c-a)(a-b)$.

$(z-x)+z^2(x-y)$, and not $x^2(y-z)+x^2(z-y)+y^2(z-x)+y^2(x-z)+z^2(x-y)+z^2(y-x)$.

We may also have **Collateral Cyclic Symmetry** with respect to two or more sets of variables. For example, $(b-c)x+(c-a)y+(a-b)z$ is cyclically symmetrical with respect to $\begin{pmatrix} x, & y, & z \\ a, & b, & c \end{pmatrix}$; the typical group may be taken to be $(b-c)x$, and we may write the function $\Sigma(b-c)x$.

## EXERCISES XXIV.

**1.** What are the degrees of the following in $x$, $y$, $z$ separately, and in $x$, $y$, $z$ together :—

$$(\alpha)\ 3x+2x^2y+6x^2y^2+z^3;\ (\beta)\ x^2y+x^2y^2z^3+xyz;$$
$$(\gamma)\ x^2(y+z)+y^2(z+x)+z^2(x+y)\,?$$

**2.** What is the degree of the lowest terms in the distributed product of

$$(2+3x+6y)(xy+x^3+y^4)(x^2-y^2+x^2y^2),$$

and what is the degree of the function ?

**3.** What is the degree of

$$\{1+x+y+x^2+y^2\}\,\{1+(x-y)^2\}\,\{1+xy\}\,?$$

Is the function homogeneous ? is it symmetrical ? is it an even or an odd function of $x$ ?

**4.** Find the terms of highest degree in $(1+x+y+x^2+y^2)(1+x-y)(1+xy)$.

**5.** Find the terms of the second degree in the functions of examples (3) and (4).

**6.** Find the terms of the second degree in $(3+2x-y+x^2+xy+y^2)(2+x-2y+xy)(1+x+y+x^2+y^2)$.

**7.** If $(a+bx+cx^2)/(a'+b'x+c'x^2)$ be an even function of $x$, and if $ac'-a'c\neq0$, then must $b=0$, $b'=0$. What possibility is covered by the excepted case where $ac'-a'c=0$ ?

**8.** Write down the most general function of $x$, $y$, $z$, $u$ of the second degree which is—(i) rational and integral ; (ii) rational, integral, and symmetrical ; (iii) rational, integral, and homogeneous ; (iv) rational, integral, symmetric, and homogeneous.

**9.** Construct the most general symmetric integral function of $x$, $y$, $z$ of the second degree which is an even function of each of its three variables.

**10.** Find the necessary and sufficient condition that $(ax+by+cz)\{(b+c)x+(c+a)y+(a+b)z\}$ be symmetric with respect to $y$ and $z$.

**11.** Write down the general form of an integral function of $x$ and $y$ which is of the first degree in $x$ and also of the first degree in $y$ ("lineo-linear function"). What modification is necessary if the function is to be symmetric ?

**12.** Give the general form of an integral function of $x$, $y$, $u$, $v$

which shall be of the first degree in $x$ and $y$ conjointly; and also of the first degree in $u$ and $v$ conjointly.

**13.** Write down the most general integral function of $x$ and $y$ which is of the second degree with respect to each of these variables taken singly. What modifications are necessary to render the function symmetric?

**14.** Write down the most general homogeneous symmetric function of $x$, $y$, $z$ of the fourth degree, which is an even function of $x$.

**15.** Write down a monotypic (*i.e.* containing terms of one type only) symmetric integral function of $x$, $y$, $z$ of the fourth degree, which is neither even nor odd with respect to any one of the variables.

**16.** How many essentially distinct monotypic symmetric functions of $x$, $y$, $z$, $u$ of the fourth degree are there which are 1st, even with respect to each of the variables; 2nd, odd; 3rd, neither even nor odd?

**17.** What are the peculiarities in the graph of a function of $x$ corresponding to evenness and oddness respectively?

**18.** Write in full the cyclically symmetric functions $\Sigma x^3y^2z$, $\Sigma(x^2-yz)$, $\Sigma(x^2-y^2z)$, $\Sigma(x+y-z)$. Are any of them absolutely symmetrical?

## Applications of the Principles of Form
## Indeterminate Coefficients

**§ 102.** The laws of homogeneity and symmetry are of the greatest use in checking and in abbreviating the work of algebraic transformations. We can often see by means of them that the presence or absence of a particular term in a result indicates some error in the calculation which leads to it. For example, we should immediately conclude that a calculation which led to $(x+y)(x^2+y^2)\equiv x^4+x^2y+2xy^2+y^3$ must be wrong, first, because the right-hand side is not homogeneous, and secondly, because it is not symmetrical.

The notion of Algebraic Form, which embraces the two laws just mentioned, leads us to the practical **Method of Indeterminate Coefficients,** which in conjunction with the laws of degree, homogeneity, and symmetry is one of the most powerful weapons of the analyst. For the present we shall confine our explanation of it to the case of integral functions. It will now be fully understood that an integral function is completely known when we know the variable parts of each of its terms, and also the values (as numbers or in terms of quantities supposed given) of its coefficients. *Since the coefficients are quite independent of the variables, it follows that when they are once determined in any way they are determined once for all.* The determination of the form of the function, and the determination of its coefficients, may, in short, be treated as perfectly separate problems. It may be repeated, for emphasis, that, when we are determining an integral function in the present sense of the word "determine," we have nothing to do with the values numerical or other of its variables; questions of that kind arise only when we make some special use of the function, *e.g.* to represent the length of a bar of iron whose temperature is given, the ordinate of a curve corresponding to a given abscissa, or the like. The variables in an integral function on the one hand, and the values that may be given to them on the other, may with advantage be compared to the pigeon holes of a desk and the documents that may be put into them. The algebraist may construct his function without reference to the arithmetical uses which it will serve, just as the cabinetmaker may construct the pigeon holes without thinking of the particular documents that may come to fill them.

*It is this process of separating the determination of the coefficients from the determination of the form of a function to which the name "method of indeterminate coefficients" is usually applied.* The name is not very descriptive, but it has the great advantage of being well established. The actual determination of the coefficients may be effected in various ways. One of the commonest is to use conditional equations obtained by asserting the existence of the identity (whose general existence has been established by considerations of form) in special cases. The whole matter will become clear from the following examples:—

Ex. 1. Since $(a+b)^3 \equiv (a+b)(a+b)(a+b)$ is a homogeneous, symmetric function of $a$ and $b$ of the third degree, being the product of three functions, each of which is homogeneous, symmetric, and of the first degree, it follows that

$$(a+b)^3 \equiv \mathrm{A}a^3 + \mathrm{B}a^2b + \mathrm{B}ab^2 + \mathrm{A}b^3 \qquad (6).$$

The form of the standard expression for $(a+b)^3$ has thus been determined.

It remains to determine the coefficients A and B. The equation (6) being an identity as regards $a$ and $b$, A and B must be such that it is an identity whatever finite values we give to $a$ and $b$. Let us suppose in the first place that $a=1$, $b=0$; then we must have $1^3 = \mathrm{A}1^3$, that is to say, $\mathrm{A}=1$. Next, suppose $a=1$, $b=1$; then we get, giving to A the value just determined, $2^3 = 1+\mathrm{B}+\mathrm{B}+1$, that is to say, $2\mathrm{B}=6$, whence $\mathrm{B}=3$. Hence $(a+b)^3 \equiv a^3+3a^2b+3ab^2+b^3$, in agreement with two other previous calculations.

The beginner will do well to convince himself by actual trial that the choice of the particular values of $a$ and $b$ actually has not, as by the general theory it ought not to have, any effect on the resulting values of A and B. This he may do by putting, say, $a=2$, $b=3$; and then $a=5$, $b=6$; and calculating A and B from the resulting equations. He will get the same values for A and B as before; and the kind of conviction that thus arises is just as necessary to one who has read a piece of algebraic theory, as is courage to a soldier who has studied the art of war.

Ex. 2. $(a+b+c)^3$ is the product of three homogeneous, symmetric, integral functions of $a$, $b$, $c$, each of the first degree; it is therefore a homogeneous symmetric integral function of $a$, $b$, $c$ of the third degree. Hence, the $\Sigma$'s having reference to $a$, $b$, $c$, we have

$$(a+b+c)^3 \equiv A\Sigma a^3 + B\Sigma a^2b + Cabc.$$

If we put $c=0$, this identity becomes

$$(a+b)^3 \equiv A(a^3+b^3) + B(a^2b+ab^2),$$

from which we see by last example that $A=1$, $B=3$.

Finally, putting $a=1$, $b=1$, $c=1$, in the original identity, and remembering that $\Sigma a^3$ has three terms and $\Sigma a^2b$ six, we get

$$3^3 = 1.3 + 3.6 + C,$$

which gives $C=6$. So that

$$(a+b+c)^3 \equiv \Sigma a^3 + 3\Sigma a^2b + 6abc.$$

Ex. 3. If we interchange any two of $a$, $b$, $c$ in the function $F \equiv (a+b+c)(-a+b+c)(a-b+c)(a+b-c)$, we merely change the positions of two of the factors. F is, therefore, a symmetric function of $a$, $b$, $c$; and it is obviously homogeneous and of the fourth degree. Hence we must have

$$F \equiv A\Sigma a^4 + B\Sigma a^3b + C\Sigma a^2b^2,$$

there being no other possible terms.

Again, if we change $a$ into $-a$, F changes into

$$\begin{aligned} F' &\equiv (-a+b+c)(a+b+c)(-a-b+c)(-a+b-c), \\ &\equiv (a+b+c)(-a+b+c)\{-(a+b-c)\}\{-(a-b+c)\}, \\ &\equiv (a+b+c)(-a+b+c)(a-b+c)(a+b-c) \equiv F. \end{aligned}$$

Hence F is an even function of $a$, and can contain no such term as $Ba^3b$. Therefore $B=0$.

Hence $$F \equiv A\Sigma a^4 + C\Sigma a^2b^2.$$

Since $a^4$ can only arise from the partial product $a.(-a).a.a = -a^4$, we see at once that $A=-1$.

Finally, putting $a=b=c=1$ in the identity last written, we get $3=-3+C.3$; whence $C=2$; and we find

$$F \equiv -\Sigma a^4 + 2\Sigma a^2b^2.$$

Ex. 4. It is required to investigate whether it is possible so to determine the coefficients $a$, $b$, $c$ that

$$(x+y)(ax^2+bxy+cy^2) \equiv x^3+y^3.$$

We are here in a different position as regards the existence of the identity; we do not know *à priori* that it exists.

Supposing, however, that it can exist, we see at once by comparing the coefficients of $x^3$ and of $y^3$ on both sides, that, if there be such an identity, then must $a=1$, $c=1$. Again, comparing the coefficients of $x^2y$ on both sides, we see that we must have $a+b=0$; so that we must have $b=-a=-1$. If there be an identity of the kind supposed, it must therefore be

$$(x+y)(x^2-xy+y^2)\equiv x^3+y^3,$$

which is, in fact, an identity, as may be readily verified.

## EXERCISES XXV.

In the following set of exercises, when a function is set down without further remark, it is required to distribute all the products contained in it, and to arrange the resulting terms according to degree and type.

When letters towards the end of the alphabet occur, they are to be taken in the first instance as variables, and the coefficients, which are functions of $a$, $b$, $c$, etc., are also to be arranged in standard form.

$\Sigma$ and $\Pi$ in all cases have reference to three variables.

**1.** $(a+b-c+d)(a+b+c-d)$.

**2.** $(a+b-c-d)(b+c-d-a)(c+d-a-b)(d+a-b-c)$.

**3.** $\{(b+c)x-by-cz\}\{x+y+z\}$.

**4.** $(x+y+z+u)^2+(x+y-z-u)^2+(x-y+z-u)^2+(x-y-z+u)^2$.

**5.** Show that $\Sigma a^2-\Sigma ab=\frac{1}{2}\Sigma(b-c)^2$.

**6.** Find the sum of the coefficients in the expansion of $(x+3y)(3x+y)(x+y)^2$.

**7.** If $s_n=x^n+y^n$, show that $s_1(s_9s_7-s_8^2)=s_{10}s_7-s_9s_8$.

**8.** $\Sigma(b-c)^3=-2\Sigma(b-c)(c-a)$. **9.** $\Sigma(cy-bz)^2+(\Sigma ax)^2=\Sigma x^2\Sigma a^2$.

**10.** $(a+b+c)(x+y+z)+(-a+b+c)(-x+y+z)+(a-b+c)(x-y+z)+(a+b-c)(x+y-z)$.

**11.** $(a^2-bc)^2(b-c)+(b^2-ca)^2(c-a)+(c^2-ab)^2(a-b)$.

**12.** $(a+b-c)^3+(a-b+c)^3+6a(a+b-c)(a-b+c)$.

**13.** $\{(y+z)/(b-c)+(z+x)/(c-a)+(x+y)/(a-b)\}\{(y-z)/a+(z-x)/b+(x-y)/c\}$.

**14.** $(x+b+c)(x^2+ax+b-c)+(x+c+a)(x^2+bx+c-a)+(x+a+b)(x^2+cx+a-b)$.

**15.** $(x+a)^2(x^2+b^2-c^2)+(x+b)^2(x^2+c^2-a^2)+(x+c)^2(x^2+a^2-b^2)$.

**16.** Show that $(x^2y+y^2z+z^2x)(xy^2+yz^2+zx^2)$ is a symmetric function of $x$, $y$, $z$.

**17.** Show that $2z(x-y)^2+2y(z-x)^2+(z+y)(x-y)(z-x)$ is symmetrical in $x$, $y$, $z$.

**18.** Show that $\{(y^2-zx)(z^2-xy)-(x^2-yz)^2\}/x$ is a symmetric function of $x$, $y$, $z$.

**19.** $(\Sigma a)^2\Sigma bc$. **20.** $\Sigma(b+c-a)(b-c)^2$.

**21.** $(b-c)(x+b+c)^2+(c-a)(x+c+a)^2+(a-b)(x+a+b)^2+(b-c)(c-a)(a-b)$.

**22.** $\Sigma a(2a-b-c)/(a-b)(a-c)=3$. **23.** $\Sigma(a^2+bc)^2-(\Sigma ab)^2$.

**24.** $\Sigma\{(y^3-z^3)/(y-z)\}^2$. **25.** $(\Sigma x)^3-\Sigma x^3=3\Pi(y+z)$.

26. $bc(b^2-c^2)+ca(c^2-a^2)+ab(a^2-b^2)+\Sigma a\Pi(b-c)=0$.
27. $(b-c)(b+c)^3+(c-a)(c+a)^3+(a-b)(a+b)^3$.
28. $\Sigma x\Sigma x(x+y-z)$.
29. $\Sigma a(b-c)^2+\Sigma(b-c)(b^2-c^2)\equiv 2\Sigma a\Sigma(a^2-bc)$;
$\equiv 2\Sigma a^3-6abc$.
30. $\Pi(x^2-yz)$.
31. $\Pi\{a^2+a(b-c)\}$.

## Integral Functions of a Single Variable

**§ 103.** When there is only one variable, $x$, say, there is only one term of any particular degree; hence the general standard form for an integral function of a single variable is

$$p_0x^n+p_1x^{n-1}+\ \dots\ +p_{n-1}x+p_n,$$

where $n$ is a positive integer denoting the degree of the function, and $p_0, p_1, \dots, p_{n-1}, p_n$ are the coefficients in the usual sense of the word. The suffixes $_0, _1, \dots _{n-1}, _n$ affixed to the letters are used partly to distinguish the coefficients from each other, partly to indicate at a glance the term to which the coefficient belongs; they must not be confounded with indices.

The theory of integral functions of a single variable has been largely developed, partly on account of its simplicity and partly because it serves as a foundation for the theory of equations.

**§ 104.** As the distribution and arrangement according to powers of the variable of the product of two integral functions of $x$ is an operation of frequent occurrence, it is worth while to have a succinct and systematic arrangement for the necessary calculations. This consists merely in arranging both multiplicand and multiplier in the standard form, *i.e.* according to descending (or ascending) powers of $x$, and placing the partial products corresponding to each term of the multiplier in a separate horizontal line in such a manner that the coefficients of the various powers of $x$ fall into vertical lines, so that they can be conveniently added. The work may be abbreviated by leaving out the powers of $x$ until the end, and merely calculating with the coefficients, a modification which is sometimes spoken of as the **method of detached coefficients.** These points will be fully understood from the following examples:—

Ex. 1. Distribute and arrange the product $(x^3-2x^2+3x-1)(3x^2-2x-3)$. The work may be arranged as follows:—

$$\begin{array}{llllll}
x^3 & -2x^2 & +3x & -1 & & \\
3x^2 & -2x & -3 & & & \\
\hline
3x^5 & -6x^4 & +9x^3 & -3x^2 & & \\
 & -2x^4 & +4x^3 & -6x^2 & +2x & \\
 & & -3x^3 & +6x^2 & -9x & +3 \\
\hline
3x^5 & -8x^4 & +10x^3 & -3x^2 & -7x & +3
\end{array};$$

or, if we omit the powers of $x$, their places being sufficiently kept by the spaces after the coefficients, the whole calculation may be represented by

$$\begin{array}{llllll}
1 & -2 & +3 & -1 & & \\
3 & -2 & -3 & & & \\
\hline
3 & -6 & +9 & -3 & & \\
 & -2 & +4 & -6 & +2 & \\
 & & -3 & +6 & -9 & +3 \\
\hline
3 & -8 & +10 & -3 & -7 & +3
\end{array}.$$

Since the highest power of $x$ in the product is $x^5$, we can be in no doubt as to the placing of the powers of $x$. The product is $3x^5 - 8x^4 + 10x^3 - 3x^2 - 7x + 3$.

Ex. 2. Distribute and arrange the product $(x^3 - 2x + 1)(x^3 - x^2 + 2)$.

The peculiarity here is that all the powers of $x$ are not present in each factor. If we are to use the method of detached coefficients, we must keep places for the missing powers. This can be done by inserting them with zero coefficients, thus: $(x^3 + 0x^2 - 2x + 1)(x^3 - x^2 + 0x + 2)$. The calculation then runs as follows:—

$$\begin{array}{lllllll}
1 & +0 & -2 & +1 & & & \\
1 & -1 & +0 & +2 & & & \\
\hline
1 & +0 & -2 & +1 & & & \\
 & -1 & +0 & +2 & -1 & & \\
 & & 0 & +0 & +0 & +0 & \\
 & & & 2 & +0 & -4 & +2 \\
\hline
1 & -1 & -2 & +5 & -1 & -4 & +2
\end{array}$$

The product is $x^6 - x^5 - 2x^4 + 5x^3 - x^2 - 4x + 2$.

In practice, the horizontal line of 0's would be omitted, care being taken to place the first coefficient of the next line in the fourth and not in the third vertical column of the scheme.

Ex. 3. Distribute and arrange $(ax^2 + bx + c)^2$. We may arrange the calculation of the coefficients thus:—

$$\begin{array}{lllll}
a & +b & +c & & \\
a & +b & +c & & \\
\hline
a^2 & +ab & +ac & & \\
 & +ab & +b^2 & +bc & \\
 & & +ac & +bc & +c^2 \\
\hline
a^2 & +2ab & +(b^2+2ac) & +2bc & +c^2
\end{array}.$$

Result, $a^2x^4 + 2abx^3 + (b^2 + 2ac)x^2 + 2bcx + c^2$.

The only special point here is the use of the bracket in the last line of the scheme to isolate the coefficient belonging to $x^2$.

Ex. 4. $(x^3 - 2x^2y + 3xy^2 - y^3)(3x^2 - 2xy - 3y^2)$. Here the coefficients, strictly speaking, are $+1, -2y, +3y^2, -y^3$, and $3, -2y, -3y^2$, but, since the powers of $y$ are in definite places, we may omit them until the end of the calculation is reached. The numerical work is the same as in Example 1; and the result of the distribution is

$$3x^5 - 8x^4y + 10x^3y^2 - 3x^2y^3 - 7xy^4 + 3y^5.$$

If the coefficient of only one power of $x$ in the distributed product be required, as is often the case, we may proceed thus :—

Ex. 5. Let the product be

$$(x^8 - 5x^7 + 4x^6 - 2x^5 + 3x^4 - 4x^3 + 2x^2 - x + 1)$$
$$\times (3x^7 - 5x^6 + 2x^5 + 3x^4 - x^3 - 3x^2 - 4x - 5);$$

and let it be required to calculate the coefficient of $x^4$ in the product. Omit all the terms above $x^4$ in both multiplicand and multiplier; reverse the order of the terms of the multiplier, writing its absolute term under the term in $x^4$ of the multiplicand, and so on; thus

$$\begin{array}{ccccc} 3x^4 & -4x^3 & +2x^2 & -x & +1 \\ -5 & -4x & -3x^2 & -x^3 & +3x^4. \end{array}$$

Multiply the pairs of terms that now stand in vertical columns; and the result is the term in $x^4$. In this case the coefficient is

$$3(-5) + (-4)(-4) + (2)(-3) + (-1)(-1) + 1.3 = -1.$$

In practice we need only write the coefficients, and arrange the calculations thus :—

$$\begin{array}{l} 3 - 4 + 2 - 1 + 1 \\ \underline{-5 - 4 - 3 - 1 + 3} \\ -15 + 16 - 6 + 1 + 3 = -1. \end{array}$$

The methods appropriate to integral functions of $x$ may be extended to fractional functions which are merely reciprocals of powers of $x$, *i.e.* $1/x$, $1/x^2$, $1/x^3$. . . . The order of the terms is then

$$\ldots ax^3 + bx^2 + cx + d + e/x + f/x^2 + g/x^3 + \ldots$$

Ex. 6. Distribute and arrange

$$(x^2 + 2x + 1 + 2/x + 1/x^2)(x + 2 + 1/x).$$

Written in full the result of the distribution is

$$\begin{array}{l} x^3 + 2x^2 + \ x + 2 + 1/x \\ \quad + 2x^2 + 4x + 2 + 4/x + 2/x^2 \\ \qquad + \ x + 2 + 1/x + 2/x^2 + 1/x^3, \\ = x^3 + 4x^2 + 6x + 6 + 6/x + 4/x^2 + 1/x^3. \end{array}$$

It is obvious that we might use detached coefficients, just as if the two factors were integral functions.

**§ 105. The Distribution of a Product of Binomial Factors,** each of which is a linear function of $x$ is, for a variety of reasons, of peculiar interest. For simplicity, we shall in the first place suppose that the coefficient of $x$ in each factor is unity. We have then to consider the product

$$(x+a_1)(x+a_2) \ . \ . \ . \ (x+a_n).$$

First take the special case of three factors

$$(x+a_1)(x+a_2)(x+a_3).$$

It is obvious from the principles already laid down in this chapter that the distributed product when arranged will be of the form

$$Ax^3+Bx^2+Cx+D,$$

where the coefficients A, B, C, D are independent of $x$, and depend on $a_1$, $a_2$, $a_3$ alone.

It will be seen at once that we can get $x^3$ in only one way, viz. from the partial product $x\times x\times x$. Hence the coefficient of $x^3$ is simply 1.

We can get terms in $x^2$ in three ways, viz.: by taking $x$ from any two brackets and an $a$ from the remaining bracket. The partial products thus arising are $a_1xx$, $xa_2x$, and $xxa_3$, *i.e.* $a_1x^2$, $a_2x^2$ and $a_3x^2$. Hence the coefficient of $x^2$ is $a_1+a_2+a_3$, say $\Sigma a_1$.

To get a term in $x$ we must take $a$'s from two brackets and an $x$ from one. The corresponding partial products are $a_1a_2x$, $a_1xa_3$, $xa_2a_3$, *i.e.* $a_1a_2x$, $a_1a_3x$, $a_2a_3x$. The coefficient of $x$ is therefore $\Sigma a_1a_2$.

Finally, the absolute term is $a_1a_2a_3$. We have, therefore—

$$(x+a_1)(x+a_2)(x+a_3)=x^3+\Sigma a_1x^2+\Sigma a_1a_2x+a_1a_2a_3,$$

where the $\Sigma$'s have reference to the $a$'s alone.

This process for calculating one by one the coefficients of $x$ is obviously general. Returning to the general case, viz.—

$$(x+a_1)(x+a_2) \ . \ . \ . \ (x+a_n),$$

let us calculate the coefficient of $x^r$, where $r<n$. In all the partial products which yield $x^r$ we must take $x$ from $r$ of the brackets, and $a$'s from the remaining $n-r$. Thus, for example, one partial product of the kind required is $a_1a_2 \ . \ . \ . \ a_{n-r}xx \ . \ . \ . \ (r \ x\text{'s})=a_1a_2 \ . \ . \ . \ a_{n-r}x^r$, and we have to form all possible partial products of the same type. Hence the coefficient of $x^r$ is simply the sum of all the products of $a_1, a_2, \ . \ . \ ., a_n$ that can be formed by taking $n-r$ of them in every possible way; this we may denote as usual by $\Sigma a_1a_2 \ . \ . \ . \ a_{n-r}$.

The absolute term of the distributed product is of course $a_1a_2 \ldots a_n$: and the highest term is $x^n$. We may therefore represent the general result at which we have arrived by writing

$$(x+a_1)(x+a_2) \ldots (x+a_n)$$
$$= x^n + \Sigma a_1 x^{n-1} + \Sigma a_1 a_2 x^{n-2} + \ldots + \Sigma a_1 a_2 \ldots a_{n-r} x^r + \ldots$$
$$+ a_1 a_2 \ldots a_n \qquad (7).$$

By substitution we may derive from this theorem a great variety of others. We may notice specially the case where we replace $a_1, a_2, \ldots, a_n$ by $-b_1, -b_2, \ldots, -b_n$. Since $\Sigma(-b_1) = -\Sigma b_1$, $\Sigma(-b_1)(-b_2) = \Sigma b_1 b_2$, $\Sigma(-b_1)(-b_2)(-b_3) = -\Sigma b_1 b_2 b_3$, and so on, we have

$$(x-b_1)(x-b_2) \ldots (x-b_n)$$
$$= x^n - \Sigma b_1 x^{n-1} + \Sigma b_1 b_2 x^{n-2} - \Sigma b_1 b_2 b_3 x^{n-2} + \ldots$$
$$\pm b_1 b_2 \ldots b_n \qquad (8).$$

where the sign of the last term is + or − according as $n$ is even or odd.

After what has been said, the apparently more general case

$$(A_1x+a_1)(A_2x+a_2) \ldots (A_nx+a_n),$$

where the coefficients of $x$ are not all unity, presents no difficulty. We have only to notice that along with every $x$ there now comes an A. Thus, for example, one of the partial products which yields $x^{n-2}$ instead of being $a_1a_2x^{n-2}$ as formerly is now $a_1a_2A_3A_4 \ldots A_nx^{n-2}$, the coefficient of which contains $n$, $a$'s and A's altogether, viz. 2 of the former and $n-2$ of the latter.

Hence the generalised formula runs—

$$(A_1x+a_1)(A_2x+a_2) \ldots (A_nx+a_n)$$
$$= A_1A_2 \ldots A_nx^n + \Sigma a_1A_2A_3 \ldots A_nx^{n-1}$$
$$+ \Sigma a_1a_2A_3 \ldots A_nx^{n-2}$$
$$\cdots\cdots\cdots$$
$$+ \Sigma a_1a_2 \ldots a_rA_{r+1} \ldots A_nx^{n-r}$$
$$\cdots\cdots\cdots$$
$$+ a_1a_2 \ldots a_n \qquad (9).$$

We have said that (9) is *apparently* more general than (7); it can, however, be readily derived from (7) by substituting $a_1/A_1$, $a_2/A_2, \ldots, a_n/A_n$ for $a_1, a_2, \ldots, a_n$ respectively, in (7) and multiplying both sides by $A_1A_2 \ldots A_n$; a fact which the reader should verify.

An important special case of (9) arises when we put $x=1$, we thus get

$$\begin{aligned}(A_1+a_1)(A_2+a_2)\ .\ .\ .\ (A_n+a_n)\\ =A_1A_2\ .\ .\ .\ A_n+\Sigma a_1A_2A_3\ .\ .\ .\ A_n+\Sigma a_1a_2A_3A_4\ .\ .\ .\ A_n\\ +\ .\ .\ .\ +\Sigma a_1a_2\ .\ .\ .\ a_rA_{r+1}A_{r+2}\ .\ .\ .\ A_n\\ +\ .\ .\ .\ +a_1a_2\ .\ .\ .\ a_n \qquad (10).\end{aligned}$$

This so-called special case might itself be made to furnish (7), (8), and (9) as special cases by substituting special values for $A_1, \ldots, A_n, a_1, \ldots, a_n$.

Ex. 1. Distribute and arrange $(x+1)(x-1)(x+2)$.

Here
$$\begin{aligned}\Sigma a_1&=+1-1+2=+2;\\ \Sigma a_1a_2&=(+1)(-1)+(+1)(+2)+(-1)(+2),\\ &=-1+2-2=-1;\\ a_1a_2a_3&=(+1)(-1)(+2)=-2;\end{aligned}$$
hence
$$(x+1)(x-1)(x+2)=x^3+2x^2-x-2.$$

Ex. 2. $(x+b-c)(x+c-a)(x+a-b)$.

$$\begin{aligned}\Sigma a_1&=(b-c)+(c-a)+(a-b)=0;\\ \Sigma a_1a_2&=(c-a)(a-b)+(b-c)(a-b)+(b-c)(c-a),\\ &=\Sigma\{-a^2+a(b+c)-bc\},\\ &=-\Sigma a^2+2\Sigma ab-\Sigma bc,\\ &=-\Sigma a^2+\Sigma ab,\\ a_1a_2a_3&=(b-c)(c-a)(a-b)=-b^2c+bc^2-c^2a+ca^2-a^2b+ab^2.\end{aligned}$$
Therefore
$$\begin{aligned}&(x+b-c)(x+c-a)(x+a-b),\\ &=x^3+(-\Sigma a^2+\Sigma ab)x+(bc^2-b^2c+ca^2-c^2a+ab^2-a^2b).\end{aligned}$$

Ex. 3. $(ax+b-c)(bx+c-a)(cx+a-b)$.

Here
$$\begin{aligned}A_1A_2A_3&=abc,\\ \Sigma a_1A_2A_3&=bc(b-c)+ca(c-a)+ab(a-b),\\ &=b^2c-bc^2+c^2a-ca^2+a^2b-ab^2;\\ \Sigma a_1a_2A_3&=a(c-a)(a-b)+b(a-b)(b-c)+c(b-c)(c-a),\\ &=-\Sigma a^3+\Sigma a^2(b+c)-3abc;\\ a_1a_2a_3&=(b-c)(c-a)(a-b),\\ &=-b^2c+bc^2-c^2a+ca^2-a^2b+ab^2.\end{aligned}$$
Therefore
$$\begin{aligned}&(ax+b-c)(bx+c-a)(cx+a-b)\\ &=abcx^3-(bc^2-b^2c+ca^2-c^2a+ab^2-a^2b)x^2\\ &\quad+(-\Sigma a^3+\Sigma a^2b-3abc)x\\ &\quad+(bc^2-b^2c+ca^2-c^2a+ab^2-a^2b).\end{aligned}$$

**§ 106.** If in the identity (7) above we put $a_1=a_2=a_3=\ .\ .\ .\ =a_n$, each $=a$, we get an important special case which we shall now consider.

Each term of the coefficient $\Sigma a_1a_2\ .\ .\ .\ a_r$ now becomes $aa\ .\ .\ .\ a$ ($r$ factors) $=a^r$. The number of separate terms in the sum $\Sigma a_1a_2\ .\ .\ .\ a_r$ is the number of different ways in which a group of $r$ letters can be selected from the $n$ different letters

$a_1, a_2, \ldots, a_n$; this number, which is evidently a perfectly definite positive integer, known when both $n$ and $r\,(n \nless r)$ are given, we may denote by ${}_nC_r$; it is usually spoken of as the number of $r$-combinations of $n$ different things. The preceding suffix denotes the whole number of things and the succeeding suffix the number selected. Methods for calculating ${}_nC_r$ will be given presently; in the meantime we assume that it is known. It is clear, therefore, that, when $a_1 = a_2 = \ldots = a_n = a$, $\Sigma a_1 a_2 \ldots a_r$ becomes ${}_nC_r a^r$. Hence the identity (7) becomes, since each factor on the left is now $x + a$, and there are $n$ factors—

$$(x+a)^n = x^n + {}_nC_1 x^{n-1}a + {}_nC_2 x^{n-2}a^2 + \ldots + {}_nC_r x^{n-r}a^r + \ldots + {}_nC_n a^n \qquad (11),$$

a result which is called **the Binomial Theorem for a positive integral index.**

Owing to the part they play in the Binomial Theorem, the positive integral numbers

$$1,\ {}_nC_1,\ {}_nC_2,\ {}_nC_3,\ \ldots,\ {}_nC_n \qquad (12)^*$$

are often spoken of as the Binomial Coefficients of the $n$th order. Two remarks are at once obvious from the definition above given of ${}_nC_r$. First, that ${}_nC_n = 1$; and that ${}_nC_1 = n$. Second, that ${}_nC_r = {}_nC_{n-r}$. This appears from the fact that for every selection of $r$ things that we make from $n$ we leave a selection of $n-r$ behind: there are therefore just as many different selections of $n-r$ things as there are of $r$ things. From this last remark it follows that the first of the numbers (12) is equal to the last; the second to the last but one; and so on.

If for $a$ we put $-a$, $+1$, $-1$ successively, we get the following special cases of the Binomial Theorem—

$$(x-a)^n = x^n - {}_nC_1 x^{n-1}a + {}_nC_2 x^{n-2}a^2 + \ldots + (-1)^r {}_nC_r x^{n-r}a^r + \ldots + (-1)^n a^n \qquad (13);$$

$$(x+1)^n = x^n + {}_nC_1 x^{n-1} + {}_nC_2 x^{n-2} + \ldots + {}_nC_r x^{n-r} + \ldots + 1 \qquad (14);$$

$$(x-1)^n = x^n - {}_nC_1 x^{n-1} + {}_nC_2 x^{n-2} + \ldots + (-1)^r {}_nC_r x^{n-r} + \ldots + (-1)^n \qquad (15).$$

**§ 107. Addition Rule for Calculating the Binomial Coefficients.** From the identity $(x+1)^{n+1} \equiv (x+1)(x+1)^n$, we have, by what has been established in last paragraph—

* It is usual to replace the 1 by ${}_nC_0$ (in itself a meaningless symbol), and write the coefficients ${}_nC_0, {}_nC_1, {}_nC_2, \ldots {}_nC_n$.

$$x^{n+1} + {}_{n+1}C_1 x^n + {}_{n+1}C_2 x^{n-1} + \ldots + {}_{n+1}C_r x^{n-r+1}$$
$$+ \ldots + {}_{n+1}C_{n+1}$$
$$\equiv (x+1)(x^n + {}_nC_1 x^{n-1} + {}_nC_2 x^{n-2} + \ldots + {}_nC_r x^{n-r}$$
$$+ \ldots + {}_nC_n),$$
$$\equiv x^{n+1} + {}_nC_1 x^n + {}_nC_2 x^{n-1} + \ldots + {}_nC_r x^{n-r+1} + \ldots$$
$$+ {}_nC_n x$$
$$+ x^n + {}_nC_1 x^{n-1} + \ldots + {}_nC_{r-1} x^{n-r+1} + \ldots$$
$$+ {}_nC_{n-1} x + {}_nC_n$$
$$\equiv x^{n+1} + (1 + {}_nC_1)x^n + ({}_nC_1 + {}_nC_2)x^{n-1} + \ldots$$
$$+ ({}_nC_{r-1} + {}_nC_r)x^{n-r+1} + \ldots$$
$$+ ({}_nC_{n-1} + {}_nC_n)x + {}_nC_n.$$

Hence, by the principle established in § 97, we must have

$$\begin{aligned} {}_{n+1}C_1 &= 1 + {}_nC_1; \\ {}_{n+1}C_2 &= {}_nC_1 + {}_nC_2; \\ &\ldots\ldots\ldots \\ {}_{n+1}C_r &= {}_nC_{r-1} + {}_nC_r; \\ &\ldots\ldots\ldots \\ {}_{n+1}C_n &= {}_nC_{n-1} + {}_nC_n; \\ {}_{n+1}C_{n+1} &= {}_nC_n. \end{aligned} \qquad (16).$$

These equations are merely expressions of the following simple rule :—

*The* r*th Binomial Coefficient of any order is the sum of the* (r − 1)*th and* r*th of the preceding order.*

Since the coefficients of $(x+1)^1$, *i.e.* $x+1$, are 1, 1, it follows by the rule that the coefficients of $(x+1)^2$ are 1, 1 + 1, 1, that is, 1, 2, 1. Hence, again, the coefficients of $(x+1)^3$ are 1, 1 + 2, 2 + 1, 1, that is, 1, 3, 3, 1. Proceeding in this way we rapidly form the following table :—

| $n$ | ${}_nC_0$ | ${}_nC_1$ | ${}_nC_2$ | ${}_nC_3$ | ${}_nC_4$ | ${}_nC_5$ | ${}_nC_6$ | ${}_nC_7$ | ${}_nC_8$ |
|---|---|---|---|---|---|---|---|---|---|
| 1 | 1 | 1 | | | | | | | |
| 2 | 1 | 2 | 1 | | | | | | |
| 3 | 1 | 3 | 3 | 1 | | | | | |
| 4 | 1 | 4 | 6 | 4 | 1 | | | | |
| 5 | 1 | 5 | 10 | 10 | 5 | 1 | | | |
| 6 | 1 | 6 | 15 | 20 | 15 | 6 | 1 | | |
| 7 | 1 | 7 | 21 | 35 | 35 | 21 | 7 | 1 | |
| 8 | 1 | 8 | 28 | 56 | 70 | 56 | 28 | 8 | 1 |

in which each number is formed by adding together the number immediately above and the number immediately above and to the left. The table may be used as a *memoria technica* for recovering the coefficients when they are wanted; in using it for this purpose one starts from the nearest line of coefficients which one happens to recollect.

Ex. 1. Expand $(x-3)^8$ in powers of $x$.

$$\begin{aligned}(x-3)^8 &= x^8 - 8x^7 3 + 28x^6 3^2 - 56x^5 3^3 + 70x^4 3^4 - 56x^3 3^5 + 28x^2 3^6 - 8x3^7 + 3^8\\ &= x^8 - 24x^7 + 252x^6 - 1512x^5 + 5670x^4 - 13608x^3 + 20412x^2\\ &\quad - 17496x + 6561.\end{aligned}$$

As a test of the correctness of this result we may put $x=1$; we then find $256=256$, as it should be.

Ex. 2. Find the term in the expansion of $(2x-3)^7$ which contains $x^4$.

The term is ${}_7C_3(2x)^4(-3)^3 = 35(2x)^4(-3)^3 = -15120x^4$.

Ex. 3. Find the coefficient of $x^5$ in $(1+x^2+3x^4)(x-1)^8$.

$$\begin{aligned}(1+x^2+3x^4)(x-1)^8 &= (1+x^2+3x^4)(1-x)^8\\ &= (1+x^2+3x^4)(1-8x+28x^2-56x^3+70x^4-56x^5+\ldots).\end{aligned}$$

We now calculate the coefficient of $x^5$ in this product, as suggested in § 104, Ex. 5. The work is—

$$\begin{array}{l}1+\ \ 0+\ \ 1+\ \ 0+\ \ 3\\ \underline{-56+70-56+28-\ \ 8}\\ -56+\ \ 0-56+\ \ 0-24=-136.\end{array}$$

Ex. 4. $(x-y)^6(x+y)^6 = (x^2-y^2)^6$,

$$\begin{aligned}&= (x^2)^6 - 6(x^2)^5(y^2) + 15(x^2)^4(y^2)^2 - 20(x^2)^3(y^2)^3 + 15(x^2)^2(y^2)^4 - 6x^2(y^2)^5\\ &\quad + (y^2)^6,\\ &= x^{12} - 6x^{10}y^2 + 15x^8y^4 - 20x^6y^6 + 15x^4y^8 - 6x^2y^{10} + y^{12}.\end{aligned}$$

Ex. 5. Expand $(1+x+x^2+x^3)^3$.

Since $1+x+x^2+x^3 = (1+x) + x^2(1+x) = (1+x)(1+x^2)$, we have

$$\begin{aligned}(1+x+x^2+x^3)^3 &= \{(1+x)(1+x^2)\}^3,\\ &= (1+x)^3(1+x^2)^3,\\ &= (1+3x+3x^2+x^3)(1+3x^2+3x^4+x^6).\end{aligned}$$

We now work out the coefficients for the distribution of this product, thus—

$$\begin{array}{l}1+3+3+\ 1\\ \underline{1+0+3+\ 0+\ 3+\ 0+\ 1}\\ 1+3+3+\ 1\\ \qquad\ \ 3+\ 9+\ 9+\ 3\\ \qquad\qquad\quad 3+\ 9+\ 9+3\\ \qquad\qquad\qquad\qquad \underline{1+3+3+1}\\ 1+3+6+10+12+12+10+6+3+1.\end{array}$$

Hence

$$(1+x+x^2+x^3)^3$$
$$=1+3x+6x^2+10x^3+12x^4+12x^5+10x^6+6x^7+3x^8+x^9.$$

Ex. 6. Expand $(1-2x+3x^2-5x^3+7x^4-9x^5)^3$ as far as $x^3$.

The simplest process is direct multiplication, if we abbreviate by means of detached coefficients and avoid useless work, as follows :—

$$\begin{array}{r}
1-2+\ 3-\ 5\\
1-2+\ 3-\ 5\\
\hline
1-2+\ 3-\ 5\\
-2+\ 4-\ 6\\
+\ 3-\ 6\\
-\ 5\\
\hline
1-4+10-22\\
1-2+\ 3-\ 5\\
\hline
1-4+10-22\\
-2+\ 8-20\\
+\ 3-12\\
-\ 5\\
\hline
1-6+21-59
\end{array}$$

Hence

$$(1-2x+3x^2-5x^3+7x^4-9x^5)^3=1-6x+21x^2-59x^3+\ .\ .\ .$$

Ex. 7. Find the coefficient of $x^5$ in $(1+3x+x^2)^6$.

In solving this example we may conveniently avail ourselves of the principle of association as follows :—

$$(1+3x+x^2)^6\equiv\{(1+3x)+x^2\}^6$$
$$\equiv(1+3x)^6+6(1+3x)^5x^2+15(1+3x)^4x^4+\text{ etc.}$$

Since the terms denoted by "etc." each contain as a factor $x^6$ or some higher power of $x$, we are not concerned with them in determining the coefficient of $x^5$. All that remains, therefore, is to pick out the terms in $x^5$ from the three composite terms of the expansion written down. $(1+3x)^6$ gives $6.3^5x^5=1458x^5$; $6(1+3x)^5x^2$ gives $6.10.3^3x^5=1620x^5$; $15(1+3x)^4x^4$ gives $15.4.3x^5=180x^5$. Hence the coefficient of $x^5$ is $1458+1620+180=3258$.

§ **107a.*** The binomial coefficients of any given order can be calculated successively by means of a **Multiplication and Division Rule**, which may be stated symbolically as follows :—

$$_nC_{r+1}={}_nC_r\times(n-r)\div(r+1) \qquad (17);$$

* This paragraph is due to the suggestion of my former pupil, Mr. J. B. Clark, who uses the demonstration here given in his class-teaching in George Heriot's School.

or, in words, *the* (r + 1)[th] *binomial coefficient of the* n[th] *order may be calculated from the* r[th] *of the same order by multiplying the latter by* n − r *and dividing by* r + 1.

The proof of this is not difficult, if we reflect that ${}_nC_r$ means the number of ways in which we can select a group of $r$ things from $n$ different things, no account being taken of the order of the things in the group.

If we take any particular $r$-group of the $n$ things, and form $(r+1)$-groups by adding to it in succession each of the remaining $n-r$ things, we shall construct $n-r$ different groups of the $n$ things, each group containing $r+1$ things. If we repeat this operation with every one of the ${}_nC_r$ groups of $r$ things, we shall construct ${}_nC_r \times (n-r)$ groups of the $n$ things, each group containing $r+1$ things.

In this way we shall get every possible group of $r+1$ things; but each of these groups will occur more than once. In fact, we can divide a group of $r+1$ things into one thing and a group of $r$ things in $r+1$ ways, viz. by selecting out in turn each of the $r+1$ things. It follows that in the above construction of the $(r+1)$-groups from the $r$-groups each $(r+1)$-group has been formed and counted $r+1$ times. Hence the whole number of different $(r+1)$-groups is ${}_nC_r \times (n-r) \div (r+1)$, which establishes the equation (17).*

By means of (17) we can find an **Expression for ${}_nC_r$ as a Function of n and r.**

In the first place, since the number of groups of $n$ things each consisting of one thing is simply $n$, we have ${}_nC_1 = n = n/1$.

Again, putting $r=1$ in (17), we get ${}_nC_2 = {}_nC_1 \times (n-1) \div 2 = n(n-1)/1.2$.

Putting $r=2$ in (17), we get ${}_nC_3 = {}_nC_2 \times (n-2) \div 3 = n(n-1)(n-2)/1.2.3$.

Proceeding step by step in this way we get finally—

$$ {}_nC_r = \frac{n(n-1)(n-2)\ \ldots\ (n-r+1)}{1.\ 2.\ \ 3\ \ldots\ldots\ r} \qquad (18), $$

which may be reduced to a verbal rule as follows:—

* The reader should verify the accuracy of the above general reasoning by setting down all the 2-combinations of the four letters $a$, $b$, $c$, $d$, forming therefrom in the manner above described the 3-combinations, and thus satisfying himself that ${}_4C_3 = {}_4C_2 \times (4-2) \div 3$.

*To get the number of* r*-combinations of* n *things, write down as denominator the product of the first* r *integers, and over these as numerator write the product of* r *successive integers in descending order beginning with* n.

We can now write **the Binomial Theorem in Newton's Form,** viz.—

$$(x+a)^n = x^n + \frac{n}{1!}x^{n-1}a + \frac{n(n-1)}{2!}x^{n-2}a^2 + \ldots$$
$$+ \frac{n(n-1)(n-2)}{3!}x^{n-3}a^3 + \ldots$$
$$+ \frac{n(n-1)\ldots(n-r+1)}{r!}x^{n-r}a^r + \ldots + a^n \qquad (19),$$

where $1!$, $2!$, $3!$, . . ., $r!$ are used to denote 1, 1.2, 1.2.3, . . ., 1.2.3 . . . $r$ respectively.

Ex. 1. To calculate the binomial coefficients of the 6th order :—

$${}_6C_1=6\ ;\quad {}_6C_2=6\div2\times5=15\ ;\quad {}_6C_3=15\div3\times4=20\ ;$$
$${}_6C_4=20\div4\times3=15\ ;\quad {}_6C_5=15\div5\times2=6\ ;\quad {}_6C_6=6\div6\times1=1.$$

Ex. 2. To calculate ${}_{20}C_7$ and ${}_{20}C_{13}$.

We observe in the first place that, by § 106, ${}_{20}C_{13}={}_{20}C_{20-13}={}_{20}C_7$ ; so that we have only to calculate ${}_{20}C_7$. Now

$${}_{20}C_7=\frac{20.19.18.17.16.15.14}{1.\ 2.\ 3.\ 4.\ 5.\ 6.\ 7},$$
$$=19.17.16.15=77520.$$

§ 108. The following table of **Standard Identities** will be found useful. Such of the results as have not already been demonstrated above may be established by the student himself as an exercise.

$$\left.\begin{aligned}(x+a)(x+b)&=x^2+(a+b)x+ab\ ;\\ (x+a)(x+b)(x+c)&=x^3+(a+b+c)x^2\\ &\quad+(bc+ca+ab)x+abc\ ;\\ \text{and generally}\qquad&\\ (x+a_1)(x+a_2)\ldots(x+a_n)&=x^n+\Sigma a_1x^{n-1}+\Sigma a_1a_2x^{n-2}+\ldots\\ &\quad+\Sigma a_1a_2\ldots a_{n-1}x+a_1a_2\ldots a_n\end{aligned}\right\}\text{(I.)}$$

$$(x \pm y)^2 = x^2 \pm 2xy + y^2 \; ;$$
$$(x \pm y)^3 = x^3 \pm 3x^2y + 3xy^2 \pm y^3 \; ;$$
etc. ;

the numerical coefficients being taken from the following table of binomial coefficients :— (II.)

| $n$ | ${}_nC_0$ | ${}_nC_1$ | ${}_nC_2$ | ${}_nC_3$ | ${}_nC_4$ | ${}_nC_5$ | ${}_nC_6$ | ${}_nC_7$ | ${}_nC_8$ | ${}_nC_9$ | ${}_nC_{10}$ | ${}_nC_{11}$ | ${}_nC_{12}$ |
|---|---|---|---|---|---|---|---|---|---|---|---|---|---|
| 1 | 1 | 1 | | | | | | | | | | | |
| 2 | 1 | 2 | 1 | | | | | | | | | | |
| 3 | 1 | 3 | 3 | 1 | | | | | | | | | |
| 4 | 1 | 4 | 6 | 4 | 1 | | | | | | | | |
| 5 | 1 | 5 | 10 | 10 | 5 | 1 | | | | | | | |
| 6 | 1 | 6 | 15 | 20 | 15 | 6 | 1 | | | | | | |
| 7 | 1 | 7 | 21 | 35 | 35 | 21 | 7 | 1 | | | | | |
| 8 | 1 | 8 | 28 | 56 | 70 | 56 | 28 | 8 | 1 | | | | |
| 9 | 1 | 9 | 36 | 84 | 126 | 126 | 84 | 36 | 9 | 1 | | | |
| 10 | 1 | 10 | 45 | 120 | 210 | 252 | 210 | 120 | 45 | 10 | 1 | | |
| 11 | 1 | 11 | 55 | 165 | 330 | 462 | 462 | 330 | 165 | 55 | 11 | 1 | |
| 12 | 1 | 12 | 66 | 220 | 495 | 792 | 924 | 792 | 495 | 220 | 66 | 12 | 1 |
| | | | | | | | etc. | | | | | | |

$$(x \pm y)^2 \mp 4xy = (x \mp y)^2. \qquad \text{(III.)}$$

$$(x + y)(x - y) = x^2 - y^2 \; ;$$
$$(x \pm y)(x^2 \mp xy + y^2) = x^3 \pm y^3 \; ;$$

and generally

$$(x - y)(x^{n-1} + x^{n-2}y + \ldots + xy^{n-2} + y^{n-1}) = x^n - y^n \; ;$$
$$(x + y)(x^{n-1} - x^{n-2}y + \ldots \mp xy^{n-2} \pm y^{n-1}) = x^n \pm y^n,$$

upper or lower sign according as $n$ is odd or even. (IV.)

$$(x^2 + y^2)(x'^2 + y'^2) = (xx' \mp yy')^2 + (xy' \pm yx')^2 \; ;$$
$$(x^2 - y^2)(x'^2 - y'^2) = (xx' \pm yy')^2 - (xy' \pm yx')^2 \; ;$$
$$(x^2 + y^2 + z^2)(x'^2 + y'^2 + z'^2) = (xx' + yy' + zz')^2 + (yz' - y'z)^2 + (zx' - z'x)^2 + (xy' - x'y)^2 \; ;$$
$$(x^2 + y^2 + z^2 + u^2)(x'^2 + y'^2 + z'^2 + u'^2) = (xx' + yy' + zz' + uu')^2 + (xy' - yx' + zu' - uz')^2 + (xz' - yu' - zx' + uy')^2 + (xu' + yz' - zy' - ux')^2$$

(V.)*

$$(x^2 + xy + y^2)(x^2 - xy + y^2) = x^4 + x^2y^2 + y^4. \qquad \text{(VI.)}$$

* These identities furnish, *inter alia*, proofs of a series of propositions in the theory of numbers, of which the following is typical :—If each of two integers be the sum of two squares, their product can be exhibited in two ways as the sum of two integral squares.

$$\left.\begin{aligned}&(a+b+c+d)^2=a^2+b^2+c^2+d^2+2ab+2ac+2ad\\&\qquad+2bc+2bd+2cd;\\&\text{and generally}\\&(a_1+a_2+\ldots+a_n)^2=\text{sum of squares of } a_1, a_2, \ldots, a_n\\&\quad+\text{twice sum of all partial products two and two.}\end{aligned}\right\}\text{(VII.)}$$

$$\left.\begin{aligned}(a+b+c)^3&=a^3+b^3+c^3+3b^2c+3bc^2+3c^2a+3ca^2+3a^2b\\&\quad+3ab^2+6abc\\&=a^3+b^3+c^3+3bc(b+c)+3ca(c+a)\\&\quad+3ab(a+b)+6abc.\end{aligned}\right\}\text{(VIII.)}$$

$$(a+b+c)(a^2+b^2+c^2-bc-ca-ab)=a^3+b^3+c^3-3abc. \quad \text{(IX.)}$$

$$\left.\begin{aligned}(b-c)(c-a)(a-b)&=-a^2(b-c)-b^2(c-a)-c^2(a-b),\\&=a(b^2-c^2)+b(c^2-a^2)+c(a^2-b^2),\\&=-bc(b-c)-ca(c-a)-ab(a-b),\\&=+bc^2-b^2c+ca^2-c^2a+ab^2-a^2b.\end{aligned}\right\}\text{(X.)}$$

$$\left.\begin{aligned}(b+c)(c+a)(a+b)&=a^2(b+c)+b^2(c+a)+c^2(a+b)+2abc,\\&=bc(b+c)+ca(c+a)+ab(a+b)+2abc,\\&=bc^2+b^2c+ca^2+c^2a+ab^2+a^2b+2abc\end{aligned}\right\}\text{(XI.)}$$

$$\left.\begin{aligned}(a+b+c)(a^2+b^2+c^2)&=bc(b+c)+ca(c+a)+ab(a+b)\\&\quad+a^3+b^3+c^3.\end{aligned}\right\}\text{(XII.)}$$

$$\left.\begin{aligned}(a+b+c)(bc+ca+ab)&=a^2(b+c)+b^2(c+a)+c^2(a+b)\\&\quad+3abc.\end{aligned}\right\}\text{(XIII.)}$$

$$\left.\begin{aligned}(b+c-a)(c+a-b)(a+b-c)&=a^2(b+c)+b^2(c+a)\\&\quad+c^2(a+b)-a^3-b^3-c^3-2abc.\end{aligned}\right\}\text{(XIV.)}$$

$$\left.\begin{aligned}(a+b+c)(-a+b+c)(a-b+c)(a+b-c)&=2b^2c^2+2c^2a^2\\&\quad+2a^2b^2-a^4-b^4-c^4.\end{aligned}\right\}\text{(XV.)*}$$

$$\left.\begin{aligned}(b-c)+(c-a)+(a-b)&=0;\\a(b-c)+b(c-a)+c(a-b)&=0;\\(b+c)(b-c)+(c+a)(c-a)+(a+b)(a-b)&=0.\end{aligned}\right\}\text{(XVI.)}$$

### EXERCISES XXVI.

The functions set down without further direction are to be simplified by distributing and arranging according to powers of $x$.

1. $(x^4-3x^2+1)(x^4+x^3-4x^2+x+1)$.
2. $(2x^3-3x^2y+3xy^2-2y^3)(x^2-2xy+y^2)$.
3. $(x^4/4-x^2+1)/(x^2/2-x+1)$.

---

* Important in connection with Hero's formula for the area of a plane triangle.

4. $(x^3+4x+1)(x^3-4x+4)(x^2+2x+1)$.

5. Find the coefficient of $x^6$ in $(5x^7+3x^6-2x^4+8x^2+5)\times(3x^6-\frac{2}{7}x^4-3x^3+7x+2)$.

6. $(a^2x^2-ax-1)(x^2/a^2-x/a+1)$.

7. Show that $a^4+(a^2+ab+b^2)^2=(a^2+b^2)\{a^2+(a+b)^2\}$.

8. $(x+1+1/x)(x-1+1/x)(x^2-1+1/x^2)$.

9. $(1+ax)(1-ax)(1+x/a)(1-x/a)$.

10. $\{(b+c-a)x^2+(c+a-b)x+(a+b-c)\}\{x^2-x+1\}$.

11. $(x+a)(x^2+ax+a^2)(x-a)(x^2-ax+a^2)$.

12. $(x^2+ax+a^2)(x^2-ax+a^2)(x^4-a^2x^2+a^4)$.

13. $\{(x-1)^3(x^2+x+1)^3+(x+1)^3(x^2-x+1)^3\}\{2x^6-6x^3+1\}$.

14. Find the coefficient of $x^4$ in $(1-x+x^2)^3$.

15. $(x^2+x+1)^3+(x^2-x+1)^3$.

16. $(-px^2+qx+r)(px^2-qx+r)(px^2+qx-r)$.

17. $\{(a+b)x^2-abxy+(a-b)y^2\}\{(a-b)x^2+abxy+(a+b)y^2\}$.

18. $(1+x)^2(1-x+x^2-x^3)^2$.

19. $(1+x+x^2+x^3)^3+(1-x+x^2-x^3)^3$.

20. Find the coefficient of $x^3$ in the product $(x-1)(x-2)(x-3)(x-4)(x-5)$.

21. Find the coefficient of $x$ in $(x+b+c-a)(x+c+a-b)(x+a+b-c)$.

22. Find the coefficient of $x^3$ and $x^4$ in $(1-ax+bx^2)^3+(1+ax-bx^2)^3$.

23. Find the coefficient of $x^2$ in $(x^2+ax+b)^3+(x^2+ax-b)^3+(x^2-ax+b)^3+(x^2-ax-b)^3$.

24. Find the coefficient of $x^2$ in $(1-ax+bx^2)^2(1-bx+ax^2)^2$.

25. Show that $x^3-4(x-1)^3+6(x-2)^3-4(x-3)^3+(x-4)^3\equiv 0$.

26. Show that $x+21(x-1)x(x+1)+14(x-2)(x-1)x(x+1)(x+2)+(x-3)(x-2)(x-1)x(x+1)(x+2)(x+3)=x^7$.

27. Show that $\{x^3-y^3+6xy(2y+x)\}^2-\{y^3-x^3+6xy(2x+y)\}^2\equiv 36xy(x+y)(x-y)^3$.

28. If $A(x+1)(x-2)+B(x-2)(x-3)\equiv C(x-4)+1$, find A, B, C.

29. Determine L, M, N so that $2x^2+3x+2=L(x-1)(x-2)+M(x-2)(x-3)+N(x-3)(x-1)$.

30. Find $l$, $m$, $n$ so that $l(x-b)(x-c)+m(x-c)(x-a)+n(x-a)(x-b)\equiv px^2+qx+r$.

31. Determine A, B, C, D so that $x^3-5x+7\equiv A+B(x-1)+C(x-1)^2+D(x-1)^3$.

32. Determine numerical values for A, B, C, D so that $2x^3-13x^2+26x-4\equiv A(x-1)(x-2)(x-5)+B(x-1)(x-2)+C(x-1)+D$.

33. Express $3x^3-4x^2+6x-3$ in the form $a+b(x-1)+c(x-1)(x+1)+d(x-1)(x+1)(x+3)$, where $a$, $b$, $c$, $d$ are constants.

34. If $f(z)=(z^4-1)/(z-1)$, and $z=x^2+x+3$, express $f(z)$ as a function of $x+1$.

## EXERCISES XXVII.

Where no other indication is given, the functions set down are to be expanded and arranged according to powers of $x$.

1. $(x+2)^5$.

2. Find the coefficient of $x^4$ in $(1-x)^8$.

3. Find the coefficient of $x^3$ in $(1+2x)^8+(1+2x)^9$.

4. Find the coefficient of $x^5$ in $(1-x)^{10}$.

5. $(x-2)^{11}+(x+2)^{11}$. 6. Find the coefficient of $x^4$ in $(1-x^2)^{15}$.

7. $(3x-2)^5(3x+2)^5$. 8. $(x^2-x+1)^5$.

9. $(a^2+b^2)^4(a^4-a^2b^2+b^4)^3$.

10. $(1-x)^5(1+x)^6$. 11. $(x^2-1)^5(x^4+x^2+1)^5$.

12. Find the coefficient of $x^{18}$ in $(x-1)^5(x^2-x+1)^5(x+1)^5(x^2+x+1)^5$.

13. $(x-1)^4(x+1)^8$.

14. $(x+y)^2(x^2-xy+y^2)^2(x^2+xy+y^2)^2(x-y)^2$.

15. Find the coefficient of $x^3$ in $(x-1/x)^7$.

16. Find the coefficient of $x^2$ in $(x^3-1/x)^{10}$.

17. Find the coefficient of $x^4$ in $(1-x)^8(1+x+3x^2)$.

18. Find the coefficients of $x^3$ and $x^5$ in $(1-x+x^2)^5-(1+x-x^2)^5$.

19. Determine $p$ and $q$ so that the terms containing $x$ and $x^2$ may disappear from the distributed product of $(1+px+qx^2+x^3)(1+x)^4$.

20. Expand $(1-x+x^2-x^3+x^4)^3$ as far as $x^3$.

21. Find the coefficient of $x^5$ in $(1-x+x^2-x^3)^4$.

22. Find the coefficient of $x^5$ in $(1+2x+3x^2)^7$.

23. $(x-y)^2(x+y)^4-(x+y)^2(x-y)^4$.

24. $x\{1+(1-x)+(1-x)^2+(1-x)^3+(1-x)^4\}\equiv 1-(1-x)^5$.

25. $(x+y)^4\equiv 2(x^2+y^2)(x+y)^2-(x^2-y^2)^2$; and deduce $(x^2-y^2)^4 \equiv \{2(x^2+y^2)^2-(x^2-y^2)^2\}^2-16x^2y^2(x^2+y^2)^2$.

26. $a^5+b^5\equiv(a+b)^5-5(a+b)^3ab+5(a+b)a^2b^2$. Deduce the corresponding equivalents of $a^5-b^5$, $(x-y)^5+(y-z)^5$, $(x-y)^5-(y-z)^5$.

27. $(x-y)(x^{n-1}+x^{n-2}y+\ .\ .\ .\ +xy^{n-2}+y^{n-1})$.

28. $(x+y)(x^{n-1}-x^{n-2}y+\ .\ .\ .\ -xy^{n-2}+y^{n-1})$ $n$ odd.

29. $(x+y)(x^{n-1}-x^{n-2}y+\ .\ .\ .\ +xy^{n-2}-y^{n-1})$ $n$ even.

30. $(x-y)^2(x^n+x^{n-1}y+\ .\ .\ .\ +xy^{n-1}+y^n)$.

31. Show by means of the identity $(1+x)^{n+3}=(1+x)^n(1+x)^3$ that ${}_{n+3}C_r={}_nC_r+3{}_nC_{r-1}+3{}_nC_{r-2}+{}_nC_{r-3}$.

32. Assign a value of $n$ such that $(1+1/1000)^n>100,000$.

33. The excess of the product of any four consecutive odd integers over 1 is always a multiple of 8.

34. Show that, if $n$ be any integer, $n^2-n+1$ is an odd integer.

35. If the sum of the squares of any three consecutive odd integers be increased by 1, show that the resulting integer is always a multiple of 12.

36. A number of six significant digits is multiplied by another of five, give an estimate of the utmost effect on the product by increasing the last digit in each of the factors by 5.

# CHAPTER XII

## THE DIVISION TRANSFORMATION

§ **109.** The quotient of one integral function of $x$ (dividend) by another (divisor) is a function of $x$ which would be classified under the scheme of § 9 as rational. By the mutual relation of multiplication and division this function has the fundamental property that if we multiply it by the divisor we reproduce the dividend. Thus, for example, $(x^2+1)/(x+1)$, or, as it is variously written, $(x^2+1)\div(x+1)$, $(x^2+1):(x+1)$, $\dfrac{x^2+1}{x+1}$, has by definition the property

$$\{(x^2+1)/(x+1)\}(x+1)\equiv x^2+1.$$

There are two distinct cases to consider :—

1. *The quotient may be an integral function of* x, *or transformable into an integral function of* x. For example, since $(x-1)(x+1)\equiv x^2-1$, it follows that $(x^2-1)/(x+1)\equiv x-1$.

In this case the dividend is said to be **exactly divisible** *by the divisor. In other words, an integral function,* A, *of* x *is said to be exactly divisible by another integral function,* D, *of* x *when there exists an integral function,* Q, *of* x *such that* $A\equiv QD$.

It will be observed that the algebraic notion of exact divisibility depends entirely on the notion of algebraic form, and has nothing whatever to do with arithmetical or absolute value either of the variable or of the function itself. Forgetfulness of this point is a frequent source of confusion and error.

Since the dividend is the product of the quotient and divisor, if the quotient be integral, it follows by the law of degree, § 97, that the degree of the dividend is equal to the sum of the degrees of the quotient and divisor. Hence—

*When the quotient of two integral functions of* x *is an integral function of* x, *its degree is the excess of the degree of the dividend over the degree of the divisor.* Thus, for example, the degree of $(x^2-1)/(x-1)$ is $2-1=1$.

2. *It may be impossible to transform the quotient of two integral functions of* x *into an integral function of* x ; the quotient is then said to be *fractional*, or, if emphasis is required, *essentially fractional.*

*Clearly this is the case when the degree of the dividend is less than the degree of the divisor.* For, as we have seen, when the division is exact, the degree of the quotient, which is positive or zero, is the excess of the degree of the dividend over the degree of the divisor ; in every case of exact divisibility the degree of the dividend must therefore be not less than the degree of the divisor.

The same may happen in other cases ; indeed, it is the exception and not the rule that the quotient is integral.

A quotient of two integral functions of $x$ in which the degree of the dividend is less than that of the divisor is called a **Proper Fraction** ; a non-integral quotient of two integral functions of $x$ in which the degree of the dividend is not less than the degree of the divisor is called an **Improper Fraction.**

It is clear from the foregoing discussion that *an integral function of* x *cannot be equal in the identical sense to an essentially fractional function of* x ; for the definition of an essentially fractional function is simply that it cannot be transformed into an integral function.

Ex. To show that $(x^2+1)/(x+1)$ is essentially fractional.

Since $$x^2+1\equiv x(x+1)-(x+1)+2,$$

we have $$(x^2+1)/(x+1)\equiv\{x(x+1)-(x+1)+2\}/(x+1)$$
$$\equiv x-1+2/(x+1).$$

If now $(x^2+1)/(x+1)$ were integral, say $\equiv \mathrm{A}x+\mathrm{B}$, then we should have

$$(\mathrm{A}x+\mathrm{B})-(x-1)\equiv 2/(x+1).$$

But $2/(x+1)$ is essentially fractional, since the degree in $x$ of its numerator is 0 and of its denominator 1. We should therefore have an integral function of $x$ identically equal to an essentially fractional function, which is impossible.

§ 110. **The Division Transformation.**—The example of last

paragraph should be carefully studied, as it illustrates much of what follows. Among other things it gives a special case of the following theorem :—

*A quotient of two integral functions of* x, *in which the degree of the dividend is not less than the degree of the divisor, can always, and that in one way only, be transformed into the sum of an integral function of* x *and a proper fraction whose denominator is the divisor of the given quotient. The integral part of the transformed function is called the* **Integral Quotient** ; *its degree is the excess of the degree of the dividend over the degree of the divisor. The numerator of the fractional part of the transformed function is called the* **Remainder** *of the given dividend with respect to the given divisor. The degree of the remainder is therefore less than the degree of the divisor.*

Ex. For the quotient $(x^2+1)/(x+1)$ the integral quotient is $x-1$ and the remainder is 2, as appears by the work at the end of § 109.

This transformation may be called the **Division Transformation.** We shall establish the possibility of the transformation by carefully discussing a particular case ; and finally, prove that it can be made in one way only.

Consider the quotient $A_6/D_4$, where

$$A_6 = 8x^6 + 8x^5 - 20x^4 + 40x^3 - 50x^2 + 30x - 10,$$
$$D_4 = 2x^4 + 3x^3 - 4x^2 + 6x - 8,$$

the suffixes in $A_6$, $D_4$, etc., being used partly for distinction, partly to indicate the degrees of the integral functions which these letters represent. Multiply the divisor $D_4$ by such a term as will make its highest term identical with the highest term of the dividend ; in other words, by the quotient of the highest term of the dividend by the highest term of the divisor (that is, multiply $D_4$ by $8x^6/2x^4 = 4x^2$), and subtract the result from the dividend $A_6$. We have

$$\begin{array}{rl} A_6 = & 8x^6 + 8x^5 - 20x^4 + 40x^3 - 50x^2 + 30x - 10 \\ 4x^2D_4 = & 8x^6 + 12x^5 - 16x^4 + 24x^3 - 32x^2 \\ \hline A_6 - 4x^2D_4 = & \quad - 4x^5 - 4x^4 + 16x^3 - 18x^2 + 30x - 10 \\ & = A_5 \text{ say ;} \end{array}$$

therefore
$$A_6 = 4x^2D_4 + A_5 \qquad (1).$$

Repeat the same process with the residue $A_5$ in place of $A_6$, and we have

$$\begin{array}{rl} A_5 = & -4x^5 - 4x^4 + 16x^3 - 18x^2 + 30x - 10 \\ -2xD_4 = & -4x^5 - 6x^4 + 8x^3 - 12x^2 + 16x \\ \hline A_5 + 2xD_4 = & 2x^4 + 8x^3 - 6x^2 + 14x - 10 \\ & = A_4 \text{ say;} \end{array}$$

therefore
$$A_5 = -2xD_4 + A_4 \qquad (2).$$

And again with $A_4$—

$$\begin{array}{rl} A_4 = & 2x^4 + 8x^3 - 6x^2 + 14x - 10 \\ 1 \times D_4 = & 2x^4 + 3x^3 - 4x^2 + 6x - 8 \\ \hline A_4 - D_4 = & 5x^3 - 2x^2 + 8x - 2 \\ & = A_3 \text{ say;} \end{array}$$

therefore
$$A_4 = D_4 + A_3 \qquad (3).$$

Here the process must stop, unless we agree to admit fractional multipliers of $D_4$; for the quotient of the highest term of $A_3$ by the highest term of $D_4$ is $5x^3/2x^4$—that is, $\frac{5}{2}/x$, which is a fractional function of $x$. Such a continuation of the process does not concern us now.

From (1) we have

$$A_6 = 4x^2D_4 + A_5 \qquad (4);$$

and, using (2) to replace $A_5$—

$$A_6 = 4x^2D_4 - 2xD_4 + A_4 \qquad (5);$$

and finally, using (3)—

$$\begin{aligned} A_6 &= 4x^2D_4 - 2xD_4 + D_4 + A_3, \\ &= (4x^2 - 2x + 1)D_4 + A_3 \qquad (6); \end{aligned}$$

Hence
$$\begin{aligned} \frac{A_6}{D_4} &= \frac{(4x^2 - 2x + 1)D_4 + A_3}{D_4}, \\ &= 4x^2 - 2x + 1 + \frac{A_3}{D_4}; \end{aligned}$$

or, replacing the capital letters by the functions they represent—

$$\begin{aligned} &\frac{8x^6 + 8x^5 - 20x^4 + 40x^3 - 50x^2 + 30x - 10}{2x^4 + 3x^3 - 4x^2 + 6x - 8} \\ &= 4x^2 - 2x + 1 + \frac{5x^3 - 2x^2 + 8x - 2}{2x^4 + 3x^3 - 4x^2 + 6x - 8} \qquad (7). \end{aligned}$$

Since $6 - 4 = 2$, it will be seen that we have established the above theorem for this special case. It so happens that the degrees of the residues $A_5$, $A_4$, $A_3$ diminish at each operation by unity only; but the student will easily see that the diminution might happen to be more rapid; and, in particular, that the degree of the first residue whose degree falls under that of the divisor might happen to be less than the degree of the divisor by more than unity. But none of these possibilities will affect the proof in any way, and the process is obviously applicable in any case whatever where the degree of the dividend is not less than that of the divisor.

The work may be arranged as follows:—

$$\begin{array}{rrrrrrr|l}
8x^6 + & 8x^5 - & 20x^4 + & 40x^3 - & 50x^2 + & 30x - & 10 & 2x^4 + 3x^3 - 4x^2 + 6x - 8 \\
\underline{8x^6 +} & \underline{12x^5 -} & \underline{16x^4 +} & \underline{24x^3 -} & \underline{32x^2} & & & \overline{4x^2 - 2x + 1} \\
 & - 4x^5 - & 4x^4 + & 16x^3 - & 18x^2 + & 30x - & 10 & \\
 & \underline{- 4x^5 -} & \underline{6x^4 +} & \underline{8x^3 -} & \underline{12x^2 +} & \underline{16x} & & \\
 & & 2x^4 + & 8x^3 - & 6x^2 + & 14x - & 10 & \\
 & & \underline{2x^4 +} & \underline{3x^3 -} & \underline{4x^2 +} & \underline{6x -} & \underline{8} & \\
 & & & 5x^3 - & 2x^2 + & 8x - & 2 &
\end{array}$$

Or, observing that the term $-10$ is not wanted till the last operation, and therefore need not be taken down from the upper line until that stage is reached, and observing further that the method of detached coefficients is clearly applicable here just as in multiplication, we may arrange the whole thus:—

$$\begin{array}{rrrrrrr|l}
8 + & 8 - & 20 + & 40 - & 50 + & 30 - & 10 & 2 + 3 - 4 + 6 - 8 \\
\underline{8 +} & \underline{12 -} & \underline{16 +} & \underline{24 -} & \underline{32} & & & \overline{4 - 2 + 1} \\
 & - 4 - & 4 + & 16 - & 18 + & 30 & & \\
 & \underline{- 4 -} & \underline{6 +} & \underline{8 -} & \underline{12 +} & \underline{16} & & \\
 & & 2 + & 8 - & 6 + & 14 - & 10 & \\
 & & \underline{2 +} & \underline{3 -} & \underline{4 +} & \underline{6 -} & \underline{8} & \\
 & & & 5 - & 2 + & 8 - & 2 &
\end{array}$$

Therefore—

$$\text{Integral quotient} = 4x^2 - 2x + 1\,;$$
$$\text{Remainder} = 5x^3 - 2x^2 + 8x - 2.$$

The process may be verbally described as follows:—

*Arrange both dividend and divisor according to descending powers of* x, *filling in missing powers with zero coefficients. Find the quotient of the highest term of the dividend by the highest term of the divisor; the result is the highest term of the "integral quotient."*

*Multiply the divisor by the term thus obtained, and subtract the result from the dividend, taking down only one term to the right beyond those affected by the subtraction; the result thus obtained will be less in degree than the dividend by one at least. Divide the highest term of this result by the highest of the divisor; the result is the second term of the "integral quotient."*

*Multiply the divisor by the new term just obtained, and subtract, etc., as before.*

*The process continues until the result after the last subtraction is less in degree than the divisor; this last result is the remainder as above defined.*

**§ 111.** We shall now show that *the division transformation is unique;* in other words, that the integral quotient and remainder for a given quotient as above defined are perfectly definite functions.* If possible, let the quotient $A/D$ be transformable in two ways into the sum of an integral function and a proper fraction whose denominator is $D$; then we should have $A/D \equiv Q + R/D$ and $A/D \equiv Q' + R'/D$, where $Q$, $Q'$, $R$, $R'$ are all integral functions of $x$, and the degrees of $R$ and $R'$ are each less than the degree of $D$. Hence, we must have $Q + R/D \equiv Q' + R'/D$. From this last identity we should have $Q - Q' \equiv (R' - R)/D$. Now, $Q$ and $Q'$ being both integral functions of $x$, $Q - Q'$ is an integral function of $x$; also $R' - R$ is an integral function of $x$, whose degree is not higher than the degree either of $R$ or of $R'$ (whichever is the higher). Hence the degree of $R' - R$ is less than the degree of $D$; and $(R' - R)/D$ is essentially fractional. But an integral function cannot be identically equal to an essentially fractional function. We are therefore driven to the conclusion that $Q - Q' \equiv 0$, and $R' - R \equiv 0$—that is, $Q = Q'$ and $R = R'$; in other words, the two transformations which we supposed to be different must be the same.

It must be carefully noticed that all that has been said supposes that a certain operand $x$ has been chosen as variable. If the dividend and divisor are functions of more than one variable, there will, in general, be a different integral quotient and corresponding remainder with respect to each of the variables (see Ex. 3, § 112 below).

* It is one of the commonest of logical errors to assume that, because a particular process has led to a certain result, therefore no other result is possible. Thus it does not follow from the first part of § 109 that every method of transforming $A_6/D_4$ into the sum of an integral function and a proper fraction whose denominator is $D_4$, will lead to the same integral function and the same proper fraction.

**§ 112.** It is now easy to give the necessary and sufficient criterion that any given integral function of $x$, say A, be exactly divisible by any other given integral function of $x$, say D.

In the first place, if the degree of A in $x$ be less than the degree of D in $x$, then, as already shown, A cannot be exactly divisible by D, unless, of course, A be identically zero, *i.e.* unless every coefficient of A, including the absolute term, vanish.

If the degree of A be not less than the degree of D, we can, always, by the division transformation put A/D into the form Q + R/D, where Q is integral and the degree of R is less than that of D. Hence we must have

$$A/D - Q = R/D.$$

Hence, if A/D be integral, A/D − Q must be integral, and R/D must be integral. Now, the degree of R being less than that of D, R/D cannot be integral, unless $R \equiv 0$. Conversely, it is obvious that, if $R \equiv 0$, then A/D − Q is integral, and therefore A/D is integral.

Therefore *the necessary and sufficient condition that any integral function of* x *be exactly divisible by another, is that the remainder of the former with respect to the latter shall vanish identically.*

Cor. Since there are $n$ coefficients in a function of the $(n-1)$th degree, it follows that, *in general,* n *conditions must be satisfied in order that any given integral function of* x *may be exactly divisible by a given integral function of* x *of the* n*th degree.*

Ex. 1. Transform the quotient

$$F \equiv \frac{abx^5 + (a^2+b^2)x^4 + (b^2+2ab)x^3 + (b^2+2ab)x^2 + (a^2+b^2)x + ab}{bx^2+ax+b}.$$

The calculation may be arranged as follows :—

$$\begin{array}{llllll|l}
ab & +(a^2+b^2) & +(b^2+2ab) & +(b^2+2ab) & +(a^2+b^2) & +ab & b+a+b \\
ab+a^2 & + ab & & & & & \overline{a+b+b+a} \\
\hline
 & b^2 & +(b^2+ab) & +(b^2+2ab) & & & \\
 & b^2 & + ab & + b^2 & & & \\
\hline
 & & b^2 & + 2ab & +(a^2+b^2) & & \\
 & & b^2 & + ab & + b^2 & & \\
\hline
 & & & ab & + a^2 & +ab & \\
 & & & ab & + a^2 & +ab & \\
\hline
 & & & & 0 & + 0. & 
\end{array}$$

The remainder vanishes identically ; and we have

$$F \equiv ax^3 + bx^2 + bx + a.$$

Ex. 2. Find the remainder when $(x^2+5x+2)^3$ is divided by $x^2+2x+3$.

Denoting $x^2+2x+3$ for the moment by D, we have $x^2+5x+2 = x^2+2x+3+3x-1 = \mathrm{D}+3x-1$.

Hence

$$(x^2+5x+2)^3 = (\mathrm{D}+3x-1)^3 = \mathrm{D}^3+3\mathrm{D}^2(3x-1)+3\mathrm{D}(3x-1)^2+(3x-1)^3.$$

Therefore

$$(x^2+5x+2)^3/\mathrm{D} = \mathrm{D}^2+3\mathrm{D}(3x-1)+3(3x-1)^2+(3x-1)^3/\mathrm{D}.$$

The proper fractional part of $(x^2+5x+2)^3/\mathrm{D}$ is therefore the same as the proper fractional part of $(3x-1)^3/\mathrm{D}$; hence the remainder when $(3x-1)^3$, *i.e.* $27x^3-27x^2+9x-1$, is divided by D is the same as the remainder when $(x^2+5x+2)^3$ is divided by D. The calculation for the former remainder is

$$\begin{array}{rrrr|l}
27 & -27+ & 9 & -1 & 1+2+3 \\
27 & +54+ & 81 & & 27-81 \\ \hline
 & -81- & 72 & -1 & \\
 & -81- & 162 & -243 & \\ \hline
 & & 90 & +242. &
\end{array}$$

The remainder required is therefore $90x+242$.

It will be a good arithmetical exercise for the beginner to verify this result by calculating out the distribution of $(x^2+5x+2)^3$ and then dividing by $x^2+2x+3$, using detached coefficients throughout.

Ex. 3. Transform the quotient

$$(x^3+px^2y+qxy^2+2y^3)/(x^2+3xy+y^2),$$

1st, taking $x$ as variable, 2nd, taking $y$ as variable, and in each case determine numerical values for $p$ and $q$, so that the quotient may be integral.

If we take $x$ as variable, the calculation runs

$$\begin{array}{rrrr|l}
1 & +p+ & & q+2 & 1+3+1 \\
1 & +3+ & & 1 & 1+(p-3) \\ \hline
 & (p-3)+ & (q-1) & +2 & \\
 & (p-3)+ & 3(p-3) & +(p-3) & \\ \hline
 & & (q-3p+8) & +(5-p). &
\end{array}$$

Hence

$$\frac{x^3+px^2y+qxy^2+2y^3}{x^2+3xy+y^2} = x+(p-3)y+\frac{(q-3p+8)xy^2+(5-p)y^3}{x^2+3xy+y^2}.$$

If we take $y$ as variable, we arrange both dividend and divisor according to descending powers of $y$. The calculation then runs

$$\begin{array}{rrrr|l}
2 & +q+ & & p+1 & 1+3+1 \\
2 & +6+ & & 2 & 2+(q-6) \\ \hline
 & (q-6)+ & (p-2) & +1 & \\
 & (q-6)+ & 3(q-6) & +(q-6) & \\ \hline
 & & (p-3q+16) & +(7-q). &
\end{array}$$

Therefore

$$\frac{x^3+px^2y+qxy^2+2y^3}{x^2+3xy+y^2}=2y+(q-6)x+\frac{(p-3q+16)yx^2+(7-q)x^3}{x^2+3xy+y^2}.$$

In order that the quotient may be integral the remainder must vanish identically. Taking the first transformation, this gives $q-3p+8=0$, $5-p=0$; from these we get $p=5$, $q=7$; so that

$$(x^3+5x^2y+7xy^2+2y^3)/(x^2+3xy+y^2)=x+2y.$$

If we take the second transformation, the conditions for the evanescence of the remainder are $p-3q+16=0$, $7-q=0$, which give $p=5$ and $q=7$, as before. In so far, therefore, as finding the condition for the integrality of the quotient is concerned, it is indifferent in the present case whether we take $x$ or $y$ as variable. The reader ought to be able to convince himself *a priori* that this must be so.

Ex. 4. Transform $\dfrac{(x-a)^4+3(x-a)^3+(x-a)^2-3(x-a)-2}{(x-a)^3+(x-a)^2-(x-a)-1}$.

Put $\xi$ for $x-a$; and we have $(\xi^4+3\xi^3+\xi^2-3\xi-2)/(\xi^3+\xi^2-\xi-1)$. If we take $\xi$ as variable, the calculation for the transformation of this quotient is

$$\begin{array}{r|l} 1+3+1-3-2 & 1+1-1-1 \\ \underline{1+1-1-1\phantom{-2}} & \overline{1+2} \\ 2+2-2-2 & \\ \underline{2+2-2-2} & \\ 0+0+0 & \end{array}$$

Hence $(\xi^4+3\xi^3+\xi^2-3\xi-2)/(\xi^3+\xi^2-\xi-1)=\xi+2$.

Replacing $\xi$ by $x-a$, we get

$$\frac{(x-a)^4+3(x-a)^3+(x-a)^2-3(x-a)-2}{(x-a)^3+(x-a)^2-(x-a)-1}=x-a+2.$$

## EXERCISES XXVIII.

A quotient set down without further direction is to be transformed into an integral function, or into the sum of an integral function and a proper fraction.

1. $(3x^4-4x^3+2x^2-3)/(x-3)$.
2. $(x^4-x^3-x+1)/(x^2+x+1)$.
3. $x^7/(x^2-x+1)$.
4. $(2x^5+3x^4-12x^3+15x^2-11x+3)/(2x^2-3x+1)$.
5. $(x^3+6x^2+3x+2)/(x-1)^2$.
6. $(x^9-2x^8+2x^6-x^5-x^3+x^2+3x-4)/(x^2-2x+1)$.
7. $(15x^6+16x^5+8x^4-9x^3-7x^2+19x-42)/(5x^2+2x-7)$.
8. $(x^7+x^6+x+1)/(x^3+x^2+x+1)$.
9. $(2x^6-7x^5+18x^4-33x^3+32x^2-36x+16)/(2x^3-x^2+3x-2)$.
10. $(x^6-3x^4+3x^2-1)/(x^2-1)(x-1)$.
11. Express $(x^4+x^3+2x^2+3x+4)/(x^2+x+1)+(x^4-x^3+2x^2-3x+4)$

$/(x^2-x+1)$ as the sum of an integral function and a proper fractional function of $x$.

**12.** $(x^6+\frac{1}{4}x^5+\frac{3}{8}x^4+1)/(x^2+\frac{1}{3}x+2)$.

**13.** $(x^6+x^2+1)/(x^3+2x^2+\frac{3}{2}x+6)$.

**14.** $(x^{16}+x^8+1)/(x^8+x^4+1)$.

**15.** $\{(x-a)^{16}+(x-a)^8+1\}/\{(x-a)^8+(x-a)^4+1\}$.

**16.** Find the remainder when $x^4-3yx^3+2y^2x^2-3y^3x+y^4$ is divided by $x+2y$, first when $x$ is regarded as the variable, second when $y$ is regarded as the variable.

**17.** Find the remainders for $(x^5-3x^4y+5x^3y^2-5x^2y^3+3xy^4-y^5)/(x^2-3xy+y^2)$ when $x$ and $y$ respectively are regarded as the variable.

**18.** $(x^8-3025x^2+8820x-5796)/(x-1)(x-2)(x-3)(x-4)$.

**19.** $(x^3-4x^2-x+12)(x^3-6x+4)/(x^2-5x+6)$.

**20.** $(x^5+bx^3+6x+1)/(x^2+ax+3)$.

**21.** $(y^2z+yz^2+z^2x+zx^2+xy^2+x^2y+3xyz)/(x+y)$, the variable being $x$.

**22.** $(x^3+x^2y-x^2z+y^2z-yz^2+xyz)/(x^2+yz)$.

**23.** $\{(x+y+z)^3+(x-y+z)^3\}/(x+z)$.

**24.** $\{px^3-(p-q)x^2-(q-r)x-r\}/(px^2+qx+r)$.

**25.** $\{(a^2+ab)x^5+(a^2+b^2)x^4+(2a^2+3ab-b^2)x^3+(2a^2-ab+b^2)x^2$
$+(a^2+2ab-b^2)x+(a^2-ab)\}/(ax^2+bx+a)$.

**26.** $\{p^2x^6-q^2x^4+(3ps-2qr)x^3+(qs-r^2)x^2+rsx+2s^2\}/(px^3+qx^2+rx+s)$.

**27.** Determine $p$ and $q$ so that $3x^4+2x^3-5x^2+px+q$ may be exactly divisible by $x^2-x+1$.

**28.** Find $p$ and $q$ so that $x^4+5x^3+6x^2+px+q$ may be exactly divisible by $x^2+2x+1$.

**29.** Find the necessary and sufficient condition or conditions that $x^5+ax^4+x^3+x^2+bx+1$ may be exactly divisible by $x^2-x+1$.

**30.** Determine $a$ and $b$ so that $(x^5+ax^4+b)/(3x^2+3x-1)$ may be integral.

**31.** What conditions must $p$ and $q$ satisfy in order that $x^2+px+q$ may divide $x^3+x^2+2x+2$ exactly?

**32.** Determine $a$ and $c$ so that $(x^3+2x^2+cx+1)/(x^2+ax+1)$ may be an integral function of $x$.

**33.** Determine $\lambda$ so that $(x-1)^2(2x+\lambda)^2-(x-1)^2(x+\lambda)^2$ may be exactly divisible by $x^3-4x^2+5x-2$; and write down the quotient.

**34.** Determine $p$ and $q$ so that $x^4+4x^3+3x^2+px+q$ and $x^3+3x^2+2x+1$ may have the same remainder when divided by $x^2+2x+1$.

## Binomial Divisor—Remainder Theorem

**§113.** The case of a binomial divisor of the first degree is of special importance. Let the divisor be $x-a$, and the dividend $p_0x^n+p_1x^{n-1}+p_2x^{n-2}+\ .\ .\ .\ +p_{n-1}x+p_n$.

Then, if we employ the method of detached coefficients, the calculation runs as follows:—

$$
\begin{array}{l|l}
p_0 + p_1 \qquad + p_2 + \cdots \qquad + p_{n-1} + p_n & 1 - a \\
\underline{p_0 - p_0 a} & \overline{p_0} \\
\quad (p_0 a + p_1) + p_2 & \; + (p_0 a + p_1) \\
\quad \underline{(p_0 a + p_1) - (p_0 a + p_1)a} & \; + (p_0 a^2 + p_1 a + p_2) \\
\qquad\qquad (p_0 a^2 + p_1 a + p_2) + p_3 & \quad \vdots \\
\qquad\qquad \underline{(p_0 a^2 + p_1 a + p_2) - (p_0 a^2 + p_1 a + p_2)a} & \quad \vdots \\
\qquad\qquad\qquad\qquad (p_0 a^3 + p_1 a^2 + p_2 a + p_3) & \quad \vdots \\
\qquad\qquad\qquad\qquad\qquad \vdots &
\end{array}
$$

The integral quotient is, therefore—

$$p_0 x^{n-1} + (p_0 a + p_1)x^{n-2} + (p_0 a^2 + p_1 a + p_2)x^{n-3} + \ldots$$

*The law of formation of the coefficients is evidently as follows:—*

*The first is the first coefficient of the dividend;*

*The second is obtained by multiplying its predecessor by* a *and adding the second coefficient of the dividend;*

*The third by multiplying the second just obtained by* a *and adding the third coefficient of the dividend; and so on.*

*It is also obvious that the remainder, which in the present case is of zero degree in* x *(that is, does not contain* x*), is obtained from the last coefficient of the integral quotient by multiplying that coefficient by* a *and adding the last coefficient of the dividend.*

The operations in any numerical instance may be conveniently arranged as follows:*—

Ex. 1. $(2x^4 - 3x^2 + 6x - 4) \div (x - 2)$.

$$
\begin{array}{l}
2 + 0 - 3 + \;\, 6 - \;\, 4 \\
\underline{0 + 4 + 8 + 10 + 32} \\
2 + 4 + 5 + 16 + 28
\end{array}
$$

Integral quotient $= 2x^3 + 4x^2 + 5x + 16$;
Remainder $= 28$.

The figures in the first line are the coefficients of the dividend.

The first coefficient in the second line is 0.

The first coefficient in the third line results from the addition of the two above it.

The second figure in the second line is obtained by multiplying the first coefficient in the third line by 2.

---

* The student should observe that this arrangement of the calculation of the remainder is virtually a handy method for calculating the value of an integral function of $x$ for any particular value of $x$, for 28 is $2 \times 2^4 - 3 \times 2^2 + 6 \times 2 - 4$, that is to say, the value of $2x^4 - 3x^2 + 6x - 4$ when $x = 2$. This method is often used, and always saves arithmetic when some of the coefficients are negative and others positive.

The second figure in the third line by adding the two over it.

And so on.

Ex. 2. If the divisor be $x+2$, we have only to observe that this is the same as $x-(-2)$; and we see that the proper result will be obtained by operating throughout as before, using $-2$ for our multiplier instead of $+2$.

$$(2x^4-3x^2+6x-4)\div(x+2)=(2x^4-3x^2+6x-4)\div(x-(-2)).$$

$$\begin{array}{l} 2+0-3+\ \ 6-4 \\ \underline{0-4+8-10+8} \\ 2-4+5-\ \ 4+4. \end{array}$$

$$\begin{aligned} &\text{Integral quotient} = 2x^3-4x^2+5x-4\,; \\ &\text{Remainder} \qquad\quad = 4. \end{aligned}$$

Ex. 3.

The following example will show the student how to bring the case of any binomial divisor of the first degree under the case of $x-a$ :—

$$\frac{3x^4-2x^3+3x^2-2x+3}{3x+2}=\frac{3x^4-2x^3+3x^2-2x+3}{3(x+\frac{2}{3})}$$
$$=\tfrac{1}{3}\left\{\frac{3x^4-2x^3+3x^2-2x+3}{x-(-\frac{2}{3})}\right\}.$$

Transforming now the quotient inside the bracket { }, we have

$$\begin{array}{l} 3-2+\ 3\ -\ \ 2\ +\ \ 3 \\ \underline{0-2+\tfrac{8}{3}-\tfrac{34}{9}+\tfrac{104}{27}} \\ 3-4+\tfrac{17}{3}-\tfrac{52}{9}+\tfrac{185}{27} \end{array}$$

$$\begin{aligned} &\text{Integral quotient} = 3x^3-4x^2+\tfrac{17}{3}x-\tfrac{52}{9}. \\ &\text{Remainder} \qquad\quad = \tfrac{185}{27}. \end{aligned}$$

Whence

$$\frac{3x^4-2x^3+3x^2-2x+3}{3x+2}=\tfrac{1}{3}\left\{3x^3-4x^2+\tfrac{17}{3}x-\tfrac{52}{9}+\frac{\frac{185}{27}}{x-(-\frac{2}{3})}\right\}$$
$$=x^3-\tfrac{4}{3}x^2+\tfrac{17}{9}x-\tfrac{52}{27}+\frac{\frac{185}{27}}{3x+2}.$$

Hence, for the division originally proposed, we have—

$$\begin{aligned} &\text{Integral quotient} = x^3-\tfrac{4}{3}x^2+\tfrac{17}{9}x-\tfrac{52}{27}\,; \\ &\text{Remainder} \qquad\quad = \tfrac{185}{27}. \end{aligned}$$

The process employed in Examples 2 and 3 above is clearly applicable in general, and the student should study it attentively as an instance of the use of a little transformation in bringing cases apparently distinct under a common treatment.

**§ 114.** Reverting to the general result of last section, we see that the remainder, when written out in full, is

$$p_0a^n+p_1a^{n-1}+\ .\ .\ .+p_{n-1}a+p_n.$$

Comparing this with the dividend

$$p_0x^n+p_1x^{n-1}+\ .\ .\ .\ \ +p_{n-1}x+p_n,$$

we have the following **Remainder Theorem** :—

*When an integral function of* x *is divided by* x − $a$, *the remainder is obtained by substituting* $a$ *for* x *in the dividend.* In other words, *the remainder is the same function of* $a$ *as the dividend is of* x.

§ **115.** Partly on account of the great importance of the Remainder Theorem, and partly as an exercise in general algebraic reasoning, we shall give an independent proof of the theorem in a slightly generalised form.

*The remainder, when any integral function of* x *is divided by* ax + b, *is the same function of* − b/a * *as the dividend is of* x.

Let us, for shortness, denote the dividend $p_0x^n + p_1x^{n-1} + \ldots + p_{n-1}x + p_n$ by $f(x)$: $f(c)$ will then, naturally, denote the result of substituting $c$ for $x$ in $f(x)$, *i.e.* $f(c)$ will denote $p_0c^n + p_1c^{n-1} + \ldots + p_{n-1}c + p_n$.

Let $\chi(x)$ denote the integral quotient, and R the remainder when $f(x)$ is divided by $ax+b$. Then $\chi(x)$ is an integral function of $x$ (of degree $n-1$), and R is a constant (that is, is independent of $x$); and we have

$$f(x)/(ax+b) \equiv \chi(x) + \mathrm{R}/(ax+b),$$

whence on multiplication by $ax+b$ we get the identity

$$f(x) \equiv (ax+b)\chi(x) + \mathrm{R}.$$

Since this holds for all values of $x$, we get, putting $x = -b/a$ throughout—

$$f(-b/a) = (-b+b)\chi(-b/a) + \mathrm{R},$$

where R remains the same as before, since it does not depend upon $x$, and is therefore not affected by giving any particular value to $x$.

Since $\chi(-b/a)$ is finite if $-b/a$ be finite, $(-b+b)\chi(-b/a) = 0 \times \chi(-b/a) = 0$; and we get finally

$$f(-b/a) = \mathrm{R},$$

which, if we remember the meaning of $f(-b/a)$, proves the Remainder Theorem for the general binomial divisor $ax+b$.

Example 3 of § 113 above is in part a special case of this theorem.

Ex. 1. Determine $l$ and $m$ so that $x^4 - 3x^3 + 5x^2 + lx + m$ shall be exactly divisible by $x^2 - 5x + 6$.

$x^2 - 5x + 6 = (x-2)(x-3)$. Now $x - 3$ is not exactly divisible by $x - 2$ (since the remainder corresponding to this division, viz. $2 - 3 \neq 0$).

* *i.e.* of the value of $x$, for which $ax+b$ vanishes.

Hence it is necessary and sufficient in order that any integral function may be exactly divisible by $(x-2)(x-3)$ that it be exactly divisible by $x-2$ and by $x-3$.

The remainders, when $x^4-3x^3+5x^2+lx+m$ is divided by $x-2$ and by $x-3$, are

$$2^4-3.2^3+5.2^2+l.2+m=2l+m+12,$$

and

$$3^4-3.3^3+5.3^2+l.3+m=3l+m+45$$

respectively. Hence the necessary and sufficient conditions are $2l+m+12=0$, $3l+m+45=0$. Solving these as a pair of equations to determine $l$ and $m$, we get the unique solution $l=-33$, $m=54$.

Ex. 2. Show by means of the remainder theorem that

$$I\equiv a^3(b-c)+b^3(c-a)+c^3(a-b)=-(b-c)(c-a)(a-b)(a+b+c).$$

First consider I as an integral function of the variable $a$. The remainder, when I is divided by $a-b$, is found by replacing $a$ by $b$ throughout I: the result is $b^3(b-c)+b^3(c-b)+c^3(b-b)$, which obviously vanishes. Hence I is exactly divisible by $a-b$.

In precisely the same way, I may be shown to be exactly divisible by $b-c$, and by $c-a$.

Hence, since no one of the three $b-c$, $c-a$, $a-b$ is divisible by any other, I must be exactly divisible by $(b-c)(c-a)(a-b)$; and, since I is of the fourth degree in each of the variables $a$, $b$, $c$, and $(b-c)(c-a)(a-b)$ of the third, the quotient must be of the first degree in $a$, $b$, $c$; and obviously also homogeneous.

We have, therefore, the identity—

$$a^3(b-c)+b^3(c-a)+c^3(a-b)=(b-c)(c-a)(a-b)(la+mb+nc),$$

where $l$, $m$, $n$ are numerical constants.

If we compare the terms in $a^3b$, $b^3c$ and $c^3a$ on the two sides of this identity, we see at once, by picking out the corresponding partial products on the right, that $l=m=n=-1$: and thus the required result is established.

Ex. 3. Show that when any integral function $f(x)$ is divided by $(x-a)(x-b)$ the remainder is

$$[\{f(b)-f(a)\}x+\{bf(a)-af(b)\}]/(b-a).$$

Since the divisor is of the second degree in $x$, the remainder will be of the first degree in $x$; and therefore of the form $Ax+B$, where A and B are constants. We may put $Ax+B\equiv A(x-a)+aA+B \equiv A(x-a)+C$, where A and C are constants which we have to determine.

We see, therefore, that there must be an identity of the form

$$f(x)=\chi(x)(x-a)(x-b)+A(x-a)+C,$$

which must hold for the particular values $x=a$ and $x=b$. Since $\chi(a)$ and $\chi(b)$ are both finite, we must therefore have

$$f(a)=C,\ \text{and}\ f(b)=A(b-a)+C.$$

Whence $\qquad C=f(a)$, and $A=\{f(b)-f(a)\}/(b-a)$.

The remainder is, therefore—

$$\frac{f(b)-f(a)}{b-a}(x-a)+f(a)\equiv\frac{\{f(b)-f(a)\}x+bf(a)-af(b)}{b-a}.$$

## EXERCISES XXIX.

1. What is the remainder when $2x^5-2x^4+3x^3-7x^2+5x-8$ is divided by $x+2$?

2. Find the integral quotient and remainder when $3x^4-3x^2+1$ is divided by $x+4$.

3. Show that $x^4-34x^2+225$ is exactly divisible by $x^2-25$.

4. Find the remainder for $(4x^4-6x^3+6x^2-1)/(2x-1)$.

5. Simplify $(x^3+2x^2+x-4)(2x^3+7x^2+8x+4)/(x-1)(x+2.)$

6. Determine $\lambda$ so that $2x^4-3x^3+\lambda x^2-9x+1$ may be exactly divisible by $x-3$.

7. For what values of $l$ and $m$ are $x^2+(l+1)x+(m+2)$ and $x^2+(l+12)x-2m$ each exactly divisible by $x-1$?

8. Determine $\lambda$ and $\mu$ so that $3x^3+\lambda x^2+\mu x+42$ may be exactly divisible by $(x-2)(x-3)$.

9. Determine $a$ and $b$ so that $x^4-ax^3+bx^2+bx+9$ may be exactly divisible by $x^2-1$.

10. Simplify $(x^n-2x+1)/(x-1)$.

11. Divide $\Pi(y+z)+xyz$ by $\Sigma x$.

12. Show that $\{\Sigma(y+z)^3-3\Pi(y+z)\}/\{\Sigma x^3-3xyz\}=2$.

13. Simplify $(\Sigma a^3-3\Sigma bcd)/\Sigma a$.
$\quad abcd$

14. Simplify $[\{(pa+b)x+(pb+c)\}^2+(ac-b^2)(x-p)^2]/(ap^2+2bp+c)$.

15. Show that $\{(2y-z-x)^3-(2z-x-y)^3\}/(y-z)$ is symmetrical in $x$, $y$, $z$.

16. Express $x^4+x^2+1$ as an integral function of $x-2$.

17. Show that $(x-1)^3\equiv A+B(x-2)+C(x-2)(x-3)+D(x-2)(x-3)(x-4)$, provided A, B, C, D have certain numerical values.

18. If $Q_1=x^2-3x+4$ and $Q_2=x^2-4x+5$, transform $x^5$ into the form $Ax+B+(Cx+D)Q_1+(Ex+F)Q_1Q_2$, where A, B, C, D, E, F are constants.

19. Transform $3x^4+4x^3+x+1$ into the form $a+b(x+1)+c(x^2-1)+d(x+1)^2(x-1)+e(x+1)^2(x-1)(x-2)$; and also into the form $a_0+a_1(x-1)+a_2(x-1)^2+a_3(x-1)^3+a_4(x-1)^4$, where $a, b, \ldots, a_0, a_1, a_2, \ldots$ are constants.

20. Express $x^5$ as an integral function of $x-1$; and $(x^2+5x+2)^3$ as an integral function of $y=x-2$ and $z=x^2+2x+3$, which shall be linear in $y$.

21. Find an integral function of $x$ of the third degree which shall vanish when $x=1$ and $x=-2$, and have the value 30 when $x=3$.

22. If A, B, Q, R be integral functions of $x$, and $A\equiv BQ+R$, show that A is or is not exactly divisible by Q according as R is or is not exactly divisible by Q.

23. If $f(x)$ denote an integral function of $x$, and $f(a)$ the result of replacing $x$ by $a$ in $f(x)$, show by means of the well-known identity $x^n-a^n\equiv(x-a)(x^{n-1}+x^{n-1}a+\ldots+a^{n-1})$, that $f(x)-f(a)$ is always exactly divisible by $x-a$: and deduce the remainder theorem.

24. Show that $1/(1+x)=1-x+x^2-x^3-\ldots+(-x)^n+(-x)^{n+1}(1+x)$; and hence show that $1/1{\cdot}00368=1-{\cdot}00368={\cdot}99632$ approximately, the error being less than ${\cdot}000014$.

# CHAPTER XIII

## RESOLUTION OF INTEGRAL FUNCTIONS INTO FACTORS

**§ 116.** In the wider sense of the word factor any two functions whose product is a given function may be said to be factors of that function, *e.g.* $(x^2+1)/(x-1)$ and $x-1$ are factors of $x^2+1$, since their product is $x^2+1$. The factorisation of a function in this sense is an absolutely indeterminate and meaningless problem.* Thus, for example, $x^2+1$ in this sense may be factorised into P and $(x^2+1)/P$, where P is any function whatever.

In the present chapter the problem we discuss is the following:—*Given an integral function of any stated variables, to find two (or it may be as many as possible) integral functions of these variables, such that their product is the given function.*

For example, $x^3-1=(x-1)(x^2+x+1)$, $x^2-\frac{1}{4}=(x+\frac{1}{2})(x-\frac{1}{2})$ $x^2-ay^2=(x+\sqrt{a}y)(x-\sqrt{a}y)$ are factorisations in the sense defined, provided the variables be $x$ in the two first cases, and $x$ and $y$ only in the third. The integrality, be it observed, has reference to the *variables* only.

On the other hand, $x^2-ay^2=(x+\sqrt{a}y)(x-\sqrt{a}y)$ is not a factorisation in our present sense if $a$ be regarded as the variable or as one of the variables. Again, $x^2+xy+y^2=\{x+y+\sqrt{(xy)}\}\{x+y-\sqrt{(xy)}\}$ and $x^2-1/x^2=(x+1/x)(x-1/x)$, although true identities, are not factorisations in the strict sense—the first because the factors are not rational functions of $x$ and $y$, the second because neither the given function nor the factors are integral.

The problem of factorising an integral function in general is difficult:

---

* A fact that seems to be occasionally forgotten by examiners and text-book writers.

(1), because we cannot in general tell beforehand whether there be any factors at all, or what are their degrees if they do exist. Thus, for example, $x^2+y^2+1$* has no factors; $x^3-y^3+x-y$ has the factors $x-y$ and $x^2+xy+y^2+1$ only; while $x^3-x^2-3x+3$ has the three linear factors $x-1$, $x+\sqrt{3}$, $x-\sqrt{3}$. (2), because the coefficients of the factors, when they do exist, may be complicated functions of the coefficients of the given function; they will, in fact, in general not be expressible by means of a finite number of the operations $+$, $-$, $\times$, $\div$, $\sqrt[n]{}$ at all. For example, $ax^5+bx^4+cx^3+dx^2+ex+f$ is undoubtedly resolvable into factors which are linear functions of $x$; but the coefficients of those factors cannot, except in special cases, be expressed in terms of $a$, $b$, $c$, $d$, $e$, $f$ by means of a finite number of the operations $+$, $-$, $\times$, $\div$, $\sqrt[n]{}$. We make these remarks, which must be taken on trust, lest the beginner should suppose that his failure to factorise a given function by elementary processes necessarily arises from his want of skill. It may be so, but it may also happen that the problem is not soluble by elementary means, or that it is insoluble.

Nevertheless, factorisation can be accomplished easily in many cases by tentative elementary processes; and, in the case of the quadratic integral function of $x$, the problem can be solved generally by such processes.

In as much as factorisation is one of the most powerful methods for abbreviating algebraic work, a certain amount of skill in it is indispensable even to a beginner, and must be acquired by practice. We now proceed to classify the artifices used, and to furnish numerous examples and exercises.

## Use of Standard Identities

**§ 117.** The use of Standard Identities in factorisation has already been pointed out in § 50. The reader has now at his disposal the extended table of § 108.

Ex. **1.** $(x+1)^4+(x^2-1)^2+(x-1)^4$.

This may be written $(x+1)^4+(x+1)^2(x-1)^2+(x-1)^4$. Hence, putting $x+1$ for $x$ and $x-1$ for $y$ in § 108, vi. we get

$$\begin{aligned}(x+1)^4+(x^2-1)^2+(x-1)^4 &= \{(x+1)^2+(x+1)(x-1)+(x-1)^2\} \\ &\quad \times \{(x+1)^2-(x+1)(x-1)+(x-1)^2\}, \\ &= (3x^2+1)(x^2+3).\end{aligned}$$

Ex. **2.** $F \equiv b^2c^4-b^4c^2+c^2a^4-c^4a^2+a^2b^4-a^4b^2$.

If we put $a^2$, $b^2$, $c^2$ in place of $a$, $b$, $c$ in § 108, x. we get

$$\begin{aligned}F &= (b^2-c^2)(c^2-a^2)(a^2-b^2), \\ &= (b-c)(b+c)(c-a)(c+a)(a-b)(a+b).\end{aligned}$$

Ex. **3.** $x^5-1$ and $x^5+1$.

---

* See A. vii. § 12.

If we put $y=1$ and $n=5$ in § 108, iv. we get

$$x^5-1=(x-1)(x^4+x^3+x^2+x+1);$$
$$x^5+1=(x+1)(x^4-x^3+x^2-x+1).$$

Ex. 4. $F\equiv xyz-yz-zx-xy+x+y+z-1.$

$$-F=1-\Sigma x+\Sigma yz-xyz$$
$$=(1-x)(1-y)(1-z), \text{ by § 108, i.}$$

Hence $$F=-(1-x)(1-y)(1-z)=(x-1)(y-1)(z-1).$$

## EXERCISES XXX.

Resolve each of the following into as many factors as you can :—

1. $a^2-b^2-c^2+d^2+2(bc-ad)$.
2. $(x-1)^2-9(x+1)^2$.
3. $(ax+by)^2-(bx-ay)^2$.
4. $(x^2+x+1)^2-(x^2-x-1)^2$.
5. $(1-xy)^2-(x-y)^2$.
6. $x^2-2px+(p^2-q^2)$.
7. $(x+1)^2+2p(x+1)+(p^2-q^2)$.
8. $(x^2+2xy+y^2)-4(x^2-2xy+y^2)$.
9. $x^4-8x^2+16$.
10. $(2x+1)^2-3(x-2)^2$.
11. $(2x-1)^2-(2x-1)(2x-4)+(x-2)^2$.
12. $x^2+(a-b)x-ab$.
13. $(x+1)^2+2(x^2-1)+(x-1)^2$.
14. $x^6-(a-b)x^3-ab$.
15. $(x+y+1)^2-2(x-y-1)^2$.
16. $x^{2m}+(a+b)x^m+ab$.
17. $(x-y)^2-(a+b)(x-y)+ab$.
18. $(xy+ab)(xy-ab)-a^2(xy-b^2)-b^2(xy-a^2)$.
19. $4(xy-ab)^2-(x^2+y^2-a^2-b^2)^2$.
20. $(2x+y-1)^2+4(2x+y)$.
21. $27x^3-27x^2+9x-1$.
22. $x^8-4x^6y^2+6x^4y^4-4x^2y^6+y^8$.
23. $x^6-(a^2+b^2+c^2)x^4+(b^2c^2+c^2a^2+a^2b^2)x^2-a^2b^2c^2$.
24. $8x^3+4(a-b+c)x^2+2(-bc+ca-ab)x-abc$.
25. $x^8-y^8$.
26. $x^7-y^7$.
27. $x^7+y^7$.
28. $x^4+9x^2+81$.
29. $x^3+y^3-z^3+3xyz$.
30. $3x^3-x^6-x^3-1$.
31. Show that $\{(z-x)^3-(x-y)^3\}/(y+z-2x)$ is a symmetric function of $x$, $y$, $z$.

### Factorisation by Grouping Terms

§ **118.** If the terms of an integral function can be associated into groups, each of which has the factor P, then, by the law of distribution, P is a factor in the function.

Ex. 1. $x^4-2x^3+2x-1$

$$=(x^4-1)-(2x^3-2x),$$
$$=(x^2-1)(x^2+1)-2x(x^2-1),$$
$$=(x^2-1)(x^2+1-2x),$$
$$=(x+1)(x-1)(x-1)^2,$$
$$=(x+1)(x-1)^3.$$

Ex. 2. $(x+y)^2+(y-z)^2-(z-u)^2-(u+x)^2$

$$\begin{aligned}&=\{(x+y)^2-(u+x)^2\}+\{(y-z)^2-(z-u)^2\},\\&=\{(x+y)-(u+x)\}\{(x+y)+(u+x)\}\\&\qquad+\{(y-z)-(z-u)\}\{(y-z)+(z-u)\},\\&=(y-u)(2x+y+u)+(y+u-2z)(y-u),\\&=(y-u)\{(2x+y+u)+(y+u-2z)\},\\&=(y-u)(2x+2y+2u-2z).\end{aligned}$$

Ex. 3. $F\equiv lx^3+\{(m-n)-l(a+b)\}x^2+\{lab-(m-n)(a+b)\}x+(m-n)ab.$

Since the function is homogeneous and of the first degree in $l, m, n$, it is clear that $l, m, n$ can only occur in one of the factors. If, therefore, we arrange the function in the form $lP+mQ+nR$, the three integral functions P, Q, R will have as a common factor the product of all the factors (if there be any) which do not contain $l, m, n$. We have

$$\begin{aligned}F&=lx\{x^2-(a+b)x+ab\}+(m-n)\{x^2-(a+b)x+ab\},\\&=(lx+m-n)\{x^2-(a+b)x+ab\},\\&=(lx+m-n)(x-a)(x-b).\end{aligned}$$

## EXERCISES XXXI.

Factorise the following as far as you can :—

1. $x^2-122x+121.$
2. $5x^2-(7+15a)x+21a.$
3. $3(x^2-y^2)-5(x-y)^2.$
4. $x^2-6xy+6yz-z^2.$
5. $x^3-x^2-x+1.$
6. $x^3+3x^2-x-3.$
7. $x^3+x^2y+xy^2+y^3.$
8. $x^3-3x^2+9x-27.$
9. $x^6-x^4y^2-x^2y^4+y^6.$
10. $2x^3-15x^2+2x-15.$
11. $x^3-ax^2+px-ap.$
12. $(x-1)(x-2)^2-(x-1)^3.$
13. $(x+2)^2+(x^2+3x+2)+(x^2-4).$
14. $1+ax-x^2-ax^3.$
15. $x^3+px^2+px+p-1.$
16. $x^3-2x^2+25x-50.$
17. $x^4y+3x^3y^2-3x^2y^3-xy^4.$
18. $x^5+x^4y-x^3y^2-x^2y^3+xy^4+y^5.$
19. $(x+z)(y-u)-(x+u)(y-z).$
20. $x^3-y^3+y(3x^2-2xy-y^2).$
21. $x^4+3x^3+2x^2+x+1.$
22. $x^4-1-4(x-1).$
23. $a(b+c)^3-b(c+a)^3-(a-b)c^3.$
24. $x^3+x^2+(p^2-1)x+p^2-1.$
25. $x^5+px^4+x^3+px^2+x+p.$
26. $(x-3)(x+1)^3+(x-3)(x+2)^3.$
27. $(l+m)x^3+(3l+2m-n)x^2+(2l-m-3n)x-2(m+n).$
28. $x(x-1)(x-2)+4(x-1)(x-2)(x-3)-(x-2)(x-3)(x-4).$
29. $lpx^4+(lp+lq+mp)x^3+(l+m)(p+q)x^2+(mp+mq+lq)x+mq.$
30. $(x+y+z)^2+(x-y-z)^2-(x+y+u)^2-(x-y-u)^2.$

### Factorisation of $ax^2+bx+c$, when $b^2-4ac$ is a Positive Perfect Square

§ **119.** We now proceed to consider in detail the factorisation of a quadratic function of $x$, say $ax^2+bx+c$, when $a, b, c$ are given algebraic quantities, which we shall suppose to be positive or negative commensurable numbers. It is obvious that we may suppose $a$ positive; for, if it were negative, we could write the given function $-(-ax^2-bx-c)$, and deal with the function inside the bracket.

The reader may take for granted, what we shall presently prove, that, *when the* **discriminant of the quadratic,** i.e. *the function* $\Delta = \mathrm{b}^2 - 4\mathrm{ac}$ *of the coefficients, is positive and the square of a commensurable number, then the quadratic can be resolved into two linear factors whose coefficients are real and commensurable.*

*In this case the factors can usually be found by inspection from the identity*

$$(x+\lambda)(x+\mu) \equiv x^2 + (\lambda+\mu)x + \lambda\mu.$$

§ **120.** Consider first the case where $a=1$. There are four sub-cases to consider according to the signs of $b$ and $c$, viz.—

| $b$ | + | − | + | − |
|---|---|---|---|---|
| $c$ | + | − | − | + |

, of which the first two and the last two should be grouped together. The method of procedure will be understood from the following four examples:—

Ex. 1. $x^2+15x+56$. Let us assume the factors to be $(x+\alpha)(x+\beta)$. Then $x^2+15x+56 \equiv x^2+(\alpha+\beta)x+\alpha\beta$. We have therefore $\alpha+\beta=15$, $\alpha\beta=56$. We have to determine two integers $\alpha$ and $\beta$ such that their sum is 15 and their product 56. Now the different ways of resolving 56 into a pair of factors are $1\times56$, $2\times28$, $4\times14$, $7\times8$. Of these only the last gives $\alpha+\beta=15$. Hence $\alpha=7$, $\beta=8$; and we have $x^2+15x+56\equiv(x+7)(x+8)$.

Ex. 2. $x^2-15x+56$. Here we must obviously assume $x^2-15x+56 \equiv(x-\alpha)(x-\beta)$ where $\alpha$ and $\beta$ are positive integers. We have $\alpha+\beta=15$, $\alpha\beta=56$, and the arithmetical problem is the same as in Example 2. We get $\alpha=7$, $\beta=8$; and $x^2-15x+56\equiv(x-7)(x-8)$.

Ex. 3. $x^2+x-56$. Since the last term is negative, if $x^2+x-56 =(x+\lambda)(x+\mu)\equiv x^2+(\lambda+\mu)x+\lambda\mu$, we see that one of the two $\lambda$, $\mu$ must be negative and the other positive; and, since $\lambda+\mu=+1$ is positive, the greater of the two quantities $\lambda$, $\mu$ must be positive. We therefore assume $x^2+x-56\equiv(x+\alpha)(x-\beta)$, where $\alpha$ and $\beta$ are positive integers, and $\alpha>\beta$. Comparing coefficients we have $\alpha-\beta=1$, $\alpha\beta=56$. As before, the decompositions of 56 are $1\times56$, $2\times28$, $4\times14$, $7\times8$; and we have to select one from among these in which the difference of the factors is 1. The last alone is suitable; we have $\alpha=8$, $\beta=7$. Hence $x^2+x-56\equiv(x+8)(x-7)$.

Ex. 4. $x^2-x-56$. The reasoning and the calculation is the same as in Example 3, only that now $\alpha-\beta=-1$, which necessitates that $\alpha<\beta$. We get $\alpha=7$, $\beta=8$. Hence $x^2-x-56\equiv(x+7)(x-8)$.

§ **121.** The case where $a\neq1$ can be brought under the case where $a = 1$ as follows:—We have $ax^2 + bx + c \equiv (a^2x^2 + abx + ac)/a \equiv \{(ax)^2 + b(ax) + ac\}/a$. We may consider $(ax)^2 + b(ax) + ac$ as a new quadratic function in which the variable is $ax \equiv z$, say. We have then merely to factorise $z^2 + bz + ac$. If $z^2 + bz + ac \equiv (z + \lambda)(z + \mu)$, then $ax^2 + bx + c \equiv (ax + \lambda)(ax + \mu)/a$.

Ex. 1. $15x^2 - 37x - 52 \equiv \frac{1}{15}\{(15x)^2 - 37(15x) - 15 \times 52\} \equiv \frac{1}{15}(15x + \alpha)(15x - \beta)$, where $\alpha$ and $\beta$ are positive integers and $\alpha < \beta$ (see Ex. 4, § 120). We have to find $\alpha$ and $\beta$ so that $\alpha - \beta = -37$, $\alpha\beta = 3 \times 2^2 \times 5 \times 13$. It is easily found that of the various ways of decomposing $3 \times 2^2 \times 5 \times 13$ into two factors the only one that gives $\alpha - \beta = -37$ is $(3 \times 5)(2^2 \times 13)$. Hence $\alpha = 15$, $\beta = 52$. Hence

$$\begin{aligned} 15x^2 - 37x - 52 &\equiv \tfrac{1}{15}(15x + 15)(15x - 52), \\ &\equiv (x + 1)(15x - 52). \end{aligned}$$

The decomposition obtained should in all cases be verified by distribution, which can generally be done mentally.

In many cases the factorisation can be obtained without passing back to the case where $a = 1$, by comparing the given quadratic with

$$(\alpha x + \beta)(\gamma x + \delta) \equiv \alpha\gamma x^2 + (\alpha\delta + \beta\gamma)x + \beta\delta.$$

Ex. 2. $6x^2 - 19x + 15 \equiv (\alpha x + \beta)(\gamma x + \delta)$. Here $\alpha\gamma = +6$, $\beta\delta = +15$. We may take $\alpha$ and $\gamma$ both positive; and it is obvious that, when $\alpha$ and $\beta$ are determined, $\gamma$ and $\delta$ are known. We might, apart from the correctness of the middle term, have any one of the 32 factorisations $(x \pm 1)(6x \pm 15)$, $(x \pm 3)(6x \pm 5)$, $(x \pm 5)(6x \pm 3)$, $(x \pm 15)(6x \pm 1)$; $(2x \pm 1)(3x \pm 15)$, etc. A glance at the middle coefficient, $-19$, at once excludes a large number of these, and we find after a few trials—

$$6x^2 - 19x + 15 \equiv (2x - 3)(3x - 5).$$

Success in this kind of work is a matter of readiness at mental arithmetic; those who are not gifted in this way may fall back on the general process of § 130, which meets all cases. If the factorisation is not immediately obvious, it is advisable before wasting time on a possibly impossible problem, to settle whether the factors really have commensurable coefficients, *i.e.* to see whether the discriminant of the quadratic is or is not a positive perfect square.

Ex. 3. Has the quadratic function $3x^2 - 31x + 9$ commensurable factors?

$\Delta = (-31)^2 - 4 \times 3 \times 9 = 853$, which is not a perfect square. Hence the function cannot be resolved into factors having commensurable coefficients.

### EXERCISES XXXII.

If the student finds it impossible to factorise any of the following by inspection he should apply the general method of § 130 :—

| | | |
|---|---|---|
| **1.** $x^2+8x+12$. | **2.** $x^2+24x+143$. | **3.** $x^2-11x+18$. |
| **4.** $x^2-28x+195$. | **5.** $x^2+5x-14$. | **6.** $x^2-5x-50$. |
| **7.** $x^2-71x-2900$. | **8.** $x^2+78x-3663$. | **9.** $x^2-7x-8$. |
| **10.** $x^6+9x^3+8$. | **11.** $x^2+x+\frac{1}{4}$. | **12.** $x^2-\frac{8}{15}x+\frac{1}{15}$. |
| **13.** $(x^2+6x+4)^2-16$. | **14.** $x^4-10x^2+9$. | **15.** $1-2a-x^2+2ax$. |
| **16.** $8x^2+6xy+y^2$. | **17.** $x^3-7x^2+14x-8$. | **18.** $2x^2+x-3$. |
| **19.** $6x^2+16xy+10y^2$. | **20.** $10x^2+x-9$. | **21.** $7x^2+44x-35$. |
| **22.** $20x^2-x-30$. | **23.** $6x^2+15x+9$. | **24.** $6x^2+59x+105$. |

### Digression on Imaginary and Complex Quantity

§ **122.** Before proceeding to the factorisation of a quadratic function in general, it is necessary to discuss briefly a fundamental point in the theory of Algebra which now arises for the first time. The special quadratic function $x^2+c$ can, as has already been seen, be factorised by means of the identity $x^2-\lambda^2 \equiv (x-\lambda)(x+\lambda)$, provided always that $c$ be a negative quantity, say $c=-d$, where $d$ is an absolute arithmetic quantity. All we have to do is to determine $\lambda$ so that $\lambda^2=d$. This can be done accurately if $d$ be the square of a commensurable number (integral or fractional), and to any required degree of approximation if $d$ is not the square of a commensurable number. In short, we may write $x^2+c \equiv x^2-(-c) \equiv (x-\sqrt{(-c)})(x+\sqrt{(-c)})$, so long as $c$ is a negative quantity.

If, however, $c$ be a positive quantity, we can no doubt write $x^2+c \equiv x^2-(-c)$; but the fundamental difficulty arises that we can no longer find a real quantity $\lambda$ such that $\lambda^2=-c$. That this is so will be obvious when we reflect that the square of any quantity in the algebraic series of real quantity

$$-\infty, \ldots, -1, \ldots, 0, \ldots, +1, \ldots, +\infty$$

is positive.

One way of meeting this difficulty would be simply to note and declare that the factorisation of $x^2+c$ by means of real operands is impossible when $c$ is a positive quantity.

§ **123.** There is, however, another course open to us. Although the laws of Algebra were derived from arithmetic, and we began by limiting the operands of Algebra to be arithmetical numbers, we have already passed beyond that limitation by introducing essentially negative quantity, the unit of which may be taken to be $-1$. The laws of Algebra have been constructed so that they are consistent with one another when the operands are either positive or negative quantities, the assemblage of which we have been in the habit of calling algebraic quantity.

Nothing hinders us from considering whether we might not still further enlarge the boundaries of Algebra by defining yet another kind of quantity having a new unit. The only point to be seen to is that any new kind of quantity must be such that we can operate with it together with the old kind of quantity by means of the laws and definitions of Algebra without landing ourselves in logical contradiction—in brief, without speaking or writing nonsense.

§ **124.** Our immediate want is an algebraic quantity whose square shall be negative. Let us take the simplest case, and define a quantity $i$ by the equation $i^2 = -1$. We call $i$ **the Imaginary Unit**; and the understanding regarding it is that it is to be an algebraic operand; in other words, it is to be obedient to all the laws of Algebra. Whether it can be introduced without turning Algebra into nonsense will be seen by, and only by, operating with it and examining the consequences.

Meantime it is obvious that $i$ cannot be any real algebraic quantity, because it has a property (viz. that its square is negative) possessed by no such quantity.

§ **125.** The introduction of the imaginary unit is sufficient for our present purpose, viz. it enables us to find a quantity whose square shall be equal to any given negative quantity $-d$, say, where $d$ is any absolute numerical quantity; for we can write $-d = (-1)d = +i^2(\sqrt{d})^2 = (i\sqrt{d})^2$, from which it appears that $i\sqrt{d}$ has for its square $-d$.

§ **126.** It should also be noted that the integral powers of $i$ are alternately real and non-real, or, as we shall say, **imaginary**, viz. $i^1 = i$; $i^2 = -1$; $i^3 = i^2 \times i = (-1)i = -i$; $i^4 = i^3 \times i = -i \times i = -i^2 = -(-1) = +1$; $i^5 = i^4 \times i = i$, $i^6 = -1$, $i^7 = -i$, etc. . . .

§ **127.** If we are to operate with $i$ just as with an ordinary algebraic quantity, we may take all possible positive or negative multiples of it, *e.g.* $+2i$, $+\frac{1}{2}i$, $+\frac{3}{4}i$, $(-1)i$, $(-\frac{1}{2})i$, $(-\frac{3}{4})i$. . . . We thus arrive at a complete series of **Purely Imaginary Quantity,** which we may symbolise by

$$-\infty i, \ldots, (-1)i, \ldots, 0i, \ldots, (+1)i, \ldots, +\infty i;$$

or, asserting the properties of 1 and 0, as heretofore—

$$-\infty i, \ldots, -i, \ldots, 0, \ldots, +i, \ldots, +\infty i.$$

It will be observed that this new series of quantity has no quantity in common with the real series of algebraic quantity except 0.

Any purely imaginary quantity may therefore be represented by $yi$, where $y$ is some real quantity positive or negative; and we see that, if $x$ and $y$ be both real and finite both ways, then $x=yi$ is an impossible equation; such an equation is possible when and only when $x=0$ and $y=0$.

§ **128.** If we are to treat purely imaginary alongside of purely real quantity, we shall arrive by addition at quantities of the form $x+yi$, which consist of two parts, viz. a purely real part or multiple (positive or negative) of the real unit 1, and a purely imaginary part or multiple (positive or negative) of the imaginary unit $i$. Such mixed quantities are called **Complex Numbers** or **Complex Quantities.**

It follows readily from § 127 that two complex quantities cannot be identically equal unless their real parts and their purely imaginary parts are separately equal. For, if $x+yi=\xi+\eta i$, ($x$, $y$, $\xi$, $\eta$ being by hypothesis all real), we have $x-\xi=\eta i-yi=(\eta-y)i$. Now, since $x-\xi$ and $\eta-y$ are real, this equation can only subsist if $x-\xi=0$ and $\eta-y=0$; that is, we must have $x=\xi$ and $y=\eta$.

§ **129.** Into the further consideration of complex quantity we do not at present enter.* All that it is necessary for the beginner to do is to familiarise himself with operations involving the imaginary unit, and to gain by practice the conviction that its introduction into Algebra leads to no logical inconsequence. In particular, he should note in the following examples and exercises particular cases of the general theorem, to be fully established later in his course, that every series of algebraic operations with complex operands leads to a complex quantity as a result, and requires no new kind of unit for its expression.

It may be repeated for emphasis that *the sources of all inferences regarding* i *are:* 1st, *its fundamental property* $i^2=-1$; 2nd, *that it obeys all the laws of algebraic operation as previously established.*

Ex. **1.** $(-3+2i)(+5+2i)=(-3)(+5)+(-3)(+2i)+(+2i)(+5)+(+2i)(+2i)=-15-6i+10i+4i^2=-15+(-6+10)i+4(-1)=-15+4i-4=-19+4i$.

---

* See A. Chap. XII.

Ex. 2. $(7-\frac{1}{2}i)(7+\frac{1}{2}i)=7^2-(\frac{1}{2}i)^2=49-(\frac{1}{2})^2i^2=49-\frac{1}{4}(-1)=49+\frac{1}{4}$ $=197/4$.

Ex. 3. $(3+2i)/(2+3i)=(3+2i)(2-3i)/(2+3i)(2-3i)=\{6+(4-9)i-6i^2\}/(2^2-3^2i^2)=\{6-5i-6(-1)\}/\{2^2-3^2(-1)\}=(12-5i)/(4+9)$ $=\frac{12}{13}-\frac{5}{13}i$.

Ex. 4. $(2+i)/(1+i)-(1+i)/(2-i)=(2+i)(2-i)/(1+i)(2-i)$ $-(1+i)(1+i)/(1+i)(2-i)=\{(4-i^2)-(1+2i+i^2)\}/\{2+i-i^2\}=\{(4+1)$ $-(1+2i-1)\}/\{2+i+1\}=(5-2i)/(3+i)=(5-2i)(3-i)/(3+i)(3-i)$ $=(15-11i+2i^2)/(3^2-i^2)=(15-11i-2)/(3^2+1)=\frac{13}{10}-\frac{11}{10}i$.

Ex. 5. Verify that $5+7i$ is one of the values of the square root of $-24+70i$.

This amounts to proving that the square of $5+7i$ is $-24+70i$. Now $(5+7i)^2=5^2+2\times5\times(7i)+(7i)^2=25+70i+7^2i^2=25+70i+49(-1)=25-49+70i=-24+70i$.

## EXERCISES XXXIII.

Reduce the following to the form $x+yi$, where $x$ and $y$ are real; any letters used, unless otherwise defined, denote real quantities:—

1. $(6-8i)(3+5i)$. 2. $(1+i)^3$. 3. $(2-3i)^2(3+2i)$.
4. $(1-i)(2+i)(3+i)$. 5. $(1+i)^5-(1-i)^5$ 6. $(a+bi)(c-di)$.
7. $(a+bi)(c+di)(e+fi)$. 8. $(1+\sqrt{3}i)^4+(1-\sqrt{3}i)^4$
9. $(1+2i+3i^2+4i^3)^2$. 10. $(1+i+i^2+i^4)^3$.
11. $(l+mi+ni^2)(n+li+mi^2)(m+ni+li^2)$.
12. Find the real part of $\Sigma x\Sigma x^2$, where $x=(1+i)-a(1-i)$, $y=1+i$, $z=(1+i)+a(1-i)$.
13. $\{(a+bi)+(b-ai)\}^5+\{(a-bi)+(b+ai)\}^5$.
14. $\{(p+q)+(p-q)i\}^6+\{(p+q)-(p-q)i\}^6$.
15. $(a+bi)/(c+di)$. 16. $(3+i)(2+i)/(3-2i)$.
17. $(2+3i)/(1-i)-(2-3i)/(1+i)$.
18. $(3-2i)/(5+i)-(3+2i)/(5-i)$.
19. $(2+i)/(4-3i)+(2-i)/(4+3i)$.
20. $(1+i)(2-3i)(4-i)/(1-i)(3-i)$.
21. $(1-i)(2-i)(3-i)/(1-3i)(3+5i)$.
22. $\{(1+i)^3+(1-i)^3\}/(2+i+i^2)^2$.
23. $(3+4i)/(2-i)-(1+3i)/(3-2i)+(1-i)/(4-7i)$.
24. $(6+4i)/(4+i)+(4+i)/(6+4i)+(3-2i)^2/(2-3i)^2$.
25. $\{(2-3i)/(1+i)\}^3\times\{(2-i)/(3-i)\}^2$.
26. $(1+i\sqrt{3})^4/(1-i\sqrt{3})^4$.
27. $(x^4-y^4)\{1/(x+yi)+1/(x-yi)\}$.
28. Show that $\sqrt{i}=\pm(1+i)/\sqrt{2}$, $\sqrt{-i}=\pm(1-i)/\sqrt{2}$.
29. $\sqrt[4]{i}=\pm[\sqrt{\{\frac{1}{2}(1\pm\sqrt{\frac{1}{2}})\}}+i\sqrt{\{\frac{1}{2}(1\mp\sqrt{\frac{1}{2}})\}}]$.
30. Find real values of $x$ and $y$ to satisfy the equation

$$(2-i)x+(3-2i)y+5-i=(1+3i)x+(5-i)y-3+7i.$$

31. Construct an integral equation whose roots are $3+i$ and $3-i$.
32. Construct an integral equation whose roots are $\sqrt{3}+i$, $\sqrt{3}-i$, $-\sqrt{3}+i$, $-\sqrt{3}-i$.
33. Construct an integral equation whose roots are $1+\sqrt{2}+i$, $1-\sqrt{2}+i$, $1+\sqrt{2}-i$, $1-\sqrt{2}-i$.

34. Construct an integral equation whose roots are $1+\sqrt{3}i$, $1-\sqrt{3}i$, $\sqrt{3}+i$, $\sqrt{3}-i$, $-\sqrt{3}+i$, $-\sqrt{3}-i$.

## Factorisation of $ax^2+bx+c$ in General

§ **130.** Let us now consider the general problem of the factorisation of the quadratic function $ax^2+bx+c$. We suppose $a \neq 0$, and for convenience denote the **discriminant** of the function, viz. $b^2-4ac$, by $\Delta$. We have

$$ax^2+bx+c = a\left\{x^2+\frac{b}{a}x+\frac{c}{a}\right\},$$

$$= a\left\{x^2+2\left(\frac{b}{2a}\right)x+\left(\frac{b}{2a}\right)^2-\left(\frac{b}{2a}\right)^2+\frac{c}{a}\right\},$$

$$= a\left\{\left(x+\frac{b}{2a}\right)^2-\frac{b^2-4ac}{4a^2}\right\},$$

$$= a\left\{\left(x+\frac{b}{2a}\right)^2-\frac{\Delta}{4a^2}\right\}.$$

There are three distinct cases to be considered.

First, **Let $\Delta$ be positive.** Then $\Delta=(\sqrt{\Delta})^2$, where $\sqrt{\Delta}$ is a real quantity; and we have

$$ax^2+bx+c = a\left\{\left(x+\frac{b}{2a}\right)^2-\left(\frac{\sqrt{\Delta}}{2a}\right)^2\right\},$$

$$= a\left\{x+\frac{b}{2a}-\frac{\sqrt{\Delta}}{2a}\right\}\left\{x+\frac{b}{2a}+\frac{\sqrt{\Delta}}{2a}\right\},$$

$$= a\left\{x+\frac{b-\sqrt{\Delta}}{2a}\right\}\left\{x+\frac{b+\sqrt{\Delta}}{2a}\right\} \qquad (1).$$

The factors are real, and will be also rational if $\Delta$ be the square of a commensurable number.

Second, **Let $\Delta=0$.** Then

$$ax^2+bx+c = a\left\{x+\frac{b}{2a}\right\}^2 = a\left\{x+\frac{b}{2a}\right\}\left\{x+\frac{b}{2a}\right\} \qquad (2);$$

that is to say, the factors are identical; and the function is the square of a linear integral function of $x$, viz. it is the square of $\sqrt{a}(x+b/2a)$.

Third, **Let $\Delta$ be negative.** Then we may write $\Delta=$

$(-1)(-\Delta)=i^2(\sqrt{-\Delta})^2$, where $i$ is the imaginary unit. Hence

$$ax^2+bx+c=a\left\{\left(x+\frac{b}{2a}\right)^2-\left(\frac{i\sqrt{-\Delta}}{2a}\right)^2\right\},$$

$$=a\left\{x+\frac{b}{2a}-i\frac{\sqrt{-\Delta}}{2a}\right\}\left\{x+\frac{b}{2a}+i\frac{\sqrt{-\Delta}}{2a}\right\},$$

$$=a\left\{x+\frac{b-i\sqrt{-\Delta}}{2a}\right\}\left\{x+\frac{b+i\sqrt{-\Delta}}{2a}\right\} \qquad (3),$$

where $\sqrt{-\Delta}$ is a real quantity, since $\Delta$ itself is negative.

The factors in this case are imaginary and unequal.

We have thus shown how to factorise $ax^2+bx+c$ in every possible case; and we can draw the following important conclusions :—

*The factors are real if the discriminant is positive.*

*The function is a perfect square as regards* x *if the discriminant vanishes.*

*The factors are imaginary if the discriminant is negative.*

*The factors are commensurable if the discriminant be the square of a commensurable number,* a, b, c *themselves being supposed commensurable.*

Ex. 1. $6x^2-19x+15=6\{x^2-\frac{19}{6}x+\frac{15}{6}\}$,

$$=6\{x-2\tfrac{19}{12}x+(\tfrac{19}{12})^2-(\tfrac{19}{12})^2+\tfrac{15}{6}\},$$
$$=6\{(x-\tfrac{19}{12})^2-(\tfrac{1}{12})^2\},$$
$$=6\{x-\tfrac{19}{12}+\tfrac{1}{12}\}\{x-\tfrac{19}{12}-\tfrac{1}{12}\}$$
$$=6(x-\tfrac{3}{2})(x-\tfrac{5}{3}),$$
$$=(2x-3)(3x-5),$$

a result obtained otherwise in § 121, Example 2.

Ex. 2. $4x^2+12x+2$. Here $a=4$, $b=12$, $c=2$. Hence $\Delta=b^2-4ac=144-32=+112$, which is not a perfect square. The factors are therefore real but not rational. We may shorten the work a little by regarding $2x$ as the variable instead of $x$, thus—

$$4x^2+12x+2=(2x)^2+6(2x)+2,$$
$$=(2x)^2+6(2x)+3^2-3^2+2,$$
$$=(2x+3)^2-7=(2x+3)^2-(\sqrt{7})^2,$$
$$=(2x+3-\sqrt{7})(2x+3+\sqrt{7}).$$

Ex. 3. $12x^2+12x+3$. Here $\Delta=144-144=0$; therefore the function is a perfect square as regards $x$. We have, in fact—

$$12x^2+12x+3=3(2x)^2+3\,.\,2(2x)+3,$$
$$=3\{(2x)^2+2(2x)+1\},$$
$$=3(2x+1)^2.$$

Hence $$12x^2+12x+3=\{\sqrt{3}(2x+1)\}^2.$$

Ex. 4. $9x^2+6x+6$. Here $\Delta=36-216=-180$. Therefore the factors must be imaginary. We have—

$$\begin{aligned}9x^2+6x+6&=(3x)^2+2(3x)+1+5\\&=(3x+1)^2-5i^2=(3x+1)^2-(i\sqrt{5})^2,\\&=(3x+1-i\sqrt{5})(3x+1+i\sqrt{5}).\end{aligned}$$

## EXERCISES XXXIV.

1. $x^2-9x-52$.
2. $29x^2+x-30$.
3. $x^2+6x-6$.
4. $4x^2+4x-2$.
5. $18x^2+24x-154$.
6. $4x^2+12x+4$.
7. $4x^2+12x+6$.
8. $78x^2-149x+11$.
9. $4x^2+6x+7$.
10. $9x^2-24x+25$.
11. $11211x^2+91x-11526$.
12. $4x^2+16x+19$.
13. $x^2+6x+13$.
14. $4x^2+4x+4$.
15. $16x^2+8x+5$.
16. $\frac{5}{21}x^2-\frac{1117}{630}x-\frac{7}{5}$.
17. $279x^2-610x+299$.
18. $4x^4+4x^2-16$.
19. $(x^2+a-1)^2-a^2x^2$.
20. $px^2+(p+q)x+q$.
21. $x^2-2(m+n)x+2(m^2+n^2)$.
22. $x^3-2bx^2-(a^2-b^2)x$.
23. $(a+b)x^2+2(a^2+b^2)x+a^3+b^3$; show that the factors are real or imaginary according as $a$ and $b$ are unlike or like in sign.
24. Find four values of $a$ for which $6x^2+ax-35$ is resolvable into linear factors whose coefficients are integral numbers. State how many more such values of $a$ could be found by your method.

§ **131.** It should be observed that the factorisation for $ax^2+bx+c$ leads at once to the factorisation of the **homogeneous function** $ax^2+bxy+cy^2$ of the second degree in two variables; for

$$\begin{aligned}ax^2+bxy+cy^2&=ay^2\left\{\left(\frac{x}{y}\right)^2+\frac{b}{a}\left(\frac{x}{y}\right)+\frac{c}{a}\right\},\\&=ay^2\left\{\frac{x}{y}+\frac{b+\sqrt{\Delta}}{2a}\right\}\cdot\left\{\frac{x}{y}+\frac{b-\sqrt{\Delta}}{2a}\right\},\\&=a\left\{x+\frac{b+\sqrt{\Delta}}{2a}y\right\}\left\{x+\frac{b-\sqrt{\Delta}}{2a}y\right\},\end{aligned}$$

if we suppose $\Delta$ positive.

By operating in a similar way any homogeneous function of two variables may be factorised, provided a certain non-homogeneous function of one variable, having the same coefficients, can be factorised.

Ex. 1. From

$$x^2+2x+3=(x+1+i\sqrt{2})(x+1-i\sqrt{2}),$$

we deduce

$$x^2+2xy+3y^2=\{x+(1+i\sqrt{2})y\}\{x+(1-i\sqrt{2})y\}.$$

Ex. 2. From

$$x^3-2x^2-23x+60=(x-3)(x-4)(x+5),$$

we deduce

$$x^3-2x^2y-23xy^2+60y^3=(x-3y)(x-4y)(x+5y).$$

§ **132. By using the principle of substitution** a great many apparently complicated cases may be brought under the case of the quadratic function, or under other equally simple forms. The following are some examples :—

Ex. 1.

$$\begin{aligned} x^4+x^2y^2+y^4 &= (x^2+y^2)^2-(xy)^2, \\ &= (x^2+y^2+xy)(x^2+y^2-xy), \\ &= \left\{\left(x+\frac{1}{2}y\right)^2+\frac{3}{4}y^2\right\}\left\{\left(x-\frac{1}{2}y\right)^2+\frac{3}{4}y^2\right\}, \\ &= \left\{\left(x+\frac{1}{2}y\right)^2-\left(\frac{\sqrt{3}}{2}yi\right)^2\right\}\left\{\left(x-\frac{1}{2}y\right)^2-\left(\frac{\sqrt{3}}{2}yi\right)^2\right\}, \\ &= \left\{x+\left(\frac{1}{2}+\frac{\sqrt{3}}{2}i\right)y\right\}\left\{x+\left(\frac{1}{2}-\frac{\sqrt{3}}{2}i\right)y\right\} \\ &\qquad \left\{x+\left(-\frac{1}{2}+\frac{\sqrt{3}}{2}i\right)y\right\}\left\{x+\left(-\frac{1}{2}-\frac{\sqrt{3}}{2}i\right)y\right\}. \end{aligned}$$

Here the student should observe that, if resolution into *quadratic* factors only is required, it can be effected with real coefficients; but, if the resolution be carried to *linear* factors, complex coefficients have to be introduced.

Ex. 2.

$$\begin{aligned} x^3+y^3 &= (x+y)(x^2-xy+y^2) \\ &= \{x+y\}\left\{x+\left(-\frac{1}{2}+\frac{\sqrt{3}}{2}i\right)y\right\}\left\{x+\left(-\frac{1}{2}-\frac{\sqrt{3}}{2}i\right)y\right\}. \end{aligned}$$

Ex. 3.

$$\begin{aligned} x^4+y^4 &= (x^2+y^2)^2-2x^2y^2 \\ &= (x^2+y^2)^2-(\sqrt{2}xy)^2 \\ &= (x^2+\sqrt{2}xy+y^2)(x^2-\sqrt{2}xy+y^2). \end{aligned}$$

Again,

$$\begin{aligned} x^2+\sqrt{2}xy+y^2 &= \left(x+\frac{\sqrt{2}}{2}y\right)^2+\frac{2}{4}y^2 \\ &= \left(x+\frac{\sqrt{2}}{2}y\right)^2-\left(\frac{\sqrt{2}}{2}iy\right)^2 \\ &= \left\{x+\frac{\sqrt{2}}{2}(1+i)y\right\}\left\{x+\frac{\sqrt{2}}{2}(1-i)y\right\}. \end{aligned}$$

The similar resolution for $x^2-\sqrt{2}xy+y^2$ will be obtained by changing the sign of $\sqrt{2}$. Hence, finally—

$$\begin{aligned} &x^4+y^4 \\ &= \left\{x+\frac{\sqrt{2}}{2}(1+i)y\right\}\left\{x+\frac{\sqrt{2}}{2}(1-i)y\right\}\left\{x-\frac{\sqrt{2}}{2}(1+i)y\right\} \\ &\qquad \left\{x-\frac{\sqrt{2}}{2}(1-i)y\right\}. \end{aligned}$$

Ex. 4.

$$\begin{aligned} x^{12}-y^{12} &= (x^6)^2-(y^6)^2 \\ &= (x^6-y^6)(x^6+y^6) \\ &= \{(x^2)^3-(y^2)^3\}\{(x^2)^3+(y^2)^3\} \\ &= (x^2-y^2)(x^4+x^2y^2+y^4)(x^2+y^2)(x^4-x^2y^2+y^4) \\ &= (x+y)(x-y)(x+iy)(x-iy)(x^4+x^2y^2+y^4)(x^4-x^2y^2+y^4), \end{aligned}$$

where the last two factors may be treated as in Example 1.

Ex. 5.

$$\begin{aligned} x^6-7x^3+10 &= (x^3)^2-7(x^3)+10, \\ &= (x^3-2)(x^3-5), \text{ etc.} \end{aligned}$$

Ex. 6. The so-called reciprocal biquadratic integral function $Ax^4+Bx^3+Cx^2+Bx+A$, in which the first and last, second and next to last coefficients are equal, may be treated thus—

$$\begin{aligned} Ax^4+Bx^3+Bx^2+Bx+A &= A(x^4+1)+B(x^2+1)x+Cx^2, \\ &= A(x^2+1)^2+B(x^2+1)x+(C-2A)x^2. \end{aligned}$$

If we now treat this last as a homogeneous quadratic function of $x^2+1$ and $x$, we can resolve it into $A(x^2+1+ax)(x^2+1+\beta x)$, where $a$ and $\beta$ are certain functions of A, B, C. Then each of the functions $x^2+ax+1$ and $x^2+\beta x+1$ can be resolved as ordinary quadratics.

## EXERCISES XXXV.

Factorise the following:—

1. $23x^2+24xy+25y^2$.
2. $(x^4-y^4)^2+(x^2-y^2)^4$.
3. $3x^4-2x^2-16$.
4. $x^4+x^2-1$.
5. $x^2y^2-4xy+25$.
6. $27(x+1)^3+8$.
7. $(2x+3)^3+(3x+2)^3$.
8. $x^{12}-x^6y^6$.
9. $x^6+64y^6$.
10. $(x-1)(x-y-z-1)+yz$.
11. $(x^2+x)^2+4(x^2+x)-12$.
12. $x^3+3x^2+3x+1$.
13. $2x^3-7x^2+7x-2$.
14. $6x^4+35x^3+62x^2+35x+6$.
15. $x^4-3x^3+x^2-3x+1$.
16. $x^4+x^3y+x^2y^2+xy^3+y^4$.
17. $x^4-x^3y+x^2y^2-xy^3+y^4$.
18. $x^5-y^5$
19. $x^5+y^5$.
20. Show that $Ax^4+Bx^3+Cx^2-Bx+A$ can be factorised by a method similar to that given in § 132, Ex. 6.
21. $2x^4+3x^3-6x^2-3x+2$.
22. $x^6-3x^5-2x^4+2x^2+3x-1$.
23. $x^5+3x^4-2x^3-2x^2+3x+1$.
24. $x^5+x^4y+x^3y^2+x^2y^3+xy^4+y^5$.
25. $x^5+x^4y+x^3y^2-x^2y^3-xy^4-y^5$.
26. $(a-b)^6+64a^3b^3$.

### Use of the Remainder Theorem

§ **133.** Inasmuch as the remainder theorem is virtually a test for the existence or non-existence of a given linear factor in any given integral function of $x$, it is very useful in factorisation. For our present purpose we may state it thus: *if $a$, $\beta$, $\gamma$, . . . be values of* x *for which any integral function of* x *vanishes, then*

$x-\alpha$, $x-\beta$, $x-\gamma$, . . . *are factors of that function.* It should be noticed that, if $x-a$ occur twice in the integral function $f(x)$, not only will $f(x)$ vanish when $x=a$, but the quotient of $f(x)$ by $x-a$ will also vanish when $x=a$.

In the present connection the reader should study § 115, Ex. 2, which is virtually a factorisation theorem.

Ex. 1. $F \equiv x^4-12x^3+51x^2-92x+60$.

Since 2 is a factor in 60, there is reason to suspect that $x-2$ may be a factor in F. Let us calculate the quotient and remainder corresponding to the divisor $x-2$ by the method of § 113, Ex. 1. The remainder is 0; hence $x-2$ is a factor; and the quotient is $x^3-10x^2+31x-30$. We may try whether this also has the factor $x-2$. The remainder is again 0; hence $x-2$ is a factor in $x^3-10x^2+31x-30$; and the quotient is $x^2-8x+15$. Since this last function is $(x-3)(x-5)$, we see finally that the given function has been resolved into $(x-2)^2(x-3)(x-5)$.

$$\begin{array}{r|l} & 1-12+51-92+60 \\ 2 & 0+\ 2-20+62-60 \\ \hline & 1-10+31-30 \mid +0 \\ 2 & 0+\ 2-16+30 \\ \hline & 1-\ 8+15 \mid +0 \end{array}$$

## EXERCISES XXXVI.

1. Given that $x^4+x^3-10x^2-4x+24$ vanishes when $x=2$ and when $x=-3$, resolve the function into linear factors.

2. Show that $x^3+15x^2+74x+120$ has the factor $x+5$, and find the other factors.

3. Show that $x^4-3x^3-7x^2+27x-18$ contains the factors $x-1$ and $x-2$, and find the other two factors.

4. For what numerical values of $p$ can the fraction $\{2px^2+(3p+4)x+7\}/(x+1)(x+2)$ be reduced to lower terms?

Factorise the following:—

5. $x^3-2x^2-x+2$.
6. $x^3-3x^2+2$.
7. $2x^3+3x^2+2x-7$.
8. $x^3-x-6$.
9. $4x^3-16x^2+9x+9$.
10. $x^4-8x^3+21x^2-20x+4$.

11. Determine $\alpha$ and $\beta$ so that $x^4+\alpha x^2y^2-4xy^3+\beta y^4$ may have the factor $(x-y)^2$.

12. $x^6-2x^3+1-(x^2-1)^2$.

13. $(a-1)x^3-ax^2-(a-3)x+(a-2)$.

14. $(1+x)^2(1+y^2)-(1+x^2)(1+y)^2$.

15. $\Sigma(y-z)^3$.

16. $\Sigma(x^2-yz)^2(z-y)=\Sigma x\Sigma x^3(z-y)=(\Sigma x)^2(y-z)(z-x)(x-y)$.

17. $x^3+(2a+1)x^2+x(a^2+2a-1)+a^2-1$.

18. Find the conditions that $(x-a)^2$ be a factor in $x^3+px+q$.

19. Show that $b+c$ is a factor in $\Sigma a^3b+2\Sigma a^2b^2+4\Sigma a^2bc$, and find all the other factors.

20. Factorise $\Sigma a^3(b^2-c^2)$ as far as you can.

## FACTORISATION OF FUNCTIONS OF MORE THAN ONE VARIABLE

§ **134.** In general an integral function of more than one variable cannot be factorised. Thus, for example, the integral function of $x$ and $y$ of the second degree and the connected homogeneous integral function of $x$, $y$, $z$, viz. $ax^2+2hxy+by^2+2gx+2fy+c$ and $ax^2+by^2+cz^2+2fyz+2gzx+2hxy$, cannot be resolved into two linear factors, unless its discriminant, $abc+2fgh-af^2-bg^2-ch^2$, vanish.* When this happens, its factors can always be found by the method employed in the following example :—

Ex. 1. Let, if possible, $6x^2-5xy+y^2+x-y-2\equiv(lx+my+n)(px+qy+r)$. Since terms of one degree cannot be transformed into terms of another degree, the terms of the second degree on the right must be identically equal to the terms of the same degree on the left. Hence $(lx+my)(px+qy)\equiv 6x^2-5xy+y^2\equiv(3x-y)(2x-y)$. We may therefore assume $lx+my\equiv 3x-y$, and $px+qy\equiv 2x-y$. We now have

$$6x^2-5xy+y^2+x-y-2\equiv(3x-y+n)(2x-y+r),$$
$$\equiv 6x^2-5xy+y^2+(2n+3r)x-(n+r)y+nr\;;$$

and the question is whether we can determine $n$ and $r$ so that this identity shall be complete.

For this it is necessary and sufficient that $2n+3r=1$, $n+r=1$, $nr=-2$. From the first two equations we get $n=2$, $r=-1$. It so happens in this case that these two values also satisfy the third equation. We can therefore complete the identity ; and the factorisation is possible, viz. $6x^2-5xy+y^2+x-y-2\equiv(3x-y+2)(2x-y-1)$.

In any case taken at random the three equations corresponding to the above would in general be inconsistent ; and this would show that the factorisation is impossible.

Ex. 2. Factorise $F\equiv b^2c^2(b-c)+c^2a^2(c-a)+a^2b^2(a-b)$. Exactly as in § 115, Ex. 2, we can show that $b-c$, $c-a$, $a-b$ are all factors in F. Hence $F/(b-c)(c-a)(a-b)$ is an integral function of $a$, $b$, $c$ of the second degree. Moreover, if we interchange any pair of $a$, $b$, $c$, both F and $(b-c)(c-a)(a-b)$ simply change sign, and therefore $F/(b-c)(c-a)(a-b)$ remains unaltered. This last is therefore a symmetric function of $a$, $b$, $c$ ; and this function must be an integral function of $a$, $b$, $c$ of the second degree. There exists therefore an identity of the form

$$b^2c^2(b-c)+c^2a^2(c-a)+a^2b^2(a-b)$$
$$\equiv(b-c)(c-a)(a-b)\{A(a^2+b^2+c^2)+B(bc+ca+ab)\},$$

where A and B are numerical coefficients.

On the right there occurs a term $-Aa^4b$, and on the left no such term ; hence we must have $A=0$. Finally, if we put $c=0$, the

---

* See A. VII. §§ 12, 13.

identity reduces to $a^2b^2(a-b)\equiv -Ba^2b^2(a-b)$, from which it is obvious that $B=-1$. Hence

$$b^2c^2(b-c)+c^2a^2(c-a)+a^2b^2(a-b)\equiv -(b-c)(c-a)(a-b)(bc+ca+ab).$$

## EXERCISES XXXVII.

Factorise the following :—

1. $xy-3y+5x-15$.
2. $2xy+7x+6y+21$.
3. $x^2+3xy+8x+18y+12$.
4. $x^2-y^2+2y-1$.
5. $x^2+xy+2x+3y-3$.
6. $x^2-y^2+x-3y-2$.
7. $2xy+7x+6y+22$.
8. $x^2-xy+x+y-2$.
9. $3x^2-12xy+12y^2-1$.
10. $x^2-4xy+4y^2-x+2y-12$.
11. $6x^2-13xy+6y^2+22x-23y+20$.
12. $3x^2-4xy+y^2+2x-1$.
13. $3x^2+2xy-y^2-x+3y-2$.
14. $6x^2+13xy+6y^2+x-y-1$.

15. Find the necessary and sufficient condition that $xy+px+qy+r$ may be resolvable into two linear factors, $p$, $q$, $r$ being numerical coefficients.

16. Show that $xy+1$ cannot be resolved into linear factors.

17. Determine $\lambda$ so that $6x^2-11xy-10y^2-19y+\lambda$ may be expressible as the product of two linear factors.

18. Factorise $ab(x^2-y^2)-(a^2-b^2)(xy+1)-(a^2+b^2)(x+y)$.

19. Show that $x^2+2xy+x+y+1$ cannot be resolved into linear integral factors.

20. Factorise $(x+y)^3+2xy(1-x-y)-1$.

21. Factorise as far as possible $x^3+3x^2y+3xy^2+y^3+x^2+3xy+2y^2$.

## EXERCISES XXXVIII.

Factorise the following :—

1. $(x-1)^3+(x-2)^3+(3-2x)^3$.

2. Show that $x^5+x^3+x^2+1$ contains the factor $x^2-x+1$ and find the other factors.

3. $(a+b)^4+(c+d)^4$.
4. $(x^2+x+1)^3+(x^2+3x+1)^3$.
5. $x^3-ax^2-(2a^2+b^2)x-ab^2$.
6. $\Sigma x^2+\Sigma(b^2+c^2)yz/bc$.
7. $(a+b)^3+(a+d)^3+(b+c)^3+(c+d)^3$.
8. $\Sigma x^2+\Sigma(a^2/bc+bc/a^2)yz$.
9. $(x+y+z)^3+(x+y-z)^3+(x+z-y)^3-(y+z-x)^3$
10. $x^3+2(p-1)x^2+(p^2-q^2-4p)x-2(p^2-q^2)$.
11. $(x-a)^3+(x-a-b+c)^3+(x-a+b-c)^3$.
12. $\Sigma(y+z)(y^3-z^3)$.
13. $a^{m+1}+b^{n+1}+a^mb+ab^n+a^3+b^3$.
14. $(\Sigma x)^5-\Sigma x^5=\frac{5}{2}\Pi(y+z)\Sigma(y+z)^2$.
15. Find the two quadratic factors of $\Sigma(yz-x^2)^2$.
16. Find two linear factors and a quadratic factor of $(x^2-yz)^2-(y^2-zx)(z^2-xy)$.
17. Show that $\Sigma yz$ is a factor in $\Sigma(y^2-zx)(z^2-xy)$; and find the other quadratic factor.
18. Factorise $\Sigma(y-z)^5$ so far as it is possible to do so; and show

that the function cannot vanish unless two of the variables have equal values.

**19.** If $a$ and $b$ be two odd integers, $a^4-b^4$ is divisible by 8.

**20.** Show that $(b-c)(c-a)(a-b)$ is a factor in $\Sigma a^n(b-c)$; and that the remaining factor is $a^{n-2}+a^{n-3}(b+c)+a^{n-4}(b^2+bc+c^2)+\ .\ .\ .\ +(b^{n-2}+b^{n-3}c+\ .\ .\ .\ +c^{n-2})$.

## Square, Cube, or other Root of an Integral Function of $x$

**§ 135.** When an integral function of $x$ happens to be the square, cube, or any other power of an integral function of $x$, its root can always be readily found by the method of indeterminate coefficients, as will be understood from the examples given below. There is a special algorithm for calculating the root,* but it is of very little practical use; and the underlying theory is too complicated to be worth giving here.

Ex. 1. Given that $4x^8+12x^7+5x^6-2x^5-x^4-14x^3+5x^2-4x+4$ is the square of an integral function of $x$, find its root. The degree of the root must be the 4th, and (§ 97) its highest term must obviously be $+2x^4$; we shall take $2x^4$ (the term $-2x^4$ will obviously belong to the value of the root which we get by changing the sign of every term). We must therefore have

$$4x^8+12x^7+5x^6-2x^5-x^4-14x^3+5x^2-4x+4$$
$$\equiv(2x^4+px^3+qx^2+rx+s)^2,$$

where $p$, $q$, $r$, $s$ are numerical coefficients, which must be determined so as to make the supposed identity complete. We calculate the coefficients of the five highest terms on the right, using detached coefficients, thus—

$$\begin{array}{lllll}
2+p\ +q & +r & +s & & \\
2+p\ +q & +r & +s & & \\
\hline
4+2p+2q & +2r & +2s & & \\
\quad 2p+p^2 & +pq & +pr & & \ldots \\
\qquad\ 2q & +pq & +q^2 & & \ldots \\
 & \ \ 2r & +pr & & \ldots \\
 & & \ \ 2s & & \ldots \\
\hline
\multicolumn{5}{l}{4+4p+(4q+p^2)+(4r+2pq)+(4s+2pr+q^2)\ \ldots}
\end{array}$$

Equating these to the first five coefficients on the left respectively, we get $4=4$, $4p=12$, $4q+p^2=5$, $4r+2pq=-2$, $4s+2pr+q^2=-1$. The second of these equations gives $p=3$; the third $4q+9=5$, whence $q=-1$; the fourth $4r-6=-2$, whence $r=+1$; the fifth $4s+6+1=-1$, whence $s=-2$. Assuming that the function is a perfect square, its root is therefore $\pm(2x^4+3x^3-x^2+x-2)$.

It will be noticed that we have only equated five pairs of co-

---

* See A. XI. § 17.

efficients; the identification of the remaining four pairs will give four more equations which the values of $p$, $q$, $r$, $s$ just found must satisfy, since the given function is a perfect square. If the function were not a perfect square, the values found by means of the first set of equations would not satisfy the rest. If, therefore, we have any doubt as to whether the radicand is a perfect square, we must either write down the remaining equations and test whether they are satisfied, or else square the function arrived at by means of the first set, and see whether the result is the given radicand.

Ex. 2. Find the cube root of $8x^6-36x^5+78x^4-99x^3+78x^2-36x+8$, which is a perfect cube as regards $x$. Confining ourselves to that root whose coefficients are all real,* we must have

$$8x^6-36x^5+78x^4-99x^3+78x^2-36x+8=(2x^2+px+q)^3.$$

Since we have two unknown coefficients $p$ and $q$ to determine, we must calculate the coefficients of the first three terms on the right. The work may be arranged thus—

$$\begin{array}{llll}
2+p & +q & & \\
\underline{2+p} & \underline{+q} & & \\
4+2p & +2q & & \\
\phantom{4+}2p & +p^2 & & \ldots \\
 & \phantom{+}2q & & \ldots \\
\hline
4+4p & +(4q & +p^2) & \ldots \\
\underline{2+p} & \underline{+q} & & \\
8+8p & +(8q & +2p^2) & \ldots \\
\phantom{8+}4p & +4p^2 & & \ldots \\
 & \phantom{+}4q & & \ldots \\
\hline
8+12p & +(12q & +6p^2) & \ldots
\end{array}$$

Equating coefficients, we have $8=8$, $12p=-36$, $12q+6p^2=78$. The second of these gives $p=-3$; the third $12q+54=78$, whence $q=2$. The cube root must therefore be $2x^2-3x+2$.

## EXERCISES XXXIX.

Find the square roots of the following functions:—

1. $(2x-3a)^2-4(2x-3a)+4$.
2. $(x^2+1)^2+4x(x^2-1)$.
3. $x^6+2x^5-5x^4+2x^3+17x^2-24x+16$.
4. $121x^6+44x^5-18x^4+18x^3+5x^2-2x+1$.
5. $1-4x+10x^2-20x^3+35x^4-44x^5+46x^6-40x^7+25x^8$.
6. $1-2x+5x^2-10x^3+18x^4-20x^5+25x^6-24x^7+16x^8$.
7. $(xy+x+y)^2-4xy(x+y)$.
8. $x^2-4xy+4y^2+6x-12y+9$.
9. $x^6+2x^3y^3+y^6+2x^5+2x^3y^2+2x^2y^3+2y^5+3x^4+2x^3y+2x^2y^2+2xy^3+3y^4+2x^3+2x^2y+2xy^2+2y^3+x^2+2xy+y^2$.

---

* The other two will be found by multiplying this one by the two imaginary cube roots of $+1$, viz. $(-1+\sqrt{3}i)/2$ and $(-1-\sqrt{3}i)/2$.

Find the cube roots of the following :—

**10.** $27x^6 - 81x^5 + 108x^4 - 81x^3 + 36x^2 - 9x + 1$.

**11.** $x^6 + 3x^5 + 6x^4 + 7x^3 + 6x^2 + 3x + 1$.

**12.** $x^6 + 6x^5 + 15x^4 + 20x^3 + 15x^2 + 6x + 1$.

**13.** $64x^6 - 192x^5 + 240x^4 - 160x^3 + 60x^2 - 12x + 1$.

**14.** Find a rational function whose cube is $x^3/8 + 1/2 + 2/3x^3 + 8/27x^6$.

Find rational functions whose squares are the following :—

**15.** $x^2/(x+1)^2 + (x+1)^2/x^2 - x/(x+1) + (x+1)/x - 7/4$.

**16.** $a^2x^4 + 2abx^3 + (b^2 + 2ab)x^2 + 2(b^2 + ab)x + (2a^2 + 3b^2)$
$+ 2(b^2 + ab)/x + (b^2 + 2ab)/x^2 + 2ab/x^3 + a^2x^4$.

**17.** $x^2 - 4x + 2 + 4/x + 1/x^2$.

**18.** Determine $\lambda$, $\mu$, $\nu$ so that $9x^2 + 2\lambda xy + 4y^2 + 2\mu x + 2\nu y + 4$ may be a complete square.

**19.** Determine $c$ and $d$ so that $x^4 + 12x^3 + 8x^2 + cx + d$ may be a complete square.

# CHAPTER XIV

## GREATEST COMMON MEASURE AND LEAST COMMON MULTIPLE

§ **136.** Two given integral functions of any given variables $x, y, z, \ldots$ have in general no common factor; in other words, there exists in general no integral function of $x, y, z, \ldots$ which divides each of the two exactly; they are then said to be **Prime** to each other.

On the other hand, two or more integral functions of $x, y, z, \ldots$ may have a common factor; in such a case *the integral function of highest degree in* x, y, z, . . . *which divides all the given functions exactly, is called the* **Greatest Common Measure** *(G.C.M.) of these functions.*

The only case of any practical importance, when more than one variable is considered, is that where all the integral functions are monomials. In this case the G.C.M. can be found by inspection. Thus, for example, the G.C.M. of $12x^2y^3zu$, $6x^3y^4z^2u^2$, and $24x^6y^2z^3$ is obviously $Ax^2y^2z$ where A is any constant; for this function will divide each of the given monomials exactly, and no monomial of higher degree in any one of the variables will. The rule for finding the variable part of the G.C.M. in such cases is obviously as follows:—*Write down the product of all the variables that occur in all of the given monomials, each raised to the lowest power in which it occurs in any one of them.* It will be observed that the coefficient is, so far as the definition is concerned, arbitrary; thus, for example, $2x^2y^2z$, $\frac{1}{2}x^2y^2z$, $6x^2y^2z$ will all divide the three monomials above exactly, and each is of the highest possible degree in $x, y, z, u$.* It is usual to make the arbitrary coefficient unity.

* Frequently the instruction is added to make the coefficient the arithmetical G.C.M. of the coefficients. It is better to omit this, because

§ **137.** By far the most important case is that where the functions considered are integral functions of a single variable. *Two integral functions of* x *which have no common factor—that is, which are not exactly divisible by any common integral function of* x *of degree other than zero—are said to be* **prime** *to each other.*

Ex. $x-1$ and $x^2+1$ are prime to each other, for the only integral function of $x$ that will divide $x-1$ exactly is $x-1$ itself (or any constant multiple of $x-1$, which for our present purpose is the same thing); and since $1^2+1 \neq 0$, $x-1$ does not divide $x^2+1$ exactly.

*The integral function of* x *of highest degree which divides each of two or more given integral functions of* x *exactly, is called the* **Greatest Common Measure** (*G.C.M.*) *of these functions.*

Ex. $x-1$, or any constant multiple of $x-1$, is the integral function of highest degree that divides $x-1$ exactly, and it divides $x^2-1$ exactly; hence $x-1$ (or any constant multiple of $x-1$) is the G.C.M. of $x-1$ and $x^2-1$.

It will be noticed that here, as in the case of monomials, the G.C.M. is arbitrary to the extent of a constant factor; this factor is usually taken to be unity, or the smallest number, or simplest function of constants that will render all the coefficients of the G.C.M. integral, but this is purely a matter of convenience.

§ **138.** We may caution the beginner against confusing the notion of the algebraic G.C.M. with the notion of the arithmetic G.C.M. He will note that in the definition of the algebraic G.C.M. no mention is made of arithmetical magnitude whatever. The question, as always in Algebra, is regarding *form.* The words "highest" and "greatest" refer merely to degree. It is not even true that the arithmetical G.C.M. of the two numerical values of two given functions of $x$, obtained by giving a particular value to $x$, is the arithmetical value of the algebraic G.C.M. when that particular value of $x$ is substituted therein. Thus $x+1$ and $x^2+1$ have no algebraic G.C.M. at all, but when $x=5$, $x+1=6$, $x^2+1=26$; and 6 and 26 have the arithmetic G.C.M. 2.* There is, in fact, no

it tends to introduce confusion in a matter where confusion is rife enough already; moreover, it is altogether inappropriate to such a case as $\frac{1}{2}x^3y^3zu$, $\frac{1}{3}x^3y^4z^2u^2$, $\frac{1}{4}x^6y^2z^3$, where the coefficients are numerical fractions.

* On account of the distinction here emphasised it has become common of late to call the G.C.M. and the L.C.M. the *Highest Common Factor* and the *Lowest Common Multiple.* There can be no objection to this; but the

fundamental connection between the two notions at all, although there are certain analogies between the two theories that are built upon these notions. The learner must, therefore, beware of confusion and looseness of statement in the demonstration of propositions in the algebraic theory.*

## G.C.M. OBTAINED BY FACTORISATION OR OTHER TENTATIVE PROCESS

§ **139.** When one of the given integral functions has been factorised into linear factors, the G.C.M. can be found without difficulty by means of the remainder theorem; and it can be written down at once when each of the functions can be factorised in such a way that every factor is prime to every other, whether in the same or in another function.

Ex. 1. $2x^2-6x+4$ and $6x^2-6x-12$. We have $2x^2-6x+4=2(x^2-3x+2)=2(x-1)(x-2)$. If we put $x=1$ in the second function, it does not vanish, hence $x-1$ is not a factor in the second function. On the other hand, it does vanish when $x=2$, hence $x-2$ is a factor in both functions; and there is no other—that is to say, $x-2$ is the G.C.M.

Ex. 2.
$$A\equiv x^5-5x^4+7x^3+x^2-8x+4;$$
$$B\equiv x^6-7x^5+17x^4-13x^3-10x^2+20x-8;$$
$$C\equiv x^5-3x^4-x^3+7x^2-4.$$

Starting with C as the simplest of the three, we notice at once that $C=0$ when $x=1$, and also when $x=-1$. Hence $x-1$ and $x+1$ are both factors; and we readily find that $x+1$ occurs twice. The remaining factor is $(x-2)^2$; hence $C=(x-1)(x+1)^2(x-2)^2$. Trying these factors in succession for A and B, we find without difficulty, $A\equiv(x-1)^2(x+1)(x-2)^2$; $B\equiv(x-1)^2(x+1)(x-2)^3$. It is now obvious that the G.C.M. is $(x-1)(x+1)(x-2)^2$, *i.e.* $x^4-4x^3+3x^2+4x-4$.

§ **140.** The following proposition is very useful in establishing conclusions regarding the G.C.M. of two integral functions:—

*If* A *and* B *be integral functions of* x, *and* m *and* n *either constants or integral functions of* x, *then any common factor of* A *and* B *is a factor of* mA + nB.†

reasons given often suggest that the reformers are not aware that there is a similar difference in the use of almost all the technical words of elementary algebra, *e.g.* "integral," "fractional," "factor," "exactly divisible," "proper fraction," etc.

* All the more because nonsense has occasionally been printed on the subject.

† The following converse is true and is often useful:—*Any factor of* mA + nB *which is also a factor of* A *and not of* n, *or a factor of* B *and not of* m, *is a common factor of* A *and* B.

To prove this let us suppose that P is a common factor of A and B, then $A \equiv aP$ and $B \equiv bP$ by hypothesis, where $a$, $b$, P are all integral functions of $x$. Hence $mA + nB \equiv maP + nbP \equiv (ma + nb)P$. Now, since $m$, $n$, $a$, $b$ are all integral functions, $ma + nb$ is an integral function. Hence P is a factor of $mA + nB$.

The reader should notice the generality of this proposition. $m$ and $n$ may be any integral functions whatsoever, or any constants—in particular, any arithmetical numbers with positive or negative signs attached. The following examples will show how the theorem can be utilised.

Ex. 1. Consider the functions $A \equiv x^2 - x + 1$ and $B \equiv x^2 + x + 1$. By the above theorem (putting $m = 1$, $n = -1$) we see that any common factor of A and B is a factor in $A - B$, that is, in $-2x$. Now $x$, the only factor in this last, is not a factor in A, and therefore cannot be a common factor of A and B. We conclude that $x^2 - x + 1$ and $x^2 + x + 1$ have no common factor.

Ex. 2. $A \equiv 2x^4 - 3x^3 - 3x^2 + 4$, $B \equiv 2x^4 - x^3 - 9x^2 + 4x + 4$. We have $A - B \equiv -2x^3 + 6x^2 - 4x \equiv -2x(x^2 - 3x + 2) \equiv -2x(x-1)(x-2)$. The common factors of A and B, if any, being factors of $A - B$, must be among the factors $x$, $x - 1$, $x - 2$ of this last. Clearly $x$ is not a factor of A or B: on the other hand, we find at once, by the remainder theorem, that both $x - 1$ and $x - 2$ are factors both of A and B. Hence the G.C.M is $(x-1)(x-2)$, that is, $x^2 - 3x + 2$.

Ex. 3. Find what relation or relations must connect $p$, $q$, $p'$, $q'$ in order that $A \equiv x^2 + px + q$ and $B \equiv x^2 + p'x + q'$ may have a common factor. Since $A - B \equiv (p - p')x + (q - q')$, the common factor must be a factor of $(p - p')x + (q - q')$, that is, a constant multiple of $(p - p')x + (q - q')$ itself. Moreover, since $A \equiv B + \{(p - p')x + (q - q')\}$, if $(p - p')x + (q - q')$ be actually a factor in B, it must also be a factor in A. The necessary and sufficient condition then is simply that $(p - p')x + (q - q')$ be a factor in B, which by the remainder theorem in its general form, § 115, is

$$\{-(q - q')/(p - p')\}^2 + p'\{-(q - q')/(p - p')\} + q' = 0\ ;$$

or, since $p - p'$ may be supposed $\neq 0$, the condition is

$$(q - q')^2 - p'(q - q')(p - p') + q'(p - p')^2 = 0\ ;$$

that is, $p$, $p'$, $q$, $q'$ must satisfy the single relation

$$q^2 - 2qq' + q'^2 - pp'q - pp'q' + p^2q' + p'^2q = 0.$$

If $p = p'$, then $A - B \equiv q - q'$. If therefore $q \neq q'$, A is prime to B. If $q = q'$, then, since $p = p'$, A and B are identical.

## EXERCISES XL.

Find the G.C.M of the following functions :—

1. $2x^3y^2$, $3x^5y$.
2. $18x^2y^2z$, $10x^3yz^3$, $12x^2y^2z^2$.
3. $\frac{2}{3}x^2y^2zu^2$, $\frac{3}{5}x^3y^3z^3u^3$, $\frac{7}{8}xyzu$.

4. $(a+b)x^2yz$, $(a^2-b^2)x^3yz^3$, $(a+b)x^2y^2z^2$.
5. $x^4y^2-x^2y^4$, $x^3y-xy^3$.
6. $6x^2(x-1)^2$, $8x^3(x-1)^3$.
7. $(x^2-1)^2$, $(x^3-1)^3$.
8. $(x^2-1)^2$, $(x^2-3x+2)^2$.
9. $x^2-y^2$, $x^4-y^4$, $x^6-y^6$.
10. $x^2y-2y^3$, $x^6-8y^6$.
11. $x^6-y^6$, $x^8-y^8$.
12. $x^2+6xy+8y^2$, $x^2+7xy+10y^2$.
13. $4x^2+4x-3$, $4x^2+8x-5$.
14. $x^2+2xy+y^2-x-y+1$, $(x+y)^4+(x+y)^2+1$.
15. $x^6-y^6$, $x^8+x^4y^4+y^8$.
16. $x^9+1$, $x^{11}+1$.
17. $x^4-2x+1$, $x^4-2x^2+1$.
18. $x^4-y^4$, $x^3+x^2y+xy^2+y^3$.
19. $x^2y^2-z^4$, $x^2y^2+5xyz^2-6z^4$.
20. $(x-2)(x-3)(x-4)$, $2x^3-13x^2+27x-18$.
21. $x^6-6x+5$, $x^3-3x+2$.
22. $x^3-6x^2+8x-3$, $x^3-7x^2+14x-8$.
23. $(x-1)^2$, $(x^2-1)^2$, $x^7-7x+6$.
24. $x^4+x^2+1$, $(x+1)^4+(x+1)^2+1$.
25. $x^3-3x+2$, $x^5-5x+4$, $x^7-7x+6$.

26. Prove that $x^2-x+1$ and $x^4-x^2+1$ have no common factor.
27. Prove that $x^2-x+1$ and $x^4+1$ are prime to each other.

### "Long Rule" for finding the G.C.M.

**§ 141.** The problem of finding the G.C.M. of two given integral functions (or of showing that they are prime to each other, as the case may be) can be solved by a direct process involving only rational operations, which is fundamental in the theory of integral functions, and also in the theory of equations.

The central point of the process is the following simple theorem :—

*If* A, B, Q, R *be all integral functions of* x, *and if* $A \equiv BQ + R$, *then the G.C.M. of* A *and* B *is the same as the G.C.M. of* B *and* R.

To prove this, it is obviously necessary and sufficient to show (1) that every factor common to A and B is common to B and R; (2) that every factor common to B and R is common to A and B.

Now, since $A \equiv BQ + R$, it follows, by the theorem of § 140 that every factor common to B and R is a factor of A. Hence every factor common to B and R is common to A and B.

Again, from $A \equiv BQ + R$ we have $R \equiv A - BQ$: from which it follows that every factor common to A and B is a factor in R, and therefore a factor common to B and R.

*Let now* A *and* B *be two integral functions whose G.C.M. is required; and let* B *be the one whose degree is not greater than that of the other. Divide* A *by* B, *and let the quotient be* $Q_1$, *and the remainder* $R_1$.

*Divide* B *by* $R_1$, *and let the quotient be* $Q_2$, *and the remainder* $R_2$.

*Divide* $R_1$ *by* $R_2$, *and let the quotient be* $Q_3$, *and the remainder* $R_3$, *and so on.*

*Since the degree of each remainder is less by unity at least than the degree of the corresponding divisor,* $R_1$, $R_2$, $R_3$, *etc., go on diminishing in degree, and the process must come to an end in one or other of two ways.*

I. *Either the division at a certain stage becomes exact, and the remainder vanishes;*

II. *Or a stage is reached at which the remainder is reduced to a constant.*

Now we have, by the process of derivation above described,

$$\left.\begin{aligned} A &= BQ_1 + R_1 \\ B &= R_1Q_2 + R_2 \\ R_1 &= R_2Q_3 + R_3 \\ &\cdot\;\cdot\;\cdot\;\cdot\;\cdot\;\cdot \\ R_{n-2} &= R_{n-1}Q_n + R_n \end{aligned}\right\} \quad (1).$$

Hence, by the fundamental proposition, the pairs of functions

$$\left.\begin{matrix} A \\ B \end{matrix}\right\} \left.\begin{matrix} B \\ R_1 \end{matrix}\right\} \left.\begin{matrix} R_1 \\ R_2 \end{matrix}\right\} \left.\begin{matrix} R_2 \\ R_3 \end{matrix}\right\} \; \cdot\;\cdot\;\cdot \; \left.\begin{matrix} R_{n-2} \\ R_{n-1} \end{matrix}\right\} \left.\begin{matrix} R_{n-1} \\ R_n \end{matrix}\right\} \text{all have the same G.C.M.}$$

In Case I. $R_n = 0$ and $R_{n-2} = Q_nR_{n-1}$. Hence the G.C.M. of $R_{n-2}$ and $R_{n-1}$, that is, of $Q_nR_{n-1}$ and $R_{n-1}$, is $R_{n-1}$, for this divides both, and no function of higher degree than itself can divide $R_{n-1}$. Hence $R_{n-1}$ is the G.C.M. of A and B.

In Case II. $R_n$ = constant. In this case A and B have no G.C.M., for their G.C.M. is the G.C.M. of $R_{n-1}$ and $R_n$, that is, their G.C.M. divides the constant $R_n$. But no integral function (other than a constant) can divide a constant exactly. Hence A and B have no G.C.M. (other than a constant, which does not count).

*If, therefore, the process ends with a zero remainder, the last divisor is the G.C.M.; if it ends with a constant, there is no G.C.M., i.e. the two functions are prime to each other.*

The following examples illustrate the process in its simplest form :—

Ex. 1. Find the G.C.M. of $x^4+4x^3+8x^2+8x+3$ and $x^3+x^2+x-3$. If we use detached coefficients the work runs as follows:—

```
             1+4+8+ 8+ 3|1+1+1-3
             1+1+1- 3   |1+3
               3+7+11+ 3
               3+3+ 3- 9
1+1+1-3|         4+ 8+12
1+2+3  |         ¼- ¼
 -1-2-3
 -1-2-3
   0+0 .
```

The last remainder vanishes identically; hence the last divisor, viz. $4x^2+8x+12$, or, rejecting the irrelevant constant factor 4, $x^2+2x+3$, is the G.C.M.

Ex. 2. Consider the functions $x^3+2x^2+3x+4$ and $x^2+x+1$. The process in this case works out as follows:—

```
        1+2+3+4|1+1+1
        1+1+1  |1+1
          1+2+4
          1+1+1
1+1+1|      1+3
1+3  |      1-2
 -2+1
 -2-6
   +7 .
```

The last remainder being a non-evanescent constant, the two functions are prime to each other.

§ **142.** It is important to remark that it follows from the nature of the above process for finding the G.C.M., which consists essentially in substituting for the original pair of functions pair after pair of others which have the same G.C.M., that *we may, at any stage of the process, multiply either the divisor or the remainder by an integral function, provided we are sure that this function and the remainder or divisor, as the case may be, have no common factor. We may similarly remove from either the divisor or the remainder a factor which is not common to both. We may remove a factor which is common to both, provided we introduce it into the G.C.M. as ultimately found. It follows of course,* a fortiori, *that a numerical factor may be introduced into or removed from divisor or remainder at any stage of the process.* This last remark is of great use in enabling us to avoid fractions and otherwise simplify the arithmetic of the process. In order to obtain the full advantage of it, the student should notice that, in what has been said, "remainder" may mean, not only the remainder properly so called at the end of each separate division, but also, if we please, the "*residue in the middle of any such division*," that is to say, the dividend less the product of the divisor by the sum of the terms of the quotient already obtained: for all that is really necessary for a step in the process is that we have a relation of the form $R_p \equiv R_{p+1}Q_{p+2} + R_{p+2}$.

Some of these remarks are illustrated in the following examples :—

Ex. 1.

To find the G.C.M. of $x^5-2x^4-2x^3+8x^2-7x+2$ and $x^4-4x+3$.

$$
\begin{array}{r|l}
x^5-2x^4-2x^3+8x^2-7x+2 & x^4-4x+3 \\
x^5 \qquad\qquad -4x^2+3x \quad & \overline{x+1} \\
\hline
-2)\ -2x^4-2x^3+12x^2-10x+2 & \\
\hline
x^4+x^3-6x^2+5x-1 & \\
x^4 \qquad\qquad -4x+3 & \\
\hline
x^3-6x^2+9x-4 &
\end{array}
$$

$$
\begin{array}{r|l}
x^4 \qquad\qquad -4x+3 & x^3-6x^2+9x-4 \\
x^4-6x^3+9x^2-4x \quad & \overline{x+2} \\
\hline
3)\ 6x^3-9x^2 \qquad +3 & \\
\hline
2x^3-3x^2 \qquad +1 & \\
2x^3-12x^2+18x-8 & \\
\hline
9)\ 9x^2-18x+9 & \\
\hline
x^2-2x+1 &
\end{array}
$$

$$
\begin{array}{r|l}
x^3-6x^2+9x-4 & x^2-2x+1 \\
x^3-2x^2+x \quad & \overline{x+1} \\
\hline
-4)\ -4x^2+8x-4 & \\
\hline
x^2-2x+1 & \\
x^2-2x+1 & \\
\hline
0. &
\end{array}
$$

Hence the G.C.M. is $x^2-2x+1$.

It must be observed that what we have written in the place of quotients are not really quotients in the ordinary sense, owing to the rejection of the numerical factors here and there. In point of fact the quotients are of no importance in the process, and need not be written down; neglecting them, carrying out the subtractions mentally, and using detached coefficients, we may write the whole calculation in the following compact form :—

| | | | |
|---|---|---|---|
| | 1 − 2 − 2 + 8 − 7 + 2 | 1 + 0 + 0 − 4 + 3 | |
| ÷ −2 | −2 − 2 + 12 − 10 + 2 | 6 − 9 + 0 + 3 | ÷ 3 |
| | 1 + 1 − 6 + 5 − 1 | 2 − 3 + 0 + 1 | |
| | 1 − 6 + 9 − 4 | 9 − 18 + 9 | ÷ 9 |
| ÷ −4 | − 4 + 8 − 4 | 1 − 2 + 1 | |
| | 1 − 2 + 1 | | |
| | 0 | | |

G.C.M., $x^2 - 2x + 1$.

Ex. 2.

Required the G.C.M. of $4x^4+26x^3+41x^2-2x-24$ and $3x^4+20x^3+32x^2-8x-32$.

Bearing in mind the general principle on which the rule for finding the G.C.M. is founded, we may proceed as follows, in order to avoid large numbers as much as possible:—

| | | | |
|---|---|---|---|
| | 4 + 26 + 41 − 2 − 24 | 3 + 20 + 32 − 8 − 32 | |
| ×2 | 1 + 6 + 9 + 6 + 8 | 2 + 5 − 26 − 56 | |
| | 2 + 12 + 18 + 12 + 16 | −53 − 318 − 424 | ÷ −53 |
| | 7 + 44 + 68 + 16 | 1 + 6 + 8 | |
| | 1 + 29 + 146 + 184 | | |
| ÷23 | 23 + 138 + 184 | | |
| | 1 + 6 + 8 | | |
| | 0 | | |

The G.C.M. is $x^2+6x+8$.

Here the second line on the left is obtained from the first by subtracting the first on the right. By the general principle referred to, the function $x^4+6x^3+9x^2+6x+8$ thus obtained and $3x^4+20x^3+32x^2-8x-32$ have the same G.C.M. as the original pair. Similarly the fifth line on the left is the result of subtracting from the line above three times the second line on the right.

If the student be careful to pay more attention to the principle underlying the rule than to the mere mechanical application of it, he will have little difficulty in devising other modifications to suit particular cases.

## G.C.M. of any Number of Integral Functions

**§ 143.** It follows at once, by the method of proof given in § 141, that *every common divisor of two integral functions* A *and* B *is a divisor of their G.C.M.*

This principle enables us at once to find the G.C.M. of any number of integral functions by successive application of the process for two. Consider, for example, four functions of $x$, A, B, C, D. Let $G_1$ be the G.C.M. of A and B, then $G_1$ is divisible by every common divisor of A and B. Find now the G.C.M. of $G_1$ and C, $G_2$ say. Then $G_2$ is the divisor of highest degree that will divide A, B, and C. Finally, find the G.C.M. of $G_2$ and D, $G_3$ say. Then $G_3$ is the G.C.M. of A, B, C, and D.

For example, consider $x^3+2x^2-1$, $x^3-x^2-3x+2$ and $x^3+4x^2+2x-3$. The G.C.M. of the first two functions is $x^2+x-1$, the G.C.M. of $x^2+x-1$ and $x^3+4x^2+2x-3$ is $x^2+x-1$. Hence the G.C.M. of the three functions is $x^2+x-1$.

§ **144.** It is important to notice that *the G.C.M. of two integral functions whose coefficients are real commensurable numbers must have real commensurable coefficients;* for it can be obtained by means of the "Long Rule," which involves only rational operations. This remark is important theoretically, and is often useful in finding the G.C.M.

Ex. Consider the functions $x^3-1$ and $2x^4+5x^3+10x^2+8x+5$. The factors of the first are $x-1$ and $x^2+x+1$; the first of which is obviously not a factor in the second function. Since the factors of $x^2+x+1$ have imaginary coefficients, it follows by the theorem just established, that one of these by itself cannot be the G.C.M. Hence either $x^2+x+1$ itself is the G.C.M., or there is no G.C.M. Dividing the second function by $x^2+x+1$ we find that it is a factor. Hence $x^2+x+1$ is the G.C.M. of the two given functions.

## EXERCISES XLI.

Find the G.C.M. of the following:—

1. $x^4-x^3-x^2-x-2$, $x^4-2x^3-2x^2-2x-3$.
2. $x^4+5x^3+11x^2+13x+6$ and $x^3-x^2-3x-9$.
3. $x^4-3x^3-x^2-3x-2$, $x^4-3x^3-3x-1$.
4. $x^3+3x^2-4x-12$ and $x^4+5x^3+5x^2-5x-6$.
5. $x^4-x^3-3x^2-19x-10$, $x^4-x^3-2x^2-17x-5$.
6. $x^6+x^4-x^2-1$, $x^8-x^6-4x^4-x^2+1$.
7. $x^4-x^3+2x^2+x+3$, and $x^4+2x^3-x-2$.
8. $x^8+4x^6+8x^4+7x^2+4$, $x^8+4x^6-x^2-4$.
9. $2x^3-x^2-16x+15$, and $x^4+3x^3-x-3$.
10. $x^4+x^3+(a^2+1)x^2+a^2x+a^2$, and $3x^3-5x^2-5x-8$.
11. $2x^4-8x^3+x^2+11x+3$, and $2x^4-4x^3+x^2+x-3$.
12. $8x^9-12x^6+6x^3-1$, $2x^9-3x^6+3x^3-1$.
13. $x^4-4x^3-x^2-x+39$, and $x^4-x^3-x^2-9x-18$.
14. $x^4+2x^3+4x^2+3x+2$, and $x^4-x^3+3x^2-x+6$.
15. $3x^4+6x^3-5x^2+1$, $6x^4+15x^3-4x^2-6x-1$.
16. $3x^4+9x^3+16x^2+11x+3$, and $2x^4+2x^3+3x^2-4x+3$.
17. $10x^4-7x^3+x^2$ and $4x^4-2x^3-2x+1$.
18. $4x^4+7x^3+5x^2-x-3$, $4x^4+5x^3+3x^2-2$.
19. $3x^5+7x^4+4x^3+10x^2+14x+4$, and $2x^5+3x^4+8x^2+7x-2$.
20. $x^5+6x^3+7x-49$, and $3x^4-2x^3+27x^2+49$.

**21.** $x^4+p^2x^2+p^4$, $x^4+2px^3+p^2x^2-p^4$.

**22.** $px^3-(p-q)x^2-(q-r)x-r$, and $qx^3-(q-r)x^2-(r-p)x-p$.

**23.** $px^3+(p+q)x^2+(q+r)x+r$, and $p^2x^3+(pr-q^2)x-qr$.

**24.** $x^3+(a+b)x^2+(ab+1)x+b$, and $bx^3+(ab+1)x^2+(a+b)x+1$.

**25.** Show that $x^3+px^2+qx+1$, and $x^3+qx^2+px+1$ cannot have a common measure unless $p=q$, or $p+q+2=0$.

**26.** If $x^2+ax+b$ and $x^2+a'x-b$ have a common linear factor, then $4b=a^2-a'^2$, and the factor is $x+\frac{1}{2}(a+a')$.

**27.** Find the value of $l$ in order that $x^2+lx+2$ and $x^2+3x+5$ may have a common factor.

**28.** Determine the constant $p$ so that $x^2+x+p$ and $x^3+x^2+x+1$ may have a common factor of the first degree.

**29.** Find the necessary and sufficient conditions that $x^3+px^2+qx+r$, and $x^3+rx^2+qx+p$ may have a common factor of the second degree.

**30.** Find the G.C.M. of $2x^4+(3+2a)x^3+(10+3a)x^2+3(5+2b)x+9b$ and $8x^3+20x^2+14x+3$: and show that it will be of the second degree if $2a+24b=21$.

**31.** Find the values of $x$ for which the equations

$$x^4-7x^3-22x^2+139x+105=0,$$
$$x^4-8x^3-11x^2+116x+70=0$$

are consistent.

## Least Common Multiple

**§ 145.** Closely allied to the problem of finding the G.C.M. of a set of integral functions is the problem of finding *the integral function of least degree which is exactly divisible by each of them.* This function is called their **least common multiple** (L.C.M.).

As in the case of the G.C.M., the degree may, if we please, be reckoned in terms of more variables than one; thus the L.C.M. of the monomials $3x^3yz^2$, $6x^2y^3z^4$, $8xyzu$, the variables being $x$, $y$, $z$, $u$, is $x^3y^3z^4u$, or any constant multiple thereof.

The general rule clearly is to *write down the product of all the variables, each raised to the highest power in which it occurs in any of the monomials.*

**§ 146.** Confining ourselves to the case of integral functions of a single variable $x$, let us investigate what are the essential factors of every common multiple of two given integral functions A and B. Let G be the G.C.M. of A and B (if they be prime to each other we may put $G=1$); then

$$A=aG, \quad B=bG,$$

where $a$ and $b$ are two integral functions which are prime to each other. Let M be any common multiple of A and B. Since M is divisible by A, we must have

$$M = PA,$$

where P is an integral function of $x$.

Therefore $$M = PaG.$$

Again, since M is divisible by B, that is, by $b$G, therefore M/$b$G, that is, P$a$G/$b$G, that is, P$a$/$b$ must be an integral function. Now $b$ is prime to $a$; hence * $b$ must divide P, that is, P = Q$b$, where Q is integral. Hence finally

$$M = QabG.$$

This is the general form of all common multiples of A and B.

Now $a$, $b$, G are given, and the part which is arbitrary is the integral function Q. Hence we get the *least* common multiple by making the degree of Q as small as possible, that is, by making Q any constant, unity say. The L.C.M. of A and B is therefore $ab$G, or ($a$G)($b$G)/G, that is, AB/G. In other words, *the L.C.M. of two integral functions is their product divided by their G.C.M.*

§ **147.** The above reasoning also shows that *every common multiple of two integral functions is a multiple of their least common multiple.*

The converse proposition, that every multiple of the L.C.M. is a common multiple of the two functions, is of course obvious.

These principles enable us to find the L.C.M. of a set of any number of integral functions A, B, C, D, etc. For, if we find the L.C.M., $L_1$ say, of A and B; then the L.C.M., $L_2$ say, of $L_1$ and C; then the L.C.M., $L_3$ say, of $L_2$ and D, and so on, until all the functions are exhausted, it follows that the last L.C.M. thus obtained is the L.C.M. of the set.

§ **148.** The process of finding the L.C.M. has neither the theoretical nor the practical importance of that for finding the G.C.M. In the few cases where the student has to solve the problem he will probably be able to use the following more direct process, the foundation of which will be obvious after what has been already said.

*If a set of integral functions can all be exhibited as powers of a set of integral factors* A, B, C, etc., *which are either all of the first degree and all different, or else are all prime to each other, then the L.C.M. of the set is the product of all these factors, each being raised to the highest power in which it occurs in any of the given functions.*

* See A. VI. § 12.

For example, let the functions be

$$(x-1)^2(x^2+2)^3(x^2+x+1),$$
$$(x-2)^2(x-3)(x^2-x+1)^2,$$
$$(x-1)^5(x-2)^3(x-3)^4(x^2+x+1)^3,$$

then, by the above rule, the L.C.M. is

$$(x-1)^5(x-2)^3(x-3)^4(x^2+2)^3(x^2+x+1)^3(x^2-x+1)^2.$$

## EXERCISES XLII.

Find the L.C.M. of the following :—

**1.** $36x^3y^2,\ 24x^6y^4,\ 18x^3y^3.$ **2.** $4x^2yz,\ 8x^3y^2z^2,\ 12x^2y^2z^3.$

**3.** $\frac{1}{2}xyzu^2,\ \frac{3}{8}x^3y^2z,\ \frac{3}{4}x^3y^3z^3u^3.$ **4.** $x^3-x,\ (x-1)^2,\ (x^2-1)^2,\ (x^3-1)^2.$

**5.** $(x^2-1)^2,\ x^2-3x+2,\ x^2-6x+9.$

**6.** $x^4-1,\ x^4+4x^2+3,\ x^4-2x^2+1.$

**7.** $x^4+16x^2+1,\ x^3-8,\ x^6-64.$

**8.** $x^4+x^2+1,\ x^2-x+1.$

**9.** $(x^4+x^2+1)^2,\ (x^3-1)^2,\ x^3+1,\ (x^2-1)^4.$

**10.** $x^2-y^2,\ x^3-y^3,\ x^4-y^4,\ x^5-y^5,\ x^6-y^6.$

**11.** $x^6-y^6,\ x^9+y^9,\ x^8+x^4y^4+y^8.$

**12.** $x^6-8,\ x^6+8,\ x^8+4x^4+16.$

**13.** $x^3+x-2,\ x^4+x^3-2x-4.$

**14.** $x^4+x^2+1,\ x^4-x^2+1.$

**15.** $x^2-2x-3,\ x^3+x^2-4x-4,\ x^3-7x-6.$

**16.** $x^3-2x^2-5x+6,\ x^3-13x+12,\ 2x^3-11x^2+18x-9.$

**17.** $x^4+x^2+1,\ x^6-1,\ x^4+x^2(1+x)^2+(x+1)^4.$

**18.** G.C.M. and L.C.M. of $14x^2+25xy-25y^2,\ 28x^2-41xy+15y^2,\ 21x^2-xy-10y^2.$

**19.** $x^4+x^2+1$ and $2x^4-5x^3+7x^2-5x+2.$

**20.** $x^4+x^2+1$ and $11x^4-x^3+10x^2+x+9.$

**21.** $x^4+x^2+1,\ x^4-x^3-x+1.$

**22.** $6x^4-11x^3+10x^2-7x+2$ and $3x^4+2x^3-2x^2+3x-2.$

**23.** If $x+c$ be the G.C.M. of $x^2+px+q$, and $x^2+rx+s$, their L.C.M. is $x^3+(p+r-c)x^2+(pr-c^2)x+(p-c)(r-c)c.$

# CHAPTER XV

## PROPERTIES OF RATIONAL FUNCTIONS—FRACTIONS

§ **149.** In this chapter, for clearness in exposition, we shall draw a momentary distinction between the quotient of two integral functions of any stated variables $x, y, z, u \ldots$, which we shall call a **Rational Fraction**, and any function whatever of these variables which, so far as $x, y, z, u \ldots$ are concerned, involves only the operations $+, -, \times, \div$ in finite number, which we shall call, as heretofore, a **Rational Function of x, y, z, u. . . .** The distinction is of no permanent practical importance, for the simple reason that we shall presently show that every Rational Function of $x, y, z, u \ldots$ can be transformed into the quotient of two integral functions of $x, y, z, u \ldots$, *i.e.* into a Rational Fraction, including for the moment under that head an integral function which may be regarded as the quotient of itself by 1, 1 being regarded as an integral function of degree 0.

### Reduction to Lowest Terms

§ **150.** We have already seen that the identity $a/b \equiv ma/mb$ results immediately from the laws of association and commutation for multiplication and division. If in view of present purposes we restrict $a$, $b$, and $m$ to be integral functions of $x, y, z, u \ldots$, we may read this as follows :—

*The numerator and denominator of any rational fraction may be multiplied by any integral function of the variables without altering its identity.*

Or, *any integral function of the variables which is a factor in both numerator and denominator of a rational fraction may be removed from both without altering its identity.*

Hence *any rational fraction may be so transformed that its numerator and denominator have no factor in common. The rational fraction is then said to be at its* **Lowest Terms.**

When a rational fraction is at its lowest terms, it may be said to be in a standard form, for it is not difficult to prove * that if two such fractions be identically equal and each at its lowest terms, then the numerator of one must be a certain constant multiple of the numerator of the other, and the denominator of that one the same constant multiple of the denominator of the other.

It may be noticed that *the identity of a rational fraction is not altered by reversing the sign of every term in both numerator and denominator;* since this is tantamount to multiplying both by $-1$.

The ultimate process for reducing any rational fraction to its lowest terms is of course to find the G.C.M. of the numerator and denominator, and to remove it by division from both. Or we may remove common factors one by one as we discover them until we are satisfied or can prove that no more exist.

The following examples, where the problem in each case is to reduce the given fraction to its lowest terms, will illustrate some of the ordinary methods of procedure :—

Ex. 1. $(x^2+x+2)/(x^2+x+1)$. No common factor is immediately obvious: indeed there is none; for $(x^2+x+2)-(x^2+x+1)$, which must contain any factor common to numerator and denominator, is equal to 1, which has no factor. The fraction is therefore already at its lowest terms.

Ex. 2. $F\equiv(x^6-2x^3+1)/(x^4-4x^3+6x^2-4x+1)$. The fraction is identically equal to $(x^3-1)^2/(x-1)^4$. Now $(x^3-1)^2\equiv\{(x-1)(x^2+x+1)\}^2 =(x-1)^2(x^2+x+1)^2$. We may therefore remove the common factor $(x-1)^2$ from numerator and denominator. The result is $F\equiv(x^2+x+1)^2/(x-1)^2$. Since $x=1$ does not cause $x^2+x+1$ to vanish, $x-1$ is not a factor in $x^2+x+1$: F is therefore now at its lowest terms.

Ex. 3. $F\equiv\{\Sigma a^2(b+c)+2abc\}/\{\Sigma a^2(b^4-c^4)\}$. By the Standard Identity, § 108, X., $\Sigma a^2(b^4-c^4)\equiv(b^2-c^2)(c^2-a^2)(a^2-b^2)\equiv(b+c)(c+a)(a+b)(b-c)(c-a)(a-b)$. It is immediately obvious that $b-c$, $c-a$, $a-b$ are not factors in the numerator of F, since it obviously does not vanish when $b=c$, etc. On the other hand, when $b=-c$, the numerator reduces to $c^2(c+a)+c^2(a-c)-2ac^2=0$; hence $b+c$ is a factor; and by like reasoning so also are $c+a$ and $a+b$. Removing these common factors, we get $F\equiv1/(b-c)(c-a)(a-b)$, obviously at lowest terms.

Ex. 4. $(x^4+2x^3-2x-1)/(x^4+x^3-3x^2-5x-2)$. We find immedi-

---

* See A. VIII. § 4.

ately by the "Long Rule" for the G.C.M. that numerator and denominator have the common factor $x^3+3x^2+3x+1$. Removing this, we reduce the fraction to $(x-1)/(x-2)$, which is obviously at lowest terms.

## EXERCISES XLIII.

Reduce the following to lowest terms :—

**1.** $27x^3yz^2/12xy^3z^3$. **2.** $144x^3y^3zu/108x^4y^4zu^2$.

**3.** $(ax^2yz+bx^2yz)/(ax^2y^2z^2-bx^2y^2z^2)$.

**4.** $xy/(x^2y+xy^2)$. **5.** $(x^2-xy)/(x^2+xy)$.

**6.** $(2x^2-2)/(x+1)^2$. **7.** $(4x-12)/(x^2-9)$.

**8.** $(x^6+2x^4)/(x^4-4)$. **9.** $(x^4-a^4)/(x^8-a^8)$.

**10.** $(x^6-a^6)/(x^4+x^2a^2+a^4)$. **11.** $(x^2y^2-9)/(x^4y^4-81)$.

**12.** $(x^2-3x+2)/\{(x-1)^2+2(x-1)\}$.

**13.** $(x^2+x-2)/(1-x^3)$.

**14.** $\{(2x+1)^2-4\}/\{(x+2)^2-(2x+3)^2\}$.

**15.** $(2-2x^3)/(x^4-1)$.

**16.** $\{(2x+1)^2-10(2x+1)+16\}/\{(2x+1)^3-8\}$.

**17.** $(15x^2+16x-15)/(15x^2+34x+15)$.

**18.** $(2x^3+6x^2)/(4x^3+8x^2-12x)$.

**19.** $(x^my-xy^m)/(x^{2m}y^2-x^2y^{2m})$. **20.** $\dfrac{(x^4-y^4)(x^3-y^3)}{(x^6-y^6)(x-y)}$.

**21.** $\dfrac{(x+y)^2(x-y)^3}{(x^6-y^6)^2}$. **22.** $\dfrac{(x+y)^2(x^2-y^2)^3(x^2-y^2)}{(x-y)^4(x+y)^3(x^3-y^3)}$.

**23.** $\dfrac{(x-1)(x-2)(x-3)}{2x^3-5x^2+x+2}$. **24.** $(x^6-6x^4+8x^2)/(x^5+5x^3-14x)$.

**25.** $\{(x+y)^3+(x-y)^3\}/\{(x+y)^5+(x-y)^5\}$.

**26.** $(xy-x-y+1)/(x^2+2xy-2y-1)$.

**27.** $(x^2+p^2x+p^3-p^2)/\{x^2+(p^2-2p)x-p^3+p^2\}$.

**28.** $\{x^3-a^3+(x+a)^2-ax\}/\{(x^2-a^2)^2+3x^2a^2\}$.

**29.** $\{x^2-(a-1)x-a\}/\{x^2-(b-1)x-b\}$.

**30.** $(x^6+2x^3+1)/(x^2-x+1)$.

**31.** $\{(x+y)^2-3(x+y)+2\}/\{3x^2+5xy+2y^2-5x-4y+2\}$.

**32.** $(x^5-3x^4-x+3)/(x^4-8x^2-9)$.

**33.** $\{(1-a^2)(1-b^2)-4ab\}/\{(1-a)(1-b)-2ab\}$.

**34.** $\dfrac{\{(ax+by)^2+(ay-bx)^2\}\{(ax+by)^2-(ay+bx)^2\}}{x^4-y^4}$.

**35.** $\{\Sigma x^3-3xyz\}/\{\Sigma(y-z)^2\}$.

**36.** $(x^8-1)/(x^2+\sqrt{2}x+1)(x^2+1)(x^2-1)$.

**37.** $\{(b-c)^2+(b-c)(c-a)-(a-b)^2\}/\{(a-b)^2+(a-b)(c-a)-(b-c)^2\}$.

**38.** $(1+2x-3x^2)/(1-3x-2x^2+4x^3)$.

**39.** $(x^4-4x^3+8x^2-8x-21)/(x^4-x^3+12x^2-7x+49)$.

**40.** $(4x^5+4x^4+3x^3+3x^2+x+1)/(2x^2+x+1)(2x^2-x+1)$.

**41.** $(2x^4+x^3-2x-1)/(3x^4+x^3-3x-1)$.

**42.** $\{(x+1)^7-x^7-1\}/\{(x+1)^5-x^5-1\}$.

**43.** $(x^4-x^3+2x^2+x+3)/(x^5+4x^4+6x^3+6x^2+3x+1)$.

**44.** $\Sigma a^2(b^3-c^3)/\Sigma a(b^4-c^4)$.

**45.** If $ad=bc$, then

$$\frac{ac(b+d)+bd(a+c)+(ac+bd+ad+bc)x}{(a+c)(b+d)+(a+b+c+d)x}$$

is independent of $x$.

## Multiplication and Division

**§ 151.** It follows from the laws of association and commutation for multiplication and division (see § 27) that

$$\left(\frac{a_1}{b_1}\right)\times\left(\frac{a_2}{b_2}\right)\times\left(\frac{a_3}{b_3}\right)\times\ \ldots\ =\frac{a_1a_2a_3\ \ldots}{b_1b_2b_3\ \ldots};$$

and that

$$\left(\frac{a}{b}\right)\div\left(\frac{c}{d}\right)=\frac{ad}{bc}=\left(\frac{a}{b}\right)\times\left(\frac{d}{c}\right).$$

In particular, if the letters all denote integral functions of any given set of variables $x$, $y$, $z$, $u$, . . ., we have the following theorems regarding rational fractions:—

*The product of any given set of rational fractions is a rational fraction, whose numerator is the product of the numerators and whose denominator is the product of the denominators of the given fractions.*

*The quotient of one rational fraction by another is a rational fraction, whose numerator is the product of the numerator of the dividend and the denominator of the divisor, and whose denominator is the product of the denominator of the dividend and the numerator of the divisor.*

Or, *the quotient of one rational fraction by another is the product of the former and the reciprocal* * *of the latter.*

From the above rules, and from what has already been said regarding the reduction of rational fractions to lowest terms, we infer that: *In transforming the product of a given set of rational fractions we may remove any factor which is common to a numerator and any denominator of the given set: and in transforming a quotient of two rational fractions we may remove any factor which is common to the numerators of the dividend and divisor, or which is common to the denominators of the dividend and divisor.*

* By the reciprocal of $c/d$ we mean $1\div(c/d)$, which is equal to $1\div c\times d$, i.e. $d/c$.

Ex. 1.
$$F \equiv \frac{x^2+x+2}{x^3+2x-3} \times \frac{x^2+x+3}{x^3+x-2}$$
$$= \frac{x^2+x+2}{(x-1)(x^2+x+3)} \times \frac{x^2+x+3}{(x-1)(x^2+x+2)}.$$

If we remove the factors $x^2+x+2$ and $x^2+x+3$, each of which is common to a numerator and a denominator of fractions in the product, we have
$$F = 1/(x-1)^2.$$

Ex. 2.
$$F \equiv \frac{x^4-16}{x^4+4x^2+16} \times \frac{x^3-8}{x^2+4} \div \frac{(x-2)^2(x+2)}{x^2-2x+4},$$
$$= \frac{(x+2)(x-2)(x^2+4)}{(x^2+2x+4)(x^2-2x+4)} \times \frac{(x-2)(x^2+2x+4)}{x^2+4} \div \frac{(x-2)^2(x+2)}{x^2-2x+4},$$
$$= \frac{(x+2)(x-2)}{x^2-2x+4} \times \frac{x-2}{1} \div \frac{(x-2)^2(x+2)}{x^2-2x+4},$$
$$= \frac{(x+2)(x-2)^2}{x^2-2x+4} \div \frac{(x-2)^2(x+2)}{x^2-2x+4},$$
$$= 1.$$

## EXERCISES XLIV.

Express the following as single fractions at lowest terms :—

1. $\dfrac{x^3+1}{x^2-9} \times \dfrac{x+3}{x+1}.$

2. $\dfrac{2x-4}{x-3} \div \dfrac{2-x}{3+x}.$

3. $\dfrac{(2x-2)^2}{x^2-4} \times \dfrac{x-2}{1-x}.$

4. $\dfrac{x^2-16}{x^2-7x+10} \div \dfrac{x^3+1}{x^2-x+1}.$

5. $\dfrac{x^3-x-6}{x^2+2x+3} \times \dfrac{x^2+2x+6}{x^3+2x-12}.$

6. $\dfrac{x^2-1}{x^2-4} \times \dfrac{x-2}{x+1} \times \dfrac{x+3}{x-1}.$

7. $\dfrac{x^3-1}{x^2-x+1} \div \dfrac{x^3+1}{x^2+x+1} \times \dfrac{x+1}{x-1}.$

8. $\dfrac{x^4-x^2}{x^4+x^2} \div \dfrac{x^2}{x^2+1} \div \dfrac{x-1}{x^2}.$

9. $\dfrac{(x+y)^3}{(x^2-y^2)^2} \times \dfrac{(x-y)^3}{x^2+y^2} \times \dfrac{2x^2+2y^2}{x^4-y^4}.$

10. $\dfrac{x^4-4x-5}{x^2-3x+2} \div \dfrac{x^2-3x-10}{2x^2-7x+3}.$

11. $\dfrac{x^4-2x^2+1}{x^4+2x^2+1} \div \dfrac{x^2-1}{x^2+1} \times \dfrac{x+1}{x-1}.$

12. $\dfrac{x^2-x-6}{x^2-3x+2} \times \dfrac{x^2+x-6}{x^2+x-2} \times \dfrac{x^2-1}{x^2+6x-9}.$

13. $\dfrac{(x+y+z)^2-4z^2}{x+y+z} \div \dfrac{x+y-z}{2x+2y+2z}.$

14. $\dfrac{x^2y^2-2xy+1}{x^2y^2+2xy+1} \div \dfrac{2xy-2}{3+3xy}.$

15. $\dfrac{a^2+b^2-c^2-d^2+2ab+2cd}{a^2+c^2-b^2-d^2+2ac+2bd} \div \dfrac{a^2+c^2-b^2-d^2-2ac-2bd}{a^2+b^2-c^2-d^2-2ab-2cd}.$

16. $\dfrac{(x+y)^4+(x+y)^2+1}{(x-y)^4+(x-y)^2+1} \div \dfrac{(x+y)^2-(x+y)+1}{(x-y)^2+(x-y)+1}.$

**17.** $\dfrac{(a-b)^2-c^2}{bc+a(b+c)}\times\dfrac{(a+b+c)^2-a^2-b^2-c^2}{a-b+c}\div\dfrac{a-b-c}{a+b+c}.$

**18.** $\dfrac{x^2-81}{x^2-5}\times\dfrac{x^4-9x^2+20}{x^3-729}\div\dfrac{x^2+8x-9}{x^2+9x+81}.$

**19.** $\dfrac{9x^2-16}{9x^2+9x-4}\times\dfrac{9x^2-1}{12x^2-x-20}\times\dfrac{12x^2+19x+5}{7x+1}\div\dfrac{9x^2+6x+1}{49x^2+14x+1}.$

**20.** $\dfrac{2a+2b+2c}{(b-c)^2+(c-a)^2+(a-b)^2}\div\dfrac{(b+c)^2+(c+a)^2+(a+b)^2-a^2-b^2-c^2}{a^2+b^2+c^2}.$

## Addition and Subtraction of Fractions

§ **152.** By means of the theorem $a/b \equiv ma/mb$ of § 150, *any number of rational fractions may be transformed so as to have a* **Common Denominator.** Hence, by the law of distribution for division, it follows that *the algebraic sum of any number of rational fractions can be expressed as a single rational fraction.*

Suppose, for example, that $a$, $b$, $c$, $d$, $e$, $f$ represent integral functions of any variables $x$, $y$, $z$, $u$, . . ., and consider the algebraic sum $a/b + c/d - e/f$.

Let L be the L.C.M. of the three integral functions $b$, $d$, $f$, so that $L/b = \beta$, $L/d = \delta$, $L/f = \phi$, where $\beta$, $\delta$, $\phi$ are integral functions of $x$, $y$, $z$, $u$, . . ., then $b\beta = L$, $d\delta = L$, $f\phi = L$; and we have $a/b = a\beta/b\beta = a\beta/L$; $c/d = c\delta/d\delta = c\delta/L$; $e/f = e\phi/f\phi = e\phi/L$. Therefore

$$\frac{a}{b}+\frac{c}{d}-\frac{e}{f}=\frac{a\beta}{L}+\frac{c\delta}{L}-\frac{e\phi}{L},$$

$$=\frac{a\beta+c\delta-e\phi}{L},$$

where $a\beta + c\delta - e\phi$ and L are integral functions. If, as will often happen, no two of the three $b$, $d$, $f$ have any factor in common, then $L = bdf$, and we have

$$\frac{a}{b}+\frac{c}{d}-\frac{e}{f}=\frac{adf}{bdf}+\frac{cbf}{bdf}-\frac{ebd}{bdf},$$

$$=\frac{adf+cbf-ebd}{bdf}.$$

The process is obviously applicable to any number of summands, its essential features being the transformation of all the fractions so that they have a common denominator and the application of the law of distribution.

In practice it may not be advisable to add all the fractions at once, but to proceed step by step, first adding one group then another, combining the results ; and so on. It may occasionally save labour to reduce certain of the summands to lowest terms, or to take advantage of the fact, proved in § 110, that any improper fraction may be expressed as the sum of an integral function and a proper fraction. Some of these artifices are illustrated in the following examples :—

Ex. 1. $2/(x-1)-3(x+1)/(x^2+1)$
$=2(x^2+1)/(x-1)(x^2+1)-3(x+1)(x-1)/(x-1)(x^2+1),$
$=\{2(x^2+1)-3(x^2-1)\}/(x-1)(x^2+1),$
$=(-x^2+5)/(x^3-x^2+x-1).$

Ex. 2. $F\equiv 1/(x-a)-1/(x+a)-a/(x^2-a^2)+a/(x^2+a^2).$
$F=\{1/(x-a)-1/(x+a)\}-a\{1/(x^2-a^2)-1/(x^2+a^2)\},$
$=\{(x+a)-(x-a)\}/(x^2-a^2)-a\{(x^2+a^2)-(x^2-a^2)\}/(x^4-a^4),$
$=2a/(x^2-a^2)-2a^3/(x^4-a^4),$
$=2a(x^2+a^2)/(x^4-a^4)-2a^3/(x^4-a^4),$
$=\{2a(x^2+a^2)-2a^3\}/(x^4-a^4),$
$=2ax^2/(x^4-a^4).$

Ex. 3. $$\frac{2x^3+5x^2+6x+2}{x^2+x+1}-\frac{2x^3+x^2+4}{x^2-x+1}$$
$$=\left\{2x+3+\frac{x-1}{x^2+x+1}\right\}-\left\{2x+3+\frac{x+1}{x^2-x+1}\right\},$$
$$=\frac{x-1}{x^2+x+1}-\frac{x+1}{x^2-x+1},$$
$$=\frac{(x-1)(x^2-x+1)-(x+1)(x^2+x+1)}{(x^2+x+1)(x^2-x+1)},$$
$$=\frac{-4x^2-2}{x^4+x^2+1}=-2\frac{2x^2+1}{x^4+x^2+1}.$$

Ex. 4. $$F\equiv\frac{2x^3+5x^2+9x+9}{x^3+2x-3}-\frac{2x^3-5x^2+9x-9}{x^3+2x+3}.$$

We have obviously $x^3+2x-3=(x-1)(x^2+x+3)$, and $x^3+2x+3=(x+1)(x^2-x+3)$. $x-1$ and $x+1$ are not factors in the numerators, as we see at once by the remainder theorem. On the other hand, we find on trial that $x^2+x+3$ and $x^2-x+3$ are factors in the respective numerators, viz. $2x^3+5x^2+9x+9=(2x+3)(x^2+x+3)$ ; and $2x^3-5x^2+9x-9=(2x-3)(x^2-x+3)$. Hence

$F=(2x+3)/(x-1)-(2x-3)/(x+1),$
$=\{(2x+3)(x+1)-(2x-3)(x-1)\}/(x^2-1),$
$=10x/(x^2-1).$

## EXERCISES XLV.

Transform each of the following into a rational fraction at its lowest terms :—

1. $1/a(x+a)+1/x(x+a)$. 2. $1/(x-1)-1/(x-3)$.
3. $(x-1)/(x+1)-(x-2)/(x+2)$.
4. $(2x-1)/(2x+3)+(2x+1)/(2x-3)$.
5. $1/(x-3)-9/(x^2-9)$.
6. $(ab-a^2)/(ab+b^2)-(ab-b^2)/(ab+a^2)$.
7. $1/(x-1)+5/(x+2)-12/(2x+3)$.
8. $(a^2+b^2)/(a-b)-(a^3-b^3)/(a^2+b^2)$.
9. $(x^3-1)/(x^4-1)-(x-1)/(x^2-1)$.
10. $(a^2+ac)/(a^2c-c^3)-(a-c)/(a+c)c-2c/(a^2-c^2)$.
11. $(x-3)^2/(x+3)^2-(x-3)/(x+3)$.
12. $1/(1-x)^2+2/(1-x^2)+1/(1+x)^2$.
13. $\dfrac{x+1}{x-1}-\dfrac{x^2+1}{x^2-1}+\dfrac{4x^3}{x^4-1}$.
14. $\dfrac{2x-y}{x(3x-y)}-\dfrac{3x-4y}{x(5x-4y)}+\dfrac{2x-3y}{(3x-y)(5x-4y)}$.
15. $(x^2+x+1)/(x^2-x+1)+(x+1)/(x-1)$.
16. $(x-4)^2/(x+4)^2-(x+4)^2/(x-4)^2$.
17. $1/(a+3b)+6b/(a^2-9b^2)+1/(3b-a)$.
18. $a/(a-b)+a/(a+b)+2a^2/(a^2+b^2)+4a^2b^2/(a^4-b^4)$.
19. $(x-a)/(x+a)+(x+a)/(x-a)+(x^2-5a^2)/(x^2-a^2)$.
20. $1/(x+2a)+1/(x-2a)+8a^2/(4a^2x-x^3)$.
21. $x^4/y^2(y^2-x^2)+x/2(x+y)-x/2(y-x)$.
22. $(x^3-1)/(x^4-1)+(x-1)/(x^2-1)-\frac{1}{2}\{(x+1)/(x^2+1)+1/(x+1)\}$.
23. $1/(x+2)-2/(x+2)(x+4)+2/(x+2)(x+4)(x+6)$.
24. $\dfrac{1}{6(x-1)}+\dfrac{1}{6(x+1)}-\dfrac{x-1}{6(x^2+x+1)}-\dfrac{x+1}{6(x^2-x+1)}$.
25. $(x+y)/(x^3-y^3)+(x-y)/(x^3+y^3)-2(x^2-y^2)/(x^4+x^2y^2+y^4)$.
26. $\dfrac{1}{x^2-1}+\dfrac{1}{x^2+1}-\dfrac{x^2-1}{x^4+x^2+1}-\dfrac{x^2+1}{x^4-x^2+1}$.
27. $1/(x^3-y^3)+2/(x^3+y^3)-(3x+y)/(x^4+x^2y^2+y^4)$.
28. $\dfrac{x+1}{x^3+x^2+x}+\dfrac{x-1}{x^3-x^2+x}+\dfrac{2(x^4+1)}{x^4+x^2+1}$.
29. $(x^2-4x+3)/(x^2+2x-3)+(x^2-6x+5)/(x^2+4x-5)+(x^2-8x+7)/(x^2+6x-7)$.
30. $1/(x^2-3x+2)-1/(x^2+2x-3)+1/(x^2+x-6)$.
31. $4/(x^2-5x+6)-3/(x^2-4x+3)+2/(x^2-3x+2)$.
32. $\dfrac{1}{2x^2+5x+2}-\dfrac{2}{2x^2+3x-2}-\dfrac{1}{2x^2-3x-2}$.
33. $\dfrac{2}{x-2}-\dfrac{2}{x-1}-\dfrac{1}{(x-1)^2}+\dfrac{1}{(x^2-3x+2)^2}$.
34. $(1+x)/(1+x+x^2)-(1-x)/(1-x+x^2)+(1-x^2-x^4)/(1+x^2+x^4)$.
35. $\dfrac{1}{(x^2+2x+1)(x+2)^2}+\dfrac{2x^2+7x+7}{(x+1)(x^2+4x+4)}-\dfrac{2}{x+2}$.
36. $(x^3-1)/(x^3+2x^2+2x+1)-(x^3+1)/(x^3-2x^2+2x-1)$.

**37.** $\dfrac{x^3-1}{x^4-1}-\dfrac{x^2-1}{x^3+x^2-x-1}-\frac{1}{2}\left\{\dfrac{x+1}{x^2+1}-\dfrac{1}{x+1}\right\}.$

**38.** $\dfrac{x+2y}{x^2+xy-6y^2}-\dfrac{x-y}{x^2+7xy+12y^2}+\dfrac{x-4y}{x^2+2xy-8y^2}.$

**39.** $\dfrac{x^3-1}{x^3+2x^2+2x+1}+\dfrac{x^3+1}{x^3-2x^2+2x-1}-2\dfrac{x^2+1}{x^2-1}.$

**40.** $\dfrac{x^2+3x+1}{x^4+x^3-4x^2+x+1}-\dfrac{x^2-3x+1}{x^4-x^3-4x^2-x+1}.$

**41.** $\dfrac{x^3+x^2-x-1}{x^3+5x^2+7x+3}-\dfrac{x^3-x^2-x+1}{x^3-5x^2+7x-3}.$

**42.** $\dfrac{x^2+(p-q-1)x-p+q}{x^2+(-p+q-1)x+p-q}-\dfrac{x^2+(p+q+1)x+p+q}{x^2+(-p-q+1)x-p-q}.$

**43.** $(x^2+y^2+x+y-xy+1)/(x-y-1)+(x^2+y^2+x-y+xy+1)/(x+y-1).$

**44.** $\dfrac{x^2-4x+3}{x^4-2x^3-2x^2-2x-3}-\dfrac{x^2-x+1}{x^4+x^2+1}+\dfrac{1}{x^2-x+1}.$

**45.** $\{(1+xy)^2-(x+y)^2\}/(1+x)(1+y)+\{(1-xy)^2-(x-y)^2\}/(1-x)(1-y).$

**46.** $\{(x+y)^3-(x-y)^3\}/\{(x+y)^3+(x-y)^3\}+\{(x+y)^3+(x-y)^3\}/\{(x+y)^3-(x-y)^3\}.$

**47.** $\dfrac{x^6-a^6+a^2x^2(x^2-a^2)}{x^8-a^8-2a^2x^2(x^4-a^4)}-\dfrac{x^6+a^6-a^2x^2(x^2+a^2)}{x^8-a^8+2a^2x^2(x^4-a^4)}.$

**48.** $\dfrac{x^2-y^2}{x^4-x^3y-xy^3+y^4}+\dfrac{x^2-y^2}{x^4+x^3y+xy^3+y^4}-2\dfrac{x^4-y^4}{x^6+x^4y^2+x^2y^4+y^6}.$

**§ 153.** The following examples illustrate the use of standard identities and the employment of the Σ-notation in dealing with fractional expressions which have cyclic symmetry (see § 101) with respect to certain of the occurring variables.

In the Exercises that follow, the beginner should endeavour to save labour by the methods just mentioned. He will probably find some difficulty with them; and in that case he should pass on, and return to them from time to time, as his skill and his grasp of the principles of algebraic form increase.

Ex. 1. $F\equiv a/(a-b)(a-c)+b/(b-c)(b-a)+c/(c-a)(c-b).$

It will be convenient to arrange the constituent fractions so that $b-c$, $c-a$, $a-b$ occur throughout, instead of $b-c$ in one place and $c-b$ in another, etc. We thus have

$$\begin{aligned}F&\equiv -a/(a-b)(c-a)-b/(b-c)(a-b)-c/(c-a)(b-c),\\ &=\{-a(b-c)-b(c-a)-c(a-b)\}/(b-c)(c-a)(a-b),\\ &=0/(b-c)(c-a)(a-b)=0;\end{aligned}$$

provided $a\neq b\neq c$.

Since F has cyclic symmetry with respect to $a$, $b$, $c$, the type-group being $a/(a-b)(a-c)=-a/(a-b)(c-a)$, we might abbreviate the above calculation as follows:—

$$\mathrm{F} = -\Sigma a/(a-b)(c-a) = -\Sigma a(b-c)/(b-c)(c-a)(a-b),$$
$$= 0/(b-c)(c-a)(a-b) = 0.$$

Ex. 2. $\mathrm{F} \equiv \Sigma a/(a^2-b^2)(a^2-c^2) = -\Sigma a/(a^2-b^2)(c^2-a^2),$
$$= -\Sigma a(b^2-c^2)/(b^2-c^2)(c^2-a^2)(a^2-b^2).$$

Now $\Sigma a(b^2-c^2) = (b-c)(c-a)(a-b)$; hence we have
$$\mathrm{F} = -(b-c)(c-a)(a-b)/(b^2-c^2)(c^2-a^2)(a^2-b^2) = -1/(b+c)(c+a)(a+b).$$

Ex. 3. $\mathrm{F} \equiv \dfrac{ax-bc}{(a-b)(a-c)} + \dfrac{bx-ca}{(b-c)(b-a)} + \dfrac{cx-ab}{(c-a)(c-b)}.$

F is cyclically symmetrical with respect to $a$, $b$, $c$ with the type-group $(ax-bc)/(a-b)(a-c)$. Hence
$$\mathrm{F} = \Sigma(ax-bc)/(a-b)(a-c) = -\Sigma(ax-bc)/(a-b)(c-a),$$
$$= -\Sigma(ax-bc)(b-c)/(b-c)(c-a)(a-b),$$
$$= \{-x\Sigma a(b-c) + \Sigma bc(b-c)\}/\Pi(b-c).$$

Now $\Sigma a(b-c) = 0$ and $\Sigma bc(b-c) = -\Pi(b-c)$; therefore $\mathrm{F} = \{0 - \Pi(b-c)\}/\Pi(b-c) = -1$. We suppose $a \neq b \neq c$.

## EXERCISES XLVI.

**Reduce the following to rational fractions at lowest terms:—**

**1.** $\Sigma 1\Big/\left(\dfrac{a}{b}-1\right)\left(\dfrac{a}{c}-1\right).$ **2.** $\Sigma(y+z)/(x^2-y^2)(x^2-z^2).$

**3.** $\Sigma bc/(a-b)(a-c).$ **4.** $\Sigma 1/\{x^2-(y-z)^2\}.$

**5.** $\sum\limits_{abc}(b-c)/(b+x)(c+x).$ **6.** $\Sigma(b-c)^2/(a-b)(a-c).$

**7.** $\sum\limits_{abc}(pa+q)/(x-a).$ **8.** $\Sigma(a^2-bc)/(a+b)(a+c).$

**9.** $\Sigma a^2(b-c)/(a^2-a(b+c)+bc).$

**10.** $\Sigma(x^2+ax+a^2)/(a-b)(a-c).$ **11.** $\Sigma(x-a)^3/(a-b)(a-c).$

**12.** $(b-c+a)/(x-a)+(c-a+b)/(x-b)+(a-b+c)/(x-c)+\Pi(b-c)/\Pi(x-a).$

**13.** $\Sigma(b-c+a)/x(x-b)(x-c) - \Sigma a/\Pi(x-a).$

**14.** $\Sigma(b+c-2a)^2/(c+a-2b)(a+b-2c).$

**15.** $\Sigma(b+c)/(a-b)(a-c)(x-b)(x-c).$

**16.** $\Sigma a^3/(a-b)(a-c).$

**17.** $\Sigma\{(b-c)x+(b+c)\}/(x-b)(x-c).$ **18.** $\Sigma(b-c)/(b+c)(x-a).$

**19.** $1/(x-a-b-c) - \Pi(b+c)/(x-a-b-c)\Pi(x-a).$

**20.** $\Sigma 1/\{(1+bc)^2-(b+c)^2\}.$

**21.** $\dfrac{(x+p)(x+q)(x+r)}{(x-a)(x-b)(x-c)} - \Sigma\dfrac{(a+p)(a+q)(a+r)}{(a-b)(a-c)(x-a)}.$

**22.** $\Sigma(b-c)/(x+b+c) + \Sigma a(b^2-c^2)/\Pi(x+b+c).$

**23.** $\Sigma(y-z)/(y+z-2x).$

**24.** $\sum\limits_{abc}(a^2+a+1)/(a-b)(a-c)(x-a).$

**25.** $\sum\limits_{abc}(x-a)(y-a)(z-a)/(a-b)(a-c).$

**26.** $\sum\limits_{lmn}1/(l-m)(l-n)(1+lx).$

**27.** $\Sigma(b-c)/(x+b+c) + \Pi(b-c)/(x+b+c).$

**28.** $\Sigma(1-a^2)/\{(1-bc)^2-(b-c)^2\}.$

**29.** $\Sigma(x^2+a^2)/\{(c-a)x+b\}\{(a-b)x+c\}.$

## RATIONAL FUNCTIONS IN GENERAL

§ **154.** Since every rational function involves a finite number of the operations $+$, $-$, $\times$, $\div$ only, it follows from §§ 151, 152 that every *rational function can, by a finite number of steps, be ultimately reduced to the quotient of two integral functions of its variables, that is to say, to what we have in this chapter called a rational fraction.*

No general rule can be laid down for the best order of procedure in such reductions; and for that very reason they form an excellent exercise for the tyro in Algebra. The most important point is to see that no step is taken which cannot be justified by reference to the fundamental laws of Algebra. Subject to this condition, the steps of the calculation should be so arranged as to avoid useless labour by removing redundant (*i.e.* mutually destroying) parts of the function at as early a stage as possible. The use of brackets and the application of standard identities will be found of great help, both in saving labour and in giving clearness and perspicuity to the work. As a general rule, brackets should not be removed or products distributed until it is clearly seen that nothing is to be gained by refraining from so doing. Frequently in a complicated piece of work it will be best to deal with a part of the function by itself, to make its reduction a separate calculation, then to restore the result in the original function, and to proceed with the main reduction. The beginner may also be here reminded that next to sound logic, neatness of arrangement is the most important part of algebraic skill, and one of the most important lessons to be derived from the study of Algebra.

Ex. 1. $$F \equiv \frac{\{x+1+1/(x-1)\}\{x-1+1/(x+1)\}}{\{x+1-1/(x-1)\}\{x-1-1/(x+1)\}}.$$

If we multiply dividend and divisor of the main quotient by $(x-1)(x+1)$, we get

$$F = \frac{\{(x+1)(x-1)+1\}\{(x-1)(x+1)+1\}}{\{(x+1)(x-1)-1\}\{(x-1)(x+1)-1\}},$$

$$= \frac{\{x^2-1+1\}\{x^2-1+1\}}{\{x^2-1-1\}\{x^2-1-1\}},$$

$$= \frac{x^4}{(x^2-2)^2}.$$

Ex. 2. $$F \equiv \left\{a+1\Big/\left(\frac{1}{b}-\frac{1}{a}\right)\right\}\left\{b-1\Big/\left(\frac{1}{b}+\frac{1}{a}\right)\right\}\left\{\frac{a}{b}-\frac{b}{a}\right\}\left\{\frac{a}{b}+\frac{b}{a}\right\}.$$

We have $a+1\Big/\left(\frac{1}{b}-\frac{1}{a}\right)=a+1/\{(a-b)/ab\}=a+ab/(a-b)=\{a(a-b)$ $+ab\}/(a-b)=(a^2-ab+ab)/(a-b)=a^2/(a-b)$.

Also

$$b-1\Big/\left(\frac{1}{b}+\frac{1}{a}\right)=b-1/\{(a+b)/ab\}=b-ab/(a+b)=\{b(a+b)-ab\}/(a+b)$$
$$=(ab+b^2-ab)/(a+b)=b^2/(a+b).$$

Hence

$$F=\frac{a^2}{a-b}\times\frac{b^2}{a+b}\times\frac{a^2-b^2}{ab}\times\frac{a^2+b^2}{ab}=a^2+b^2,$$

after removal of factors which are common to numerators and denominators of the product of the fractions.

Ex. 3. $$F\equiv\frac{\left(\dfrac{x+a}{x-a}\right)^2-2+\left(\dfrac{x-a}{x+a}\right)^2}{\left(\dfrac{x+a}{x-a}\right)^2-\left(\dfrac{x-a}{x+a}\right)^2}.$$

We observe that the dividend of the main quotient can be written

$$\left(\frac{x+a}{x-a}\right)^2-2\left(\frac{x+a}{x-a}\right)\left(\frac{x-a}{x+a}\right)+\left(\frac{x-a}{x+a}\right)^2=\left(\frac{x+a}{x-a}-\frac{x-a}{x+a}\right)^2;$$

while the denominator can be put into the form

$$\left(\frac{x+a}{x-a}+\frac{x-a}{x+a}\right)\left(\frac{x+a}{x-a}-\frac{x-a}{x+a}\right).$$

Removing the common factor $(x+a)/(x-a)-(x-a)/(x+a)$, we have

$$F=\frac{\dfrac{x+a}{x-a}-\dfrac{x-a}{x+a}}{\dfrac{x+a}{x-a}+\dfrac{x-a}{x+a}}.$$

If we now multiply dividend and divisor of the quotient denoted by the longer line by $(x-a)(x+a)$, we get

$$F=\{(x+a)^2-(x-a)^2\}/\{(x+a)^2+(x-a)^2\},$$
$$=4xa/(2x^2+2a^2)=2ax/(x^2+a^2).$$

Rational functions of the form

$$F\equiv\cfrac{a}{b+\cfrac{c}{d+\cfrac{e}{f+\cfrac{g}{h}}}},$$

commonly called **Continued Fractions**, are of frequent occurrence, and deserve special attention. The function in question may also be written

$$a/(b+c/(d+e/(f+g/h)));$$

or more conveniently still, with the special notation

$$\frac{a}{b+}\,\frac{c}{d+}\,\frac{e}{f+}\,\frac{g}{h},$$

where the $+$'s are written under the fraction lines to prevent confusion with $a/b+c/d+e/f+g/h$.

Ex. 4. In reducing continued fractions it is usually convenient to begin at the bottom and proceed upwards. Thus $f+g/h=(fh+g)/h$. Therefore

$$\frac{e}{f+\dfrac{g}{h}}=\frac{e}{(fh+g)/h}=\frac{eh}{fh+g}.$$

Therefore

$$\frac{c}{d+\dfrac{e}{f+\dfrac{g}{h}}}=\frac{c}{d+eh/(fh+g)},$$

$$=\frac{c(fh+g)}{d(fh+g)+eh},$$

$$=\frac{cfh+cg}{dfh+dg+eh}.$$

Finally—

$$F=\frac{a}{b+(cfh+cg)/(dfh+dg+eh)},$$

$$=\frac{a(dfh+dg+eh)}{b(dfh+dg+eh)+(cfh+cg)},$$

$$=\frac{adfh+adg+ach}{bdfh+bdg+beh+cfh+cg},$$

which is the final expression of F as the quotient of two integral functions of $a$, $b$, $c$, $d$, $e$, $f$, $g$, $h$. Where, as often happens, $a$, $b$, $c$, etc., are themselves functions of other operands, simplifications may occur before the final stage of the reduction is reached.

## EXERCISES XLVII.

Transform each of the following into a rational fraction at lowest terms:—

1. $\left(\dfrac{a+b}{a-b}-\dfrac{a-b}{a+b}\right)\left(\dfrac{a+b}{a-b}+\dfrac{a-b}{a+b}\right).$
2. $(x/c+c/x-2)/(x-c)+(x/c+c/x+2)/(x+c).$
3. $\{2x/(1+2x)+(1-2x)/2x\}\div\left(\dfrac{2x}{1+2x}-\dfrac{1-2x}{2x}\right).$
4. $\dfrac{a+x^2/(a+x)}{x+a^2/(a+x)}\Big/\dfrac{(x+a)^2-x^2}{x^2+ax+a^2}.$
5. $(1/x^2-1/y^2)/(1/x^4-1/y^4)\{(1/x+1/y)^2+(1/x-1/y)^2\}.$

6. $\dfrac{(x+y)^2-(x-y)^2}{(x+y)^2+(x-y)^2}\left(\dfrac{x}{y}+\dfrac{y}{x}\right)$.

7. $\{(ax+by)^2+(ay-bx)^2\}/\{(a/x+b/y)^2+(a/y-b/x)^2\}$.

8. $\left\{\dfrac{x-1}{(x-2)(x-3)}-\dfrac{x-2}{(x-1)(x-3)}\right\}\Big/\left\{\dfrac{x-2}{(x-1)(x-3)}-\dfrac{x-3}{(x-1)(x-2)}\right\}$.

9. $\left[\dfrac{1/x^2+1}{1/x^2-1}+\dfrac{1/x^2-1}{1/x^2+1}\right]\div\left[\dfrac{x+1}{x-1}+\dfrac{x-1}{x+1}\right]$.

10. Show that $1+\frac{1}{2}(x/y+y/x)=1\Big/\left(1+\dfrac{x-y}{x+y}\right)+1\Big/\left(1-\dfrac{x-y}{x+y}\right)$.

11. $\left\{\left(\dfrac{x+p}{x+q}\right)^2-\left(\dfrac{x-p}{x-q}\right)^2\right\}\Big/\left\{\left(\dfrac{x-p}{x+q}\right)^2-\left(\dfrac{x+p}{x-q}\right)^2\right\}$.

12. $\left\{\dfrac{a-b}{1+ab}+\dfrac{b-c}{1+bc}\right\}\Big/\left\{1-\dfrac{(a-b)(b-c)}{(1+ab)(1+bc)}\right\}$.

13. $\dfrac{1/(a-b)-1/(a+b)}{1/(a-b)+1/(a+b)}+\dfrac{1/(b-c)-1/(b+c)}{1/(b-c)+1/(b+c)}+\dfrac{1/(c-a)-1/(c+a)}{1/(c-a)+1/(c+a)}$.

14. $\left(\dfrac{x+y}{x^3+y^3}+\dfrac{x-y}{x^3-y^3}\right)\Big/\left(\dfrac{x^3+y^3}{x+y}+\dfrac{x^3-y^3}{x-y}\right)$.

15. $\dfrac{1-1/x}{1+1/x}\Big/\dfrac{x-1/x}{x+1/x}+\dfrac{2/x}{1+1/x^2}$.

16. $\dfrac{a/b-b/a}{a/b+1+b/a}+\dfrac{1+b/a-b^2/a^2}{a/b-b^2/a^2}$.

17. $(a^2/c^2-c^2/a^2)(a/c+c/a-1)/\{a^2c^2(1/c^3+1/a^3)(1/c-1/a)\}$.

18. $\{(x/y+y/x+1)(1/x-1/y)^2\}/\{x^2/y^2+y^2/x^2-(x/y+y/x)\}$.

19. $\left(1-\dfrac{b^2+c^2-a^2}{2bc}\right)\left(1+\dfrac{(a+b+c)(b+c-a)}{(a-b+c)(a+b-c)}\right)$.

20. $\left\{\dfrac{(x-1)(x^3-1)}{x^2-x+1}-\dfrac{(x+1)(x^3+1)}{x^2+x+1}\right\}\times\left\{1+\dfrac{3}{2(x^2-1)}-\dfrac{1}{2(x^2+1)}\right\}$.

21. $1/(1-1/(1-1/x))-1/(1+1/(1+1/x))\equiv -x/(1+1/2x)$.

22. $\left(\dfrac{1}{a}+\dfrac{1}{b}\right)\Big/\left(\dfrac{1}{a}-\dfrac{1}{b}\right)+\left(\dfrac{1}{a^2}+\dfrac{1}{b^2}\right)\Big/\left(\dfrac{1}{a^2}-\dfrac{1}{b^2}\right)+\left(\dfrac{1}{a^3}+\dfrac{1}{b^3}\right)\Big/\left(\dfrac{1}{a^3}-\dfrac{1}{b^3}\right)$.

23. $\left\{\dfrac{(a-b)(a^3-b^3)}{a^2-ab+b^2}-\dfrac{(a+b)(a^3+b^3)}{a^2+ab+b^2}\right\}\times\left\{1+\dfrac{3b^2}{2(a^2-b^2)}-\dfrac{b^2}{2(a^2+b^2)}\right\}$.

24. $\left(1+\dfrac{c}{a+b}+\dfrac{c^2}{(a+b)^2}\right)\left(1-\dfrac{c^2}{(a+b)^2}\right)\Big/\left(1-\dfrac{c^3}{(a+b)^3}\right)\left(1+\dfrac{c}{a+b}\right)$.

25. $x^3-x^2/\{1+(1-x)/(x+1/x)\}$.

26. $x/[1-x/\{1+x+x/(1-x+x^2)\}]$.

27. $(x+1)/\{x+1+1/[x-1+1/(x+1)]\}$.

28. $\dfrac{y^2}{x-}\,\dfrac{y^2}{x+}\,\dfrac{y^2}{x-}\,\dfrac{y^2}{x}-\dfrac{y^2}{x+}\,\dfrac{y^2}{x-}\,\dfrac{y^2}{x+}\,\dfrac{y^2}{x}$.

## Rational Functions of a Single Variable

**§ 155.** We have already explained that a rational fraction involving (or regarded as involving) a single variable $x$ is called a **Proper** or an **Improper Fraction** according as the degree in

$x$ of its numerator is less or not less than the degree in $x$ of its denominator; and it has been shown (§ 115) that every improper fraction can be resolved, and that in one way only, into the sum of an integral function of $x$ and a proper fractional function of $x$.

Partly on account of its intrinsic importance, and partly to remind the beginner of the fundamental difference between algebraic and arithmetic fractionality, we here prove the following theorem:—

*The algebraic sum of two (and therefore of any number of) algebraic proper fractions is an algebraic proper fraction.**

Let the proper fractions be $A_\alpha/B_\beta$, $C_\gamma/D_\delta$ where $A_\alpha$, $B_\beta$, $C_\gamma$, $D_\delta$ are integral functions of $x$ of degrees $\alpha$, $\beta$, $\gamma$, $\delta$; so that $\alpha<\beta$, $\gamma<\delta$. Then

$$\pm A_\alpha/B_\beta \pm C_\gamma/D_\delta = (\pm A_\alpha D_\delta \pm C_\gamma B_\beta)/B_\beta D_\delta.$$

Now the degree of $B_\beta D_\delta$ is $\beta+\delta$, and the degree of $\pm A_\alpha D_\delta \pm C_\gamma B_\beta$ cannot exceed the greater of $\alpha+\delta$ and $\beta+\gamma$; but, since $\alpha<\beta$ and $\gamma<\delta$, neither of these can be so great as $\beta+\delta$, and our theorem is proved.

From the theorem just established we can readily deduce the following:—

*If two rational improper fractions be identically equal, their integral and proper fractional parts must be separately equal.*

Suppose that the two improper fractions have been reduced to $Q+F$ and $Q'+F'$, where $Q$ and $Q'$ are integral and $F$ and $F'$ proper fractional functions of $x$. Then, by hypothesis, $Q+F\equiv Q'+F'$, and therefore $Q-Q'\equiv F'-F$. Suppose now $F'\not\equiv F$, and therefore by last equality $Q'\not\equiv Q$. $F'-F$, being the algebraic sum of two proper fractions, is a proper fractional function of $x$, and therefore essentially fractional. $Q'-Q$, being the difference of two integral functions, is an integral function of $x$. We have thus the absurdity that an essentially integral function is transformable into an essentially fractional function of $x$ (see § 109). It follows that we must have $F\equiv F'$, and therefore $Q\equiv Q'$.

§ **156.** It may be shown by direct elementary reasoning connected with the process for finding the G.C.M. of two integral functions of $x$ that

* There is of course no analogous theorem for arithmetic fractions, *e.g.* $\frac{2}{3}+\frac{3}{4}=\frac{17}{12}$, an improper arithmetic fraction.

*If the denominator of a proper rational fraction be the product of two integral functions of* x *which are prime to each other, then the fraction can be decomposed, and that in one way only, into the sum of two proper fractions whose denominators are the two integral functions in question.** The component fractions are spoken of as **Partial Fractions.**

Given *a priori* the possibility of this decomposition, it is always easy in practice to effect it by means of indeterminate coefficients or otherwise.

Ex. 1. Decompose $(3x-4)/(x-1)(x-2)$ into the sum of two fractions whose denominators are $x-1$ and $x-2$. Since the denominators are of the first degree, the fractions will be $A/(x-1)$ and $B/(x-2)$, where A and B are constants to be determined. We have

$$(3x-4)/(x-1)(x-2) \equiv A/(x-1)+B/(x-2) \qquad (1).$$

Hence, multiplying both sides by $(x-1)(x-2)$, we get

$$3x-4 \equiv A(x-2)+B(x-1) \qquad (2).$$

We may determine A and B by equating the coefficients of $x$ in this identity. Then we get $A+B=3$, $2A+B=4$; whence $A=1$, $B=2$.

Or, better thus:—since the equation (2) is true for all values of $x$, we may put therein $x=1$ and $x=2$. We thus get $-1=-A$, and $2=B$; that is, $A=1$ and $B=2$, as before.

Ex. 2. $F \equiv (x^2-13x+26)/(x-2)(x-3)(x-4)$.

Since $x-2$ and $(x-3)(x-4)$ are obviously prime to each other, we may choose these for denominators of the partial fractions. The numerator for $x-2$ will be a constant, say A, as before; but the numerator corresponding to $(x-3)(x-4)$, which is of the second degree in $x$, may be of the first degree in $x$, and must therefore be written $Bx+C$, where B and C are constants to be determined. We must therefore have

$$(x^2-13x+26)/(x-2)(x-3)(x-4) \equiv A/(x-2)+(Bx+C)/(x-3)(x-4).$$

Hence there must exist an identity of the form

$$x^2-13x+26 \equiv A(x-3)(x-4)+(Bx+C)(x-2).$$

The coefficients may be most readily determined as follows:—

Put $x=2$ and we get $4=A(-1)(-2)$; therefore $A=2$.

There must therefore be an identity of the form

$$x^2-13x+26-2(x-3)(x-4) \equiv (Bx+C)(x-2),$$

that is—

$$-x^2+x+2 \equiv (Bx+C)(x-2).$$

Since $x-2$ is a factor on the right, it must also be a factor on the left; removing it we have

---

* See A. VIII. § 6.

$$-x-1 \equiv Bx+C.$$

Hence we must have $B=-1$, $C=-1$. Therefore

$$\frac{x^2-13x+26}{(x-2)(x-3)(x-4)}=\frac{2}{x-2}-\frac{x+1}{(x-3)(x-4)}.$$

§ 157. **Resolution into Ultimate Partial Fractions.**—It can be proved that every integral function of $x$ can be resolved into a product of real prime factors, each of which will be a linear or else a quadratic function of $x$, each, it may be, repeated several times. Corresponding hereto we have the following theorem :—

*Every proper fractional function of* x *can be decomposed uniquely into a sum of proper partial fractions, containing a partial fraction of the form* $A/(x-a)$ *corresponding to every non-repeated linear factor of the denominator, a group of partial fractions of the form*

$$B_1/(x-\beta)+B_2/(x-\beta)^2+\ \ldots\ +B_r/(x-\beta)^r$$

*corresponding to every linear factor of the denominator which is repeated* r *times ;*

*and also a partial fraction of the form* $(Cx+D)/(x^2+px+q)$ *for every single quadratic factor of the denominator ; and a group of partial fractions of the form*

$$(C_1x+D_1)/(x^2+px+q)+(C_2x+D_2)/(x^2+px+q)^2+\ \ldots$$
$$+(C_sx+D_s)/(x^2+px+q)^s$$

*corresponding to every quadratic factor of the denominator which is repeated* s *times. Here* A, B, C, D, *etc., denote constants.*

Every integral function of $x$ can be resolved into a product of factors, single or repeated, which are all linear, provided we do not require the coefficients of the factors to be real. In correspondence with this fact, we may restate the above theorem for the decomposition of a proper rational fraction into partial fractions, leaving out all that follows the words "*and also.*" In that case, however, it must be kept in view that $\alpha$, $\beta$, etc., A, $B_1$, $B_2$, etc., may not be real constants.

The decomposition just described may be spoken of as *the decomposition into ultimate partial fractions.* Given its possibility, it is not difficult to carry it out in any particular case. Some of the methods for determining the coefficients are illustrated in the examples given below.

It follows from what has just been set forth that every

rational function of $x$ can be expressed uniquely as the sum of an integral function of $x$ (which vanishes when the function reduces to a proper fraction) and a number of ultimate partial fractions. Since this expression is unique, it is a standard form by means of which the identity of the function can be established at a glance. We have thus in this chapter established two standard forms for a rational function of $x$: one the Rational-Fraction- (*i.e.* quotient of two integral functions) Form, the other the Ultimate-Partial-Fraction-Form. It is the former that is usually most insisted upon in Elementary books, and the phrase "Simplify" applied to a rational function usually means—reduce to the first of these forms. For many purposes, however, the latter is the more suitable or "simpler" form; and often in establishing complicated fractional identities the method of partial fractions proves very powerful.

Ex. 1. Decompose $x/(x-1)(x-2)^2$ into ultimate partial fractions.

The proper general form is

$$x/(x-1)(x-2)^2 \equiv A/(x-1)+B/(x-2)+C/(x-2)^2.$$

Hence we must have

$$x \equiv A(x-2)^2+B(x-1)(x-2)+C(x-1).$$

Put $x=1$, and we get at once $A=1$. Subtract from both sides the part of the right which is now determined, viz. $(x-2)^2$, and suppress the factor $x-1$ which occurs on both sides, and we get

$$-x+4 \equiv B(x-2)+C.$$

Put $x=2$, and we get $C=2$; equate the coefficients of $x$, and we get $B=-1$. Hence

$$x/(x-1)(x-2)^2 \equiv 1/(x-1)-1/(x-2)+2/(x-2)^2.$$

Ex. 2. Decompose $1/(x^2+1)$.

$$1/(x^2+1) \equiv A/(x+i)+B/(x-i)$$

$i$ being the imaginary unit. Hence

$$1 \equiv A(x-i)+B(x+i).$$

If we put successively $x=-i$ and $x=i$, we get

$$A=-1/2i=i/2, \quad B=1/2i=-i/2.$$

Therefore $\qquad 1/(x^2+1) \equiv \frac{1}{2}i/(x+i)-\frac{1}{2}i/(x-i).$

Ex. 3. Decompose $(3x^2+2x+1)/(x+2)(x^2+x+1)^2$ into ultimate partial fractions.

$$\frac{3x^2+2x+1}{(x+2)(x^2+x+1)^2} \equiv \frac{A}{x+2}+\frac{Bx+C}{x^2+x+1}+\frac{Dx+E}{(x^2+x+1)^2}.$$

Hence

$$3x^2+2x+1 \equiv A(x^2+x+1)^2+(Bx+C)(x+2)(x^2+x+1)+(Dx+E)(x+2).$$

Put $x=-2$, and we get $9=A9$, whence $A=1$. Now subtract $(x^2+x+1)^2=x^4+2x^3+3x^2+2x+1$ from both sides, and suppress the factor $x+2$. We then have

$$-x^3\equiv(Bx+C)(x^2+x+1)+Dx+E.$$

Now divide both sides by $x^2+x+1$, and transform the improper fraction on the left by division into the sum of an integral function and a proper fraction. Then we have

$$-x+1+\frac{-1}{x^2+x+1}\equiv Bx+C+\frac{Dx+E}{x^2+x+1}.$$

In this identity the integral and proper fractional parts must be equal separately (by § 155). Therefore $Bx+C\equiv-x+1$ and $Dx+E\equiv-1$. Hence, finally—

$$\frac{3x^2+2x+1}{(x+2)(x^2+x+1)^2}\equiv\frac{1}{x+2}-\frac{x-1}{x^2+x+1}-\frac{1}{(x^2+x+1)^2}.$$

B, C, D, E might also be obtained very readily by equating coefficients of powers of $x$.

## EXERCISES XLVIII.

Resolve the following into ultimate partial fractions :—

**1.** $(8x-87)/(x-9)(x-11)$.

**2.** Transform $x^2/(x-a)(x-b)$ into the form $A+B/(x-a)+C/(x-b)$, where A, B, C are constants.

**3.** $(px+q)/(x-a)(x-b)(x-c)$.

**4.** $x/(x^2-11x+30)$.

**5.** $(x^2+1)/(x-1)(x-2)(x-3)$.

**6.** $(x^2+1)/(2+x)(2-x)(x-1)$.

**7.** $(5x^2-8x-1)/(x-1)^2(x-2)$.

**8.** $(x^2+2x-1)/(x-2)^2(x-3)$.

**9.** $x/(x-1)(x-2)$.

**10.** $1/(x-1)^2(x^2+1)$.

**11.** $(4x^2-2x+2)/(x-1)^2(x^2+1)$.

**12.** $(x^4+x^2-1)/(x^3-1)$.

**13.** $(x^3+1)/(x^3-1)$.

**14.** $(3x^2-3x+2)/(x-1)^2(x^2-x+1)$.

**15.** $(x^2+2x+5)/(x^2+x+2)(x^2+2x+3)$.

**16.** $x/(x-1)(x-2)(x^2+3)$.

**17.** $1/(x^6-1)$.

**18.** $2x/(x^4+x^2+1)$.

**19.** $5x^2/(x-1)(2x+3)(x^2+1)$.

**20.** $(3x^2+4x+5)/(x-1)(x+2)^2(x^2+2)$.

**21.** $x^4/(x^6-1)$.

**22.** $1/(1+x+x^2)$ (into imaginary partial fractions).

**23.** Decompose $1/(x^4+x^2+1)^2$ into two proper fractions whose denominators are $(x^2+x+1)^2$ and $(x^2-x+1)^2$.

# CHAPTER XVI

## IRRATIONAL FUNCTIONS

§ **158.** We have already (§ 5) defined $\sqrt[n]{a}$ ($n$ being a positive integer) as the quantity whose $n$th power is $a$; in short, the defining property of $\sqrt[n]{a}$ is $(\sqrt[n]{a})^n = a$. It has also been explained that, so long as we confine the radicand $a$ to have only positive or purely arithmetical values, there is always one, and only one, real positive or arithmetical value of $\sqrt[n]{a}$; this we shall call for distinction the **Principal Value** of the $n$th root. The $n$th root has other values besides the principal value (for example, there are three quantities, viz. $+1$, $(-1+i\sqrt{3})/2$, $(-1-i\sqrt{3})/2$, which have the property that their cube is $+1$); but with these we are not concerned in this chapter. We also refuse for the present to consider any case where the radicand has a negative value or an imaginary value; the theory of such cases falls quite naturally under the theory of the radication of complex numbers.*

### FUNDAMENTAL RULES OF OPERATION WITH RADICAL SYMBOLS

§ **159.** We may lay down the following five fundamental rules for operating with radical symbols; we shall see, however, that in reality only the first two are independent principles.

$$\sqrt[m]{(\sqrt[n]{a})} = \sqrt[mn]{a} \qquad (1);$$

or, *the* m*th root of the* n*th root of any radicand is the* mn*th root of that radicand.*

* See A. VII. § 17-20.

Ex. 1. $\sqrt{(\sqrt[3]{729})} = \sqrt[6]{(729)} = \sqrt[3]{(\sqrt{729})}$, which may be verified arithmetically, for $\sqrt[3]{729}=9$, $\sqrt{729}=27$, $\sqrt{9}=3$, $\sqrt[3]{27}=3$, $\sqrt[6]{729}=3$.

$$\sqrt[n]{(abc\ .\ .\ .)} = \sqrt[n]{a}\sqrt[n]{b}\sqrt[n]{c}\ .\ .\ . \qquad (2);$$

or, *the* n*th root of a product is the product of the* n*th roots of the factors*

Ex. 2. $\sqrt[3]{(27 \times 8)} = \sqrt[3]{27} \times \sqrt[3]{8}$—that is, $\sqrt[3]{216}=6=3 \times 2$.

$$\sqrt[n]{(a/b)} = \sqrt[n]{a}/\sqrt[n]{b} \qquad (3);$$

or, *the* n*th root of a quotient is the quotient of the* n*th roots of the dividend and divisor.*

Ex. 3. $\sqrt[3]{(27/8)} = \sqrt[3]{27}/\sqrt[3]{8} = 3/2$.

$$\sqrt[n]{a} = \sqrt[mn]{a^m} \qquad (4);$$

or, *the* n*th root of any radicand is the* mn*th root of the* m*th power of that radicand.*

Ex. 4. $\sqrt[2]{4} = \sqrt[6]{4^3} = \sqrt[6]{64}$.

$$(\sqrt[n]{a})^m = \sqrt[n]{a^m} \qquad (5);$$

or, *the* m*th power of the* n*th root of any radicand is the* n*th root of the* m*th power of that radicand.*

Ex. 5. $(\sqrt[3]{8})^2 = \sqrt[3]{8^2} = \sqrt[3]{64}$.

These five identities are simply translations of the laws of integral indices into a new language or symbolism; they can all be established by the same method, which we shall illustrate by application to (1) and (2).

First consider (1). Since we restrict $a$ to be positive, and consider only the principal value of the $n$th root, $\sqrt[n]{a}$ is a real positive quantity, and $\sqrt[m]{(\sqrt[n]{a})}$ denotes the principal value of the $m$th root of this real positive quantity $(\sqrt[n]{a})$. If, therefore, $x \equiv \sqrt[m]{(\sqrt[n]{a})}$, $x$ is a real positive quantity. Also $x^m = \{\sqrt[m]{(\sqrt[n]{a})}\}^m = \sqrt[n]{a}$, by the definition of $\sqrt[m]{\ }$, and $(x^m)^n = (\sqrt[n]{a})^n = a$, by the definition of $\sqrt[n]{\ }$. Hence $x^{mn} = a$, by the laws of integral indices. It follows that $x$ is one of the $mn$th roots of $a$, and, being real and positive, it is the principal $mn$th root, viz. that which we have agreed to denote by $\sqrt[mn]{a}$. Hence $\sqrt[mn]{a} = \sqrt[m]{(\sqrt[n]{a})}$, as was to be shown.

To prove (2), let us consider three radicands $a$, $b$, $c$, and denote $\sqrt[n]{a}\sqrt[n]{b}\sqrt[n]{c}$ by $x$. Then, since $a$, $b$, $c$ are supposed to be all real and positive, and only principal values of the $n$th roots are considered, $x$ is a real positive quantity. Also we have $x^n = (\sqrt[n]{a}\sqrt[n]{b}\sqrt[n]{c})^n = (\sqrt[n]{a})^n(\sqrt[n]{b})^n(\sqrt[n]{c})^n$, by the laws of positive integral indices. Therefore $x^n = abc$, by the definition of $\sqrt[n]{\ }$. Hence $x$ is one of the $n$th roots of the real positive quantity $abc$, and, being real and positive, it is the principal value of the $n$th root, which is what we have agreed to denote by $\sqrt[n]{(abc)}$. Therefore $\sqrt[n]{a}\sqrt[n]{b}\sqrt[n]{c} = \sqrt[n]{(abc)}$. The proof will obviously apply to any number of radicands.

As an exercise the beginner should apply the same direct method to establish the identities (3), (4), (5). These can, however, be derived from (1) and (2).

To prove (3). We have, $x$ and $y$ being any two real positive quantities—

$$\sqrt[n]{(xy)} = \sqrt[n]{x}\sqrt[n]{y}.$$

Now if $a$ and $b$ be any two real positive quantities, and $b \neq 0$, $x = a/b$ and $y = b$ will be two real positive quantities. Hence the last equation gives

$$\sqrt[n]{\left(\frac{a}{b} \times b\right)} = \sqrt[n]{\left(\frac{a}{b}\right)}\sqrt[n]{b},$$

that is—

$$\sqrt[n]{a} = \sqrt[n]{\left(\frac{a}{b}\right)}\sqrt[n]{b};$$

whence

$$\sqrt[n]{a}/\sqrt[n]{b} = \sqrt[n]{(a/b)}\sqrt[n]{b}/\sqrt[n]{b},$$

that is—

$$\sqrt[n]{a}/\sqrt[n]{b} = \sqrt[n]{(a/b)},$$

which is (3).

To prove (4). By the definition of $\sqrt[m]{\ }$ we have, since $a$ is real and positive, $a = \sqrt[m]{a^m}$. Hence $\sqrt[n]{a} = \sqrt[n]{(\sqrt[m]{a^m})} = \sqrt[nm]{a^m}$, by (1).

To prove (5). We have, as in last demonstration, $a = \sqrt[m]{a^m}$. Hence

$$\sqrt[n]{a} = \sqrt[n]{(\sqrt[m]{a^m})} = \sqrt[nm]{a^m} = \sqrt[m]{(\sqrt[n]{a^m})},$$

by (1).

Therefore

$$(\sqrt[n]{a})^m = \{\sqrt[m]{(\sqrt[n]{a^m})}\}^m$$
$$= \sqrt[n]{(a^m)},$$

by the definition of $\sqrt[m]{}$.

It thus appears that (3) is but a case of (2); and that (4) and (5) might be looked upon as particular cases of (1).

§ **160.** For the sake of comparison with the theory of indices, we deduce from the above fundamental laws the following :—

$$\sqrt[q]{x^p}\sqrt[s]{x^r} = \sqrt[qs]{x^{ps+qr}} \qquad \text{(A)};$$
$$\sqrt[s]{\{\sqrt[q]{x^p}\}^r} = \sqrt[qs]{x^{pr}} \qquad \text{(B)};$$
$$\sqrt[q]{(xy)^p} = \sqrt[q]{x^p}\sqrt[q]{y^p} \qquad \text{(C)}.$$

To deduce (A) we have merely to notice that, by (4), $\sqrt[q]{x^p} = \sqrt[qs]{x^{ps}}$ and $\sqrt[s]{x^r} = \sqrt[qs]{x^{qr}}$. Hence $\sqrt[q]{x^p}\sqrt[s]{x^r} = \sqrt[qs]{x^{ps}}\sqrt[qs]{x^{qr}}$ $= \sqrt[qs]{(x^{ps}x^{qr})}$, (by (2)), $= \sqrt[qs]{x^{ps+qr}}$, by the laws of positive integral indices.

To prove (B). We have, by (5), $\{\sqrt[q]{x^p}\}^r = \sqrt[q]{(x^p)^r} = \sqrt[q]{x^{pr}}$, by laws of positive integral indices. Hence $\sqrt[s]{\{\sqrt[q]{x^p}\}^r} = \sqrt[s]{\{\sqrt[q]{x^{pr}}\}}$ $= \sqrt[qs]{x^{pr}}$, by (1).

We deduce (C) from (2) at once by putting $a = x^p$, $b = y^p$, which will be perfectly legitimate if $x$, $y$, $z$, . . . be all real and positive.

§ **161.** It should be noticed that, if the restrictions on the value of the roots and radicands (§ 158) be departed from, some of the above propositions will not hold. For example, if we admit that $\sqrt{4^2}$ may be either $\pm 4$, we should have $(\sqrt{4})^2 = (\pm 2)^2$ $= 4$; but $\sqrt{4^2} = \pm 4$; so that we could not assert the identity $(\sqrt{4})^2 = \sqrt{4^2}$. Another example of paradox arising from the same cause is $-1 = \sqrt{(-1)} \times \sqrt{(-1)} = \sqrt{\{(-1)(-1)\}} = \sqrt{1} = 1$.

§ **162.** The following examples are given in order to familiarise the beginner with transformations involving the direct use of radical symbols. The references (1), (2), (3), (4), (5) are to the theorems of § 159.

Ex. 1. Express $\sqrt{32}$ and $\sqrt{(350/21)}$ as rational multiples of the smallest possible square root.

$\sqrt{32} = \sqrt{(16 \times 2)} = \sqrt{16} \times \sqrt{2}$, by (2), $= 4\sqrt{2}$.

$\sqrt{(350/21)} = \sqrt{(50/3)} = \sqrt{50}/\sqrt{3}$, by (3), $= \sqrt{25} \times \sqrt{2}/\sqrt{3}$, by (2), $= 5 \times \sqrt{2} \times \sqrt{3}/(\sqrt{3})^2 = \frac{5}{3}\sqrt{6}$, by (2).

Ex. 2. $\sqrt{108} + \sqrt{27} - \sqrt{75} = \sqrt{(36 \times 3)} + \sqrt{(9 \times 3)} - \sqrt{(25 \times 3)}$, $= \sqrt{36}\sqrt{3} + \sqrt{9}\sqrt{3} - \sqrt{25}\sqrt{3}$, by (2), $= 6\sqrt{3} + 3\sqrt{3} - 5\sqrt{3}$, $= (6+3-5)\sqrt{3}$, by the law of distribution, $= 4\sqrt{3}$.

Ex. 3. Express $2\sqrt{3}$ and $3\sqrt[3]{2}$ as roots of the same order.

$$2\sqrt{3} = \sqrt{4}\sqrt{3} = \sqrt{12}, \text{ by (2)},$$
$$3\sqrt[3]{2} = \sqrt[3]{27}\sqrt[3]{2} = \sqrt[3]{54}, \text{ by (2)}.$$

Also, by (4), $\sqrt{12} = \sqrt[6]{12^3}$; and $\sqrt[3]{54} = \sqrt[6]{54^2}$. Hence $2\sqrt{3} = \sqrt[6]{1728}$; and $3\sqrt[3]{2} = \sqrt[6]{2916}$, which are the required expressions.

Ex. 4. To arrange $\sqrt{5}$, $\sqrt[3]{4}$, and $\sqrt[4]{10}$ in order of magnitude.

We can express all three as roots of the order 12. Thus $\sqrt{5} = \sqrt[12]{5^6}$; $\sqrt[3]{4} = \sqrt[12]{4^4}$; $\sqrt[4]{10} = \sqrt[12]{10^3}$. Hence $\sqrt{5} = \sqrt[12]{15625}$; $\sqrt[3]{4} = \sqrt[12]{256}$; $\sqrt[4]{10} = \sqrt[12]{1000}$. It is now evident that the order of ascending magnitude is $\sqrt[3]{4}$, $\sqrt[4]{10}$, $\sqrt{5}$.

Ex. 5. $\sqrt[m]{x^{pm+q}} = \sqrt[m]{(x^{pm} \times x^q)} = \sqrt[m]{x^{pm}}\sqrt[m]{x^q}$, by (2), $= x^p\sqrt[m]{x^q}$, by (4).

Ex. 6. $\sqrt{(yx+x^2)}\sqrt{(yz+zx)} = \sqrt{x}\sqrt{(y+x)}\sqrt{z}\sqrt{(y+x)}$, by (2), $= \{\sqrt{(y+x)}\}^2\sqrt{(zx)}$, by (2), $= (y+x)\sqrt{(zx)}$.

Ex. 7. Simplify $2\sqrt{2}\sqrt[4]{2}\sqrt[8]{4}$.

We can, by (4), express each factor as an eighth root; thus $2\sqrt{2}\sqrt[4]{2}\sqrt[8]{4} = \sqrt[8]{2^8}\sqrt[8]{2^4}\sqrt[8]{2^2}\sqrt[8]{2^2} = \sqrt[8]{(2^8 2^4 2^2 2^2)}$, by (2), $= \sqrt[8]{2^{16}} = (\sqrt[8]{2})^{16}$, by (5), $= \{(\sqrt[8]{2})^8\}^2 = 2^2 = 4$.

Ex. 8. $\{\sqrt{(1-x)} + \sqrt{(1+x)}\}^4 = \{\sqrt{(1-x)}\}^4 + 4\{\sqrt{(1-x)}\}^3\{\sqrt{(1+x)}\} + 6\{\sqrt{(1-x)}\}^2\{\sqrt{(1+x)}\}^2 + 4\{\sqrt{(1-x)}\}\{\sqrt{(1+x)}\}^3 + \{\sqrt{(1+x)}\}^4$. Now $\{\sqrt{(1-x)}\}^4 = [\{\sqrt{(1-x)}\}^2]^2 = (1-x)^2$; $\{\sqrt{(1-x)}\}^3 = \{\sqrt{(1-x)}\}^2\{\sqrt{(1-x)}\} = (1-x)\sqrt{(1-x)}$, etc. Hence

$$\begin{aligned}\{\sqrt{(1-x)} + \sqrt{(1+x)}\}^4 &= (1-x)^2 + 4(1-x)\sqrt{(1-x)}\sqrt{(1+x)} + 6(1-x)\\ &\quad (1+x) + 4(1+x)\sqrt{(1-x)}\sqrt{(1+x)} + (1+x)^2\\ &= (8-4x^2) + 8\sqrt{(1-x^2)}, \text{ by (2)}.\end{aligned}$$

Ex. 9. Prove that

$$\sqrt{\left\{\frac{a+\sqrt{(a^2-b)}}{2}\right\}} + \sqrt{\left\{\frac{a-\sqrt{(a^2-b)}}{2}\right\}} = \sqrt{(a+\sqrt{b})},$$

where $a$ and $b$ are both real positive quantities and $a^2 > b$.

First we remark that, since $a^2 - b < a^2$, it follows that $\sqrt{(a^2-b)} < a$. Hence, since $a$ is positive, $a - \sqrt{(a^2-b)}$ is a real positive quantity; and $a + \sqrt{(a^2-b)}$ is obviously also real and positive. Therefore, if

$$x = \sqrt{\left\{\frac{a+\sqrt{(a^2-b)}}{2}\right\}} + \sqrt{\left\{\frac{a-\sqrt{(a^2-b)}}{2}\right\}},$$

$x$ is a real positive quantity. Now we have

$$x^2=\left[\sqrt{\left\{\frac{a+\sqrt{(a^2-b)}}{2}\right\}}\right]^2+\left[\sqrt{\left\{\frac{a-\sqrt{(a^2-b)}}{2}\right\}}\right]^2$$
$$+2\sqrt{\left\{\frac{a+\sqrt{(a^2-b)}}{2}\right\}}\sqrt{\left\{\frac{a-\sqrt{(a^2-b)}}{2}\right\}},$$
$$=\frac{a+\sqrt{(a^2-b)}}{2}+\frac{a-\sqrt{(a^2-b)}}{2}+2\sqrt{\left\{\frac{a^2-(\sqrt{(a^2-b)})^2}{4}\right\}},$$

by (2). Therefore $x^2=a+2\sqrt{(b/4)}=a+2\sqrt{b}/2$, by (3).

Hence $x^2=a+\sqrt{b}$. Therefore, since $x$ is real and positive, $x=\sqrt{(a+\sqrt{b})}$.

## EXERCISES XLIX.*

Express each of the following functions, first, as the simplest rational multiple of a single root of the simplest possible rational radicand; second, as a single root of the simplest possible rational radicand:—

1. $(x/y)/\sqrt{(x^2/y^3)}$.
2. $(\sqrt{a^3}\sqrt{b})/\sqrt{(b^3/a^2)}$.
3. $\sqrt{(x^3/y^3)}\times\sqrt[3]{(x^2/y^2)}$.
4. $\sqrt{\left(\frac{a}{b}\sqrt{\frac{a}{b}}\right)}\sqrt{\left(\frac{b}{a}\sqrt{\frac{b}{a}}\right)}$.
5. $\sqrt{x}\times 3\sqrt[3]{x}\times 2\sqrt[6]{x}$.
6. $\sqrt[3]{(x^2y^2)}\times\sqrt{(x^3y^3)}\times\sqrt{x}\times\sqrt{y}$.
7. $\sqrt[3]{x}\times\sqrt[9]{x^3}\times\sqrt[6]{x^2}$.
8. $\sqrt[m]{x^n}\times\sqrt[n]{x^m}\times\sqrt[mn]{x}$.
9. $\sqrt{(x\sqrt{(x\sqrt{(x\sqrt{x})})})}$.
10. $\sqrt[q]{x^p}/\sqrt[s]{y^r}$.
11. $\dfrac{\sqrt{(xy)}}{\sqrt{(xy)}-y}\Big/\dfrac{x+\sqrt{(xy)}}{x}$.
12. $\dfrac{\sqrt{x}}{\sqrt{x}-\sqrt{y}}\Big/\dfrac{\sqrt{x}+\sqrt{y}}{\sqrt{x^3}}$.
13. $\sqrt{(x^2+2xy+y^2-1)}/\sqrt{(x+y+1)}$.
14. $\sqrt{(x+y)}\sqrt[3]{(x+y)}\sqrt{(x-y)}\sqrt[3]{(x-y)}$.
15. $\sqrt{(x+y)}\sqrt{(x-y)}\sqrt{(x^2+y^2)}\sqrt{(x^4-y^4)}$.
16. $x\sqrt{x^5}+3\sqrt{x^7}+\sqrt{x^3}$.
17. $\sqrt{(a^3+a^2x)}+\sqrt{(x^3+ax^2)}+\sqrt{(a+x)}$.
18. $\sqrt{(4x^3+4x^2+x)}-\sqrt{(x^3+2x^2+x)}-\sqrt{(x^3-2x^2+x)}$.
19. $(x+y)\sqrt{\dfrac{x-y}{x+y}}+(x-y)\sqrt{\dfrac{x+y}{x-y}}$.
20. $(a-x)\sqrt{(a^2-x^2)}-(a^2+x^2)\sqrt{\{(a+x)/(a-x)\}}$.
21. $(3\sqrt[3]{x}+\sqrt[3]{x^4})/(\sqrt{x^3}+3\sqrt{x})$.
22. $\sqrt[3]{(27x^4)}+2\sqrt[3]{x^7}+6\sqrt[3]{x}$.
23. $\dfrac{xy}{x-y}+\sqrt{\left\{\dfrac{x^2y^2}{(x-y)^2}+\dfrac{x^2y}{x-y}\right\}}\div\dfrac{x\sqrt{y}}{\sqrt{x}-\sqrt{y}}$, $x>y$.

Arrange each of the following functions as an algebraic sum of rational multiples of single roots of the simplest possible rational radicands:—

24. $(x+\sqrt{x}+1)(x-\sqrt{x}+1)$.
25. $\{x+\sqrt{(a^2-x^2)}\}^2$.

---

* If the beginner finds difficulty with these exercises, he may postpone some of them until he has finished the next chapter; but he should not solve them by using the index notation for radicals.

26. $(\sqrt{x}+1+1/\sqrt{x})(\sqrt{x}-1+1/\sqrt{x})$.
27. $\{\sqrt{(x+1)}+\sqrt{(x-1)}+1\}\{\sqrt{(x+1)}-\sqrt{(x-1)}+1\}$.
28. $(\sqrt{x}+\sqrt[4]{x}+1)(\sqrt{x}-\sqrt[4]{x}+1)$.
29. $\{\sqrt[4]{(x/y)}+\sqrt[4]{(y/x)}\}^4$. 30. $(\sqrt[3]{x}-1)(\sqrt[3]{x^2}+\sqrt[3]{x}+1)$.
31. $(\sqrt[3]{x^2}+\sqrt[3]{x}+1)(\sqrt[3]{x^2}-\sqrt[3]{x}+1)$.
32. Arrange $\sqrt[3]{3}$, $\sqrt{5}$, $\sqrt[4]{7}$ in order of magnitude.

Prove the following irrational identities :—

33. $\sqrt{(a+b)}+\sqrt{(a-b)}=\sqrt{2}\sqrt{\{a+\sqrt{(a^2-b^2)}\}}$.

34. $\sqrt{\dfrac{a+\sqrt{b}}{2}}+\sqrt{\dfrac{a-\sqrt{b}}{2}}=\sqrt{\{a+\sqrt{(a^2-b)}\}}$.

35. $\sqrt{(a+b)}+\sqrt{(a-b)}=\sqrt[3]{\{(4a-2b)\sqrt{(a+b)}+(4a+2b)\sqrt{(a-b)}\}}$.

36. $\sqrt[4]{(a+\sqrt{b})}+\sqrt[4]{(a-\sqrt{b})}=\sqrt[4]{[2a+6\sqrt{(a^2-b)}+4\sqrt{\{2a\sqrt{(a^2-b)}+2(a^2-b)\}}]}$.

37. Show that $x=\sqrt[3]{\{a+\sqrt{(a^2-b^3)}\}}+\sqrt[3]{\{a-\sqrt{(a^2-b^3)}\}}$ is a root of the equation $x^3-3bx-2a=0$.

38. If $x=\sqrt[3]{(a+\sqrt{b})}+\sqrt[3]{(a-\sqrt{b})}$, show that $(x^3-2a)^3=27(a^2-b)x^3$.

39. Show that $x=2+\sqrt{(2+\sqrt{(2+\sqrt{2})})}$ is a root of the equation $x^8-16x^7+104x^6-352x^5+660x^4-672x^3+336x^2-64x+2=0$. What other roots can you perceive the equation to have ?

## Representation of Radication by Means of Fractional Indices

**§ 163.** The radical symbol $\sqrt[n]{\ }$ in conjunction with the fundamental rules which we have established for its use is quite sufficient for all the purposes of Algebra, in so far as Algebra deals with synthetic irrational functions; but there is another way of representing root extraction which is for many purposes more convenient than the radix notation, and which is also interesting in connection with the theory of exponential and logarithmic functions, which are the simplest kind of function that cannot be described as a synthetic algebraic function (see § 8).

**§ 164.** Up to this point we have restricted the index or exponent $n$ in $x^n$ to be a positive integer; indeed, the original definition of $x^n$ becomes meaningless if we attribute any other kind of value to $n$. It is therefore open to us to give to $x^n$, when $n$ is fractional or zero or negative, any meaning we please, *provided always that such meaning consorts with the fundamental laws of Algebra and all their consequences, and also with the laws of indices as already established.*

Let us confine our attention for the moment to positive

fractional (commensurable) indices. If such a symbol as $x^{\frac{3}{4}}$ is to be admitted as an algebraic operand, subject, *inter alia*, to the laws of indices, we must have, by the first law, $x^{\frac{3}{4}} \times x^{\frac{3}{4}} \times x^{\frac{3}{4}} \times x^{\frac{3}{4}} = x^{\frac{3}{4}+\frac{3}{4}+\frac{3}{4}+\frac{3}{4}} = x^3$, or, what is virtually the same thing, by the second law, $(x^{\frac{3}{4}})^4 = x^3$. That is to say, $x^{\frac{3}{4}}$ is an operand whose fourth power is $x^3$; or $x^{\frac{3}{4}}$ is one of the fourth roots of $x^3$. Introducing, for the sake of clearness and simplicity, the same restrictions on the radicand and the value of the root as heretofore (§ 158), we are thus led to define as follows:—

$x^{p/q}$, *where* p *and* q *are finite positive integers, is defined to mean the principal* (i.e. *real positive*) *value of the* q*th root of* $x^p$, x *being restricted to be a real positive quantity;* or, in symbols—

$$x^{p/q} = \sqrt[q]{x^p}.$$

It will be observed that, since $p$ is a positive integer, and $x$ is real and positive, $x^p$ is real and positive, so that there is no ambiguity or indefiniteness in the meaning of $\sqrt[q]{x^p}$.

Since, by § 159 (5), under the restrictions introduced, $\sqrt[q]{x^p} = (\sqrt[q]{x})^p$, it follows that $x^{p/q}$ *also means the* p*th power of the principal value of the* q*th root of* x. Sometimes this is made the primary definition of $x^{p/q}$; but that is not a convenient arrangement.

Again, it would appear from our definition and the fundamental property of a fraction that we ought to have $x^{p/q} = x^{mp/mq}$, where $m$ is any positive integer; and this is as it ought to be; for the identity just written is in radical symbols $\sqrt[q]{x^p} = \sqrt[mq]{x^{mp}}$, which is a case of § 159 (4).

**§ 165.** It remains to show that the definition of $x^{p/q}$ just given introduces no contradiction into Algebra; and in particular that it agrees with the laws of indices.

As regards Algebra generally, there is no more difficulty regarding $x^{p/q}$ than there is regarding its equivalent $\sqrt[q]{x^p}$, which, when it is not commensurable, we regard as being replaced by a commensurable approximation of sufficient accuracy.

Before proceeding to discuss the application of the laws of indices (§ 29) to the new definition, it will be an advantage to reduce them to their simplest independent elements; these are

$$x^m \times x^n = x^{m+n} \qquad (A);$$

$$(x^m)^n = x^{mn} \qquad (B);$$

$$(xy)^m = x^m y^m \qquad (C).$$

For the general case of I. ($\alpha$), viz. $x^m \times x^n \times x^p \times \ldots = x^{m+n+p+\ldots}$, follows by repeated application of the special case $x^m \times x^n = x^{m+n}$. Again, I. ($\beta$), viz. $x^m \div x^n = x^{m-n}$, if $m>n$, $= 1 \div x^{n-m}$ if $m<n$, follows from I. ($\alpha$). For, if $m>n$, we have, by I. ($\alpha$), $x^{m-n} \times x^n = x^{(m-n)+n} = x^m$, whence $x^{m-n} \times x^n \div x^n = x^m \div x^n$; that is, $x^m \div x^n = x^{m-n}$; and again, if $m<n$, $x^{n-m} \times x^m = x^n$, by I. ($\alpha$), whence $x^{n-m} \times x^m \div x^{n-m} \div x^n = x^n \div x^{n-m} \div x^n$; that is, $x^m \div x^n = 1 \div x^{n-m}$.

Lastly, the general case of III. ($\alpha$) follows from the particular case $(xy)^m = x^m y^m$ by repetition; and III. ($\beta$) may be derived from III. ($\alpha$), thus: $\{(x/y)y\}^m = (x/y)^m y^m$, by III. ($\alpha$); that is, $x^m = (x/y)^m y^m$. Hence $x^m \div y^m = (x/y)^m y^m \div y^m$; that is, $x^m/y^m = (x/y)^m$, which is III. ($\beta$).

In these deductions no appeal has been made to the definition of an index, we have merely used the laws themselves supposed to be valid. If, therefore, the simple laws (A), (B), (C), as written in this paragraph, have been proved for any definition of an index, the more extended laws of § 29 will follow.

Suppose now $m$ and $n$ to be any two positive commensurable quantities, say $p/q$ and $r/s$, where $p$, $q$, $r$, $s$ are positive integers, then we have to deduce from our definition of a positive fractional power that

$$x^{p/q}x^{r/s} = x^{p/q+r/s},$$

that is to say—

$$x^{p/q}x^{r/s} = x^{(ps+qr)/qs} \qquad \text{(A)};$$

$$\left(x^{\frac{p}{q}}\right)^{\frac{r}{s}} = x^{\frac{p}{q}\times\frac{r}{s}},$$

that is to say—

$$\left(x^{\frac{p}{q}}\right)^{\frac{r}{s}} = x^{\frac{pr}{qs}} \qquad \text{(B)};$$

and

$$(xy)^{\frac{p}{q}} = x^{\frac{p}{q}}y^{\frac{p}{q}} \qquad \text{(C)}.$$

Now, bearing in mind the meaning attached to $x^{p/q}$, etc., we see that these identities expressed in radical symbols are

$$\sqrt[q]{x^p} \times \sqrt[s]{x^r} = \sqrt[qs]{x^{ps+qr}} \qquad \text{(A')};$$

$$\sqrt[s]{(\sqrt[q]{x^p})^r} = \sqrt[qs]{x^{pr}} \qquad \text{(B')};$$

$$\sqrt[q]{(xy)^p} = \sqrt[q]{x^p} \times \sqrt[q]{y^p} \qquad \text{(C')};$$

which are simply the identities (A), (B), (C) of § 160, already established.

It should be remarked that the case where fractional and integral indices are mixed is covered by the above demonstration, because nothing in the reasoning used prevents either $p/q$ or $r/s$ from being actually integral in value; all that is required is that they be positive and commensurable.

## Interpretation of $x^0$ and $x^{-m}$

**§ 166.** Having extended the definition of $x^n$ to cover all cases where $n$ is any positive commensurable number, it is natural to endeavour to complete the extension by including the cases where $n$ is 0 or any negative commensurable number. As before, we may get suggestions by supposing that the laws of indices are to hold without restriction on the value of the index and tracing the consequences of this assumption.

Let $m$ be any positive commensurable number and $x$ a base *differing from* 0. Then, if the laws of indices are to hold without restriction on the index, we should have, *inter alia*, $x^n \div x^m = x^{n-m}$, by I. ($\beta$), $= x^{n+(-m)} = x^n \times x^{-m}$, by I. ($\alpha$).

Hence $$x^n \div x^m = x^n \times x^{-m},$$

therefore $$x^n \div x^m \div x^n = x^n \times x^{-m} \div x^n;$$

whence $$1 \div x^m = x^{-m}.$$

*We are thus led to define* $\mathbf{x}^{-\mathbf{m}}$ *as the reciprocal of* $\mathbf{x}^{\mathbf{m}}$, *with the understanding that, if* m *be fractional,* x *is to be restricted to be real and positive, and that the principal value of the root indicated is to be taken.* The base $x$ must of course be different from 0.

Again, we should have $x^0 = x^{m-m} = x^m \div x^m$ by II. ($\beta$), if the values of indices are to be unrestricted. *We are thus led to define* $\mathbf{x}^0$, *where* x *is any non-evanescent base, as meaning* 1.

To complete the theory, it now requires to be shown that these meanings of $x^{-m}$ and $x^0$ are consistent with all the laws of indices, there being obviously no difficulty as regards the laws of Algebra generally.

This means that we have to re-establish the laws of indices in the form (A), (B), (C), § 165, for the newly-introduced meanings.

In the case of laws (A) and (B), where two different indices appear, there are a good many cases to consider, viz. $m$ positive, $n$ negative; $m$ negative, $n$ negative; $m$ positive, $n$ zero; $m$ negative, $n$ zero; $m$ zero, $n$ zero. It will be sufficient to take two cases, and leave the rest as exercises for the reader.

Case $$m = -p, \quad n = -q,$$

where $p$ and $q$ are positive commensurable numbers. To prove

$$x^{-p}x^{-q} = x^{-p-q}.$$

By definition— $x^{-p} = 1/x^p, \quad x^{-q} = 1/x^q.$
Hence $x^{-p}x^{-q} = (1/x^p)(1/x^q) = 1/x^px^q.$

Now, since the laws are already established for any positive commensurable indices, we have $x^px^q = x^{p+q}$. Hence $1/x^px^q = 1/x^{p+q}$; but this last is $x^{-(p+q)}$ or $x^{-p-q}$ by the definition of a negative index.

Again, to prove $(x^{-p})^{-q} = x^{(-p)(-q)}$,

we have, by the definition of a negative index, $(x^{-p})^{-q} = (1/x^p)^{-q} = 1/(1/x^p)^q = 1/\{1^q/(x^p)^q\} = 1/\{1/x^{pq}\}$, the last two steps by the laws for positive commensurable indices already established. Hence $(x^{-p})^{-q} = x^{pq} = x^{(-p)(-q)}$, which was to be proved.

Case $$m = -p, \quad n = 0,$$

where $p$ is positive and commensurable. To prove

$$x^{-p}x^0 = x^{-p+0}.$$

Since, by definition, $x^0 = 1$, this amounts to proving that $x^{-p} \times 1 = x^{-p}$, which is obviously true.

Next to prove

$$(x^{-p})^0 = (x^0)^{-p} = x^{(-p \times 0)}.$$

This is tantamount to proving that $1 = 1^{-p} = x^0 = 1$, which is true, provided always $x \neq 0$, and we adhere to our restrictions on the value of any root that may be represented by $p$.

Finally, as to the law (C), we have to prove that

$$(xy)^{-p} = x^{-p}y^{-p},$$

where $p$ is any positive commensurable number, and $x$ and $y$ non-evanescent bases. Now $(xy)^{-p} = 1/(xy)^p$, by definition, $= 1/x^py^p$ by the laws established for positive commensurable indices. Hence $(xy)^{-p} = (1/x^p)(1/y^p) = x^{-p}y^{-p}$, by the definition of a negative index.

To prove that $$(xy)^0 = x^0y^0,$$

where $x$ and $y$ are non-evanescent bases, simply amounts to proving that $1=1\times 1$, which is obvious.

§ **167.** We can now arrange the powers of a single base in an ascending series $x^{-\infty}, \ldots, x^{-1}, \ldots, x^{-\frac{1}{2}}, \ldots, x^{0}, \ldots, x^{+\frac{1}{2}} \ldots, x^{+1}, \ldots, x^{+\infty}$ corresponding to the series of real commensurable algebraic quantity. Thus we speak of $x^{\frac{3}{4}}$ as a higher power than $x^{\frac{1}{2}}$; $x^{-\frac{1}{2}}$ as a higher power than $x^{-\frac{3}{4}}$, and so on.

§ **168.** In working with the laws of indices as now extended, the only new points are to attend to the interpretations of $x^{p/q}$, $x^{-m}$, and $x^0$. The following examples are intended to familiarise the beginner with the subject. The first eight examples are the first eight of § 162 worked out with the index notation for radication. The references I. ($\alpha$), I. ($\beta$); II. III. ($\alpha$); III. ($\beta$) are to the laws of indices as stated in § 29.

Ex. **1.** $\sqrt{32}=(32)^{\frac{1}{2}}=(16\times 2)^{\frac{1}{2}}=16^{\frac{1}{2}}2^{\frac{1}{2}}$, by III. ($\alpha$), $=4\sqrt{2}$.

$\sqrt{(350/21)}=(50/3)^{\frac{1}{2}}=50^{\frac{1}{2}}/3^{\frac{1}{2}}$, by III. ($\beta$),

$=(25\times 2)^{\frac{1}{2}}/3^{\frac{1}{2}}=25^{\frac{1}{2}}2^{\frac{1}{2}}/3^{\frac{1}{2}}$, by III. ($\alpha$), $=5\times 2^{\frac{1}{2}}3^{\frac{1}{2}}/3^{\frac{1}{2}+\frac{1}{2}}$

$=5(2\times 3)^{\frac{1}{2}}/3$, by III. ($\alpha$), $=\frac{5}{3}\sqrt{6}$.

Ex. **2.** $\sqrt{108}+\sqrt{27}-\sqrt{75}=(36\times 3)^{\frac{1}{2}}+(9\times 3)^{\frac{1}{2}}-(25\times 3)^{\frac{1}{2}}$

$=36^{\frac{1}{2}}3^{\frac{1}{2}}+9^{\frac{1}{2}}3^{\frac{1}{2}}-25^{\frac{1}{2}}3^{\frac{1}{2}}$, by III. ($\alpha$), $=(6+3-5)3^{\frac{1}{2}}=4\sqrt{3}$.

Ex. **3.** $2\sqrt{3}=2.3^{\frac{1}{2}}=2^{\frac{2}{2}}3^{\frac{1}{2}}=(2^2)^{\frac{1}{2}}3^{\frac{1}{2}}$, by II., $=(2^2 3)^{\frac{1}{2}}$, by III. ($\alpha$), $=12^{\frac{1}{2}}$.

$3\sqrt[3]{2}=3^{\frac{3}{3}}2^{\frac{1}{3}}=(3^3)^{\frac{1}{3}}2^{\frac{1}{3}}$, by II., $=(3^3 2)^{\frac{1}{3}}$, by III. ($\alpha$), $=54^{\frac{1}{3}}$.

Hence $2\sqrt{3}=12^{\frac{1}{2}}=12^{\frac{3}{6}}=\sqrt[6]{12^3}=\sqrt[6]{1728}$;

$3\sqrt[3]{2}=54^{\frac{1}{3}}=54^{\frac{2}{6}}=\sqrt[6]{54^2}=\sqrt[6]{2916}$.

Ex. **4.** $\sqrt{5}=5^{\frac{1}{2}}=5^{\frac{6}{12}}=\sqrt[12]{5^6}=\sqrt[12]{15625}$;

$\sqrt[3]{4}=4^{\frac{1}{3}}=4^{\frac{4}{12}}=\sqrt[12]{4^4}=\sqrt[12]{256}$;

$\sqrt[4]{10}=10^{\frac{1}{4}}=10^{\frac{3}{12}}=\sqrt[12]{10^3}=\sqrt[12]{1000}$.

Ex. **5.** $\sqrt[m]{x^{pm+q}}=x^{(pm+q)/m}=x^{p+q/m}=x^p x^{q/m}$, by I. ($\alpha$), $=x^p\sqrt[m]{x^q}$.

Ex. **6.** $\sqrt{(yx+x^2)}\sqrt{(yz+zx)}=(yx+x^2)^{\frac{1}{2}}(yz+zx)^{\frac{1}{2}}=\{(y+x)x\}^{\frac{1}{2}}\{(y+x)z\}^{\frac{1}{2}}=(y+x)^{\frac{1}{2}}x^{\frac{1}{2}}(y+x)^{\frac{1}{2}}z^{\frac{1}{2}}$, by III. ($\alpha$), $=(y+x)^{\frac{1}{2}+\frac{1}{2}}(zx)^{\frac{1}{2}}$, by I. ($\alpha$) and III. ($\alpha$), $=(y+x)\sqrt{(zx)}$.

Ex. **7.** $2\sqrt{2}\sqrt[4]{2}\sqrt[8]{4}=2\times 2^{\frac{1}{2}}\times 2^{\frac{1}{4}}\times 4^{\frac{1}{8}}=2\times 2^{\frac{1}{2}}\times 2^{\frac{1}{4}}\times(2^2)^{\frac{1}{8}}=2\times 2^{\frac{1}{2}}\times 2^{\frac{1}{4}}\times 2^{\frac{2}{8}}$, by II., $=2^1\times 2^{\frac{1}{2}}\times 2^{\frac{1}{4}}\times 2^{\frac{1}{4}}=2^{1+\frac{1}{2}+\frac{1}{4}+\frac{1}{4}}$, by I. ($\alpha$), $=2^2=4$.

Ex. **8.** $\{\sqrt{(1-x)}+\sqrt{(1+x)}\}^4=\{(1-x)^{\frac{1}{2}}+(1+x)^{\frac{1}{2}}\}^4=\{(1-x)^{\frac{1}{2}}\}^4+4\{(1-x)^{\frac{1}{2}}\}^3\{(1+x)^{\frac{1}{2}}\}+6\{(1-x)^{\frac{1}{2}}\}^2\{(1+x)^{\frac{1}{2}}\}^2+4\{(1-x)^{\frac{1}{2}}\}\{(1+x)^{\frac{1}{2}}\}^3+\{(1+x)^{\frac{1}{2}}\}^4=(1-x)^2+4(1-x)^{\frac{3}{2}}(1+x)^{\frac{1}{2}}+6(1-x)(1+x)+4(1-x)^{\frac{1}{2}}(1+x)^{\frac{3}{2}}+(1+x)^2$, by II. Now $(1-x)^{\frac{3}{2}}(1+x)^{\frac{1}{2}}=(1-x)^1(1-x)^{\frac{1}{2}}(1+x)^{\frac{1}{2}}$, by I. ($\alpha$),

$=(1-x)(1-x^2)^{\frac{1}{2}}$, by III. ($\alpha$), etc. Hence $\{\sqrt{(1-x)}+\sqrt{(1+x)}\}^4$ $=(1-x)^2+4(1-x)(1-x^2)^{\frac{1}{2}}+6(1-x^2)+4(1+x)(1-x^2)^{\frac{1}{2}}+(1+x)^2=$ $(8-4x^2)+8\sqrt{(1-x^2)}$.

Ex. 9. Simplify $\sqrt[3]{(x^{-2}\sqrt{y^3})}/\sqrt{(y\sqrt[3]{x^{-4}})}$.

$\sqrt[3]{(x^{-2}\sqrt{y^3})}/\sqrt{(y\sqrt[3]{x^{-4}})}=(x^{-2}y^{\frac{3}{2}})^{\frac{1}{3}}/(yx^{-\frac{4}{3}})^{\frac{1}{2}}=(x^{-2})^{\frac{1}{3}}(y^{\frac{3}{2}})^{\frac{1}{3}}/y^{\frac{1}{2}}(x^{-\frac{4}{3}})^{\frac{1}{2}}$, by III. ($\alpha$), $=x^{-\frac{2}{3}}y^{\frac{1}{2}}/y^{\frac{1}{2}}x^{-\frac{2}{3}}$, by II., $=x^{-\frac{2}{3}+\frac{2}{3}}y^{\frac{1}{2}-\frac{1}{2}}$, by I. ($\beta$), $=x^0y^0=1\times1=1$.

Ex. 10. Show that $\sqrt[n+3]{\left\{\sqrt[n-1]{x^2}\Big/\sqrt[n+1]{x}\right\}^{n^2-1}}=x$.

$$\sqrt[n+3]{\left\{\sqrt[n-1]{x^2}\Big/\sqrt[n+1]{x}\right\}^{n^2-1}}=\{x^{2/(n-1)}/x^{1/(n+1)}\}^{(n^2-1)/(n+3)},$$
$$=\{x^{2/(n-1)-1/(n+1)}\}^{(n^2-1)/(n+3)},$$
$$=\{x^{(n+3)/(n^2-1)}\}^{(n^2-1)/(n+3)},\text{ by I. }(\beta),$$
$$=x^{[(n+3)/(n^2-1)][(n^2-1)/(n+3)]},\text{ by II.},$$
$$=x^1=x.$$

## EXERCISES L.

Simplify the following as much as you can :—

1. $8^{\frac{4}{3}}$.
2. $16^{-\frac{3}{2}}$.
3. $\sqrt{(25^{-1})}$.
4. $\sqrt[5]{\left\{\left(\frac{21}{28}\right)^2\times\frac{(21\times2^2)^3}{28\times7^2}\right\}}$.
5. $\frac{\sqrt{x^4}\sqrt[4]{x^3}\sqrt{x^5}}{x^{\frac{1}{3}}(\sqrt{x})^{\frac{1}{2}}x^{\frac{3}{4}}}\div(x^{\frac{1}{2}})^{\frac{2}{3}}$.
6. $\left\{\left(\frac{\sqrt{x}}{\sqrt[4]{x}}\right)^2(\sqrt[4]{x\sqrt{x}})^4\right\}\Big/\left\{\left(\frac{\sqrt[4]{x}}{\sqrt{x}}\right)^2\left(\frac{\sqrt{x}}{x}\right)^3\right\}^2$.
7. $\sqrt[4]{(x^3y^2z^5)}/\{\sqrt[3]{(x^2yz^3)}\sqrt{(x^2y^2z)}\}$.
8. $\sqrt[5]{(x^3y^2z)}\sqrt[5]{(y^3z^2x)}\sqrt[5]{(z^3x^2y)}\Big/\sqrt[3]{\frac{x^5}{yz}}\sqrt[3]{\frac{y^5}{zx}}\sqrt[3]{\frac{z^5}{xy}}$.
9. $\sqrt[3]{2}\sqrt{(\sqrt{2}\sqrt[3]{2})}/\sqrt[3]{\{2(\sqrt{2}\sqrt[3]{2})^3\}}$.
10. $\sqrt[3]{x^{-4}}\sqrt{x^{-5}}x^{\frac{7}{5}}/x^{-\frac{6}{5}}\sqrt[4]{x^{10}}x^{-\frac{1}{3}}$.
11. Show that $\sqrt[3]{(\sqrt[4]{x})}/\sqrt[4]{(\sqrt[5]{x})}=\sqrt[3]{x}\sqrt[4]{x^{-2}}\sqrt[5]{x}$.
12. $\{(a^{-3}/b^{-3})^{\frac{3}{4}}\sqrt{(a/b)}\}^{\frac{2}{3}}\sqrt{(a^{\frac{2}{3}}b^{\frac{2}{3}})}$.
13. $(x^{\frac{2}{3}}y^{-\frac{1}{2}}z^2)^{\frac{3}{2}}(x^{-\frac{1}{2}}yz^{\frac{1}{4}})/(xy^2)^{-\frac{3}{4}}(y^2z)^{\frac{1}{2}}(z^{\frac{1}{4}}x^{-\frac{1}{3}})^{\frac{1}{2}}$.
14. $\frac{\sqrt[3]{(x^2y^5)}\{\sqrt[4]{(x^2y^6)}\}^{-2}}{(x^{\frac{2}{3}}y^{\frac{1}{3}})^{-2}(x^2y)^{\frac{2}{3}}}\div\frac{(x^{-2}y^{\frac{2}{3}})^{-2}}{\{\sqrt[3]{(x^{-2}y)}\}^2}$.
15. $\{\sqrt[5]{2^2}(\sqrt[7]{3})^3(6-\frac{1}{10})^{29}\}^{35}$, calculate to two places of decimals.
16. $\frac{(a^{\alpha+\beta}b^{\beta+\gamma})^{\alpha+\beta}}{(a^{\alpha-\beta}b^{\beta-\gamma})^{\beta-\gamma}}\times\frac{(a^{\beta+\gamma}b^{\gamma+\alpha})^{\gamma+\alpha}}{(a^{\beta-\gamma}b^{\gamma-\alpha})^{\gamma-\alpha}}$.

Find the value when $\alpha=4$, $\beta=2$, $\gamma=1$, $a=2$, $b=4$.

**17.** $\{x^{1/m}(xy)^{1/n}/y^{1/m}\}^m / \{x^{1/n}(xy)^{1/m}/y^{1/n}\}^n$.

**18.** Show that $(x^l/x^m)^l(x^m/x^n)^m(x^n/x^l)^n(x^{mn}x^{nl}x^{lm})^3 = (x^{l+m+n})^{l+m+n}$.

**19.** $x^{l(m^2-n^2)}x^{m(n^2-l^2)}x^{n(l^2-m^2)}\left\{\left(\dfrac{x^m}{x^n}\right)^{n-l}\right\}^m\left\{\left(\dfrac{x^n}{x^m}\right)^{n-l}\right\}^l$.

**20.** $\left\{\dfrac{(x^m x^n)^{lm}}{(x^n x^m)^{ln}}\right\}^l\left\{\dfrac{(x^n x^l)^{mn}}{(x^l x^n)^{ml}}\right\}^m\left\{\dfrac{(x^l x^m)^{nl}}{(x^m x^l)^{nm}}\right\}^n$.

**21.** Show that $\sqrt[n-1]{x} \div \sqrt[n+1]{\left(\sqrt[n+1]{x^2}\right)} \div \sqrt[n+1]{x}$

$$= \left[\sqrt[n+1]{\left\{\sqrt[n-1]{\left(\sqrt[n+1]{x}\right)}\right\}}\right]^4.$$

**22.** $\sqrt[a]{\left(\dfrac{\sqrt[b]{x}}{\sqrt[c]{x}}\right)}\sqrt[b]{\left(\dfrac{\sqrt[c]{x}}{\sqrt[a]{x}}\right)}\sqrt[c]{\left(\dfrac{\sqrt[a]{x}}{\sqrt[b]{x}}\right)}$.

**23.** $\{(x^a)^{1/n}(x^m)^{1/b}\}^{bn} / \{\sqrt[m]{y^b}\sqrt[a]{y^n}\}^{ma}\left(\dfrac{x}{y}\right)^{mn}$.

**24.** $\{(x^b)^b/(x^c)^c\}^{1/(b-c)} \times \{(x^c)^c/(x^a)^a\}^{1/(c-a)} \times \{(x^a)^a/(x^b)^b\}^{1/(a-b)}$.

**25.** Show that $2^{1/5} > 3^{1/8}$.

**26.** Is $(\frac{1}{2})^{\frac{1}{2}} >< (\frac{2}{3})^{\frac{1}{3}}$?

**27.** Prove that $x^{x\sqrt{x}} = (x\sqrt{x})^x$ is satisfied by $x = 2\frac{1}{4}$.

**28.** If $x$, $y$, $z$ be finite positive integers, and $(\sqrt[x]{a^{y-z}})(\sqrt[y]{a^{z-x}})(\sqrt[z]{a^{x-y}}) = 1$, where $a \neq 0$, show that at least two of $x$, $y$, $z$ are equal.

## Irrational Terms and Linear Irrational Functions

**§ 169.** Any function of $x, y, z, \ldots$ which is the product of a coefficient A (which may be constant or a rational function of $x, y, z, \ldots$), and a series of fractional powers of $x, y, z, \ldots$ we may call an **Irrational Term.** If A be a constant we may distinguish when necessary by speaking of an **Irrational Term with a Constant Coefficient.**

Ex. $3x^{\frac{2}{3}}y^{\frac{3}{4}}z^{\frac{1}{5}}$ and $\left(\dfrac{x}{x^2+y^2}\right)x^{\frac{1}{2}}y^{\frac{2}{3}}$ are irrational terms, the first having a constant, the second a variable coefficient.

*Any irrational term can be expressed as a rational multiple of a single root of a rational function of the variables, or as a single root pure and simple.*

Ex. 1. $3x^{\frac{2}{3}}y^{\frac{3}{4}}z^{\frac{1}{5}} = 3x^{\frac{40}{60}}y^{\frac{45}{60}}z^{\frac{12}{60}} = 3(x^{40}y^{45}z^{12})^{\frac{1}{60}} = 3\sqrt[60]{(x^{40}y^{45}z^{12})}$, also $= \sqrt[60]{(3^{60}x^{40}y^{45}z^{12})}$, as will be seen from the laws of indices or the equivalent theorems regarding radication given in § 159.

Ex. 2. $\left(\dfrac{x}{x^2+y^2}\right)x^{\frac{1}{2}}y^{\frac{2}{3}} = \dfrac{x}{x^2+y^2}(x^3y^4)^{\frac{1}{6}},$

$$= \left(\frac{x}{x^2+y^2}\right)\sqrt[6]{(x^3y^4)},$$

$$= \sqrt[6]{\left\{\frac{x^6(x^3y^4)}{(x^2+y^2)^6}\right\}},$$

$$= \sqrt[6]{\left\{\frac{x^9y^4}{(x^2+y^2)^6}\right\}}.$$

§ **170.** The algebraic sum of a number of irrational terms is called a **Linear Irrational Function**, binomial, trinomial, etc., according to the number of separate terms.

For example, $Ax^{\frac{2}{3}}y^{\frac{3}{4}} - Bx^{\frac{1}{5}}y^{\frac{1}{2}} + Cx^{\frac{1}{3}}y^{\frac{1}{2}}$, where A, B, C are either constant or rational functions of $x$ and $y$, is a linear irrational function of $x$ and $y$.

If the coefficients of the terms are independent of the variables, the function may be described as a **Linear Irrational Function with Constant Coefficients.**

Ex. $x^{\frac{2}{3}} + 2x^{\frac{1}{3}}y^{\frac{1}{3}} - y^{\frac{2}{3}}$ is a linear irrational function of $x$ and $y$ with constant coefficients.

It follows from § 169 that *every linear irrational function can be expressed as an algebraic sum of rational multiples of roots of rational functions of its variables, or, if we choose, as an algebraic sum of roots of rational functions pure and simple.*

Ex. $Ax^{\frac{2}{3}}y^{\frac{3}{4}} - Bx^{\frac{1}{5}}y^{\frac{1}{2}} + Cx^{\frac{1}{3}}y^{\frac{1}{2}}$

$$= A\sqrt[12]{(x^8y^9)} - B\sqrt[10]{(x^2y^5)} + C\sqrt[6]{(x^2y^3)},$$

$$= \sqrt[12]{(A^{12}x^8y^9)} - \sqrt[10]{(B^{10}x^2y^5)} + \sqrt[6]{(C^6x^2y^3)}.$$

It is this property which leads us to characterise the form as linear; it is linear in the sense that it can be expressed as a sum of first powers of certain roots, *e.g.* $Ax^{\frac{2}{3}}y^{\frac{3}{4}} - Bx^{\frac{1}{5}}y^{\frac{1}{2}} + Cx^{\frac{1}{3}}y^{\frac{1}{2}}$ is linear when we regard $\sqrt[12]{(x^8y^9)}$, $\sqrt[10]{(x^2y^5)}$, and $\sqrt[6]{(x^2y^3)}$ as variables.

The linear irrational function is *a standard form to which can be reduced all rational functions of any roots of the variables* $x, y, z, \ldots$

It is this proposition that gives to the linear irrational form its importance; to prove it completely would transcend our present limits; but it is established, so far as square roots alone are concerned, in §§ 172, 173.

## SPECIAL PROPERTIES OF THE LINEAR IRRATIONAL FUNCTION WITH CONSTANT COEFFICIENTS

**§ 171.** The linear irrational form with constant coefficients has many properties analogous to those of an integral function. We may speak of its degree, meaning the sum of the indices of the variables in its highest term (*i.e.* the term for which the algebraic sum of the indices is highest); and we may speak of such a function as being homogeneous or the contrary; and so on.

For example, $x^{\frac{3}{4}}+3x^{\frac{1}{4}}+2$, $x^{\frac{2}{3}}+x^{\frac{1}{3}}+1+x^{-\frac{1}{3}}+x^{-\frac{2}{3}}$, $x^{-\frac{1}{2}}+x^{-\frac{1}{4}}$, are linear irrational functions of $x$ with constant coefficients of degrees $\frac{3}{4}$, $\frac{2}{3}$, $-\frac{1}{4}$ respectively; and $x^{\frac{2}{3}}+2x^{\frac{1}{3}}y^{\frac{1}{3}}+y^{\frac{2}{3}}$ is a homogeneous linear irrational function of $x$ and $y$ with constant coefficients of degree $\frac{2}{3}$.

It is obvious from the laws of indices as now extended, combined with the generalised form of the law of distribution, that the product of any number of given linear irrational functions with constant coefficients is a linear irrational function with constant coefficients whose highest term is the product of the highest terms of the given functions, and whose degree is the sum of their degrees.

When one linear irrational function with constant coefficients, A, can be produced by multiplying together two other such functions B and C, then B and C may be said in a certain sense to be factors or exact divisors of A; and processes for testing divisibility of this sort may be used analogous to some of those employed for integral functions.

There are, however, important contrasts. Thus, for example, we have

$$\begin{aligned} x^{\frac{1}{2}}-y^{\frac{1}{2}} &= (x^{\frac{1}{4}}+y^{\frac{1}{4}})(x^{\frac{1}{4}}-y^{\frac{1}{4}}) = (x^{\frac{1}{4}}+y^{\frac{1}{4}})(x^{\frac{1}{8}}+y^{\frac{1}{8}})(x^{\frac{1}{8}}-y^{\frac{1}{8}}) \\ &= (x^{\frac{1}{4}}+y^{\frac{1}{4}})(x^{\frac{1}{8}}+y^{\frac{1}{8}})(x^{\frac{1}{16}}+y^{\frac{1}{16}})(x^{\frac{1}{16}}-y^{\frac{1}{16}}) = \text{etc.}, \end{aligned}$$

from which we see that a linear irrational form with constant coefficients may have an infinite number of factors in the present sense of the word factor.

We follow the fashion of English text-books* in giving some examples of this kind of Algebra; its practical importance is,

* In our own opinion the subject has no business in an elementary text-book at all. Introduced carelessly, as it is for the most part, it tends to confuse the ideas of the beginner about much more important things.

however, next to nothing ; and its theoretical importance appears only in recondite branches of the subject never reached by the great majority of mathematical students.

Ex. 1. Distribute $(x^{\frac{3}{4}} - x^{\frac{1}{4}} + x^{-\frac{1}{4}} - x^{-\frac{3}{4}})(x^{\frac{1}{2}} + 1 + x^{-\frac{1}{2}})$, and arrange the product according to fractional powers of $x$.

$$\begin{aligned}
&(x^{\frac{3}{4}} - x^{\frac{1}{4}} + x^{-\frac{1}{4}} - x^{-\frac{3}{4}})(x^{\frac{1}{2}} + 1 + x^{-\frac{1}{2}}) \\
&\qquad = x^{\frac{5}{4}} - x^{\frac{3}{4}} + x^{\frac{1}{4}} - x^{-\frac{1}{4}} \\
&\qquad\qquad + x^{\frac{3}{4}} - x^{\frac{1}{4}} + x^{-\frac{1}{4}} - x^{-\frac{3}{4}} \\
&\qquad\qquad\qquad + x^{\frac{1}{4}} - x^{-\frac{1}{4}} + x^{-\frac{3}{4}} - x^{-\frac{5}{4}}, \\
&\qquad = x^{\frac{5}{4}} + x^{\frac{1}{4}} - x^{-\frac{1}{4}} - x^{-\frac{5}{4}}.
\end{aligned}$$

Ex. 2. Express in linear form $(x + x^{\frac{1}{2}} + 1)/(x^{\frac{1}{4}} + x^{\frac{1}{8}} + 1)$. We may use the process for dividing one integral function by another, and proceed thus—

$$\begin{array}{l|l}
x + x^{\frac{1}{2}} + 1 & x^{\frac{1}{4}} + x^{\frac{1}{8}} + 1 \\
\underline{x + x^{\frac{7}{8}} + x^{\frac{3}{4}}} & x^{\frac{3}{4}} - x^{\frac{5}{8}} + x^{\frac{3}{8}} - x^{\frac{1}{8}} + 1 \\
\quad - x^{\frac{7}{8}} - x^{\frac{3}{4}} & \\
\quad \underline{- x^{\frac{7}{8}} - x^{\frac{3}{4}} - x^{\frac{5}{8}}} & \\
\qquad\qquad x^{\frac{5}{8}} + x^{\frac{1}{2}} & \\
\qquad\qquad \underline{x^{\frac{5}{8}} + x^{\frac{1}{2}} + x^{\frac{3}{8}}} & \\
\qquad\qquad\qquad - x^{\frac{3}{8}} & \\
\qquad\qquad\qquad \underline{- x^{\frac{3}{8}} - x^{\frac{1}{4}} - x^{\frac{1}{8}}} & \\
\qquad\qquad\qquad\qquad x^{\frac{1}{4}} + x^{\frac{1}{8}} + 1 & \\
\qquad\qquad\qquad\qquad \underline{x^{\frac{1}{4}} + x^{\frac{1}{8}} + 1}\ ; &
\end{array}$$

hence $(x + x^{\frac{1}{2}} + 1)/(x^{\frac{1}{4}} + x^{\frac{1}{8}} + 1) = x^{\frac{3}{4}} - x^{\frac{5}{8}} + x^{\frac{3}{8}} - x^{\frac{1}{8}} + 1$. We might also use the identity $X^4 + X^2 + 1 = (X^2 + X + 1)(X^2 - X + 1)$ ; thus $x + x^{\frac{1}{2}} + 1 = (x^{\frac{1}{2}} - x^{\frac{1}{4}} + 1)(x^{\frac{1}{2}} + x^{\frac{1}{4}} + 1) = (x^{\frac{1}{2}} - x^{\frac{1}{4}} + 1)(x^{\frac{1}{4}} - x^{\frac{1}{8}} + 1)(x^{\frac{1}{4}} + x^{\frac{1}{8}} + 1)$. Hence the given function reduces to $(x^{\frac{1}{2}} - x^{\frac{1}{4}} + 1)(x^{\frac{1}{4}} - x^{\frac{1}{8}} + 1) = (x^{\frac{4}{8}} - x^{\frac{2}{8}} + 1)(x^{\frac{2}{8}} - x^{\frac{1}{8}} + 1)$. If we arrange according to powers of $x^{\frac{1}{8}}$, and use detached coefficients, the work for the distribution of this product is

$$\begin{array}{r}
1 + 0 - 1 + 0 + 1 \qquad\qquad \\
\underline{1 - 1 + 1 \qquad\qquad\qquad\quad} \\
1 + 0 - 1 + 0 + 1 \qquad\qquad \\
- 1 + 0 + 1 + 0 - 1 \qquad \\
\underline{1 + 0 - 1 + 0 + 1} \\
1 - 1 + 0 + 1 + 0 - 1 + 1 .
\end{array}$$

Hence the final result is $x^{\frac{6}{8}} - x^{\frac{5}{8}} + x^{\frac{3}{8}} - x^{\frac{1}{8}} + 1$, which is the same as before.

Ex. 3. Simplify $\{1 - 3x - 2x^{\frac{1}{3}}(1 + x^{\frac{1}{3}})\}/\{1 + 3x + 4x^{\frac{1}{3}}(1 + x^{\frac{1}{3}})\}$. For convenience of calculation, let us put $x^{\frac{1}{3}} = y$, then

$$\frac{1-3x-2x^{\frac{1}{3}}(1+x^{\frac{1}{3}})}{1+3x+4x^{\frac{1}{3}}(1+x^{\frac{1}{3}})}=\frac{1-2y-2y^2-3y^3}{1+4y+4y^2+3y^3}.$$

It is readily found that numerator and denominator of this last fraction have the G.C.M. $1+y+y^2$. Removing this factor, we get for the value of the fraction $(1-3y)/(1+3y)$.

Hence the given irrational function of $x$ reduces to $(1-3x^{\frac{1}{3}})/(1+3x^{\frac{1}{3}})$.

It may be noted that we can reduce this to a linear form by multiplying numerator and denominator by $1-3x^{\frac{1}{3}}+9x^{\frac{2}{3}}$. The result is

$$\{(1-27x)-6x^{\frac{1}{3}}+18x^{\frac{2}{3}}\}/\{1+27x\};$$

but the coefficients of the form are no longer constant.

## EXERCISES LI.

Simplify the following irrational functions, and reduce them to linear form wherever you can :—

1. $(x^{\frac{1}{2}}-2+x^{-\frac{1}{2}})^3$.
2. $\{(x+y)^{\frac{3}{2}}-(x+y)^{-\frac{3}{2}}\}^2$.
3. $(x+x^{-1})^2(x^{\frac{1}{2}}-x^{-\frac{1}{2}})^3$.
4. $(x^{\frac{1}{4}}-1+x^{-\frac{1}{4}})(x^{\frac{1}{4}}+1+x^{-\frac{1}{4}})(x^{\frac{1}{2}}-1+x^{-\frac{1}{2}})$.
5. $(x^3+2x+3x^{-1}+4x^{-3})(x^3-2x+3x^{-1}-4x^{-3})$.
6. $(3^{\frac{2}{3}}+3^{\frac{1}{3}}+1)(3^{\frac{2}{3}}+3^{\frac{1}{3}}-1)(3^{\frac{2}{3}}-3^{\frac{1}{3}}+1)(3^{\frac{2}{3}}-3^{\frac{1}{3}}-1)$.
7. $1/(x-y)+1/(\sqrt{x}-\sqrt{y})+1/(\sqrt[4]{x}-\sqrt[4]{y})$.
8. $(a-b)/(a^{\frac{1}{3}}-b^{\frac{1}{3}})$.
9. $(x^{-1}-y^{-1})/(x^{-\frac{1}{3}}-y^{-\frac{1}{3}})$.
10. $(x^2-y)/(x^{\frac{2}{3}}-y^{\frac{1}{3}})$.
11. $(x^{\frac{5}{4}}+y^{\frac{5}{2}})/(x^{\frac{1}{4}}+y^{\frac{1}{2}})$.
12. $(x-y)/(x^{\frac{1}{3}}-x^{\frac{1}{6}}y^{\frac{1}{6}}+y^{\frac{1}{3}})+(x-y)/(x^{\frac{1}{3}}+x^{\frac{1}{6}}y^{\frac{1}{6}}+y^{\frac{1}{3}})$.
13. $(x^2-y^2)/(x^{\frac{2}{3}}-x^{\frac{1}{3}}y^{\frac{1}{3}}+y^{\frac{2}{3}})+(x^2-y^2)/(x^{\frac{2}{3}}+x^{\frac{1}{3}}y^{\frac{1}{3}}+y^{\frac{2}{3}})$.
14. $(x^{\frac{3}{2}}+x^{\frac{1}{2}}+x^{-\frac{1}{2}}+x^{-\frac{3}{2}})/(x^{\frac{1}{2}}+x^{-\frac{1}{2}})-(x^{\frac{3}{2}}-x^{\frac{1}{2}}+x^{-\frac{1}{2}}-x^{-\frac{3}{2}})/(x^{\frac{1}{2}}-x^{-\frac{1}{2}})$.
15. $(x^{\frac{2}{5}}-2+x^{-\frac{2}{5}})/(x^{\frac{1}{5}}-2+x^{-\frac{1}{5}})$.
16. $1/(\sqrt[3]{3}+1)+1/(\sqrt[3]{3}-1)$.
17. $\{(1-x^2)^{\frac{1}{2}}+(1+x)^{\frac{1}{2}}+(1-x)^{\frac{1}{2}}+1\}/\{(1+x)^{\frac{1}{4}}-(1+x)^{-\frac{1}{4}}\}\{(1-x)^{\frac{1}{4}}-(1-x)^{-\frac{1}{4}}\}$.
18. $(x-y)/(x^{1/n}-y^{1/n})$.
19. $(x^{3n/2}-x^{-3n/2})/(x^{n/2}-x^{-n/2})$.
20. $(1+2x^{\frac{1}{2}}+2x^{\frac{3}{4}}+x)(1+3x^{\frac{1}{4}}+3x^{\frac{3}{4}}+2x)$.
21. $\{5x^{\frac{2}{3}}+(x+4)x^{\frac{1}{3}}+(4x+1)\}/\{x^{\frac{2}{3}}-(x-2)x^{\frac{1}{3}}+(2x-1)\}$.
22. $\sqrt{\{(x+x^{-1})^2-4(x-x^{-1})\}}$.
23. $\sqrt{(4a^5-12a^{\frac{5}{2}}b^{\frac{1}{3}}+9b^{\frac{2}{3}}+16a^{\frac{5}{2}}c^{\frac{3}{4}}-24b^{\frac{1}{3}}c^{\frac{3}{4}}+16c^{\frac{3}{2}})}$.
24. $\sqrt{\left\{\frac{x+y}{x}+\frac{x+y}{y}+\frac{\sqrt{x}+\sqrt{y}}{\sqrt{x}}+\frac{\sqrt{x}-\sqrt{y}}{\sqrt{y}}+\frac{1}{4}\right\}}$.
25. $\sqrt{(9x^4-6x^2+24x^{\frac{3}{2}}+1-8/\sqrt{x}+16/x)}$.
26. Show that $\dfrac{x^{\frac{2}{3}}+3x^{\frac{1}{3}}+1}{x^{\frac{2}{3}}-3x^{\frac{1}{3}}+1}=\dfrac{(6x+18)x^{\frac{2}{3}}+(18x+6)x^{\frac{1}{3}}+(x^2+30x+1)}{x^2-18x+1}$.

**27.** Find by means of the identity $x^m - y^m = (x-y)(x^{m-1} + x^{m-2}y + \ . \ . \ . \ + y^{m-1})$ a rationalising factor for $\sqrt[3]{x} - \sqrt[5]{y}$.

**28.** Find a rationalising factor for $x^{\frac{2}{3}} + y^{\frac{3}{2}}$.

## Reduction of Rational Functions of Square Roots to Linear Forms

§ **172.** *Every rational function of* $\sqrt{x}$ *can be reduced to the form* $A + B\sqrt{x}$, *where* A *and* B *do not contain* $\sqrt{x}$.

In the first place, it is obvious that every integral function of $\sqrt{x}$ can be reduced to the form in question; for every term of the function is of the form $A_n(\sqrt{x})^n$, and if $n$ be even this is rational, if $n$ be odd $= 2m+1$, say, $A_n(\sqrt{x})^{2m+1} = A_n(\sqrt{x})^{2m}\sqrt{x} = A_n x^m \sqrt{x}$, which is a rational multiple of $\sqrt{x}$. All the even terms therefore may be collected into a rational part A, say; and all the odd terms will furnish a rational multiple of $\sqrt{x}$, say $B\sqrt{x}$. The whole function thus reduces to $A + B\sqrt{x}$, where A and B are rational (see Ex. 1 below).

Consider now any rational function of $\sqrt{x}$; it may, see § 154, be reduced to the form P/Q, where P and Q are integral functions of $\sqrt{x}$. Now, as we have seen, we may reduce P and Q to $C + D\sqrt{x}$ and $E + F\sqrt{x}$, where C, D, E, and F are rational in so far as $\sqrt{x}$ is concerned. Then we have, multiplying numerator and denominator by $E - F\sqrt{x}$—

$$\frac{P}{Q} = \frac{C + D\sqrt{x}}{E + F\sqrt{x}} = \frac{(C + D\sqrt{x})(E - F\sqrt{x})}{(E)^2 - (F\sqrt{x})^2},$$
$$= \frac{CE - DFx}{E^2 - F^2x} + \frac{DE - CF}{E^2 - F^2x}\sqrt{x},$$
$$= A + B\sqrt{x},$$

where A and B are rational, in the sense that they do not contain $\sqrt{x}$ (see Ex. 2 below).

Ex. 1.

$$(1+2\sqrt{x})^3 + (1-\sqrt{x})^2 + 1 = 1 + 6\sqrt{x} + 12x + 8x\sqrt{x} + 1 - 2\sqrt{x} + x + 1,$$
$$= (13x+3) + (8x+4)\sqrt{x}.$$

Ex. 2. Reduce $(1+\sqrt{x})^3/(1-\sqrt{x})$ to linear form.

$$(1+\sqrt{x})^3/(1-\sqrt{x}) = (1+\sqrt{x})^4/\{1-(\sqrt{x})^2\}$$
$$= (1 + 4\sqrt{x} + 6x + 4x\sqrt{x} + x^2)/(1-x),$$
$$= \frac{1+6x+x^2}{1-x} + \frac{4(1+x)}{1-x}\sqrt{x}.$$

**§ 173.** *Every rational function of two square roots* $\sqrt{x}$ *and* $\sqrt{y}$ *can be reduced to the linear form* $A + B\sqrt{x} + C\sqrt{y} + D\sqrt{x}\sqrt{y}$, *where* A, B, C, D *are rational so far as* $\sqrt{x}$, $\sqrt{y}$ *are concerned; every rational function of three square roots* $\sqrt{x}$, $\sqrt{y}$, $\sqrt{z}$ *to the linear form* $A + B\sqrt{x} + C\sqrt{y} + D\sqrt{z} + E\sqrt{x}\sqrt{y} + F\sqrt{x}\sqrt{z} + G\sqrt{y}\sqrt{z} + H\sqrt{x}\sqrt{y}\sqrt{z}$, *where* A, B, C, D, E, F, G, H *are rational so far as* $\sqrt{x}$, $\sqrt{y}$, $\sqrt{z}$ *are concerned; and so on.*

Let $\Phi$ be a rational function of $\sqrt{x}$ and $\sqrt{y}$; and let us first consider $\sqrt{x}$ alone. Then $\Phi$ is a rational function of $\sqrt{x}$, and may, as we have seen, be reduced to the form $A' + B'\sqrt{x}$, where $A'$ and $B'$ will no longer contain $\sqrt{x}$, but will in general contain $\sqrt{y}$. Since $A'$ and $B'$ are arrived at by rational operations, they will be rational functions of $\sqrt{y}$; hence we can reduce them to the forms $A' = A'' + B''\sqrt{y}$, $B' = A''' + B'''\sqrt{y}$, where $A''$, $B''$, $A'''$, $B'''$ do not now contain either $\sqrt{x}$ or $\sqrt{y}$. Hence

$$\begin{aligned}\Phi &= A'' + B''\sqrt{y} + (A''' + B'''\sqrt{y})\sqrt{x},\\ &= A'' + A'''\sqrt{x} + B''\sqrt{y} + B'''\sqrt{x}\sqrt{y},\end{aligned}$$

which is the form in question.

If $\Phi$ contain a third square root, say $\sqrt{z}$, then each of the coefficients $A''$, $A'''$, etc., will be a rational function of $\sqrt{z}$, and therefore reducible to the form $P + Q\sqrt{z}$, $P' + Q'\sqrt{z}$, etc. Substituting these we shall arrive at the form

$$A + B\sqrt{x} + C\sqrt{y} + D\sqrt{z} + E\sqrt{x}\sqrt{y} + F\sqrt{x}\sqrt{z} + G\sqrt{y}\sqrt{z} + H\sqrt{x}\sqrt{y}\sqrt{z},$$

where A, B, C, etc., now contain neither $\sqrt{x}$, nor $\sqrt{y}$, nor $\sqrt{z}$.

This process can evidently be continued until all the square roots, however many there may be in the function, have been dealt with.

Ex. 1. Reduce $F \equiv (1+\sqrt{x}-\sqrt{y})/(1+\sqrt{x}+\sqrt{y})$ to linear form.

First we get rid of $\sqrt{y}$ in the denominator by multiplying both denominator and numerator by $1+\sqrt{x}-\sqrt{y}$ (which is often called the conjugate of $1+\sqrt{x}+\sqrt{y}$ with respect to $\sqrt{y}$). Thus

$$\begin{aligned}F &= (1+\sqrt{x}-\sqrt{y})^2/((1+\sqrt{x})^2-(\sqrt{y})^2),\\ &= (1+x+y+2\sqrt{x}-2\sqrt{y}-2\sqrt{x}\sqrt{y})/(1+x+2\sqrt{x}-y).\end{aligned}$$

We next get rid of $\sqrt{x}$ in the denominator by multiplying denominator and numerator by the conjugate of the former with respect to $\sqrt{x}$, viz. by $1+x-y-2\sqrt{x}$. Thus

$$\begin{aligned}F &= \frac{(1+x+y+2\sqrt{x}-2\sqrt{y}-2\sqrt{x}\sqrt{y})(1+x-y-2\sqrt{x})}{(1+x-y)^2-(2\sqrt{x})^2},\\ &= \frac{\{(1-x)^2-y^2\}-4y\sqrt{x}-2(1-x-y)\sqrt{y}+2(1-x+y)\sqrt{x}\sqrt{y}}{1-2(x+y)+(x-y)^2}.\end{aligned}$$

Ex. 2. $\sqrt{\frac{x+1}{x-1}}+\sqrt{\frac{x-1}{x+1}}=\frac{\sqrt{(x+1)}}{\sqrt{(x-1)}}+\frac{\sqrt{(x-1)}}{\sqrt{(x+1)}},$

$$=\frac{\{\sqrt{(x+1)}\}^2+\{\sqrt{(x-1)}\}^2}{\sqrt{(x^2-1)}},$$

$$=\frac{(x+1)+(x-1)}{\sqrt{(x^2-1)}}=\frac{2x}{\sqrt{(x^2-1)}}=\frac{2x}{x^2-1}\sqrt{(x^2-1)}.$$

**§ 174.** It appears from the theory and examples of §§ 172, 173, that any linear form F containing the square roots $\sqrt{x}$, $\sqrt{y}$, $\sqrt{z}$ . . . can be rendered rational so far as $\sqrt{x}$ is concerned by multiplying it by a factor $R_1$ obtained from F itself by changing the sign of $\sqrt{x}$ wherever it occurs. This factor is called a **Rationalising Factor with respect to** $\sqrt{x}$. The product will be reducible to a linear form containing only $\sqrt{y}$ and $\sqrt{z}$ . . . If we now multiply the product by a factor $R_2$, obtained by changing the sign of $\sqrt{y}$ therein, the resulting product will be free from $\sqrt{y}$. We can proceed in this way until all the square roots are removed. Hence *there exists a factor which will completely rationalise any linear form and therefore any rational function of a given set of square roots.*

Ex. To find a rationalising factor for $F=\sqrt{x}-\sqrt{y}+\sqrt{z}$, and also the corresponding rational product.

$$F(\sqrt{x}-\sqrt{y}-\sqrt{z})=(\sqrt{x}-\sqrt{y})^2-(\sqrt{z})^2,$$
$$=x+y-z-2\sqrt{x}\sqrt{y}.$$
$$F(\sqrt{x}-\sqrt{y}-\sqrt{z})(x+y-z+2\sqrt{x}\sqrt{y})=(x+y-z)^2-4(\sqrt{x}\sqrt{y})^2,$$
$$=x^2+y^2+z^2-2yz-2zx-2xy.$$

Hence a rationalising factor is $(\sqrt{x}-\sqrt{y}-\sqrt{z})(x+y-z+2\sqrt{x}\sqrt{y})$ $=(\sqrt{x}-\sqrt{y}-\sqrt{z})((\sqrt{x}+\sqrt{y})^2-(\sqrt{z})^2)=(\sqrt{x}-\sqrt{y}-\sqrt{z})(\sqrt{x}+\sqrt{y}+\sqrt{z})(\sqrt{x}+\sqrt{y}-\sqrt{z})$, and the corresponding rational product is $\Sigma x^2-2\Sigma yz$.

It should be noticed that the rationalising factor is the product of all the functions that can be derived from F by keeping the sign of $\sqrt{x}$ fixed and taking every possible arrangement of signs for $\sqrt{y}$ and $\sqrt{z}$ except that which occurs in F itself.

This result might have been deduced at once from the identity $(a+b+c)(-a+b+c)(a-b+c)(a+b-c)=-\Sigma a^4+2\Sigma b^2c^2$ by putting $a=\sqrt{x}$, $b=\sqrt{y}$, $c=\sqrt{z}$.

## EXERCISES LII.

Reduce the following to linear form :*—

1. $(\sqrt{x}+\sqrt{y})(\sqrt[4]{x}+\sqrt[4]{y})(\sqrt[4]{x}-\sqrt[4]{y})$.
2. $\{x+\sqrt{(a^2+x^2)}\}^2\{x-\sqrt{(a^2+x^2)}\}^2$.

* Some of the more difficult of these exercises may be postponed until the beginner has finished the next chapter.

**3.** $\{\sqrt{(\sqrt{x}+1)}+1\}\{\sqrt{(\sqrt{x}+2)}+2\}\{\sqrt{(\sqrt{x}+1)}-1\}$
$\{\sqrt{(\sqrt{x}+2)}-2\}$.

**4.** $\{y+\sqrt{(x^2-y^2)}\}^3+\{y-\sqrt{(x^2-y^2)}\}^3$.

**5.** $(\sqrt{x}+\sqrt{y}+\sqrt{z})^3+\Sigma(-\sqrt{x}+\sqrt{y}+\sqrt{z})^2$.

**6.** $(\sqrt{x}+1)^6$. **7.** $(\sqrt[3]{x}-1)^6$.

**8.** Show that $\{x+y+\sqrt{(1+x^2)}\}\{1+y^2-2y\sqrt{(1+x^2)}\}\{x-y-\sqrt{(1+x^2)}\}=(2xy-y^2+1)(y^2+2xy-1)$.

**9.** $(x^2-x+1)/(x^2+x+1)$, where $x=a-\sqrt{b}$.

**10.** $\{\sqrt{(a^2+x^2)}+x\}/\{\sqrt{(a^2+x^2)}-x\}$, where $x=(b-c)a/2\sqrt{(bc)}$.

**11.** $\{\sqrt{(x^2+y^2)}+x+y\}/\{\sqrt{(x^2+y^2)}-x-y\}$.

**12.** $\dfrac{1}{4(a+\sqrt{(ax)})}+\dfrac{1}{2(a+x)}+\dfrac{1}{4(a-\sqrt{(ax)})}+\dfrac{x}{a^2-x^2}$.

**13.** $\{\sqrt{(1-x)}-1/\sqrt{(1+x)}\}/\{\sqrt{(1-x)}+1/\sqrt{(1+x)}\}$.

**14.** $\Sigma x/(y-z)$, where $x=\sqrt{(b+c)}$, $y=\sqrt{(c+a)}$, $z=\sqrt{(a+b)}$.

**15.** $\left\{\dfrac{\sqrt{(x+a)}}{\sqrt{(x^2+ax)}+\sqrt{(x^2-a^2)}}\right\}^2+\left\{\dfrac{\sqrt{(x-a)}}{\sqrt{(x^2-ax)}-\sqrt{(x^2-a^2)}}\right\}^2$.

**16.** $(1+\sqrt{p}+\sqrt{q})/(1+\sqrt{p}+\sqrt{q}+\sqrt{p}\sqrt{q})-(1+\sqrt{p}+\sqrt{q})/$
$(1-\sqrt{p}-\sqrt{q}+\sqrt{p}\sqrt{q})$.

**17.** $1/\{2+\sqrt{(1-x)}+\sqrt{(1+x)}\}$.

**18.** $1/\{1+\sqrt{(1-x)}+\sqrt{(1+x)}\}^2$.

**19.** $1/\{\sqrt{(p-q)}+\sqrt{(p+q)}-\sqrt{(2p)}\}$.

**20.** $\dfrac{2-\sqrt{(1-x)}+\sqrt{(1+x)}}{2+\sqrt{(1-x)}-\sqrt{(1+x)}}-\dfrac{2-\sqrt{(1-x)}-\sqrt{(1+x)}}{2+\sqrt{(1-x)}+\sqrt{(1+x)}}$.

**21.** $\dfrac{\sqrt{(x-1)}}{\sqrt{x}-\sqrt{\{x-\sqrt{(x^2-1)}\}}}+\dfrac{\sqrt{(x-1)}}{\sqrt{x}+\sqrt{\{x+\sqrt{(x^2-1)}\}}}$.

**22.** $1/[a-\sqrt{\{a^2-\sqrt{(a^4-b^4)}\}}]+1/[a+\sqrt{\{a^2+\sqrt{(a^4-b^4)}\}}]$.

**23.** $\Pi(y+z-x)$, where $x=\sqrt{b}-\sqrt{c}$, $y=\sqrt{c}-\sqrt{a}$, $z=\sqrt{a}-\sqrt{b}$.

**24.** If $x=\sqrt{b}+\sqrt{c}-\sqrt{a}$, $y=\sqrt{c}+\sqrt{a}-\sqrt{b}$, $z=\sqrt{a}+\sqrt{b}-\sqrt{c}$, $u=\sqrt{a}+\sqrt{b}+\sqrt{c}$, show that $(xu+yz)(yu+xz)(zu+xy)=64abc$.

**25.** Show that

$$\frac{\sqrt{x}+\sqrt[4]{x}+1}{\sqrt{x}-\sqrt[4]{x}+1}+\frac{\sqrt{x}-\sqrt[4]{x}+1}{\sqrt{x}+\sqrt[4]{x}+1}=2\frac{x^2-x+1}{x^2+x+1}+4\frac{x+1}{x^2+x+1}\sqrt{x}.$$

# CHAPTER XVII

## ARITHMETICAL THEORY OF SURDS

### Algebraical and Arithmetical Irrationality

**§ 175.** In last chapter we discussed the properties of irrational functions in so far as they depend merely on outward form; in other words, we considered them merely from the algebraical point of view. We have now to consider certain peculiarities of a purely arithmetical nature. Let $p$ denote any *commensurable number;* that is, either an integer, or a proper or improper vulgar fraction with a finite number of digits in its numerator and denominator; or, what comes to the same thing, let $p$ denote a number which is either a terminating or a repeating decimal. Then, if $n$ be any positive integer, $\sqrt[n]{p}$ will not be commensurable unless $p$ be the $n$th power of a commensurable number;* for if $\sqrt[n]{p}=k$, where $k$ is commensurable, then, by the definition of $\sqrt[n]{p}$, $p=k^n$, that is, $p$ is the $n$th power of a commensurable number.

If therefore $p$ be not a perfect $n$th power, $\sqrt[n]{p}$ is incommensurable. For distinction's sake $\sqrt[n]{p}$ is then called a **surd number.** In other words, we define a *surd number as the incommensurable root of a commensurable number.*

Surds are classified according to the index, $n$, of the root to be extracted, as quadratic, cubic, biquadratic or quartic, quintic, . . . $n$-tic surds.

The student should attend to the last phrase of the definition of a surd; because incommensurable roots might be conceived which do

---

* This is briefly put by saying that $p$ is a perfect $n$th power.

not come under the above definition; and to them the demonstrations of at least some of the propositions in this chapter would not apply. For example, $\sqrt{(\sqrt{2}+1)}$ is not a surd in the exact sense of the above definition, because the radicand $\sqrt{2}+1$ is incommensurable. On the other hand, $\sqrt{(\sqrt{2})}$, which can be expressed in the form $\sqrt[4]{2}$, does comé under that definition, although not as a quadratic but as a *biquadratic* surd.

He should also observe that an algebraically irrational function, say $\sqrt{x}$, may or may not be arithmetically irrational—that is, surd, strictly so called, according to the value of the variable $x$. Thus $\sqrt{4}$ is not a surd, but $\sqrt{2}$ is.

## Rational Functions of Surds

§ **176.** A single surd number, or, what comes to the same, a rational multiple of a single surd, is spoken of as a **Simple Monomial Surd Number**; the sum of two such surds, or of a rational number and a simple monomial surd number, as a **Simple Binomial Surd Number**, and so on. A simple surd number is therefore merely a particular case of a linear irrational function where all the roots indicated happen to be incommensurable.

The propositions stated in last chapter amount to a proof of the statement that *every rational function of surd numbers can be expressed as a simple surd number, monomial, binomial, trinomial*, etc., as the case may be.

The arithmetical importance of this reduction lies in the fact that the linear- or simple - surd - number - form is most convenient for computation, especially when a table of roots is available.

§ **177.** *Two surds are said to be* **similar** *when they can be expressed as rational multiples of one and the same surd; dissimilar when this is not the case.*

For example, $\sqrt{18}$ and $\sqrt{8}$ can be expressed respectively in the forms $3\sqrt{2}$ and $2\sqrt{2}$, and are therefore similar; but $\sqrt{6}$ and $\sqrt{2}$ are dissimilar.

Again, $\sqrt[3]{54}$ and $\sqrt[3]{16}$, being expressible in the forms $3\sqrt[3]{2}$ and $2\sqrt[3]{2}$, are each similar to $\sqrt[3]{2}$.

*The product or quotient of two similar quadratic surds is rational, and if the product or quotient of the two quadratic surds is rational they are similar.*

For, if the surds are similar, they are expressible in the

forms $A\sqrt{p}$ and $B\sqrt{p}$, where A and B are rational; therefore $A\sqrt{p} \times B\sqrt{p} = ABp$; and $A\sqrt{p}/B\sqrt{p} = A/B$, which proves the proposition, since $ABp$ and $A/B$ are rational.

Again, if $\sqrt{p} \times \sqrt{q} = A$, or $\sqrt{p}/\sqrt{q} = B$, where A and B are rational, then in the one case $\sqrt{p} = (A/q)\sqrt{q}$, in the other $\sqrt{p} = B\sqrt{q}$. But $A/q$ and B are rational. Hence $\sqrt{p}$ and $\sqrt{q}$ are similar in both cases.

From this it follows that *the product or quotient of two dissimilar quadratic surds cannot be rational.*

§ **178.** The following examples illustrate some of the commonest operations with surd numbers:—

Ex. **1.** Show that $\sqrt{45}$, $\sqrt{245}$, and $\sqrt{(252/35)}$ are similar quadratic surds.

We have $\sqrt{45} = \sqrt{(9 \times 5)} = 3\sqrt{5}$, $\sqrt{(245)} = \sqrt{(49 \times 5)} = 7\sqrt{5}$, $\sqrt{(252/35)} = \sqrt{(36/5)} = 6/\sqrt{5} = 6\sqrt{5}/5 = \frac{6}{5}\sqrt{5}$. Hence all three surds are rational multiples of $\sqrt{5}$.

Ex. **2.** Calculate $6\sqrt{7} + 5\sqrt[3]{3}$ in the simplest possible way by means of Barlow's tables, which give, *inter alia,* the square and cube roots of all integral numbers from 1 to 10,000.

$$\begin{aligned} 6\sqrt{7} + 5\sqrt[3]{3} &= \sqrt{(36 \times 7)} + \sqrt[3]{(125 \times 3)} = \sqrt{252} + \sqrt[3]{375}, \\ &= 15{\cdot}8745079 + 7{\cdot}2112479, \text{ by the table,} \\ &= 23{\cdot}0857558. \end{aligned}$$

Ex. **3.** Express $N = \sqrt{18} - \sqrt{147} + \sqrt{72} + \sqrt{108} - \sqrt{128}$ in terms of the minimum number of the smallest possible surds.

$$\begin{aligned} N &= \sqrt{(9 \times 2)} - \sqrt{(49 \times 3)} + \sqrt{(36 \times 2)} + \sqrt{(36 \times 3)} - \sqrt{(64 \times 2)}, \\ &= 3\sqrt{2} - 7\sqrt{3} + 6\sqrt{2} + 6\sqrt{3} - 8\sqrt{2}, \\ &= (3 + 6 - 8)\sqrt{2} - (7 - 6)\sqrt{3} = \sqrt{2} - \sqrt{3}. \end{aligned}$$

Ex. **4.** Given $\sqrt{2} = 1{\cdot}4142$, calculate $N \equiv \sqrt{18} \times 5\sqrt{2} \times \sqrt{24} \times 3\sqrt{147}$.

$$\begin{aligned} N &= \sqrt{(9 \times 2)} \times 5\sqrt{2} \times \sqrt{(4 \times 2 \times 3)} \times 3\sqrt{(49 \times 3)}, \\ &= 3\sqrt{2} \times 5\sqrt{2} \times 2\sqrt{2}\sqrt{3} \times 3 \times 7\sqrt{3}, \\ &= 3^3 \times 5 \times 2^2 \times 7\sqrt{2} = 3780 \times \sqrt{2}, \\ &= 3780 \times 1{\cdot}4142 = 5345{\cdot}676. \end{aligned}$$

*N.B.*—Since the value of $\sqrt{2}$ errs in defect by a quantity lying between ·00001 and ·00002, the value obtained for N errs in defect by a quantity lying between ·0378 and ·0756, and we ought in strictness to write merely $N = 5345{\cdot}6$.

Ex. **5.** To calculate $N = (3 + \sqrt[3]{3})/\sqrt[3]{5}$ by means of Barlow's tables.

$$\begin{aligned} N &= (3 + 3^{\frac{1}{3}})/5^{\frac{1}{3}} = (3 + 3^{\frac{1}{3}})5^{\frac{2}{3}}/5^{\frac{1}{3}+\frac{2}{3}} = (3 \times 5^{\frac{2}{3}} + 3^{\frac{1}{3}} \times 5^{\frac{2}{3}})/5, \\ &= \{(3^3 \times 5^2)^{\frac{1}{3}} + (3 \times 5^2)^{\frac{1}{3}}\}/5 = \tfrac{1}{5}(\sqrt[3]{675} + \sqrt[3]{75}), \\ &= \tfrac{1}{5}(8{\cdot}7720532 + 4{\cdot}2171633), \text{ by the table,} \\ &= \tfrac{1}{5}(12{\cdot}9892165) = 2{\cdot}5978433. \end{aligned}$$

*N.B.*—The object of rationalising the denominator $5^{\frac{1}{3}}$ is to avoid division by the seven-place decimal which represents $5^{\frac{1}{3}}$.

Ex. 6. Reduce $1/(\sqrt{2}-\sqrt{3}+\sqrt{5})$ to a simple surd number (linear form).

We first partially rationalise the denominator with respect to $\sqrt{5}$, by multiplying numerator and denominator of the given quotient by $\sqrt{2}-\sqrt{3}-\sqrt{5}$; thus

$$\begin{aligned}1/(\sqrt{2}-\sqrt{3}+\sqrt{5}) &= (\sqrt{2}-\sqrt{3}-\sqrt{5})/\{(\sqrt{2}-\sqrt{3})^2-(\sqrt{5})^2\},\\ &= (\sqrt{2}-\sqrt{3}-\sqrt{5})/(2+3-2\sqrt{2}\sqrt{3}-5),\\ &= (\sqrt{2}-\sqrt{3}-\sqrt{5})/(-2\sqrt{2}\sqrt{3}).\end{aligned}$$

We next rationalise the denominator entirely by multiplying it and also the numerator by $-\sqrt{2}\sqrt{3}$. We then get

$$\begin{aligned}1/(\sqrt{2}-\sqrt{3}+\sqrt{5}) &= -(\sqrt{2}-\sqrt{3}-\sqrt{5})\sqrt{2}\sqrt{3}/12,\\ &= \tfrac{1}{12}(-2\sqrt{3}+3\sqrt{2}+\sqrt{2}\sqrt{3}\sqrt{5}),\end{aligned}$$

which is the required linear form.

If it were intended to use a table of square roots to evaluate the surd number, we should further transform it into $\frac{1}{12}(-\sqrt{12}+\sqrt{18}+\sqrt{30})$.

Note the choice of the rationalising factor in the first step. We preferred to rationalise first with respect to $\sqrt{5}$, because (since $2+3=5$) the rational part of the partially rationalised denominator vanishes. If, instead, we had multiplied by $\sqrt{2}+\sqrt{3}+\sqrt{5}$, we should no doubt have got rid of $\sqrt{3}$; but the denominator would then have become $4+2\sqrt{2}\sqrt{5}$, so that the final step would not have been so simple.

Ex. 7. Discuss, without extracting the roots, whether $2\sqrt{5}+\sqrt{7}$ is greater or less than $\sqrt{5}+\sqrt{23}$.

Obviously $2\sqrt{5}+\sqrt{7} >$ or $< \sqrt{5}+\sqrt{23}$ according as $\sqrt{5}+\sqrt{7} >$ or $< \sqrt{23}$. Since both of these last are real positive quantities, the one is obviously greater or less than the other, according as the square of the one is greater or less than the square of the other. Hence the reasoning may proceed thus :—

$$2\sqrt{5}+\sqrt{7} >< \sqrt{5}+\sqrt{23},$$

according as $(\sqrt{5}+\sqrt{7})^2 >< 23$;

that is— $12+2\sqrt{35} >< 23$;

according as $2\sqrt{35} >< 11$;

and, since both of these are positive, according as

$$(2\sqrt{35})^2 >< 121,$$

that is— $140 >< 121$.

Now $140 > 121$, therefore we infer that $2\sqrt{5}+\sqrt{7} > \sqrt{5}+\sqrt{23}$.

## EXERCISES LIII.

Express each of the following as a rational multiple of the smallest possible single surd :—

1. $2\sqrt{\frac{1}{2}}$.
2. $3\sqrt{\frac{1}{27}}$.
3. $\frac{1}{5}\sqrt{125}$.
4. $\sqrt{2048}$.
5. $(1024)^{\frac{3}{8}}$.
6. $\sqrt{(6/8)}$.
7. $\sqrt[3]{(27/16)}$.
8. $\sqrt{21}\times\sqrt{(3/7)}$.
9. $\sqrt{15}\times\sqrt{(8/5)}$.
10. $\sqrt{(5\sqrt{6})}\sqrt{(6\sqrt{5})}\sqrt{(\sqrt{30})}$.
11. $\sqrt{3}\sqrt[5]{2}\sqrt[3]{5}$.
12. $\sqrt[3]{2}\sqrt[5]{8}\sqrt[3]{4}$.
13. $\sqrt{2}+\sqrt{8}-\sqrt{128}$.
14. $\sqrt{12}+\sqrt{300}-\sqrt{432}$.
15. $5\sqrt{24}-2\sqrt{54}-\sqrt{6}$.

16. $\sqrt{(\frac{1}{3})} + \sqrt{(1\frac{1}{3})} - \sqrt{(8\frac{1}{3})}$.
17. $\frac{1}{2}\sqrt{32} - \frac{7}{3}\sqrt{162} + \frac{5}{2}\sqrt{288} - \frac{1}{5}\sqrt{200}$.
18. $9\sqrt{80} - 2\sqrt{125} - 5\sqrt{245} + \sqrt{320}$.
19. $\sqrt{2} + \sqrt{\frac{1}{2}} - \sqrt{\frac{1}{8}}$.
20. $2\sqrt[3]{(1/5)} + 3\sqrt[3]{(8/320)}$.

Calculate each of the following to three places of decimals, using Barlow's tables if accessible :—

21. $16/\sqrt{2}$.
22. $3/\sqrt[3]{7}$.
23. $2/5^{\frac{2}{3}}$.
24. $3 + 9/\sqrt{3}$.
25. $1/(\sqrt{3} - 1)$.
26. $(\sqrt{3} - 1)/(\sqrt{2} + 1)$.
27. $1/(1 + \sqrt{2} + \sqrt{3})$.

Reduce the following to linear form :—

28. $(\sqrt{5} - \sqrt{2})(\sqrt{2} + 1)(\sqrt{5} + \sqrt{2})(\sqrt{2} - 1)$.
29. $(x - 1 + \sqrt{2})(x + 1 - \sqrt{2})(x + 1 + \sqrt{3})(x - 1 - \sqrt{3})$.
30. $\left(\frac{\sqrt{5}-1}{2}\right)^5 - \left(\frac{\sqrt{5}+1}{2}\right)^5 + 5\left\{\left(\frac{\sqrt{5}-1}{2}\right)^2 + \left(\frac{\sqrt{5}+1}{2}\right)^2\right\}$.
31. $2\sqrt{3}/(\sqrt{5} + 2\sqrt{3}) + \sqrt{5}/(2\sqrt{3} - \sqrt{5})$.
32. $(4 + \sqrt{2})/(2 + \sqrt{2}) + (4 - \sqrt{2})/(3 + \sqrt{2}) + (9\sqrt{2} - 13)/(3 - \sqrt{2})$.
33. $1/(1 + \sqrt{2}) + 1/(1 + \sqrt{3}) + 1/(1 - \sqrt{2}) + 1/(1 - \sqrt{3})$.
34. $\{(1 + \sqrt{3})/(1 + \sqrt{2})\}^2 + \{(1 - \sqrt{3})/(1 - \sqrt{3})\}^2$.
35. $\{(1 - \sqrt{2})/(2 + \sqrt{3})\}^2 + \{(1 + \sqrt{2})/(2 - \sqrt{2})\}^2$.
36. $(2 - \sqrt{3})/(\sqrt{2} - \sqrt{3})^2 + (2 - \sqrt{3})/(\sqrt{2} + \sqrt{3})^2$. Given $\sqrt{3} = 1{\cdot}7320508$, calculate to six decimal places.
37. $1/(\sqrt{2} + 3\sqrt{3} - 5)$.
38. $1/(\sqrt{6} - \sqrt{10} - \sqrt{35})$.
39. $(1 + \sqrt{2} + \sqrt{3})/(1 + \sqrt{2} - \sqrt{3})$.
40. $(\sqrt{3} + \sqrt{2} - \sqrt{5})/(\sqrt{3} + \sqrt{2} + \sqrt{5})$.
41. $(1 + 3\sqrt{2} - \sqrt{3})/(\sqrt{2} + \sqrt{10} + \sqrt{12})$.
42. $(2\sqrt{2} + \sqrt{3} + \sqrt{5})/(2\sqrt{2} - \sqrt{3} - \sqrt{5})$.
43. $(\sqrt{3} - \sqrt{5})^2/(\sqrt{2} + \sqrt{5} - \sqrt{7})$.
44. $\dfrac{1}{\sqrt{2} + 1/(\sqrt{3} + 1/\sqrt{5})} + \dfrac{1}{\sqrt{2} - 1/(\sqrt{3} - 1/\sqrt{5})}$.
45. $(x^6 + 3\sqrt{6}x^3 + 1)/(x^2 + \sqrt{6}x + 1)$.
46. $(x^6 + x^4 + x^2 + 1)/(x^2 + \sqrt{2}x + 1)$.
47. $1/(\sqrt[4]{3} + \sqrt{3} + 1) + 1/(\sqrt[4]{3} + \sqrt{3} - 1) + 1/(\sqrt[4]{3} - \sqrt{3} + 1)$.
48. Show that $\sqrt[3]{(1 + \frac{1}{3}\sqrt{2})} + \sqrt[3]{(1 - \frac{1}{3}\sqrt{2})} = \sqrt[3]{\{2 + \sqrt[3]{(21 + 7\sqrt{2})} + \sqrt[3]{(21 - 7\sqrt{2})}\}}$.

Discuss the following inequalities without extracting the roots :—

49. $2\sqrt{3} + \sqrt{5} >< \sqrt{2} + \sqrt{5}$.
50. $\sqrt{3} + \sqrt{7} >< \sqrt{6} + 2$.
51. $\sqrt{10} + \sqrt{7} >< \sqrt{19} + \sqrt{3}$.
52. $\sqrt[3]{5} + 1 >< 2\sqrt{2}$.

## Independence of Surd Numbers *

§ **179.** *If* $p$, $q$, A, B *be all commensurable, and none of them zero, and* $\sqrt{p}$ *and* $\sqrt{q}$ *incommensurable, then we cannot have*

$$\sqrt{p} = \mathrm{A} + \mathrm{B}\sqrt{q}.$$

* The theorems given under this head are merely small fragments of a theory which belongs partly to the higher Theory of Equations, partly to that special branch of Algebra which is called the Theory of Numbers.

For, squaring, we should have as a consequence—

$$p = A^2 + B^2q + 2AB\sqrt{q};$$

whence, since $2AB \neq 0$—

$$\sqrt{q} = (p - A^2 - B^2q)/2AB,$$

which asserts, contrary to our hypothesis, that $\sqrt{q}$ is commensurable.

Since every rational function of $\sqrt{q}$ may (Chap. X. § 15) be expressed in the form $A + B\sqrt{q}$, the above theorem is equivalent to the following:—

*One quadratic surd cannot be expressed as a rational function of another which is dissimilar to it.*

§ **180.** If x, y, z, u *be all commensurable, and* $\sqrt{y}$ *and* $\sqrt{u}$ *incommensurable, and if* $x + \sqrt{y} = z + \sqrt{u}$, *then must* $x = z$ *and* $y = u$.

For if $x \neq z$, but $= a + z$ say, where $a \neq 0$, then by hypothesis

$$a + z + \sqrt{y} = z + \sqrt{u},$$

whence

$$a + \sqrt{y} = \sqrt{u},$$

$$a^2 + y + 2a\sqrt{y} = u,$$

$$\sqrt{y} = (u - a^2 - y)/2a,$$

which asserts that $\sqrt{y}$ is commensurable. But this is not so. Hence we must have $x = z$; and, that being so, we must also have $\sqrt{y} = \sqrt{u}$, that is, $y = u$.

This theorem might also be deduced from the theorem of last paragraph.

## Square Roots of Simple Surd Numbers

§ **181.** Since the square of every simple binomial surd number takes the form $p + \sqrt{q}$, it is natural to inquire whether $\sqrt{(p + \sqrt{q})}$ can always be expressed as a simple binomial surd number—that is, in the form $\sqrt{x} + \sqrt{y}$, where $x$ and $y$ are rational numbers. Let us suppose that such an expression exists; then

$$\sqrt{(p + \sqrt{q})} = \sqrt{x} + \sqrt{y},$$

whence

$$p + \sqrt{q} = x + y + 2\sqrt{(xy)}.$$

If this equation be true, we must have, by § 180, since $\sqrt{x}$ and $\sqrt{y}$ obviously cannot be similar, and therefore $\sqrt{(xy)}$ must be a surd,—

$$x + y = p \qquad (1),$$

$$2\sqrt{(xy)} = \sqrt{q} \qquad (2);$$

and, from (1) and (2), squaring and subtracting, we get

$$(x+y)^2 - 4xy = p^2 - q,$$

that is—

$$(x-y)^2 = p^2 - q \qquad (3).$$

Now (3) gives either

$$x - y = +\sqrt{(p^2-q)} \qquad (4),$$

or

$$x - y = -\sqrt{(p^2-q)} \qquad (4^*).$$

Taking, meantime, (4) and combining it with (1), we have

$$(x+y)+(x-y) = p + \sqrt{(p^2-q)} \qquad (5),$$

$$(x+y)-(x-y) = p - \sqrt{(p^2-q)} \qquad (6);$$

whence

$$2x = p + \sqrt{(p^2-q)}, \quad 2y = p - \sqrt{(p^2-q)};$$

that is—

$$x = \tfrac{1}{2}\{p + \sqrt{(p^2-q)}\} \qquad (7),$$

$$y = \tfrac{1}{2}\{p - \sqrt{(p^2-q)}\} \qquad (8).$$

If we take (4*) instead of (4), we simply interchange the values of $x$ and $y$, which leads to nothing new in the end.

Using the values of (7) and (8) we obtain the following result:—

$$\sqrt{x} = \pm\sqrt{\left\{\frac{p+\sqrt{(p^2-q)}}{2}\right\}}, \quad \sqrt{y} = \pm\sqrt{\left\{\frac{p-\sqrt{(p^2-q)}}{2}\right\}}.$$

Since, by (2), $2\sqrt{x} \times \sqrt{y} = +\sqrt{q}$, we must take either the two upper signs together or the two lower; and, if we agree as usual that $\sqrt{(p+\sqrt{q})}$ is to represent the principal value of the root, we must take the two upper signs.

If we had started with $\sqrt{(p-\sqrt{q})}$, it would have been necessary to choose $\sqrt{x}$ and $\sqrt{y}$ with opposite signs.

Finally, therefore, we have

$$\sqrt{(p+\sqrt{q})} = \sqrt{\left\{\frac{p+\sqrt{(p^2-q)}}{2}\right\}} + \sqrt{\left\{\frac{p-\sqrt{(p^2-q)}}{2}\right\}} \quad (9),$$

$$\sqrt{(p-\sqrt{q})} = \sqrt{\left\{\frac{p+\sqrt{(p^2-q)}}{2}\right\}} - \sqrt{\left\{\frac{p-\sqrt{(p^2-q)}}{2}\right\}} \quad (9^*).$$

The identities (9) and (9*) are certainly true; we have, in fact, already verified one of them (see § 162, Example 9). They will not, however, furnish a solution of our problem, unless the values of $x$ and $y$ are real and rational. For this it is necessary and sufficient that $p^2 - q$ be a positive perfect square, and that $p$ be positive. Hence *the square root of* p + $\sqrt{q}$ *can be expressed as a simple binomial surd number, provided* p *be positive and* $p^2 - q$ *be a positive perfect square.*

Ex. Simplify $\sqrt{(19-4\sqrt{21})}$.

Let $\sqrt{(19-4\sqrt{21})}=\sqrt{x}+\sqrt{y}$.

Then
$$x+y=19,$$
$$2\sqrt{x}\sqrt{y}=-4\sqrt{21},$$
$$(x-y)^2=361-336$$
$$=25,$$
$$x-y=+5 \text{ say},$$
$$x+y=19;$$
whence
$$x=12, \quad y=7,$$
$$\sqrt{x}=\pm\sqrt{12}, \quad \sqrt{y}=\mp\sqrt{7},$$
so that
$$\sqrt{(19-4\sqrt{21})}=\pm(\sqrt{12}-\sqrt{7}).$$
That is, since the principal value of the root is in question, and $\sqrt{12}>\sqrt{7}$—
$$\sqrt{(19-4\sqrt{21})}=\sqrt{12}-\sqrt{7}.$$

This example, like many others, might have been solved by inspection; for, since $19-4\sqrt{21}=19-2\sqrt{84}$, we have simply to find two numbers whose sum is 19 and whose product is 84. The numbers required are 12 and 7; and we have $19-2\sqrt{84}=12+7-2\sqrt{(12\times 7)}=(\sqrt{12}-\sqrt{7})^2$. Hence $\sqrt{(19-4\sqrt{21})}=\sqrt{12}-\sqrt{7}$.

§ **182.** It is obvious that $\sqrt{(\sqrt{p}+\sqrt{q})}$, where $\sqrt{p}$ and $\sqrt{q}$ are dissimilar quadratic surds, cannot be expressible as a simple quadratic surd number, among other reasons, because the square of every simple quadratic surd number has a rational part and $\sqrt{p}+\sqrt{q}$ has no such part.

$\sqrt{(\sqrt{p}+\sqrt{q})}$ can, however, under certain conditions, be expressed in the linear form $(\sqrt{x}+\sqrt{y})\sqrt[4]{p}$, where $p>q$. We have, in fact, $\sqrt{(\sqrt{p}+\sqrt{q})}=\sqrt[4]{p}\sqrt{\{1+\sqrt{(q/p)}\}}$. By § 181, if $1-q/p$ be a positive perfect square, $\sqrt{\{1+\sqrt{(q/p)}\}}$ is reducible to the form $\sqrt{x}+\sqrt{y}$; and in that case the linear expression proposed is possible.

Ex.
$$\sqrt{(5\sqrt{7}+2\sqrt{42})}=\sqrt[4]{7}\sqrt{(5+2\sqrt{6})},$$
$$=\sqrt[4]{7}(\sqrt{3}+\sqrt{2}),$$
$$=\sqrt[4]{63}+\sqrt[4]{28}.$$

§ **183.** $\sqrt{(p+\sqrt{q}+\sqrt{r})}$, where $p$ is rational and $\sqrt{q}$ and $\sqrt{r}$ are dissimilar quadratic surds, cannot be reduced to a simple quadratic surd number; for $p+\sqrt{q}+\sqrt{r}$ has too many irrational terms to be the square of a binomial, and too few to be the square of a trinomial.

On the other hand, since
$$(\sqrt{x}+\sqrt{y}+\sqrt{z})^2=x+y+z+2\sqrt{(yz)}+2\sqrt{(zx)}+2\sqrt{(xy)},$$
it is clear that in certain cases it must be possible to reduce $\sqrt{(p+\sqrt{q}+\sqrt{r}+\sqrt{s})}$, where $p$ is rational and $\sqrt{q}$, $\sqrt{r}$, $\sqrt{s}$ quadratic surds no two of which are similar, to the form

$\sqrt{x}+\sqrt{y}+\sqrt{z}$. The process of reduction and the conditions necessary for its success will be understood from the following examples :—

Ex. 1. Let, if possible—

$$\sqrt{(10+\sqrt{24}+\sqrt{40}+\sqrt{60})}=\sqrt{x}+\sqrt{y}+\sqrt{z} \qquad (1).$$

Then

$$10+\sqrt{24}+\sqrt{40}+\sqrt{60}=x+y+z+2\sqrt{(yz)}+2\sqrt{(zx)}+2\sqrt{(xy)} \qquad (2).$$

Let us suppose

$$2\sqrt{yz}=\sqrt{24},\quad 2\sqrt{(zx)}=\sqrt{40},\quad 2\sqrt{(xy)}=\sqrt{60} \qquad (3);$$

then we must also have

$$x+y+z=10 \qquad (4).$$

From the two last of (3) we have, by multiplication, $4x\sqrt{(yz)} = \sqrt{(40\times 60)}$; whence, by the first of (3), $2x\sqrt{24}=\sqrt{(24\times 100)}$, so that $2x=10$—that is, $x=5$. Using the first and third and the second of (3) in the same way, we get $y=3$; and, from the first, second, and third of (3), $z=2$.

These values of $x$, $y$, $z$ are all rational; and they also *happen* to satisfy (4). Hence

$$\sqrt{(10+\sqrt{24}+\sqrt{40}+\sqrt{60})}=\sqrt{5}+\sqrt{3}+\sqrt{2}.$$

Ex. 2. $\sqrt{(6+\sqrt{24}+\sqrt{40}+\sqrt{60})}=\sqrt{x}+\sqrt{y}+\sqrt{z}$.

We find rational values $x=5$, $y=3$, $z=2$ as before; but these values do not satisfy the equation $x+y+z=6$, which corresponds to (4) of last example. Hence in this case the expression as a simple quadratic surd number is impossible.

Ex. 3. $\sqrt{(6+\sqrt{2}+\sqrt{3}+\sqrt{5})}=\sqrt{x}+\sqrt{y}+\sqrt{z}$.

Here the process fails because the values found for $x$, $y$, $z$ are not rational.

**§ 184.** $\sqrt[4]{(a+\sqrt{b})}$ may be expressible in one or other of the two linear forms $\sqrt{x}+\sqrt{y}$, $(\sqrt{x}+\sqrt{y})\sqrt[4]{z}$, or it may not be expressible in linear form at all according to circumstances. When it can be expressed in linear form, the reduction is most readily accomplished by extracting the square root twice, *i.e.* by considering $\sqrt[4]{(a+\sqrt{b})}$ as $\sqrt{\{\sqrt{(a+\sqrt{b})}\}}$.

Ex. 1. $\sqrt[4]{(49+20\sqrt{6})}=\sqrt{(5+2\sqrt{6})}$,
$=\sqrt{2}+\sqrt{3}$.

Ex. 2. $\sqrt[4]{(56+32\sqrt{3})}=\sqrt{(\sqrt{32}+\sqrt{24})}$,
$=(1+\sqrt{3})\sqrt[4]{2}$.

Ex. 3. $\sqrt{(49\sqrt{3}+60\sqrt{2})}=\sqrt[8]{3}\sqrt[4]{(49+20\sqrt{6})}$,
$=\sqrt[8]{3}\sqrt{(5+2\sqrt{6})}$,
$=\sqrt[8]{3}(\sqrt{3}+\sqrt{2})$.

## EXERCISES LIV.

**1.** Prove that a quadratic surd cannot be the sum of two dissimilar quadratic surds. Why is the adjective "dissimilar" used in the statement of the theorem?

Express the following in linear form, where such expression is possible:—

**2.** $\sqrt{(4+\sqrt{15})}$. **3.** $\sqrt{(7+2\sqrt{6})}$. **4.** $\sqrt{(4+\sqrt{18})}$.

**5.** $\sqrt{(3\frac{1}{10}-\sqrt{6})}$. **6.** $\sqrt{(37+4\sqrt{78})}$. **7.** $\sqrt{\{7(3+2\sqrt{2})\}}$.

**8.** $\sqrt{(12+6\sqrt{3})/(\sqrt{3}+1)}$.

**9.** $\{1+\sqrt{(23+4\sqrt{15})}\}/\{1-\sqrt{(23-4\sqrt{15})}\}$.

**10.** $1/\sqrt{(3-2\sqrt{2})}+1/\sqrt{(3+2\sqrt{2})}$.

**11.** $2/\sqrt{(16-6\sqrt{7})}-1/(17+6\sqrt{8})$.

**12.** $(1+x)/\{1+\sqrt{(1+x)}\}+(1-x)/\{1-\sqrt{(1-x)}\}$, where $x=\sqrt{3}/2$.

**13.** $1/\{1+\sqrt{(2+\sqrt{3})}-\sqrt{(3+\sqrt{5})}\}$.

**14.** $\dfrac{\sqrt{(3+\sqrt{2})}+\sqrt{(3-\sqrt{2})}}{\sqrt{(3+\sqrt{2})}-\sqrt{(3-\sqrt{2})}}+\dfrac{\sqrt{(3+\sqrt{2})}-\sqrt{(3-\sqrt{2})}}{\sqrt{(3+\sqrt{2})}+\sqrt{(3-\sqrt{2})}}$.

**15.** $(3-2\sqrt{2})^{\frac{3}{2}}+(3+2\sqrt{2})^{\frac{3}{2}}$.

**16.** $(14+6\sqrt{5})^{\frac{3}{2}}+(14-6\sqrt{5})^{\frac{3}{2}}$.

**17.** $\sqrt{\{a+\sqrt{(a^2+2bc-b^2-c^2)}\}}$.

**18.** $\sqrt{[(2p-1)+\sqrt{\{(2p-1)^2-1\}}]}$.

**19.** $\sqrt{(6+\sqrt{8}-\sqrt{12}-\sqrt{24})}$.

**20.** $\sqrt{(48+12\sqrt{5}+12\sqrt{7}+2\sqrt{35})}$.

**21.** $\sqrt{(51-6\sqrt{3}+8\sqrt{5}-12\sqrt{15})}$.

**22.** $\sqrt{(21-\sqrt{192}-\sqrt{168}+\sqrt{224})}$.

**23.** $\sqrt{\left(\frac{17}{12}+\dfrac{1}{\sqrt{2}}+\dfrac{2}{\sqrt{3}}+\dfrac{\sqrt{2}}{\sqrt{3}}\right)}$.

**24.** $\sqrt{\{4+\sqrt{5}+\sqrt{(17-4\sqrt{15})}\}}$.

**25.** Show that the necessary and sufficient conditions that $\sqrt[4]{(a+\sqrt{b})}$ may be expressible in the form $\sqrt{x}+\sqrt{y}$ are that $a$ be positive, $\sqrt[4]{(a^2-b)}$ real and rational, and $\sqrt{\{2a+2\sqrt{(a^2-b)}\}}$ rational.

Show also that $\sqrt[4]{(a+\sqrt{b})}$ can be expressed in the form $(\sqrt{x}+\sqrt{y})/\sqrt[4]{p}$ provided $p$ can be found so as to make both $\sqrt[4]{(p^2a^2-p^2b)}$ and $\sqrt{\{2pa+2p\sqrt{(a^2-b)}\}}$ real and rational.

**26.** $\sqrt{(184-40\sqrt{21})}$. **27.** $\sqrt[4]{(14+8\sqrt{3})}$. **28.** $\sqrt[4]{(5\frac{2}{3}-4\sqrt{2})}$.

**29.** $\sqrt{(168+72\sqrt{5})}$. **30.** $\sqrt[4]{(137\sqrt{2}+36\sqrt{28})}$.

**31.** If $a$, $b$, $c$, $d$ be all rational, and $\sqrt{a}$, $\sqrt{b}$, $\sqrt{c}$, $\sqrt{d}$ surds of which $\sqrt{a}$ and $\sqrt{b}$ are dissimilar, then if $\sqrt{a}+\sqrt{b}=\sqrt{c}+\sqrt{d}$, it follows that either $a=c$ and $b=d$, or else $a=d$ and $b=c$.

# CHAPTER XVIII

## RATIO AND PROPORTION

### Ratio

§ **185.** *If* A *and* B *be two concrete quantities of the same kind which are expressible in terms of one and the same unit by the commensurable numbers* a *and* b *respectively, then the ratio of* A *to* B *is defined to be the quotient of these abstract numbers,* viz. $a/b$, *or in the notation specially appropriated to the present purpose* $a:b$.

Thus, for example, if A and B be lengths of $3\frac{1}{2}$ and $4\frac{3}{4}$ feet respectively, the ratio of A to B is $3\frac{1}{2}/4\frac{3}{4}$. If, instead of choosing a foot as the common unit, we choose a quarter of a foot, the numbers representing A and B would be 14 and 19, and the ratio 14/19. This might also be seen by the transformation $3\frac{1}{2}/4\frac{3}{4} \equiv (3\frac{1}{2}) \times 4/(4\frac{3}{4}) \times 4 \equiv 14/19$. It is obvious, indeed, from the definition that the value of a ratio is independent of the unit chosen; and this is reflected in the algebraical theorem that a quotient is unaltered by multiplying dividend and divisor by the same quantity. The ratio of A to B indicates, to use arithmetical language, the number of times that B is contained in A.

It is also obvious from what has been said, that by properly choosing to begin with the common unit in which A and B are measured, or by suitable transformation of the quotient $a/b$ afterwards, *we can always so arrange that the ratio is represented by the quotient of two integral numbers.*

The definition, however, supposes that there exists a finite unit in terms of which A and B can be expressed by commensurable numbers; in short, that A and B are commensurable. When A and B are incommensurable, the definition cannot be applied directly, and A and B must be replaced by commensurable approximations of accuracy sufficient for the purpose in hand. Any required degree of accuracy can be attained by choosing the common unit sufficiently small. For example, if we take any two lengths A and B at random, and measure them with any given rule, divided say to millimetres, neither will be an exact number of millimetres; but we take for $a$ and $b$ respectively the nearest number of millimetres in each case. If this degree of accuracy is not sufficient, we use a rule divided to tenths of a millimetre or to hundredths, etc., as the case may require.

§ **186.** The considerations just set forth lead us to *define the* **Ratio of two Abstract Quantities** a *and* b *as the quotient* a/b or $a:b$. $a$ is spoken of as the **Antecedent** and $b$ as the **Consequent** of the ratio; and the ratio is said to be a **Ratio of Greater Inequality**, the **Unit Ratio** or a **Ratio of Less Inequality** according as $a >=< b$.

Algebraically considered, therefore, a ratio is simply a quotient, and has all the properties of a quotient; indeed, $a$ and $b$ may be algebraic quantities and not merely arithmetical or positive quantities.

We may therefore multiply two ratios $a:b$ and $a':b'$ together. The ratio $aa':bb'$ which thus arises is said to be **compounded** *of the ratios* a : b *and* a': b'.

The ratio $a^2:b^2$, which is compounded of $a:b$, and $a:b$ again, is called the **duplicate** of $a:b$; $a^3:b^3$ is called the **triplicate** of $a:b$; and so on.

§ **187.** If we measure the relative value of two quantities by means of their difference, the relative value is obviously unaltered by increasing or diminishing each by the same amount. The case is otherwise if we measure the relative value of two quantities by means of their ratio. We have on this subject the following important proposition :—

*The ratio of two positive quantities* a *and* b *is brought nearer to unity by increasing each of them by the same positive quantity* x, *and can be brought as near to unity as we please by making* x *sufficiently large.*

For we have

$$1 - a/b = (b-a)/b;$$
$$1 - (a+x)/(b+x) = (b-a)/(b+x).$$

Therefore, since $b$ and $x$ are both positive and consequently

$b+x>b$, we see that $(b-a)/b$ is numerically greater than $(b-a)/(b+x)$. It follows that whether $a/b$ is greater or less than unity $(a+x)/(b+x)$ is nearer to unity than $a/b$.

Moreover, since $b-a$ is independent of $x$, we can, by making $x$ (and therefore $b+x$) larger and larger, that is, by dividing the fixed quantity $b-a$ by a larger and larger quantity, make $(b-a)/(b+x)$ as small as we like. In other words, we can, by adding to the antecedent and consequent of the ratio $a:b$ a sufficiently large positive quantity, bring the value of that ratio as near to unity as we please.

COR. *It follows also from the above reasoning that the ratio of two positive quantities is removed further from unity by subtracting from each of them the same positive quantity which does not exceed either of them.*

It would be easy to extend this theorem so as to include the ratios of algebraic quantities; but the results are not of much practical importance.

Ex. 1. If the ratio formed from $a:b$ by subtracting the same quantity $x$ from both antecedent and consequent be the duplicate of the ratio formed from $a:b$ by adding the same quantity $x$ to both antecedent and consequent, find $x$.

We have $$(a-x)/(b-x)=(a+x)^2/(b+x)^2.$$
Hence $$(a-x)(b+x)^2=(b-x)(a+x)^2,$$
since $x=\pm b$ are obviously not solutions. On reducing this equation to normal form we find

$$3(a-b)x^2+(a^2-b^2)x-ab(a-b)=0.$$

If we suppose $a\neq b$, this last equation is equivalent to

$$3x^2+(a+b)x-ab=0.$$

This last equation can (see § 130) be written

$$3[x+\tfrac{1}{6}(a+b)-\tfrac{1}{6}\sqrt{\{(a+b)^2+12ab\}}]$$
$$\times[x+\tfrac{1}{6}(a+b)+\tfrac{1}{6}\sqrt{\{(a+b)^2+12ab\}}]=0,$$

which has two real roots, one positive the other negative, viz. :—

$$x=\tfrac{1}{6}[-(a+b)+\sqrt{\{(a+b)^2+12ab\}}],$$
$$x=\tfrac{1}{6}[-(a+b)-\sqrt{\{(a+b)^2+12ab\}}].$$

If we restrict $x$ to be positive, only the former solution will be available. It will be a good exercise for the beginner to discuss whether, on the hypothesis that $a$ and $b$ are both positive, the positive value of $x$ is such as to make both $a-x$ and $b-x$ positive.

Ex. 2. If $a_1, a_2, \ldots, a_n, b_1, b_2, \ldots, b_n$ be all positive, show that the ratio $(a_1+a_2+\ldots+a_n):(b_1+b_2+\ldots+b_n)$ is not greater than the greatest and not less than the least of $a_1:b_1, a_2:b_2, \ldots, a_n:b_n$.

Let $\rho$ be the least and $\sigma$ the greatest of the ratios $a_1 : b_1$, $a_2 : b_2$, . . . , $a_n : b_n$; then $a_1/b_1 \nless \rho$, $a_2/b_2 \nless \rho$, . . . , $a_n/b_n \nless \rho$.

Hence, since $b_1$, $b_2$, . . . , $b_n$ are all positive quantities, we infer that

$$a_1 \nless \rho b_1,\ a_2 \nless \rho b_2,\ \ldots,\ a_n \nless \rho b_n.$$

Hence, by addition—

$$a_1+a_2+\ \ldots\ +a_n \nless \rho(b_1+b_2+\ \ldots\ +b_n);$$

whence, finally, since $b_1+b_2+\ \ldots\ +b_n$, being a sum of positive quantities, is positive, we have

$$(a_1+a_2+\ \ldots\ +a_n)/(b_1+b_2+\ \ldots\ +b_n) \nless \rho.$$

In exactly the same way we prove that

$$(a_1+a_2+\ \ldots\ +a_n)/(b_1+b_2+\ \ldots\ +b_n) \ngtr \sigma.$$

## EXERCISES LV.

**1.** Find the ratio of £3 : 6 : $8\frac{1}{2}$ to £2 : 3 : $4\frac{1}{2}$.

**2.** Given that a mètre is 39·37 inches, find the ratio of 5·34 m. to 6 yds. 3 ft. 2 in.

**3.** What is the ratio compounded of 2 : 3 and 6 : 8?

**4.** If $c$ be added to the antecedent of the ratio $a : b$, what quantity must be added to its consequent so that the value of the ratio may be unaltered?

**5.** Show that $x^2 : y^2 >< x : y$ according as $x >< y$.

**6.** If $x$ and $y$ be real positive quantities and $x>y$, show that $x^2 : y^2 > x-y : x+y$.

**7.** Find the number which must be taken from each term of the ratio 38 : 31 so that it may become equal to 4 : 3.

**8.** What quantity must be subtracted from each term of the ratio 5 : 6 in order that the ratio thus formed may be one-half the original ratio?

**9.** Two numbers are in the ratio 3 : 4, and if 8 be added to each, they will be in the ratio of 7 : 9. Find the numbers.

**10.** A is 24 years old, B is 15 years old. What is the least number of years after which the ratio of their ages will be less than 7 : 5?

**11.** What common quantity must be added to each term of $a^2 : b^2$ to make it equal to $a : b$?

**12.** Divide 112 into two parts whose ratio shall be 2 : 3.

**13.** If $0<x<1$, arrange the ratios $1 : 1+x$, $1+x : 1+2x$, $1-x : 1$ in order of magnitude.

**14.** Given that $x$ and $y$ are both positive, find when $x-y : x+y$ is greater than $x-2y : x+3y$, and when less.

**15.** Find $x : y$, given $3x-2y : x-5y = 1/3$.

**16.** Find the relation between $a$, $b$, $c$ in order that the ratio $a-c : b-c$ may be the duplicate of $a : b$.

**17.** If $3x^2-10xy+3y^2=0$, find the ratio $x : y$.

**18.** Find the ratio of a velocity of 3 miles per minute to a velocity of 650 yards per second.

**19.** If $x$, $y$, $a$ be all positive and $xy > a^2$, show that $(x+a)^2 : (y+a)^2 >=< x : y$ according as $x >=< y$.

**20.** Find the ratio of an acceleration of 3 feet per second per second to an acceleration of 2 yards per minute per minute.

**21.** If $a_1, a_2, \ldots, a_n, b_1, b_2, \ldots, b_n$ be all positive real quantities, show that the ratio $\sqrt[n]{(a_1^n + a_2^n + \ldots + a_n^n)} : \sqrt{(b_1^n + b_2^n + \ldots + b_n^n)}$ is intermediate in value between the greatest and the least of $a_1 : b_1$, $a_2 : b_2, \ldots, a_n : b_n$.

**22.** Two armies whose numbers were as 20 to 3 met in battle. The respective losses were as 40 : 3, and the number of survivors as 5 : 1. The number of survivors in the larger army being 1200, find their original numbers.

**23.** Three vessels contain volumes A, B, C respectively of mixtures of salt and water. In A the ratio of salt to water is $1 : a$, in B, $1 : b$, in C, $1 : c$. If the three vessels be emptied into one, what will be the ratio of salt to water in the resulting mixture?

## Proportion

**§ 188. Concrete Proportion.**—*When two concrete quantities* A *and* B *of the same kind, and two others also of the same kind, but not necessarily of the same kind as* A *and* B, *are such that the ratio of* A *to* B *is the same as the ratio of* C *to* D, *then* A, B, C, D *are said to be in proportion.* In other words, if $a : b$ be the ratio of A to B, and $c : d$ the ratio of C to D, then A, B, C, D are said to be in proportion when $a : b = c : d$.* This is sometimes expressed by saying that A is to B as C is to D.

The above definition of concrete proportion depends on the notion of ratio as previously defined, and can therefore be only indirectly applied to incommensurables. In this respect our definition differs from that given by *Euclid* (Bk. V. Def. 5; Todhunter's or Mackay's edition), which defines proportion and does not directly define ratio, using terms that are applicable alike to commensurables and incommensurables, since the notion of measurement in terms of a unit is not employed. There can be no doubt, however, that the majority of minds reach the notion of ratio through the idea of scale or measurement; and this appears to be the reason for the familiar fact that, although beginners can be taught to repeat the propositions of the Euclidian

* Some conservative writers still use in stating proportions the old sign of equality : : introduced by Oughtred. In Algebra at least there is no reason whatever for maintaining this alternative sign of equality.

theory, they rarely grasp its logical cogency, much less appreciate its elegance as an abstract theory.*

There is no doubt that a definition of ratio (and consequently of proportion) might be given without introducing the notion of a unit; but it would be necessary to deduce from this definition a series of propositions to show that a ratio has all the characteristics of a quotient and can be admitted as such and operated with under the fundamental laws of Algebra.†

§ **189.** The discussion of last paragraph leads us to give a purely **Abstract Definition of Proportion**, as follows:—

*Four algebraic quantities* a, b, c, d *are said to be in proportion when* a/b = c/d, *that is, when the ratio of the first to the second is equal to the ratio of the third to the fourth.* We may write this relation with the special notation $a:b=c:d$ whenever it is desirable or convenient to do so.

In abstract proportion there is of course no question as to the *kind* of magnitudes represented by $a$ and $b$ and by $c$ and $d$; $a$ and $d$ are spoken of as the extremes, and $b$ and $c$ as the means of the proportion.

*If* a, b, c, d, e . . . *be such that* a : b = b : c = c : d = d : e = *etc.*, a, b, c, d, e . . . *are said to be in* **Continued Proportion** (or to be in **Geometric Progression**, or to form a **Geometric Series**).

*When* a, b, c *are in continued proportion,* b *is said to be a* **Mean Proportional,** *or* **Geometric Mean,** *between* a *and* c.

*When* a, b, c, d *are in continued proportion,* b *and* c *are spoken of as* **Two Mean Proportionals,** *or* **Two Geometric Means,** *between* a *and* d, *and so on.*

§ **190.** *Expression of a set of proportionals in terms of the minimum number of independent variables.*

If $a$, $b$, $c$, $d$ be proportionals, we have $a/b=c/d=\rho$, say, where we may call $\rho$ the **Common Ratio of the Proportion.** Since there is only one relation connecting the four variables $a$, $b$, $c$, $d$, we can express them all in terms of three and no fewer. We might take as these three independent variables any three of the four quantities themselves; but for most purposes it is more convenient to select two of the four, say $b$ and $d$, and the common ratio $\rho$. Thus we have from $a/b=\rho$, $a=\rho b$; from $c/d=\rho$, $c=\rho d$. Hence $\rho b$, $b$, $\rho d$, $d$ is a perfectly general representation for four

* For a comparison of the two theories, see A. Ch. XIII. § 16.

† Professor Hill has recently shown the writer an interesting sketch of a theory of this kind which he uses in his elementary instruction.

quantities in proportion. It follows that, if three of the terms of a proportion are given, or any three independent relations connecting its terms are given, all its terms are determined.

Similarly if $a, b, c, d, e \dots$ be in continued proportion, we have $a/b = b/c = c/d = d/e = \dots = 1/\rho$, say. (It is usual to call $\rho$ in this case the common ratio.) Hence $b/a = \rho$, $c/b = \rho$, $d/c = \rho$, $e/d = \rho \dots$ Whence $b = \rho a$, $c = \rho b = \rho\rho a = \rho^2 a$, $d = \rho c = \rho\rho^2 a = \rho^3 a$, $e = \rho d = \rho\rho^3 a = \rho^4 a \dots$

It appears therefore that *any continued proportion is expressible in terms of two independent variables in the form*

$$a,\ a\rho,\ a\rho^2,\ a\rho^3,\ \dots,\ a\rho^{n-1},\ \dots$$

**§ 191.** *If* a, b, c, d *be proportional, then* ad = bc ; *and, conversely, if* ad = bc, *then the four quantities* a, b, c, d *form a proportion in any one of the eight orders* a, b, c, d ; a, c, b, d ; d, b, c, a ; d, c, b, a ; b, a, d, c ; b, d, a, c ; c, a, d, b ; c, d, a, b.

For, if $a/b = c/d$, multiplying both sides of this equation by $bd$ we deduce $ad = bc$.

Again, if $ad = bc$, and we divide by $bd$, we deduce $ad/bd = bc/bd$, that is, $a/b = c/d$. If we divide by $cd$, we deduce $ad/cd = bc/cd$, that is, $a/c = b/d$. If we divide by $ab$, we deduce $ad/ab = bc/ab$, that is, $d/b = c/a$, and so on.

Cor. *If* a, b, c, d *form a proportion in one of the eight orders given above, they also form a proportion in each of the other seven orders.*

Ex. 1. If $\frac{3}{8} : 2 = x : 1\frac{1}{2}$, find $x$.

We have $2x = \frac{3}{8} \times 1\frac{1}{2}$, whence $x = \frac{9}{32}$.

This is an example of the "Rule of Three," so familiar in arithmetic. The fact that the unknown term of the proportion is not the last is an immaterial difference, since it follows from $\frac{3}{8} : 2 = x : 1\frac{1}{2}$ that $2 : \frac{3}{8} = 1\frac{1}{2} : x$.

**§ 192.** If $a : b = c : d$, then

$$a + b : b = c + d : d \qquad (1);$$
$$a - b : b = c - d : d \qquad (2);$$
$$a + b : a - b = c + d : c - d \qquad (3);$$
$$la + mb : pa + qb = lc + md : pc + qd \qquad (4);$$
$$a^r : b^r = c^r : d^r \qquad (5);$$
$$la^r + mb^r : pa^r + qb^r = lc^r + md^r : pc^r + qd^r \qquad (6);$$

*where* l, m, p, q *are any quantities, of which* l *and* m *do not both vanish, and* p *and* q *do not both vanish, and* r *is any commensurable real number.*

It will be observed that the first five of these results are particular cases of the last. For example, (3) is obtained from (6) by putting $l=1$, $m=1$, $p=1$, $q=-1$, $r=1$. They can all be proved by a uniform method, which we shall exemplify by proving (3) and (6).

Let the common ratio of the given proportion be denoted by $\rho$, so that $a:b=\rho$, $c:d=\rho$, and therefore, as in § 190, $a=b\rho$, $c=d\rho$.

Hence we have

$$(a+b)/(a-b)=(b\rho+b)/(b\rho-b)=(\rho+1)/(\rho-1).$$

Also

$$(c+d)/(c-d)=(d\rho+d)/(d\rho-d)=(\rho+1)/(\rho-1).$$

Therefore

$$(a+b)/(a-b)=(c+d)/(c-d),$$

which proves (3).

Again—

$$\begin{aligned}(la^r+mb^r)/(pa^r+pb^r) &= (l(b\rho)^r+mb^r)/(p(b\rho)^r+qb^r),\\ &= (lb^r\rho^r+mb^r)/(pb^r\rho^r+qb^r),\\ &= (l\rho^r+m)/(p\rho^r+q),\\ (lc^r+md^r)/(pc^r+qd^r) &= (l(d\rho)^r+md^r)/(p(d\rho)^r+qd^r),\\ &= (ld^r\rho^r+md^r)/(pd^r\rho^r+qd^r),\\ &= (l\rho^r+m)/(p\rho^r+q).\end{aligned}$$

Hence, etc.

**§ 193.** By the method of last paragraph it is easy to prove the following useful theorem :—

*If* $a_1:b_1=a_2:b_2= \ldots =a_n:b_n$, *then each of these ratios is equal to*

$$a_1+a_2+ \ldots +a_n : b_1+b_2+ \ldots +b_n,$$

*provided*

$$b_1+b_2+ \ldots +b_n \neq 0\,;$$

*and also to*

$$\sqrt[r]{(l_1a_1^r+l_2a_2^r+ \ldots +l_na_n^r)} : \sqrt[r]{(l_1b_1^r+l_2b_2^r+ \ldots +l_nb_n^r)},$$

*where* $l_1, l_2, \ldots, l_n$ *are any real quantities which do not all vanish, and* $r$ *is any positive integer:* the principal value of the $r$th root, as usual, is to be taken, and the radicand is supposed to be positive.

The reader should compare this theorem with the inequality of § 187, Example 2.

**§ 194.** The method of proof used in § 192 owes its power

and directness to the following important principle, which is of course nothing fresh, but merely an assertion of the fundamental principles of Algebra :—

*If in any equation connecting conditioned variables we express by means of the given conditions all the variables involved in the equation in terms of the smallest possible number of variables, then the equation, if true, is an identity.*

This, which we may for convenience call the **Principle of the Minimum Number of Variables,** is merely the assertion of a truth frequently insisted upon already, viz. that all algebraic equality is in the last resort identity in the sense defined in § 41.

**§ 195.** Deduction of consequences from given proportions is merely a simple example of the deduction of conditional equations from given equations of condition; and we are by no means tied up to the special method above illustrated. For example, we might prove (3) of § 192 as follows :—

From $a/b=c/d$, we have $a/b+1=c/d+1$; whence $(a+b)/b=(c+d)/d$. (7).

Again from $a/b=c/d$, we have $a/b-1=c/d-1$; whence $(a-b)/b=(c-d)/d$ (8).

From (7) and (8) we have

$$\frac{(a+b)/b}{(a-b)/b}=\frac{(c+d)/d}{(c-d)/d},$$

whence $(a+b)/(a-b)=(c+d)/(c-d)$, which is (3).

**§ 196.** *If* a, b, c *be in continued proportion, then*

$$a:c=a^2:b^2=b^2:c^2;$$

*and* $$b=\sqrt{(ac)}.$$

*If* a, b, c, d *be in continued proportion, then*

$$a:d=a^3:b^3=b^3:c^3=c^3:d^3;$$

*and* $$b=\sqrt[3]{(a^2d)},\qquad c=\sqrt[3]{(ad^2)};$$

*and so on.*

It will be sufficient to prove the second of these theorems.

Let $a:b=b:c=c:d=1/\rho$, so that $b=\rho a$, $c=\rho^2a$, $d=\rho^3a$, as in § 190. It follows that $d/a=\rho^3=(b/a)^3=b^3/a^3$; whence $a/d=a^3/b^3$, etc., which proves the first part of the theorem.

Again, since $d/a=\rho^3$, we have $\rho=(\sqrt[3]{d/a})$. To avoid confusion and needless complication we assume that $a$, $b$, $c$, $d$ are all real and positive, and we consider only principal values of the cube root. We therefore have

$$b = a\rho = a\sqrt[3]{(d/a)} = \sqrt[3]{(a^3 d/a)},$$
$$= \sqrt[3]{(a^2 d)};$$

and

$$c = a\rho^2 = a\{\sqrt[3]{(d/a)}\}^2 = a\sqrt[3]{(d^2/a^2)}$$
$$= \sqrt[3]{(a^3 d^2/a^2)} = \sqrt[3]{(ad^2)}.$$

COR. From the above it appears that *the insertion of two mean proportionals between two given positive quantities depends on the extraction of the cube root.*

This is the famous **Delian Problem** of antiquity.

Ex. 1. Find a mean proportional between 3 and 12.

If $x$ be the required mean, we have $3 : x = x : 12$, and therefore $x^2 = 3 \times 12$; whence, if we confine ourselves to positive values of $x$, $x = 6$.

Ex. 2. Calculate to two places of decimals the values of three geometric means inserted between 1 and 2.

Let $\rho$ be the common ratio of the geometric progression, then its terms may be written $1, \rho, \rho^2, \rho^3, \rho^4$. The last of these must be 2, hence $\rho^4 = 2$. Extracting the square root we get $\rho^2 = 1{\cdot}414$; and again extracting the square root we get $\rho = 1{\cdot}189$. Also $\rho^3 = \rho\rho^2 = 1{\cdot}189 \times 1{\cdot}414 = 1{\cdot}681$. Hence the three means are 1·19, 1·41, 1·68.

Ex. 3. If $a : b = c : d = e : f$, prove that

$$(a + 2c + 3e)^2 : (b + 2d + 3f)^2 = (ac + ce) : (bd + df).$$

If we denote the common ratio of the given proportions by $\rho$, we have $a = b\rho$, $c = d\rho$, $e = f\rho$. Expressing all the quantities in terms of the minimum number, viz. $b, d, f, \rho$, we have

$$(a + 2c + 3e)^2/(b + 2d + 3f)^2 = (b\rho + 2d\rho + 3f\rho)^2/(b + 2d + 3f)^2,$$
$$= \rho^2(b + 2d + 3f)^2/(b + 2d + 3f)^2,$$
$$= \rho^2.$$

Also

$$(ac + ce)/(bd + df) = (b\rho d\rho + d\rho f\rho)/(bd + df),$$
$$= \rho^2(bd + df)/(bd + df),$$
$$= \rho^2.$$

Hence the theorem follows.

Ex. 4. If $a : (b + c) = b : (c + a) = c : (a + b)$, show that each of these ratios is either $\frac{1}{2}$ or $-1$, it being given that none of the three $b + c$, $c + a$, or $a + b$ vanish.

If we denote the common value of the ratios by $\rho$, we have

$$a = (b + c)\rho, \quad b = (c + a)\rho, \quad c = (a + b)\rho.$$

Therefore by addition $a + b + c = 2(a + b + c)\rho$. If now $a + b + c \neq 0$, it follows from the last equation that $1 = 2\rho$, *i.e.* $\rho = \frac{1}{2}$.

If $a + b + c = 0$, this inference fails but then $b + c = -a$, $c + a = -b$,

$a+b=-c$, from which it follows, since $b+c\neq 0$, that $a/(b+c)=-1$, etc.

Ex. 5. If $x:(a+2b+c)=y:(2a+b+c)=z:(4a-4b+c)$, show that $a:(-5x+6y-z)=b:(-2x+3y-z)=c:(12x-12y+3z)$.

Denote each of the three given ratios by $1/\sigma$, then we have

$$a+2b+c=\sigma x,\quad 2a+b+c=\sigma y,\quad 4a-4b+c=\sigma z.$$

From the first two of these equations, by eliminating $c$, we get

$$a-b=-\sigma x+\sigma y.$$

From the last two—

$$2a-5b=-\sigma y+\sigma z;$$

from these last we get

$$a=\tfrac{1}{3}\sigma(-5x+6y-z),\quad b=\tfrac{1}{3}\sigma(-2x+3y-z).$$

The equation $a+2b+c=\sigma x$ now gives

$$c=\tfrac{1}{3}\sigma(12x-12y+3z).$$

Hence

$$a/(-5x+6y-z)=b/(-2x+3y-z)=c/(12x-12y+3z),$$

each of these being equal to $\frac{1}{3}\sigma$.

## EXERCISES LVI.

1. The first and fourth terms of a proportion are 5 and 54; the sum of the second and third terms is 51. Find the terms.

Find $x$ from the following proportions:—

2. $3:6=x:8$.
3. $x:3=6:8$.
4. $3:4=12:x$.
5. $3:x=x:48$.
6. If $7x-y:3x-y=11x-5y:y-x$, find $x:y$.
7. Given $x-y:x-2y=3x-y:x-3y$, find $x:y$.

Find mean proportionals between

8. $3a^2b$ and $12a^4b$.
9. $(a-b)(a+b)^2$ and $(a+b)(a-b)^2$.
10. $\sqrt{12}$ and $\sqrt{27}$.
11. $5+7\sqrt{2}$ and $(29+47\sqrt{2})/73$.
12. Insert four mean proportionals between 6 and 192.
13. Given the third and fifth of a series of quantities in continued proportion, find the first and fourth of the series.
14. Find two numbers such that their sum, their difference, and the sum of their squares are in the ratio $5:3:51$.
15. Three positive quantities are in continued proportion. The difference between the greatest and least is 624, and the other number is 91. What are the numbers?
16. Three numbers are in continued proportion; the middle number is 12 and the sum of the others 25; find the numbers.
17. If $x-x':y-y'=x+x':y+y'$, show that $x:x'=y:y'$.
18. If $x+x':y+y'=xx':yy'$, show that $xy:x'y'=y-x:x'-y'$.
19. If $(a+b+c+d)(a-b-c+d)=(a-b+c-d)(a+b-c-d)$ then $a$, $b$, $c$, $d$ are proportionals.

**20.** If $a:b=c:d$, show that $ab+cd$ is a mean proportional between $a^2+c^2$ and $b^2+d^2$.

**21.** If $(a+bx)/(b+cy)=(b+cx)/(c+ay)=(c+ax)/(a+by)$, then in general each is equal to $(1+x)/(1+y)$. Why is this not true when $a=2$, $b=-5$, $c=3$, $x=2/3$, $y=3$?

**22.** If $a, b, c, d$ be proportionals and $a+d=b+c$, then either $a=b$ and $c=d$, or else $a=c$ and $b=d$.

**23.** If $b(a-c):c(b-d)=a-b:c-d$, prove that either $b=c$ or $a:b=c:d$.

**24.** If $a^2-ab+b^2:c^2-cd+d^2=ab:cd$, then either $a:b=c:d$ or $a:b=d:c$.

**25.** If $a:b=c:d$, then $(a/b)^2+(c/d)^2=2ac/bd$.

**26.** If $a/b=c/d$, then $(1/a+1/d)-(1/b+1/c)=(a-b)(a-c)/abc$.

**27.** If $a/b=b/c$, then $(a+b+c)/(a-b+c)=(a+b+c)^2/(a^2+b^2+c^2)$.

**28.** If $a:b=c:d$, then $(a^2+c^2)(b^2+d^2)=(ab+cd)^2$.

**29.** If $a:b=c:d=e:f$, then $a^2/(a^2+ce)=b^2/(b^2+df)$.

**30.** If $a:b=c:d$, then $(a^3+c^3)/(b^3+d^3)=(a+c)^3/(b+d)^3$.

**31.** If $\dfrac{a-b}{d-c}=\dfrac{b-c}{c-f}$, then each $=\dfrac{b(f-d)+cd-af}{cf-de}$.

**32.** If $\dfrac{(ax+by)(c+d)+(cx+dy)(a+b)}{(ax+by)(c+d)-(cx+dy)(a+b)}=\dfrac{(ax-by)(c-d)+(cx-dy)(a-b)}{(ax-by)(c-d)-(cx-dy)(a-b)}$ determine $x:y$.

**33.** If $a$, $b$, $c$ be in continued proportion, then $(\Sigma a)^2+\Sigma a^2=2(a+c)\Sigma a$.

**34.** If $a$, $b$, $c$, $d$ be in continued proportion, then $(a+d)(b+c)-(a+b)(c+d)=b(a^2-b^2)^2/a^3$.

**35.** If $a$, $b$, $c$, $d$ be in continued proportion, then $(a+d)/(a-d)=(a^2-b^2+ac+bc)/(a^2-b^2+ac-bc)$.

**36.** If $a, b, c$ be in continued proportion, so also are $\alpha^2a+2\alpha\beta b+\beta^2c$, $\alpha\gamma a+(\beta\gamma+\alpha\delta)b+\beta\delta c$, $\gamma^2a+2\gamma\delta b+\delta^2c$.

**37.** If $(a^2+ab+b^2)/(a^2-ab+b^2)=(c^2+cd+d^2)/(c^2-cd+d^2)$, then either $a:b=c:d$ or $a:b=d:c$.

**38.** If $a:b=c:d$, and $x+a$, $x+b$, $y+c$, $y+d$ are in continued proportion, show that $x:y=a:c$, and find an equation for $y$.

**39.** If $ab:cd=ef:gh$, and $ac:bd=eg:fh$, prove that $ah=\pm ed$ and $bg=\pm fc$.

**40.** If $a, b, c, d$ be such that on adding a certain quantity to each they are in continued proportion, then $a-b$, $b-c$, $c-d$ must also be in continued proportion.

**41.** A sum of money is divided into two parts in the ratio of $x:y$. A and B divide between themselves the first part in the ratio $a:b$, and the second in the ratio $c:d$. What fraction of the whole does each receive? If they receive equal amounts, find the ratio of $x:y$.

**42.** If $a, b, c$ be all positive, and $a+b+c:c+b-a=c+a-b:a+b-c$, show that $a, b, c$ will form the sides of a right-angled triangle; and, if $a+b+c$, $c+b-a$, $c+a-b$, $a+b-c$ be in continued proportion, show that if $\rho$ be the common ratio then $\rho^3+\rho^2+\rho=1$. Show also that $\rho=a/c$.

**43.** The weights of two men are as 12 to 11; a year later they have gained additional weight in the proportion of 18 to 1, and their total weights are now as 29 to 24. If the heavier man weighs 12 stone 6 lbs., find his original weight.

**44.** A man who has been making an income of £1000 a year has to draw upon his capital, and his income is thus reduced to £800. Next year he again loses the same amount of capital; but getting a higher rate of interest his income remains at £800. Show that the rates of interest are as 3 : 4.

**45.** A merchant deals in three commodities A, B, C, the numbers of tons of A, B, and C sold in a year are as 3 : 5 : 7, and the prices of A, B, C per ton are as 7 : 8 : 9. If the total number of tons sold in a year be 1500, and the total value of the material sold be £37,200, find the price of each commodity per ton.

## The Functional Relation of Proportionality

**§ 197.** Under this head we propose to discuss from the point of view of the theory of proportion certain of the simpler ways in which one variable quantity $y$, which we call for distinction the **Dependent Variable**, may be a function of another variable $x$, or several others, which we call the **Independent Variable** or **Independent Variables.**

The following is the fundamental theorem :—*If* y *depend on* x, *and on* x *alone, in such a way that,* (x, y), (x′, y′) *being two pairs of corresponding values, we have always*

$$y : y' = x : x' \qquad (1),$$

*then* y *is a constant multiple of* x.

To prove this, we may suppose the pair of values $(x', y')$ fixed for reference. Then, since we have $y/y' = x/x'$, it follows that $y/x = y'/x'$. Now $x'$ and $y'$ being fixed, $y'/x'$ is a constant which we may denote by $a$, and we have

$$y = ax \qquad (1'),$$

which proves our theorem.

When $y$ depends in this simple manner upon $x$, $y$ is said to be **proportional to** $x$; and the relation is often denoted by

$$y \propto x \qquad (1''),$$

the constant $a$ being omitted for shortness.*

* It used to be the practice to use the phrase, "$y$ varies as $x$," instead of "$y$ is proportional to $x$." This usage ought certainly to be dropped, as it involves a strain upon the meaning of the word "vary," and introduces a useless and confusing technical phrase where none is needed. The notation $\propto$ might also be dropped with advantage.

Obviously *the constant* a *in the equation* (1′) *is known as soon as we know the value of* y *corresponding to any given value of* x.

Ex. The extension per unit of length of an elastic string is proportional to the weight attached to it. Given that the extension is one-tenth of an inch per inch of length when the weight attached is 2 lbs., find the extension per unit of length when the weight attached is 5 lbs.

Let $y$ denote the extension in inches per inch of length corresponding to a weight of $x$ lbs.; then by hypothesis $y=ax$. Now, when $x=2$, $y=1/10$, hence $1/10=a2$; therefore $a=1/20$; hence generally $y=x/20$. From this again, putting $x=5$, we deduce $y_5=5/20=1/4$, where we use $y_5$ to denote the extension per inch corresponding to 5 lbs. Our result then is that, when a weight of 5 lbs. is attached, every inch of the string is lengthened by a quarter of an inch.

§ **198.** In the equation (1′) of § 197 we might replace $x$ by some less simple function of $x$; for example, by $x^2$, or by $1/x$, or by $x^2+x$, etc. We should thus have

$$y=ax^2 \qquad (2');$$
$$y=a/x \qquad (3');$$
$$y=a(x^2+x) \qquad (4');$$

with the corresponding equations

$$y:y'=x^2:x'^2 \qquad (2);$$
$$y:y'=1/x:1/x' \qquad (3);$$
$$y:y'=x^2+x:x'^2+x' \qquad (4);$$

which are analogous to (1). These functional relations we might describe by saying that $y$ is proportional to $x^2$, proportional to $1/x$,* proportional to $x^2+x$, and so on.

In all these cases there is only a single constant to determine, so that the functional relation is completely determined when a corresponding pair of values of the dependent and independent variables are known.

Ex. 1. If the value of diamonds is proportional to the squares of their weights, and a diamond worth £50 be divided into two pieces whose weights are as 2 : 3, find the values of the pieces.

Let £$y$ be the value of a diamond weighing $x$ units, then by hypothesis $y=ax^2$. If $w$ be the weight of a diamond worth £50, we have $50=aw^2$. The two pieces weigh $\frac{2}{5}w$ and $\frac{3}{5}w$ respectively; if, therefore, the values of these be $y'$ and $y''$, we have

$$y'=a(\tfrac{2}{5}w)^2=\tfrac{4}{25}aw^2=\tfrac{4}{25}\times 50=8.$$
$$y''=a(\tfrac{3}{5}w)^2=\tfrac{9}{25}aw^2=\tfrac{9}{25}\times 50=18.$$

Hence the pieces are worth £8 and £18 respectively.

* In this case $y$ is sometimes said to be *inversely proportional to* x.

It will be noticed that the whole value of the divided diamond is only £26. As a contrast, the beginner should work out the problem on the hypothesis that the value of a diamond is proportional to its weight simply.

§ **199.** We might suppose that $y$ is the sum of several parts, each part being proportional to a different function of $x$. For example, one part of $y$ might be proportional to $x$, and another part to $x^2$. We should then have $y=ax+bx^2$. Here we have two constants, $a$ and $b$, to determine, so that the functional relation is not completely determined until we know two pairs of corresponding values of $x$ and $y$.

Ex. If income tax were made up of two parts, one part proportional to income, another to the square of income, and if the whole tax on £100 were £1, and on £1000, £55, find the income tax on £10,000; and show that no man's net income could exceed £5000.

If £$y$ be the income tax on an income of £$x$, we have, by hypothesis, $y=ax+bx^2$. Our data give us

$$a100+b100^2=1, \quad a1000+b1000^2=55.$$

From these equations we find $a=1/200$, $b=1/20,000$. Hence the general formula for calculating the tax is $y=x/200+x^2/20,000$.

The tax on £10,000 is therefore $10,000/200+10,000^2/20,000$, that is, $50+5000=$ £5050.

The net income corresponding to a nominal income of £$x$ is

$$x-x/200-x^2/20,000=\{19900x-x^2\}/20,000,$$
$$=\{9950^2-(9950-x)^2\}/20,000.$$

The largest possible net income is obtained by so choosing $x$ that $(9950-x)^2$ is the least possible—that is to say, by putting $x=9950$. Hence the largest possible net income is

$$9950^2/20,000=995\times 4{\cdot}975=\text{£}4950 \text{ odd};$$

so that no man's net income could reach £5000. It would be to a man's advantage to prevent his nominal income from exceeding £9950, inasmuch as any increase beyond that point, although it would benefit the state, would be an actual loss to him. It is also obvious, from the above calculation, that a man whose nominal income reached £19,900 would be a pauper; beyond that point the tax would be more than his income.

§ **200.** We might also consider functions of more than one independent variable. Thus, for example, we might replace $x$ in § 197 (1′) by $xz$, by $x/z$, by $x+z^2$, by $xzu$, and so on. We should thus have

$$y=axz, \quad y=ax/z, \quad y=a(x+z^2), \quad y=axzu, \text{ etc.,}$$

and corresponding thereto—

$$y:y'=xz:x'z', \text{ etc.}$$

When $y = axz$ it is usual to say that $y$ *is* **proportional** *to* x *and* z **conjointly**; and when $y = ax/z$, that $y$ is **proportional** to $x$ **directly** *and to* z **inversely.**

In this connection it is usual to prove the following theorem :—*If* y *depend on* x *and* z, *and on these only, and be proportional to* x *when* z *is constant, and proportional to* z *when* x *is constant, then* y *is proportional to* x *and* z *conjointly when both* x *and* z *vary.*

Let $(x, z, y)$, $(x', z, y_1)$, $(x', z', y')$ be three sets of corresponding values of the variables, then, by hypothesis, since $z$ is the same for $(x, z, y)$ and $(x', z, y_1)$, we have

$$y/y_1 = x/x'.$$

Also, since $x'$ is the same for $(x', z, y_1)$ and $(x', z', y')$, we have

$$y_1/y' = z/z'.$$

Hence

$$(y/y_1)(y_1/y') = (x/x')(z/z'),$$

that is—

$$y/y' = zx/z'x'.$$

In other words—

$$y : y' = zx : z'x' ;$$

and, if we fix $y'$, $z'$, $x'$ for reference, so that $y'/z'x'$ is a constant, say $a$, we have

$$y = azx.$$

Our theorem is thus proved.

Ex. It is found that the quantity of work done by a man in an hour is directly proportional to his pay per hour, and inversely proportional to the square root of the number of hours he works per day. He can finish a piece of work in 6 days when working 9 hours a day at 1s. per hour. How many days will he take to finish the same piece of work when working 16 hours a day at 1s. 6d. per hour?

Let $y$ be the work done per hour when the pay is $x$ shillings per hour and the working day is $z$ hours. Then, by hypothesis, $y = ax/\sqrt{z}$. When $x = 1$, $z = 9$, then $y = a/3$; and when $x = 1\frac{1}{2}$, $z = 16$, then $y = a\frac{3}{2}/4 = 3a/8$. If, therefore, $d$ be the number of days required, we have $6 \times 9 \times a/3 = d \times 16 \times 3a/8$, that is, $18a = 6ad$, whence $d = 3$.

## EXERCISES LVII.

1. $y \propto x$, and, when $x = 3$, $y = 4$, find the value of $y$ when $x = 5$.
2. $y \propto 1/x^2$, and, when $x = 8$, $y = 10$, find the value of $y$ when $x = 9$.
3. If $x \propto p + q$, $p \propto y$, $q \propto 1/y$, and if when $y = 1$, $x = 18$, when $y = 2$, $x = 19\frac{1}{2}$, find $x$ when $y = 11$.

**4.** If $u$ is directly proportional to $x$, and inversely proportional to $y^2$, and $u=10$ when $x=2$ and $y=3$, find the value of $u$ when $x=3$ and $y=2$.

**5.** If $y$ is proportional to $z$ and $z$ is inversely proportional to $x^2$, show that $xy$ is inversely proportional to $x$.

**6.** If $x+y \propto x-y$, show that $x^2+y^2 \propto xy$.

**7.** If $a^2-b^2$ is proportional to $ab$, show that $a$ is proportional to $b$ in one or other of two ways.

**8.** Given that the area of a circle is proportional to the square of its radius, find the radius of the circle whose area is the sum of the areas of two circles whose radii are 3·82 and 2·75 respectively. Work to two places of decimals.

**9.** The whole pressure on the horizontal head of a cylindrical drum immersed in a liquid is proportional to the depth of the head under the surface, and to the square of the radius of the head. If the pressure be 1500 lbs. when the depth is 15 feet and the radius of the head 3 feet, find the pressure when the depth is 20 feet and the radius 8 feet.

**10.** The price of a passenger's ticket on a French railway is proportional to the distance he travels; he is allowed 25 kilogrammes of luggage free, but on every kilogramme beyond this amount he is charged a sum proportional to the distance he goes. If a journey of 200 miles with 50 kilogrammes of luggage cost 35 francs, and a journey of 150 miles with 35 kilogrammes cost 24 francs, what will a journey of 100 miles with 100 kilogrammes of luggage cost?

**11.** The distance from its starting-point of a particle which moves from rest with uniformly accelerated velocity is proportional to the square of the time measured from the start. The particle is observed at three successive points A, B, C. The distance between A and B is 1 foot, and between B and C 3 feet; the time between A and B is 1 second, and between B and C also 1 second. How far will the particle have moved from A in $4\frac{1}{2}$ seconds?

**12.** The distance from its starting-point of a particle which starts with a given initial velocity is the sum of two parts—one proportional to the time from starting, the other to its square. If the distances from the starting-point at the end of 3 and 5 seconds are 159 feet and 425 feet respectively, find the distance at the end of 10 seconds.

**13.** It is found that the quantity of work done by a man in an hour varies directly as his pay per hour and inversely as the number of hours he works per day. He can finish a piece of work in 6 days, when working 9 hours a day at 1s. per hour. How many days will he take to finish the same piece of work when working 16 hours a day at 1s. 6d. per hour?

**14.** Assuming the cost of digging a trench to be proportional to the product of the length, the width, and the square of the depth, find the depth of one which costs £44 : 2s. and is 90 ft. long and 10 ft. wide, when the total cost of two trenches—one 60 ft. long, 7 ft. wide, and 5 ft. deep, and the other 80 ft. long, 10 ft. wide, and 6 ft. deep—is £39 : 6s.

**15.** Given that the volume of a right circular cone varies conjointly as its height and the square of the radius of its base, solve the following problem :—The height of a certain cone is equal to the radius of its base. If the height were increased by an inch and the radius of the base were unaltered, 462 cubic inches would be added to the volume; or, if the radius were decreased by an inch and the height were unchanged, 902 cubic inches would be taken from its volume. Find the height and volume of the cone.

**16.** Owing to taxes and other unavoidable outlay, a man's expenditure is proportional to the square of his income. In a certain year he found his nominal income was £500 and his expenditure £300; what is the largest income he could have without incurring a deficit?

**17.** Two hemispherical bowls of gold of the same thickness $t$, whose external radii are $a$ and $b$ respectively, are melted down and formed into a single hemispherical bowl of the same thickness. Find the external radius of the single bowl (the volume of a sphere is proportional to the cube of its radius).

**18.** The distance of the horizon as seen from a balloon 2 miles above the earth's surface is 126·507 miles. Find the radius of the earth and also the distance of the horizon as seen from a balloon 5 miles above the earth's surface.

# CHAPTER XIX

## ARITHMETIC, GEOMETRIC, AND OTHER SERIES

### ARITHMETICAL PROGRESSION

**§ 201.** A succession of quantities (spoken of in the present connection as **terms** *) each of which exceeds (in the algebraic sense) the preceding by the same **common difference** are said to be in **Arithmetic Progression,** *or to form an* **Arithmetic Series.**

*E.g.* $1, 3, 5, 7, \ldots,$ common difference $2$ (1);
$\frac{3}{2}, 2, \frac{5}{2}, 3, \ldots,$ ,, $\frac{1}{2}$ (2);
$7, 4, 1, -2, \ldots,$ ,, $-3$ (3);
are examples of arithmetic progressions.

If the first term of an A. P. be $a$, and the common difference $d$, the successive terms are

$$a, a+d, a+2d, a+3d, \ldots,$$

the $n$th term being $a+(n-1)d$.

For example, the $n$th terms of the series (1), (2), (3) above are $1+(n-1)2=2n-1$, $\frac{3}{2}+(n-1)\frac{1}{2}=\frac{1}{2}(n+2)$, $7+(n-1)(-3)=-3n+10$ respectively.

**§ 202.** Since all the terms of an A. P. are known when the first term and the common difference are given, it is obvious that an A. P. depends essentially on two variables; and is in general determined when any two conditions on its terms are given; for example, it is determined if the values of any two terms of named orders are given.

Ex. Find the A. P. whose second and fifth terms are 3 and 10 respectively.

---

* The beginner will note that the word "term" is here used in a sense different from that defined in § 28.

Let the first term be $a$ and the common difference $d$. Then $a+d=3$, $a+4d=10$. From these we have $3d=7$, $d=7/3$, $a=3-7/3=2/3$. Hence the required A. P. is 2/3, 9/3, 16/3, 23/3, 30/3. . . .

It follows also from what has just been pointed out that

$$\left.\begin{array}{l} \alpha-\beta,\ \alpha,\ \alpha+\beta\ ; \\ \alpha-3\beta,\ \alpha-\beta,\ \alpha+\beta,\ \alpha+3\beta\ ; \\ \alpha-2\beta,\ \alpha-\beta,\ \alpha,\ \alpha+\beta,\ \alpha+2\beta\ ; \\ \qquad\qquad \text{etc.} \end{array}\right\} \qquad (4)$$

are perfectly general expressions for arithmetical progressions of 3, 4, 5, etc., terms respectively; for each of them is an A. P., and each contains two independent variables $\alpha$ and $\beta$.

The expressions (4) are often convenient in establishing theorems regarding quantities in A. P.

Ex. 1. Show that three quantities in arithmetic progression cannot also be in geometric progression unless all three are equal. We may represent any three quantities in A. P. by $\alpha-\beta$, $\alpha$, $\alpha+\beta$. In order that these may also be in G. P. we must have, by § 196, $(\alpha-\beta)(\alpha+\beta)=\alpha^2$—that is, $\alpha^2-\beta^2=\alpha^2$; whence $\beta=0$. Hence the three quantities must be $\alpha$, $\alpha$, $\alpha$, which are all equal.

Ex. 2. If $a$, $b$, $c$ be in A. P., show that $a^3+4b^3+c^3=3b(a^2+c^2)$.

We may put $a=\alpha-\beta$, $b=\alpha$, $c=\alpha+\beta$.

Hence
$$\begin{aligned} a^3+4b^3+c^3 &= (\alpha-\beta)^3+4\alpha^3+(\alpha+\beta)^3, \\ &= 6\alpha^3+6\alpha\beta^2, \\ &= 6\alpha(\alpha^2+\beta^2). \end{aligned}$$

Also $a^2+c^2=(\alpha-\beta)^2+(\alpha+\beta)^2=2(\alpha^2+\beta^2)$. The relation to be proved is therefore

$$6\alpha(\alpha^2+\beta^2)=3\alpha\times 2(\alpha^2+\beta^2),$$

which is an identity.

The beginner will here note the use of the principle of the minimum number of variables (see § 194).

§ 203. **Summation of an A. P.**—If we denote the sum of $n$ terms of an A. P. by $S_n$, we have

$$S_n = a + \{a+d\} + \{a+2d\} + \ldots + \{a+(n-1)d\}\ ;$$

also

$$S_n = \{a+(n-1)d\} + \{a+(n-2)d\} + \{a+(n-3)d\} + \ldots + a.$$

If we now add the two expressions for $S_n$, associating the terms that stand over each other in pairs, we get

$$2S_n = \{2a+(n-1)d\} + \{2a+(n-1)d\} + \{2a+(n-1)d\} + \ldots + \{2a+(n-1)d\}.$$

Since the terms on the right are now all equal, and there are $n$ of them, we have $2S_n = n\{2a+(n-1)d\}$; whence

$$S_n = \tfrac{1}{2}n\{2a+(n-1)d\} \qquad (5).$$

This expression for the sum of $n$ terms of the A. P. we speak of as *the* sum for a reason which we shall explain more fully presently. Meantime, it is obvious that the formula (5) enables us to calculate the sum of a large number of terms more conveniently than we could do by simply adding the terms together.

Ex. 1. Find the sum of 100 terms of the series $1+3+5+\ .\ .\ .$

Here $a=1$, $d=2$. Hence $S_{100}=\frac{1}{2}100(2+99\times 2)=50\times 200=10,000$.

If we denote the last or $n$th term of the A. P. by $l$, we have $l=a+(n-1)d$, and $a+l=2a+(n-1)d$. Hence the formula (5) may be written

$$S_n = n(a+l)/2 \qquad (6),$$

which may be expressed in words thus:—*The sum of* n *terms in A. P. is* n *times half the sum of the first and last.*

Ex. 2. Sum the series

$$1-3+5-7+9-\ .\ .\ .\ -(4n-1).$$

The series (of $2n$ terms) as it stands is not an A. P. We may, however, write it as follows:—

$$\{1+5+\ 9+\ .\ .\ .\ +(4n-3)\}$$
$$-\{3+7+11+\ .\ .\ .\ +(4n-1)\}.$$

Hence it is the difference of two A. P.'s, the common difference in each of which is 4. Using (6) we have

$$1+5+\ 9+\ .\ .\ .\ +(4n-3)=n\{1+(4n-3)\}/2=n(2n-1);$$
$$3+7+11+\ .\ .\ .\ +(4n-1)=n\{3+(4n-1)\}/2=n(2n+1).$$

Hence

$$1-3+5-\ .\ .\ .\ -(4n-1)=n(2n-1)-n(2n+1)=-2n.$$

This result might also have been obtained as follows:—

$$\begin{aligned}1-3+5-7+9-\ .\ .\ .\ -(4n-1)&\\ =(1-3)+(5-7)+(9-11)+\ .\ .\ .\ &+(\overline{4n-3}-\overline{4n-1});\\ =-2-2-2+\ .\ .\ .\ -2;&\\ =-2n.&\end{aligned}$$

Ex. 3. How many terms of the A. P.

$$20+17+14+11+8+\ .\ .\ .$$

must be taken in order that the sum may be 70?

Let the number of terms be $n$. Since the common difference is $-3$, we must have

$$\tfrac{1}{2}n\{40+(n-1)(-3)\}=70.$$

This gives

$$3n^2 - 43n + 140 = 0.$$

Since $3n^2 - 43n + 140 \equiv (n-5)(3n-28)$, the roots of this quadratic are $n=5$ and $n=9\frac{1}{3}$. The first of these values, being positive and integral, is immediately available; and we find, in fact, that the sum of the first five terms, viz. $20+17+14+11+8$, is 70.

The value $n=9\frac{1}{3}$, not being integral, does not furnish a solution. We might, however, infer that 70 will lie between the sum of 9 terms of the series and the sum of 10 terms. We have, in fact—

$$20+17+14+11+8+5+2-1-4 \quad = 72;$$
$$20+17+14+11+8+5+2-1-4-7 = 65.$$

**§ 204.** If $n$ quantities $u_1, u_2, \ldots, u_n$ be found such that

$$a, u_1, u_2, \ldots, u_n, c \qquad (7)$$

form an A.P., $u_1, u_2, \ldots, u_n$ are spoken of as $n$ **Arithmetic Means inserted between** $a$ and $c$.

Since (7) is an A.P. of $n+2$ terms, if $d$ be its common difference, we must have $a+(n+1)d=c$. Therefore $d=(c-a)/(n+1)$. Hence

$$u_1 = a + (c-a)/(n+1), \quad u_2 = a + 2(c-a)/(n+1), \text{ etc.}$$

Therefore

$$u_1 + u_2 + \ldots + u_n = \left(a + \frac{c-a}{n+1}\right) + \left(a + 2\frac{c-a}{n+1}\right) + \ldots$$

to $n$ terms;

$$= \frac{n}{2}\left\{2a + 2\frac{c-a}{n+1} + (n-1)\frac{c-a}{n+1}\right\};$$

$$= \frac{n}{2}\left\{2a + (n+1)\frac{c-a}{n+1}\right\};$$

$$= n\frac{a+c}{2}.$$

Hence the following interesting theorem:—*The sum of* n *arithmetic means inserted between any two quantities is* n *times the single arithmetic mean inserted between the two.*

It should be noted that when we speak of the arithmetic mean of two quantities $a$ and $c$, we mean $\frac{1}{2}(a+c)$, *i.e.* the single arithmetic mean inserted between them. On the other hand, by the arithmetic mean of $n$ quantities $a_1, a_2, \ldots, a_n$ is meant $(a_1+a_2+\ldots+a_n)/n$, or what is ordinarily called their **average**, which has nothing to do with the $n$ arithmetic means inserted between two quantities.

## EXERCISES LVIII.*

**1.** Find the tenth term of the A. P. whose first three terms are 9, 13, 17.

**2.** If the sixth term of an A. P. be 7, and the eleventh term 13, find the first term and the $n$th term.

**3.** Sum to 10 terms the series $5+10+15+20+ \dots$

**4.** Find the sum of thirteen terms of the arithmetic series whose first three terms are $\frac{1}{2}$, $-\frac{2}{3}$, $-\frac{11}{6} \dots$

**5.** Sum the A. P. $1{\cdot}35+1{\cdot}50+1{\cdot}65+ \dots$ to 150 terms.

**6.** Sum to $n$ terms the A. P. $(a-3b)+(5a+5b)+(9a+13b)+ \dots$

**7.** Sum the arithmetic series 81, 79, 77 . . . to 11 terms. How many terms will amount to 160?

**8.** Find the sum of all the numbers between 100 and 2000 which have the remainder 7 when divided by 9.

**9.** Find the first four terms of an A. P. whose ninth term is 4, and whose fifteenth term is $-14$.

**10.** If the first two terms of an A. P. be $u_1$ and $u_2$, find the $n$th term.

**11.** The sum of 10 terms of an A. P. is 100, and the second term is zero, find the first term.

**12.** Find the sum of $n$ terms of an A. P. whose third term is 5, and whose seventh term is 20.

**13.** In any A. P. of $2n$ terms, show that the sum of the odd terms is to the sum of the even terms as the $n$th term is to the $(n+1)$th.

**14.** Show that the sum of $2n+1$ consecutive integers of which the smallest is $n^2+1$ is $n^3+(n+1)^3$.

**15.** Find the sum of all the integers between 1 and 200, excluding those that are multiples of 3 or of 7.

**16.** If $s_1$ be the sum of the first three terms of an A. P., $s_2$ the sum of the next three, and so on, show that $s_1, s_2, s_3, \dots$ form another A. P.

**17.** If the first term of an A. P. be 1, and the sum of the first $n$ terms be $1/n$th of the sum of the second $n$ terms, show that the sum of the first $n$ terms is $-2n^2/(n^2-4n+1)$.

**18.** The sum of 30 terms of an A. P. whose common difference is 5 is 2355; find the first term.

**19.** The first term of an A. P. is 3 and the sum of 20 terms is 155; find the common difference.

**20.** How many terms of the series 5, 8, 11, 14, . . . must be added together so that the sum may be 493?

**21.** The first term and common difference of an A. P. are each equal to $\frac{1}{4}$; the sum of the terms is $2\frac{1}{2}$; find the number of the terms.

**22.** Find the last term in the A. P. 201, 204, 207, . . . when the sum is 8217.

**23.** Insert three arithmetic means 4 and 324.

---

* When quadratic equations occur in the working of Exercises LVIII.-LX., they are to be solved by inspection, by factorisation (see § 62), or by the graphic method, as in Exercise XI. **26**.

**24.** Find the sum of the 10 arithmetic means which can be inserted between 2 and 20.

**25.** Any uneven cube $n^3$ is the sum of $n$ consecutive uneven numbers of which $n^2$ is the middle one.

**26.** Sum to 40 terms $(x+y)+(x-2y)+(x+3y)+(x-4y)+\ldots$

**27.** Show that the sum of an A. P. whose first term is $a$, whose second term is $b$, and whose last term is $c$, is equal to $(a+c)(b+c-2a)/2(b-a)$.

**28.** If $a$, $b$, $c$, $d$ be in A. P., then $\frac{1}{2}(b^2+d^2)+a^2+c^2=2(ab+cd)-bd$.

**29.** Four positive numbers are in A. P. The product of the second and third exceeds the product of the other two by 32; and the product of the second and fourth exceeds the product of the other two by 72. Find the numbers.

**30.** Three quantities are in A. P.; their sum is 21, and the sum of their squares is 165; find them.

**31.** If the sum of $n$ terms of a series be $n^2+n$, show that it is an A. P., and find the first term and the common difference.

**32.** If $a^2$, $b^2$, $c^2$ be in A.P., $1/(b+c)$, $1/(c+a)$, $1/(a+b)$ are also in A. P.

**33.** Find the common difference of an A. P. whose first term is unity, when the sum of $n$ terms is $n^2$.

**34.** Find three positive numbers in A. P. such that their product multiplied by their sum is equal to 5880; and the sum of their squares to 165.

**35.** An exploring party numbering 819 men set out with a stock of provisions sufficient to last to the end of the expedition. After 20 days disease broke out and carried off 2 men every night; and, although at the same time there was an unexpected delay of 10 days, each man received his full daily allowance to the end. How long did the expedition last, and what was the number of the survivors?

## Geometric Progression

**§ 205.** We have already (§ 189) defined a geometric progression to be a succession of quantities each of which bears to the preceding the same **common ratio**; and we have seen that, if $a$ be the first term and $r$ the common ratio, we may represent any geometric progression whatever by

$$a,\ ar,\ ar^2,\ \ldots,\ ar^{n-1},\ \ldots,$$

$ar^{n-1}$ being the $n$th term.

It is obvious that a G. P. like an A. P. depends essentially on two independent variables; and is in general completely determined when two conditions upon its terms are given. Examples of relations connecting quantities in G. P. have already been given in Chapter XVIII.

§ **206. Summation of a G. P.**—If $S_n$ denote the sum of $n$ terms of a G. P. whose first term is $a$ and whose common ratio is $r$, we have

$$S_n = a + ar + ar^2 + \ . \ . \ . \ + ar^{n-1} \qquad (8);$$

therefore

$$rS_n = \qquad ar + ar^2 + \ . \ . \ . \ + ar^{n-1} + ar^n \qquad (9).$$

Hence by subtraction we have

$$S_n - rS_n = a - ar^n.$$

If, therefore, $1 - r \neq 0$, we have

$$S_n = a(1 - r^n)/(1 - r) \qquad (10).$$

The expression $a(1 - r^n)/(1 - r)$ is spoken of as the sum of the G. P.; it is obviously in general more convenient for calculating the sum of $n$ terms than the primary form of the sum, viz. $a + ar + ar^2 + \ . \ . \ . \ + ar^{n-1}$.

Ex. 1. Sum the geometric series $3 + 6 + 12 + \ . \ . \ .$ to 10 terms.
Here $a=3$, $r=2$. Hence $S_{10} = 3(2^{10} - 1)/(2 - 1) = 3.2^{10} - 3 = 3069$.

Ex. 2. Sum the geometric series $1 - \frac{1}{2} + \frac{1}{4} - \frac{1}{8} + \ . \ . \ .$ to 8 terms.
Here $a=1$, $r= -\frac{1}{2}$. Hence $S_{10} = 1\{1 - (-\frac{1}{2})^8\}/\{1 - (-\frac{1}{2})\} = (1 - 1/2^8)/\frac{3}{2} = \frac{85}{128}$.

§ **207.** I. *Any definite quantity, however small, can be made as large as we please by multiplying it by a quantity sufficiently large.* We take this as an axiom; but we may illustrate by supposing a case. Consider the quantity 1/1,000,000 which is very small. If we multiply 1/1,000,000 by 1,000,000, we produce 1; if we multiply it by 1,000,000,000,000, we produce 1,000,000, and so on.

II. *Any definite quantity, however large, may be made as small as we please by dividing it by a quantity sufficiently large.* This may be illustrated by starting with 1,000,000 and dividing by 1,000,000, then by 1,000,000,000,000, and so on.

To these two propositions should be added the two following:—

III. *Any definite quantity, however large, can be made as small as we please by multiplying it by a quantity sufficiently small.*

*E.g.* 1,000,000 × (1/1,000,000,000,000) = 1/1,000,000.

IV. *Any definite quantity, however small, can be made as large as we please by dividing it by a quantity sufficiently small.*

*E.g.* (1/1,000,000) ÷ (1/1,000,000,000,000) = 1,000,000.

From the above principles we can deduce the following important theorem :—

V. *If* r *be a positive quantity exceeding* 1 *by any definite quantity, however small, and* n *a positive integer, then by making* n *sufficiently large we can make* $r^n$ *as large as we please. If* r *be a positive quantity which is less than* 1 *by any definite quantity, however small, and* n *a positive integer, then by making* n *sufficiently large we can make* $r^n$ *as small as we please.*

If $r$ exceed 1 by a definite quantity, we may write $r=1+\rho$, where $\rho$ is a definite positive quantity. Then, by the Binomial Theorem, § 106—

$$r^n=(1+\rho)^n=1+{}_nC_1\rho+{}_nC_2\rho^2+\ \ldots\ +\rho^n.$$

Now ${}_nC_1=n$, and ${}_nC_2$, ${}_nC_3$, etc., are all positive. Also $\rho^2$, $\rho^3$, . . ., $\rho^n$ are all positive. Hence $r^n>1+n\rho$. Now, however small $\rho$ may be, it follows from (I.) that we can make $n\rho$ as large as we please by sufficiently increasing $n$. The same follows therefore regarding $r^n$, which is always greater than $1+n\rho$.

Next, suppose that $r$ is positive and less than 1; then, if $r'=1/r$, $r'$ is positive and greater than 1. Also we have $r=1/r'$. Hence $r^n=(1/r')^n=1/r'^n$. By what has been shown it follows, since $r'$ is positive and greater than 1, that we can make $r'^n$ as large as we please by making $n$ sufficiently large. Therefore, by (II.), we can make $1/r'^n$, that is $r^n$, as small as we please by making $n$ sufficiently large.

§ **208.** By means of the principles established in last paragraph, we can establish some interesting theorems regarding the sum of a G. P. when the number of its terms is made very large. We shall consider the following cases:—First, $r$ positive or negative and $<1$; second, $r$ positive and $>1$, or $r=1$; third, $r=-1$; fourth, $r$ negative and $>1$. To save unnecessary detail, we suppose throughout that $a$ is positive. The effect on the result of $a$ being negative will easily be understood.

*Case* 1. We may write

$$S_n=a/(1-r)-r^na/(1-r).$$

If $r$ be positive then, by (V.) of § 207, by making $n$ sufficiently large we can make $r^n$, and therefore, by (III.) of § 207, $r^na/(1-r)$ as small as we please. Since the sign of $r$ merely affects the sign of $r^na/(1-r)$, it follows that, whether $r$ is positive or negative, provided only that its numerical value be less than 1, we can make $r^n a/(1-r)$ as small as we please by sufficiently increasing $n$. Hence the following important theorem :—

*If the common ratio* r *of a G. P. be numerically less than* 1, *we can, by sufficiently increasing* n, *make the sum of* n *terms differ as little as we choose from* a/(1 − r).

This is commonly expressed by saying that the series **converges** to $a/(1-r)$ ; and $a/(1-r)$ is often spoken of as the **sum to infinity** of the series.

*Case* 2. If $r$ be positive and $>1$, by sufficiently increasing $n$ we can make $r^n$ and therefore $r^n a/(r-1)$ as large as we please (see § 207, V. and I.). Hence, since

$$S_n = r^n a/(r-1) - a/(r-1),$$

*if the common ratio of a G. P. be positive and greater than* 1, *we can, by sufficiently increasing* n, *make the sum of* n *terms exceed any positive quantity, however large.*

The same conclusion obviously follows when $r=1$ ; for then $S_n = na$, which can be made as large as we please by sufficiently increasing $n$.

The result in case 2 is expressed by saying that the series **diverges to** $+\infty$.

*Case* 3. If $r=-1$, it is obvious that the sum of any odd number of terms is $a$ ; and the sum of any even number is 0.

In this case the series is said to **oscillate.**

*Case* 4. *If* r *be negative and greater than* 1, *it is easy to see that the sum of an even number of terms can be made to exceed any positive quantity, however great, by making the number of terms sufficiently large; and that the sum of an odd number of terms is negative and may be made to exceed numerically any quantity, however large.*

This might be expressed by saying that the series **both diverges and oscillates.**

Ex. 1. Sum the series $1-\frac{1}{2}+\frac{1}{4}-\frac{1}{8}+ \ldots$ to infinity.

In this case $a=1$, $r=-\frac{1}{2}$. Hence

$$S_\infty = 1/\{1-(-\tfrac{1}{2})\} = 1/\tfrac{3}{2} = \tfrac{2}{3}.$$

Ex. 2. The first terms of an endless series of G. P.'s, having the same common ratio $f_1<1$, themselves form a G. P. having the common ratio $f_2<1$. Show that the sum of all possible terms of all the progressions is $a/(1-f_1)(1-f_2)$, $a$ being the first term of the first of the G. P.'s.

The first terms of the G. P.'s are $a, af_2, af_2^2, \ldots, af_2^{n-1}, \ldots$

The sum to infinity of the $n$th of them is therefore $af_2^{n-1}/(1-f_1)$. Hence we have to sum the series

$$\frac{a}{1-f_1}+\frac{a}{1-f_1}f_2+\frac{a}{1-f_1}f_2^2+ \ . \ . \ . \ \text{ad } \infty.$$

Now this series is a G. P. whose first term is $a/(1-f_1)$, and whose common ratio is $f_2$. Since $f_2<1$ the series converges to $\{a/(1-f_1)\}/(1-f_2)=a/(1-f_1)(1-f_2)$, which is therefore the sum of all possible terms of all the progressions in the sense that by taking a sufficient number of terms of each of the progressions, and a sufficient number of the progressions, we can make the sum differ from $a/(1-f_1)(1-f_2)$ by as little as we please.

Ex. 3. Show that any repeating decimal, *e.g.* 3·1614141414 . . ., commonly written $3\cdot16\dot{1}\dot{4}$, can be expressed as a vulgar fraction.

$$3\cdot16\dot{1}\dot{4}=3+\frac{16}{100}+\left\{\frac{14}{10^4}+\frac{14}{10^6}+\frac{14}{10^8}+\frac{14}{10^{10}}+ \ . \ . \ .\right\}.$$

Now the part within the bracket is a convergent G. P., whose first term is $14/10^4$, and whose common ratio is $1/10^2$. Hence by taking a sufficient number of the terms within the bracket, we can make the sum of these terms differ as little as we please from $(14/10^4)/(1-1/10^2)$ $=14/(100-1)100$.

Therefore—

$$3\cdot16\dot{1}\dot{4}=3+\frac{16}{100}+\frac{14}{(100-1)100}=3+\frac{1600-16+14}{(100-1)100}$$
$$=3+\frac{1614-16}{9900}=3\frac{1598}{9900},$$

from which the rule commonly given in books on Arithmetic for evaluating a repeating decimal can be readily deduced.

## EXERCISES LIX.

1. Sum $81+108+144+ \ . \ . \ .$ to 5 terms.

2. Sum $5-10+20-40+ \ . \ . \ .$ to $n$ terms.

3. Sum to seven terms the G. P. $2+1\frac{1}{3}+\frac{8}{9}+ \ . \ . \ .$, and show that the sum of a very large number of terms is very nearly equal to 6.

4. Sum to seven terms the G. P. $112-84+63- \ . \ . \ .$ Find also the sum to infinity.

5. Sum the G. P. $3\frac{3}{8}-2\frac{1}{4}+1\frac{1}{2}- \ . \ . \ .$ to infinity.

6. Sum to $n$ terms the G. P. $3\sqrt{2}+6+6\sqrt{2}+ \ . \ . \ .$

7. If the fourth term of a geometric progression be 1, and its seventh term be $\frac{1}{8}$, find the sum of 10 terms.

8. Sum to infinity the G. P. whose two first terms are $1+2\sqrt{3}$ and $1+\sqrt{3}$.

9. Evaluate ·141414 . . .

10. Evaluate ·315151515 . . .

11. Sum the series whose $n$th term is $a^n(b+a)^{n+1}$ to $n$ terms.

12. Sum the series whose $n$th term is $(1+r^n)(1-1/r^n)$ to $n$ terms.

13. Sum to $n$ terms the series whose two first terms are $1+4+ \ . \ . \ .$, first, when the series is arithmetic; second, when the series is geometric.

**14.** Insert three geometric means between 3 and 243.

**15.** Insert four geometric means between 5 and 135.

**16.** Insert five geometric means between 3 and 1875.

**17.** The fourth term of a G. P. is 36, and the seventh $-10\frac{2}{3}$. Find the sum of the series to infinity.

**18.** Find the first term of a G. P. of which the second term is 2, and the sum to infinity $-1\frac{1}{8}$.

**19.** The first term of a G. P. of positive terms is 27, and the third is 48. Find the sum of 6 terms.

**20.** The sum of five terms of a G. P. is 242, and the common ratio is 3 : find the first term.

**21.** Between each consecutive pair of terms of the series $a$, $ar$, $ar^2$, . . ., $ar^{n-1}$ $m$ arithmetic means are inserted : find the sum of the whole series thus obtained.

**22.** If $s_1$ be the sum of the first three terms of a G.P., $s_2$ the sum of the next three terms, and so on, show that $s_1$, $s_2$, $s_3$, . . . form another G. P.

**23.** Find the geometrical progression whose second term is $-21$; and whose sum to infinity is 16.

**24.** Two infinite G. P.'s each beginning with unity have the sums $\delta$ and $\delta'$ respectively. Show that the sum of the series formed by multiplying their corresponding terms is $\delta\delta'/(\delta+\delta'-1)$.

**25.** If the $(p-q)$th and $(p+q)$th terms of a G. P. be P and Q, find the first term and the common ratio.

**26.** If $x : y = (x+z)^2 : (y+z)^2$, and $x \neq y$, show that $z$ is the G. M. of $x$ and $y$.

**27.** If $a$, $b$, $c$ be in G. P. and $x$ and $y$ be the arithmetic means between $a$, $b$ and $b$, $c$ respectively, prove that $a/x + c/y = 2$.

**28.** If $a$, $b$, $c$, $d$ are in G. P., then $(a^2-c^2)(b^2+d^2) = a^3c - bd^3$.

**29.** If $a$, $b$, $c$ be in G. P., $\Sigma a^2 = (a-b+c)\Sigma a$.

**30.** If $a$, $b$, $c$ are in A. P., and $b$, $c$, $d$ in G. P., show that $(2b-a)^2 = bd$, $2bc^2 = (2b-a)^2(a+c)$.

**31.** If $1/(b-a)$, $1/2b$, $1/(b-c)$ are in A. P., prove that $a$, $b$, $c$ are in G. P.

**32.** If P, Q, R be the $p$th, $q$th, and $r$th terms of a G. P., then $P^{q-r}Q^{r-p}R^{p-q} = 1$.

**33.** If $a$, $b$, $c$ be in A. P., and $a$, $b-a$, $c-a$ in G. P., prove that $a = b/3 = c/5$.

**34.** The sum of the first four terms of a G. P. is 20, and the sum of the next two terms is 1. Find the progression.

**35.** If $r$ $(<1)$ be the common ratio of a G. P., $s$ the sum of the series to infinity, and $\sigma$ the sum to infinity of the series formed by taking the squares of the terms of the first series, prove that $(1-r)s^2 = (1+r)\sigma$.

**36.** $A_1B_1C_1$ are the middle points of the sides of the triangle ABC ; $A_2B_2C_2$ the middle points of the sides of $A_1B_1C_1$, and so on : find the sum of the areas of $A_1B_1C_1$, $A_2B_2C_2$, $A_3B_3C_3$, . . . ad$\infty$.

**37.** If the 3rd, the $(p+2)$th, and $3p$th terms of an A. P. be in G. P., prove that the $(p-2)$th term of the A. P. is double the first term.

**38.** If $a_1$, $a_2$, . . ., $a_n$ are in G. P., so also are $1/(a_1^3 - a_2^3)$,

$1/(a_2^3 - a_3^3)$, . . ., $1/(a_{n-1}^3 - a_n^3)$. Find the sum of the latter series in terms of $a_1$, $a_2$, and $n$.

**39.** If $s$ be the sum of $n$ terms of a G. P., $p$ the product of all the terms of the series, and $\sigma$ the sum of the reciprocals of its terms, show that $p^2\sigma^n = s^n$.

**40.** If $s_n$ denote the sum of $n$ terms of the G. P. $a + ar + ar^2 + \ldots$, where $r < 1$, find the sum of the series $s_1 + s_2 + s_3 + \ldots + s_n$.

**41.** If $a$, $b$, $c$ are in G. P., so also are $a^3/c$, $b^2$, $c^3/a$.

**42.** If 1, $x$, $y$ be in A. P., and 1, $y$, $x$ in G. P., find unequal values of $x$ and $y$.

**43.** The sum of four quantities in G. P. is 170, and the third exceeds the first by 30: find the quantities.

**44.** An A. P. and a G. P. have the same first and third terms. If the second term of the A. P. exceeds the second term of the G. P. by 2, and the fourth term of the G. P. exceeds the fourth term of the A. P. by 8; find the two series.

**45.** If

$$\begin{matrix} x, & y & z \\ x', & y', & z' \\ x'', & y'', & z'' \end{matrix}$$

be such that the rows are in A. P. and the columns in G. P., show that the common ratios of the three geometrical progressions are the same.

## Harmonic Progression

§ **209.** *A succession of quantities are said to be in* **Harmonic Progression,** *or to form a* **Harmonic Series,** *when their reciprocals are in arithmetic progression.*

For example, $\frac{1}{1}, \frac{1}{2}, \frac{1}{3}, \frac{1}{4}, \ldots$ is a H. P., since 1, 2, 3, 4, . . . is an A. P.

The most general form for a H. P. may be taken to be

$$1/a,\ 1/(a+d),\ 1/(a+2d),\ \ldots,\ 1/(a+(n-1)d),\ \ldots$$

A harmonic progression depends therefore essentially upon two independent variables; and it follows from § 202 that

$$\left.\begin{matrix} 1/(\alpha-\beta), \quad 1/\alpha, \quad 1/(\alpha+\beta); \\ 1/(\alpha-3\beta), \quad 1/(\alpha-\beta), \quad 1/(\alpha+\beta), \quad 1/(\alpha+3\beta); \\ \text{etc.} \end{matrix}\right\} \quad (11)$$

are perfectly general representations of H. P.'s of 3, 4, etc. terms.

§ **210.** If $a$, $b$, $c$ be any three consecutive terms of a H. P., by definition, $1/a$, $1/b$, $1/c$ are in A. P. Hence

$$1/b - 1/a = 1/c - 1/b.$$

From this we readily deduce

$$(a-b)/(b-c)=a/c \quad . \qquad (12);$$

and also

$$b=2ac/(a+c) \qquad (13).$$

The relation (12) is often given as the definition of a H. P.

In (13) we have an expression for what is called the **Harmonic Mean between** a *and* c.

We may also speak of $n$ harmonic means inserted between $a$ and $c$, meaning thereby $n$ quantities $u_1, u_2, \ldots, u_n$ such that $a, u_1, u_2, \ldots, u_n, c$ form a H. P. The problem to find $u_1, \ldots, u_n$ may be solved by remarking that $1/a, 1/u_1, \ldots, 1/u_n, 1/c$ is an A. P. and proceeding as in § 204.

§ **211.** There is no summation theorem for a H. P. such as is given in §§ 203, 206 for an A. P. and for a G. P.

The following are examples of problems relating to quantities in H. P.:—

Ex. 1. To insert three harmonic means between 2 and 8.

Let the means be $u_1, u_2, u_3$. Then $1/2, 1/u_1, 1/u_2, 1/u_3, 1/8$ are in A. P. If $d$ be the common difference of this A.P., we have $1/2+4d=1/8$, whence $d=-3/32$. Hence the A. P. is 16/32, 13/32, 10/32, 7/32, 4/32. The corresponding H. P. is 32/16, 32/13, 32/10, 32/7, 32/4. The three means required are 32/13, 32/10, 32/7.

Ex. 2. Show that the Arithmetic, Geometric, and Harmonic means between two given unequal positive quantities are in G. P., and in descending order of magnitude.

Let the two quantities be $a$ and $c$, then the three means are $(a+c)/2$, $\sqrt{(ac)}$, $2ac/(a+c)$ respectively.

Now

$$\{\sqrt{(ac)}\}/\{(a+c)/2\}=2\sqrt{(ac)}/(a+c);$$

and

$$\{2ac/(a+c)\}/\{\sqrt{(ac)}\}=2\sqrt{(ac)}/(a+c).$$

Hence the means form a G. P. whose common ratio is $2\sqrt{(ac)}/(a+c)$.

It remains to prove that this common ratio is less than 1; that is, to show that $a+c>2\sqrt{(ac)}$; in other words, that $a+c-2\sqrt{(ac)}>0$. This is tantamount to showing that $(\sqrt{a}-\sqrt{c})^2>0$. Now, since $a$ and $c$ are positive, and $\sqrt{a}$ and $\sqrt{c}$ therefore real quantities, $\sqrt{a}-\sqrt{c}$ is real and must have a positive square; therefore $(\sqrt{a}-\sqrt{c})^2>0$, and our theorem is established.

Ex. 3. If $a, b, c, d, e$ be five consecutive terms of a H. P., prove that $(a+e)(b+c)(c+d)=2c(a+d)(b+e)$.

We may put $a=1/(\alpha-2\beta)$, $b=1/(\alpha-\beta)$, $c=1/\alpha$, $d=1/(\alpha+\beta)$, $e=1/(\alpha+2\beta)$. Hence the given relation expressed in terms of the minimum number of variables is

$$\left(\frac{1}{\alpha-2\beta}+\frac{1}{\alpha+2\beta}\right)\left(\frac{1}{\alpha-\beta}+\frac{1}{\alpha}\right)\left(\frac{1}{\alpha}+\frac{1}{\alpha+\beta}\right)$$
$$=\frac{2}{\alpha}\left(\frac{1}{\alpha-2\beta}+\frac{1}{\alpha+\beta}\right)\left(\frac{1}{\alpha-\beta}+\frac{1}{\alpha+2\beta}\right).$$

The reader will have no difficulty in verifying that each side of this equation reduces to $2(4\alpha^2-\beta^2)/\alpha(\alpha^2-\beta^2)(\alpha^2-4\beta^2)$. Hence the equation is an identity.

## EXERCISES LX.

**1.** Find the $n$th term of the H. P. whose first two terms are 3 and 7.

**2.** Insert a harmonic mean between 7 and 9.

**3.** Insert three harmonic means between 2 and 8.

**4.** Insert four harmonic means between $\frac{1}{2}$ and $\frac{1}{12}$.

**5.** Find the 6th term of the H. P. whose first two terms are 3 and 5.

**6.** If $a$, $b$, $c$ be in G. P., and if $p$, $q$ be the arithmetic means between $a$, $b$ and $b$, $c$ respectively, then $b$ will be the harmonic mean between $p$ and $q$.

**7.** If $a+b$, $b+c$, $c+a$ are in H. P., then $b^2$, $a^2$, $c^2$ are in A. P.

**8.** If $a$, $b$, $c$ are positive quantities in H. P., prove that $a^2+c^2>2b^2$.

**9.** If $a$, $b$, $c$, $d$ are in H. P., then $(a^2-d^2)b^2c^2=3(b^2-c^2)a^2d^2$.

**10.** O is a point outside a circle whose centre is P. OP meets the circle in A and B: the tangents from O meet the circle in Q and Q′, and QQ′ meets OP in R. Show that OP, OQ, OR are the arithmetic, geometric, and harmonic means respectively between OA and OB.

**11.** If $\dfrac{1}{a}+\dfrac{1}{c}=\dfrac{1}{b-a}+\dfrac{1}{b-c}$, $a$, $b$, $c$ are in H.P., provided $b\neq a+c$.

**12.** If $a$, $b$, $c$ be in G. P., and $a+x$, $b+x$, $c+x$ in H. P., then $x=b$.

**13.** If the common ratio of the G. P. formed by the A. M., G. M., and H. M. of $a$ and $b$ $(a>b)$ be $r$, show that $a/b=\{1+\sqrt{(1-r^2)}\}/\{1-\sqrt{(1-r^2)}\}$.

**14.** If three unequal numbers are in H. P., and their squares in A. P., they are in the ratios $1\pm\sqrt{3}:-2:1\mp\sqrt{3}$.

**15.** Find two numbers whose A. M. is greater by 12 than their G. M., and whose G. M. is greater by 7·2 than their H. M.

**16.** If $x$, $y$, $a$ be in A. P., $x$, $y$, $b$ in G. P., and $x$, $y$, $c$ in H. P., show that $4b^2c+4ac^2-3bc^2+a^2b-6abc=0$.

**17.** Find three numbers in G. P. such that if each is increased by 18 they are in H. P., and that if the last is diminished by 24 they are in A. P.

### Summation in General

§ **212.** In order fully to understand the significance of the summation formulæ of §§ 203, 206, it is necessary to attend to some distinctions which we have not yet pointed out. Consider the identity

$$1+2+3+\ \ldots\ +n\equiv\tfrac{1}{2}n(n+1),$$

which is a particular case of (5). The function on the left is distinguished by two peculiarities from its equivalent on the right. In the first place, it is meaningless unless $n$ be an integral number; on this account we might describe it as a **Function of an Integral Variable.** In the second place, *the number of operations in the construction of the function depends on the value of* n ; thus when $n=3$, we have two additions, $1+2+3$; when $n=4$, three additions, $1+2+3+4$; and so on; for this reason we call $1+2+3+\ .\ .\ .\ +n$ an **Unclosed Function of** $n$. It is this second peculiarity on which we wish to lay stress at present. The function $\frac{1}{2}n(n+1)$ or $\frac{1}{2}n^2+\frac{1}{2}n$, on the other hand, is a **Closed Function** of $n$, because the number of steps in its construction, *i.e.* the number of operations required to calculate its value when $n$ is given, does not depend upon $n$.

Certain functions, such as $r^n$, which are not according to their definition closed functions of $n$ in the strict sense above given (*e.g.* $r^2=rr$, $\times$ $r^3=r\times r\times r$, etc.), are included in the category of closed functions, because their values have been tabulated or can be calculated from numerical tables by a number of operations not depending on the variable.

Also, we may for any temporary purpose include in the category of closed functions any functions we choose to name or define as closed functions. A *closed function* in such cases *means a function which is a closed function of* n *and of any functions of* n *which are, or are for the moment regarded as closed.* Thus, for example, $a(r^n-1)/(r-1)$ is a closed function of $n$, if we regard $r^n$ as a closed function of $n$.

§ **213.** We can now explain what is meant by saying that a series is **Summable to** $n$ **Terms,** or admits of **Finite Summation,** in the special sense of the word "**summable.**"

*A series is said to be summable when the sum of* n *terms can be transformed into a closed function of* n.*

It is of course a matter of exception when a series admits of summation in the present sense; series in general are not "summable" in the special sense of the word. The following may be regarded as the fundamental theorem on the subject:—

*The* n*th term of every summable series can be expressed in the form* f(n) − f(n − 1), *where* f(n) *is a closed function of* n ; *and every series whose* n*th term is so expressible is summable.*

* It will thus be seen that "summable" may have different senses according to the functions of $n$ which we take as closed functions.

The proof is simple. Let the terms of the series be denoted by $u_1, u_2, \ldots, u_n$. Then if we have

$$u_1 + u_2 + \ldots + u_n \equiv f(n) \qquad (14),$$

where $f(n)$ is a closed function of $n$, we must also have

$$u_1 + u_2 + \ldots + u_{n-1} \equiv f(n-1) \qquad (15).$$

From (14) and (15) by subtraction we have $u_n = f(n) - f(n-1)$.

Again, suppose we have

$$u_n = f(n) - f(n-1) \qquad (16_n).$$

Then, putting in succession $n-1, n-2, \ldots, 2, 1$, in place of $n$, we have

$$u_{n-1} = f(n-1) - f(n-2) \qquad (16_{n-1});$$
$$u_{n-2} = f(n-2) - f(n-3) \qquad (16_{n-2});$$
$$\cdot \quad \cdot \quad \cdot \quad \cdot$$
$$u_2 = f(2) - f(1) \qquad (16_2);$$
$$u_1 = f(1) - f(0) \qquad (16_1).$$

From $(16_1)$, $(16_2)$, . . ., $(16_n)$, by addition, we deduce

$$u_1 + u_2 + \ldots + u_n = f(n) - f(0).$$

Hence $u_1 + u_2 + \ldots + u_n$ is transformable into a closed function of $n$.

COR. *If* $u_n = f(n) - f(n-1) + v_n$, *where* $f(n)$ *is a closed function of* $n$, *and* $v_n$ *is the* $n$*th term of a summable series, then the series whose* $n$*th term is* $u_n$ *is summable.*

Ex. 1. We have

$$(n+1)^3 - n^3 \equiv 3n^2 + 3n + 1;$$

hence also

$$n^3 - (n-1)^3 \equiv 3(n-1)^2 + 3(n-1) + 1;$$
$$(n-1)^3 - (n-2)^3 \equiv 3(n-2)^2 + 3(n-2) + 1;$$
$$\cdot \quad \cdot \quad \cdot \quad \cdot \quad \cdot$$
$$3^3 - 2^3 \equiv 3 . 2^2 \quad + 3 . 2 \quad + 1;$$
$$2^3 - 1^3 \equiv 3 . 1^2 \quad + 3 . 1 \quad + 1.$$

By addition from these identities we have

$$(n+1)^3 - 1^3 \equiv 3(1^2 + 2^2 + \ldots + n^2) + 3(1 + 2 + \ldots + n) + n.$$

Now $1 + 2 + \ldots + n \equiv \frac{1}{2}n(n+1)$; hence

$$1^2 + 2^2 + \ldots + n^2 \equiv \tfrac{1}{3}\{n^3 + 3n^2 + 2n - \tfrac{3}{2}n^2 - \tfrac{3}{2}n\},$$
$$\equiv \tfrac{1}{6}\{2n^3 + 3n^2 + n\} \equiv \tfrac{1}{6}n(n+1)(2n+1) \qquad (17).$$

In like manner, by means of the identity $(n+1)^4 - n^4 \equiv 4n^3 + 6n^2 + 4n + 1$, and the summations $1+2+\ldots+n=\frac{1}{2}n(n+1)$, $1^2+2^2+\ldots+n^2=\frac{1}{6}n(n+1)(2n+1)$, we can deduce

$$1^3+2^3+3^3+\ldots+n^3\equiv\{\tfrac{1}{2}n(n+1)\}^2 \qquad (18);$$

and so on.

Finally, by means of the summations of $1+2+\ldots+n$, $1^2+2^2+\ldots+n^2$, $1^3+2^3+\ldots+n^3$, . . ., we can sum any series whose $n$th term is an integral function of $n$.*

Ex. 2. Sum the series

$$\frac{1^3+1^2+1}{1.2}+\frac{2^3+2^2+1}{2.3}+\ldots+\frac{n^3+n^2+1}{n(n+1)}.$$

Here
$$\begin{aligned}
u_n&\equiv n+1/n(n+1)=n \quad +1/n \qquad -1/(n+1);\\
u_{n-1}&\equiv n-1+1/(n-1)-1/n;\\
&\ldots\ldots\\
u_2&\equiv 2 \quad +1/2 \quad -1/3;\\
u_1&\equiv 1 \quad +1/1 \quad -1/2.
\end{aligned}$$

Hence, by addition—

$$\begin{aligned}
u_1+u^2+\ldots+u_n&\equiv 1+2+\ldots+n+1/1-1/(n+1),\\
&\equiv\tfrac{1}{2}n(n+1)+1-1/(n+1).
\end{aligned}$$

Ex. 3. Sum the series whose $n$th term is $(n+6)/n(n+2)(n+3)$.

Decomposing into partial fractions we have

$$u_2\equiv 1/n-2/(n+2)+1/(n+3);$$

hence also

$$\begin{aligned}
u_{n-1}&\equiv 1/(n-1)-2/(n+1)+1/(n+2);\\
u_{n-2}&\equiv 1/(n-2)-2/n+1/(n+1);\\
u_{n-3}&\equiv 1/(n-3)-2/(n-1)+1/n;\\
u_{n-4}&\equiv 1/(n-4)-2/(n-2)+1/(n-1);\\
&\ldots\ldots\\
u_5&\equiv 1/5-2/7+1/8;\\
u_4&\equiv 1/4-2/6+1/7;\\
u_3&\equiv 1/3-2/5+1/6;\\
u_2&\equiv 1/2-2/4+1/5;\\
u_1&\equiv 1/1-2/3+1/4.
\end{aligned}$$

If now we add the right and left hand sides of all these identities, and observe that on the right all the fractions which have denominators between $n$ and 4 completely destroy each other, we get

$$\begin{aligned}
u_1+u_2+\ldots+u_n\equiv&\,1/(n+3)-1/(n+2)-1/(n+1)\\
&-1/3+1/2+1/1,
\end{aligned}$$

which is the summation of the given series to $n$ terms.

It may be observed that as $n$ is increased more and more, the sum of this series converges to $-1/3+1/2+1/1=1\frac{1}{6}$.

*N.B.*—The artifice used in this example will effect the summation of any series whose $n$th term is a proper fractional function of $n$, the

---

* See A. XX. §§ 4-8.

degree of whose numerator is less by 2 at least than the degree of its denominator, and whose denominator is of the form $(n+a)(n+b)$ . . . $(n+k)$, where $a$, $b$, . . ., $k$ differ by integers.

## EXERCISES LXI.

1. A sum of money is distributed among a certain number of persons. The first receives 1d. more than the second; the second 2d. more than the third; the third 3d. more than the fourth; and so on. If the first receive £1 and the last 4s. 2d., what sum was distributed and how many persons were there?

2. Find A, B, C so that

$$\{Ax^3+Bx^2+Cx\}-\{A(x-1)^3+B(x-1)^2+C(x-1)\}\equiv ax^2+bx+c.$$

Hence sum the series whose $n$th term is $an^2+bn+c$. Show that this result includes the summation of the series $1+2+\ldots+n$, $1^2+2^2+\ldots+n^2$, and of the arithmetic series.

3. The sum of all the products in a multiplication table going up to $n$ times $n$ is $\frac{1}{4}n^2(n+1)^2$.

4. Sum $1.3+2.5+3.7+4.9+\ldots$ to $n$ terms.

5. Sum the series $1^2+3^2+5^2+7^2+\ldots$ to $n$ terms.

6. Sum the series $a^2+(a+b)^2+(a+2b)^2+\ldots+(a+\overline{n-1}b)^2$.

7. Sum $1.2.3+2.3.4+3.4.5+4.5.6+\ldots$ to $n$ terms.

8. In a pile of timber each horizontal layer contains three beams more than the one above it. If on the top there are 70 beams, and on the ground 376, how many beams and how many layers are there?

9. Sum the series $1^3-2^3+3^3-4^3+\ldots+(2n-1)^3-(2n)^3$.

10. Sum the series whose $n$th term is $(n-b)(n-c)+(n-c)(n-a)+(n-a)(n-b)$.

11. Sum the series whose $n$th term is $(a+bn)(\alpha+\beta n)$.

12. Show that it is impossible to construct a series, all of whose terms are formed according to the same law, such that the sum of $n$ terms is always $1/(n+1)$, but that a series can be constructed such that the sum of $n$ terms is always $C+1/(n+1)$, where C is independent of $n$. If the series be complete determine C.

13. Sum $\frac{1}{1.2}+\frac{1}{2.3}+\frac{1}{3.4}+\ldots$ to $n$ terms.

14. Sum to $n$ terms the series whose $n$th term is $1/(5n-2)(5n+3)$.

15. Sum $\frac{2^2+1}{2^2-1}+\frac{3^2+1}{3^2-1}+\frac{4^2+1}{4^2-1}+\ldots$ to $n$ terms.

16. Sum the series $\frac{1^2.10}{3.4.5}+\frac{2^2.11}{4.5.6}+\ldots+\frac{(n-1)^2(n+8)}{(n+1)(n+2)(n+3)}$.

17. Sum the series whose $n$th term is $(10n^3-33n^2+36n-1)/n^2(n-1)^2(n-2)^2$.

18. Sum to $n$ terms and also to infinity the series

$$\frac{2}{3\sqrt{1}+1\sqrt{3}}+\frac{2}{4\sqrt{2}+2\sqrt{4}}+\ldots+\frac{2}{(n+1)\sqrt{(n-1)}+(n-1)\sqrt{(n+1)}}.$$

19. Sum to $n$ terms

$$\frac{\sqrt{2}-\sqrt{1}}{1+\sqrt{2}+\sqrt{1}+\sqrt{(2^2-2)}}+\ldots+\frac{\sqrt{n}-\sqrt{(n-1)}}{1+\sqrt{n}+\sqrt{(n-1)}+\sqrt{(n^2-n)}}.$$

# CHAPTER XX

## QUADRATIC EQUATIONS

### Solution of a Quadratic Equation

§ **214.** By a quadratic equation is meant an integral equation in which no higher power of the unknown quantity or variable occurs than the second. The most general form of such an equation would be $Ax^2 + Bx + C = A'x^2 + B'x + C'$, where A, B, C, A′, B′, C′ are constants. We could, however, by subtracting $A'x^2 + B'x + C'$ from both sides, reduce this to the form

$$ax^2 + bx + c^* = 0 \qquad (1),$$

where $a$, $b$, $c$ are constants; and this we shall take to be *the* **Standard Form of a Quadratic Equation.**

*It is supposed in what follows that* $a \neq 0$; *but* b *or* c *or both may vanish. We shall also suppose that* a, b, c *are all real.*

§ **215.** It follows at once from § 130 that every quadratic equation has two roots either real and distinct, real and equal, or imaginary.

For, if the discriminant $\Delta \equiv b^2 - 4ac \neq 0$, we have

$$ax^2 + bx + c \equiv a\{x - (-b + \sqrt{\Delta})/2a\}\{x - (-b - \sqrt{\Delta})/2a\}.$$

Hence the equation (1) is equivalent to

$$a\{x - (-b + \sqrt{\Delta})/2a\}\{x - (-b - \sqrt{\Delta})/2a\} = 0 \quad (2),$$

which, since $a \neq 0$, is satisfied by

$$x = (-b + \sqrt{\Delta})/2a, \quad x = (-b - \sqrt{\Delta})/2a \qquad (3).$$

In the particular case where $\Delta = 0$ (3) reduces to

* $ax^2 + bx + c$ we shall speak of as the **Characteristic Function**, or simply the **Characteristic**, of the equation (1).

$$x = -b/2a, \quad x = -b/2a \qquad (4);$$

that is to say, the two roots are each equal to $-b/2a$.

If $\Delta \equiv b^2 - 4ac < 0$, we have

$$ax^2 + bx + c \equiv a\{x - [-b + i\surd(-\Delta)]/2a\} \\ \times \{x - [-b - i\surd(-\Delta)]/2a\};$$

hence in this case the roots are

$$x = [-b + i\surd(-\Delta)]/2a, \quad x = [-b - i\surd(-\Delta)]/2a \qquad (5);$$

that is to say, the roots are imaginary. It should be noticed that *when the roots are imaginary they are conjugate complex numbers;* that is to say, they differ only in the sign of the purely imaginary part.

In practice, with numerical examples it is usually convenient first to calculate $\Delta \equiv b^2 - 4ac$, then extract the square root of $\Delta$ or $-\Delta$ and substitute the values of $b$, $\Delta$, $a$ in (3) or (5). When the coefficients are complicated, it is often convenient, on account of simplifications that occur by the way, to go through the process of completing the square by which the factorisation of $ax^2 + bx + c$ was originally obtained (see Ex. 9 below).

Since the object is to factorise the characteristic, it would of course be absurd to quote the general formula for this factorisation or for the roots of the quadratic when the factorisation is obvious, as in Ex. 1, 2, 3, 4, 5 below.

Ex. 1. $x^2 + 2x = 0$. Since $x^2 + 2x \equiv x(x+2)$, the roots are given by $x = 0$, $x + 2 = 0$, *i.e.* they are 0 and $-2$.

Ex. 2. $x^2 = 0$. Since $x^2 \equiv xx$, the roots are $x = 0$, $x = 0$.

Ex. 3. $2x^2 - 3 = 0$. $2x^2 - 3 \equiv 2(x - \surd(3/2))(x + \surd(3/2))$; hence the roots are $x = \surd(3/2)$, $x = -\surd(3/2)$.

Ex. 4. $2x^2 + 3 = 0$. $2x^2 + 3 \equiv 2(x^2 - i^2 3/2) \equiv 2(x - i\surd(3/2))(x + i\surd(3/2))$; hence the roots are $x = i\surd(3/2)$, $x = -i\surd(3/2)$.

Ex. 5. $x^2 - 999x - 1000 = 0$. By inspection $x^2 - 999x - 1000 \equiv (x - 1000)(x + 1)$. Hence the roots are $x = 1000$, $x = -1$.

Ex. 6. $7x^2 + 6x - 1 = 0$. Here $D = 36 + 4 \times 7 = 64$; $\surd D = 8$; hence, by formula (3), the roots are $(-6 \pm 8)/14$, *i.e.* $-1$ and $1/7$.

Ex. 7. $2x^2 - 8x + 5 = 0$. $D = 64 - 4 \times 2 \times 5 = 24$; $\surd D = 2\surd 6$; the roots are $(8 \pm 2\surd 6)/4$, that is, $(4 \pm \surd 6)/2$.

Ex. 8. $3x^2 + 2x + 7 = 0$. $D = 4 - 4 \times 3 \times 7 = -80$; $\surd - D = 4\surd 5$; hence the roots are $(-2 \pm 4\surd 5i)/6$, *i.e.* $(-1 \pm 2\surd 5i)/3$.

Ex. 9. $(p^2 - q^2)x^2 + 2(p^2 + q^2)x + p^2 - q^2 = 0$. The quadratic is equivalent to

$$x^2+2\frac{p^2+q^2}{p^2-q^2}x=-1.$$

Adding $(p^2+q^2)^2/(p^2-q^2)^2$ to both sides in order to make the left a complete square as regards $x$, we have

$$\left(x+\frac{p^2+q^2}{p^2-q^2}\right)^2=\frac{4p^2q^2}{(p^2-q^2)^2}.$$

This last is equivalent to

$$x+\frac{p^2+q^2}{p^2-q^2}=\pm\frac{2pq}{p^2-q^2};$$

whence

$$\begin{aligned}x&=-(p^2+q^2\mp 2pq)/(p^2-q^2)\\&=-(p-q)/(p+q)\quad\text{or}\quad -(p+q)/(p-q).\end{aligned}$$

This equation might also be solved thus. It is equivalent to

$$p^2(x^2+2x+1)=q^2(x^2-2x+1),$$

*i.e.*

$$\{p(x+1)\}^2=\{q(x-1)\}^2;$$

which gives

$$p(x+1)=\pm q(x-1),$$

from which we get the same two values of $x$ as before.

### EXERCISES LXII.

1. $x^2-3x+2=0$.
2. $x^2+x-2=0$.
3. $x^2-13x+30=0$.
4. $x^2-20x+91=0$.
5. $x^2+x-132=0$.
6. $6x^2+7x=75$.
7. $12x^2+13x-35=0$.
8. $6x^2-19x+15=0$.
9. $10x^2+19x-15=0$.
10. $3x^2-13x+10=0$.
11. $x^2-2x-2=0$.
12. $x^2-6x+2=0$.
13. $100x^2+220x+108=0$.
14. $9x^2+30x+23=0$.
15. $100=100x-6x^2$.
16. $x^2-4x+5=0$.
17. $x^2-6x+34=0$.
18. $4x^2+4x+17=0$.
19. $x^2-14x+52=0$.
20. $4x^2+20x+32=0$.
21. $(x-1)^2+4(x-2)^2=0$.
22. $77(x^2-1)=72x$.
23. $6x^2-535x+1881=0$.
24. $3x^2-236x+333=0$.
25. $(x-1)(x-2)+(x-2)(x-3)=2$.
26. $(x-1)(2x-11)+(x+2)(2x-8)=(x+1)(x+2)$.
27. $ax^2-(a-b)x-b=0$.
28. $a(x^2-1)=(a^2-1)x$.
29. $x^2+a(x+b)=b^2$.
30. $x^2-2(p+q)x+2pq=0$.
31. $(x+a)^2+2(x+a)(2x+c)+(2x+c)^2=b^2$.
32. $(a+b)^2x^2+(a+b)(a^2+ab+b^2)x+ab(a^2+b^2)=0$.
33. $pqx^2+(p^2+q^2)x+pq=0$.
34. $a^2x^2+(a+c)^2x+2b^2=2c^2+(a-c)^2x-a^2x^2$.
35. $(l^2x+m^2)(p^2x+q^2)=(lqx+m^2)(p^2x+lq)$.

§ **216.** We have shown that *every quadratic equation has two roots.* We now complete the theory of the solution of such equations by showing that *a quadratic equation cannot have more than two roots.*

This follows from the Remainder Theorem (§ 114), for to every value of $x$ which makes $ax^2+bx+c\equiv 0$, *i.e.* to every root, $\alpha$, of the equation (1), corresponds a factor, $x-\alpha$, of the characteristic function $ax^2+bx+c$. Now this function, being of the second degree, cannot have more than two linear factors. Hence the equation (1) cannot have more than two roots.

§ **217.** The theorem just proved can be extended at once to integral equations of any degree, viz. *an integral equation of the* n*th degree cannot have more than* n *roots;* the proof by means of the remainder theorem is exactly the same.

The proof that *an integral equation of the* n*th degree always actually has* n *roots* cannot be furnished by a method like that employed in § 215 in the case of a quadratic, because it has been shown * that, if $n>4$, the roots are not, save in exceptional cases, expressible in terms of the coefficients of the equation by means of a finite number of the operations $+$, $-$, $\times$, $\div$, $\sqrt[m]{}$.† The proof of this fundamental theorem depends on the simpler theorem that *every integral equation has at least one root real or imaginary;* but the demonstration is beyond the scope of an elementary work. Granted that every integral equation has at least one real root, it follows very easily by the remainder theorem that *every integral equation of the* n*th degree has* n *roots real or imaginary, but not necessarily all different.*

## Relations between Roots and Coefficients

§ **218.** *If* $\alpha$ *and* $\beta$ *be the roots of the quadratic* $ax^2+bx+c=0$, *then*

$$\alpha+\beta=-b/a,\quad \alpha\beta=c/a \qquad (6).\ddagger$$

For, since $a\alpha^2+b\alpha+c\equiv 0$ and $a\beta^2+b\beta+c\equiv 0$, we must have, by the remainder theorem—

$$ax^2+bx+c\equiv a(x-\alpha)(x-\beta) \qquad (7),$$

* Originally by the Norwegian mathematician, Abel.

† This is usually expressed by saying that integral equations of degree higher than the fourth do not in general admit of algebraic (or formal) solution.

‡ This may be proved directly by substituting for $\alpha$ and $\beta$ the values (3); but this method of proof is not applicable to equations of any degree.

which is an identity between two integral functions. Now (7) may be written

$$ax^2 + bx + c \equiv a\{x^2 - (\alpha + \beta)x + \alpha\beta\}.$$

Hence, comparing coefficients, we have

$$b = -a(\alpha + \beta), \quad c = a\alpha\beta.$$

Since $a \neq 0$, the equations (6) immediately follow.

§ **219.** By exactly the same methods we can show that, *if* $\alpha_1, \alpha_2, \ldots, \alpha_n$ *be the roots of the equation*

$$a_0x^n + a_1x^{n-1} + a_2x^{n-2} + \ldots + a_{n-1}x + a_n = 0,$$

then

$$\Sigma\alpha_1 = -a_1/a_0, \quad \Sigma\alpha_1\alpha_2 = a_2/a_0, \quad \Sigma\alpha_1\alpha_2\alpha_3 = -a_3/a_0, \ldots$$
$$\Sigma\alpha_1\alpha_2 \ldots \alpha_{n-1} = (-1)^{n-1}a_{n-1}/a_0, \ \alpha_1\alpha_2 \ldots \alpha_n = (-1)^n a_n/a_0.$$

§ **220.** By means of the relations (6) *any* **Symmetric Integral Function of the Roots** *of the quadratic* $ax^2 + bx + c = 0$ *can be expressed as an integral function of* b/a *and* c/a ; *and any alternating* integral function of the roots as an integral function of* b/a *and* c/a *multiplied by* $\sqrt{(b^2 - 4ac)}/a$.†

Ex. 1. If $\alpha$ and $\beta$ be the roots of $ax^2+bx+c=0$, express $\alpha^3+\alpha^2\beta+\alpha\beta^2+\beta^3$ in terms of $a$, $b$, $c$.

$$\alpha^3+\alpha^2\beta+\alpha\beta^2+\beta^3 = (\alpha+\beta)^3 - 2\alpha\beta(\alpha+\beta),$$
$$= \left(-\frac{b}{a}\right)^3 - 2\frac{c}{a}\left(-\frac{b}{a}\right),$$
$$= (-b^3+2abc)/a^3.$$

Ex. 2. Express $\alpha^5 - \beta^5$ in terms of $a$, $b$, $c$.

$$\alpha^5-\beta^5 = (\alpha-\beta)(\alpha^4+\alpha^3\beta+\alpha^2\beta^2+\alpha\beta^3+\beta^4),$$
$$= \sqrt{\{(\alpha+\beta)^2-4\alpha\beta\}}\{(\alpha+\beta)^4-3\alpha\beta(\alpha^2+\beta^2)-5\alpha^2\beta^2\},$$
$$= \sqrt{\left(\frac{b^2}{a^2}-4\frac{c}{a}\right)} \times \left\{\frac{b^4}{a^4}-3\frac{c}{a}\left(\frac{b^2}{a^2}-2\frac{c}{a}\right)-5\frac{c^2}{a^2}\right\},$$
$$= (b^4+a^2c^2-3ab^2c)\sqrt{(b^2-4ac)}/a^5.$$

Ex. 3. If $\alpha$ and $\beta$ be the roots of $x^2-px+q=0$, find (in terms of $h$, $p$, $q$) the equation whose roots are $(\alpha-h)/(\alpha+h)$, $(\beta-h)/(\beta+h)$.

---

* A function of $\alpha$ and $\beta$ is said to be an **Alternating Function**, when its value is merely changed in sign by interchanging $\alpha$ and $\beta$, *e.g.* $\alpha^2\beta-\alpha\beta^2$, $\alpha^3-\beta^3$ are alternating functions of $\alpha$ and $\beta$. Every such function is the product of $\alpha-\beta$, and a symmetric function of $\alpha$ and $\beta$.

† It should be noticed that

$$D = b^2 - 4ac = a^2(\alpha-\beta)^2.$$

For proofs of the theorems stated in § 220, see A. XVIII. 1-4.

Let the required equation, evidently a quadratic, be $x^2 - p'x + q' = 0$; then

$$p' = \frac{\alpha - h}{\alpha + h} + \frac{\beta - h}{\beta + h} = \frac{2(\alpha\beta - h^2)}{\alpha\beta + (\alpha + \beta)h + h^2},$$
$$= 2(q - h^2)/(q + ph + h^2).$$
$$q' = \frac{(\alpha - h)(\beta - h)}{(\alpha + h)(\beta + h)} = \frac{\alpha\beta - (\alpha + \beta)h + h^2}{\alpha\beta + (\alpha + \beta)h + h^2},$$
$$= (q - ph + h^2)/(q + ph + h^2).$$

Hence the required equation, always supposing that $q + ph + h^2 \neq 0$, is

$$(q + ph + h^2)x^2 - 2(q - h^2)x + (q - ph + h^2) = 0.$$

## Discrimination of the Roots

§ **221.** By means of the discriminant and the rational relations (6) between the roots and the coefficients, we can express conditions upon the roots of a quadratic, and obtain a variety of useful information regarding the roots without actually solving the quadratic. The methods by which this is done are very important in the applications of Algebra; and we shall therefore give a few specimens.

§ **222.** We have already seen that the roots of $ax^2 + bx + c = 0$ are real and distinct, real and equal, or imaginary, according as $b^2 - 4ac$ is positive, zero, or negative. If $b^2 - 4ac$ is positive, so that the roots $\alpha$ and $\beta$ are real, it is obvious that $\alpha$ and $\beta$ will have the same sign if $\alpha\beta$ be positive, opposite signs if $\alpha\beta$ be negative. Now, by § 218, $\alpha\beta = c/a$. Hence *the roots of* $ax^2 + bx + c = 0$, *if real, will have the same or opposite signs according as* $c/a$ *is positive or negative.*

§ **223.** Again, if $\alpha\beta$ be positive, and $\alpha + \beta$ also positive, $\alpha$ and $\beta$ will evidently both be positive; and, if $\alpha\beta$ be positive, and $\alpha + \beta$ negative, $\alpha$ and $\beta$ will both be negative. Hence, since $\alpha + \beta = -b/a$, $\alpha\beta = c/a$, we see that *the roots of* $ax^2 + bx + c = 0$, *if real, will both be positive if* $b/a$ *be negative and* $c/a$ *positive; and both negative if* $b/a$ *be positive and* $c/a$ *positive.* These conditions are obviously necessary as well as sufficient.

§ **224.** *The necessary and sufficient condition that the roots of* $ax^2 + bx + c = 0$ *be numerically equal and of opposite sign is* $b = 0$. For $\alpha + \beta = -b/a$, and, since $a$ is finite both ways, $b = 0$ is tantamount to $\alpha + \beta = 0$, *i.e.* $\alpha = -\beta$.

§ **225.** *The necessary and sufficient condition that one root of*

$ax^2 + bx + c = 0$ *be zero is* $c = 0$; *the necessary and sufficient conditions that both roots of* $ax^2 + bx + c = 0$ *be zero are* $b = 0$, $c = 0$.

For, since $a$ is finite both ways, $b = 0$ and $c = 0$ are tantamount to $\alpha + \beta = 0$, $\alpha\beta = 0$. Now, if $\alpha\beta = 0$, either $\alpha = 0$ or $\beta = 0$; and the converse is clearly true. Again, if $\alpha\beta = 0$ and $\alpha + \beta = 0$, then either $\alpha = 0$ and $\alpha + \beta = 0$, or $\beta = 0$ and $\alpha + \beta = 0$. But $\alpha = 0$ and $\alpha + \beta = 0$ are equivalent to $\alpha = 0$ and $\beta = 0$; and $\beta = 0$ and $\alpha + \beta = 0$ are equivalent to $\beta = 0$ and $\alpha = 0$. Hence if $c = 0$, one root is zero; if both $b = 0$ and $c = a$, both roots are zero.

Ex. 1. $5x^2 - 7x - 1 = 0$. $\Delta = +69$, $\alpha + \beta = 7/5$, $\alpha\beta = -1/5$. The roots are real, one positive the other negative, the former being numerically greater.

Ex. 2. $5x^2 + 7x - 1 = 0$. $\Delta = +69$, $\alpha + \beta = -7/5$, $\alpha\beta = -1/5$. Result as before, only the negative root is numerically greater.

Ex. 3. $5x^2 - 7x + 1 = 0$. $\Delta = +29$, $\alpha + \beta = 7/5$, $\alpha\beta = 1/5$. The roots are real and both positive.

Ex. 4. $5x^2 + 7x + 1 = 0$. $\Delta = +29$, $\alpha + \beta = -7/5$, $\alpha\beta = 1/5$. The roots are real and both negative.

Ex. 5. Examine whether there is any restriction on the values which $2x^2 - x + 1$ can assume when all possible real values are given to $x$. Let $y = 2x^2 - x + 1$; then the value of $x$ corresponding to any given value of $y$ is given by the quadratic

$$2x^2 - x + (1 - y) = 0 \qquad (8);$$

and the question is whether the value or values of $x$ given by this quadratic are real or not. If $\Delta$, as usual, denote the discriminant of (8), we have

$$\Delta = 1 - 8(1 - y) = 8y - 7 = 8(y - 7/8) \qquad (9);$$

from which we see that the roots of (8) will be real if, and only if, $y =$ or $> 7/8$. It appears, therefore, that for real values of $x$ the quadratic function $2x^2 - x + 1$ cannot have any value less than $+7/8$, but may have any value greater than $7/8$. In other words, $7/8$ is a minimum value of the function.

Ex. 6. Find the condition that the equations $x^2 + px + q = 0$, $x^2 + p'x + q' = 0$ may have one root in common.

Let the roots of the quadratics be $\alpha$, $\beta$ and $\alpha'$, $\beta'$. Then the necessary and sufficient condition that one of the two $\alpha$, $\beta$ be equal to one of the two $\alpha'$, $\beta'$ is obviously $(\alpha - \alpha')(\alpha - \beta')(\beta - \alpha')(\beta - \beta') = 0$. Now the characteristic function of this equation is a symmetric integral function of $\alpha$ and $\beta$, and also of $\alpha'$ and $\beta'$; it can therefore be expressed as an integral function of $p$, $q$, $p'$, $q'$.

We have, in fact—

$$\begin{aligned}(\alpha - \alpha')(\alpha - \beta')(\beta - \alpha')(\beta - \beta') &= \{\alpha^2 - \alpha(\alpha' + \beta') + \alpha'\beta'\}\{\beta^2 - \beta(\alpha' + \beta') + \alpha'\beta'\}, \\ &= \{\alpha^2 + p'\alpha + q'\}\{\beta^2 + p'\beta + q'\},\end{aligned}$$

[since $\alpha'+\beta'=-p', \quad \alpha'\beta'=q'$, by § 218,]

$$=p'^2\alpha\beta+p'q'(\alpha+\beta)+q'^2+p'\alpha\beta(\alpha+\beta)+q'(\alpha^2+\beta^2)+\alpha^2\beta^2,$$
$$=p'^2q-p'q'p+q'^2-p'qp+q'(p^2-2q)+q^2,$$

[since $\alpha+\beta=-p, \quad \alpha\beta=q$]

$$=(q-q')^2+(p-p')(pq'-p'q).$$

Hence the required condition is

$$(q-q')^2+(p-p')(pq'-p'q)=0.$$

This problem may also be solved as follows:—

Since $x^2+px+q=0$ and $x^2+p'x+q'=0$ have a root in common

$$(x^2+px+q)-(x^2+p'x+q')=0,$$

and

$$q(x^2+p'x+q')-q'(x^2+px+q)=0$$

must have a root in common and conversely. Hence we have to find the condition that

$$(p-p')x+(q-q')=0 \quad \text{and} \quad x\{(q-q')x+(p'q-pq')\}=0$$

have a root in common, *i.e.* since in general $x=0$ is not the common root, that

$$(p-p')x+(q-q')=0, \quad (q-q')x+(p'q-pq')=0$$

have the same solution. This leads (see § 64) at once to the above result.*

## EXERCISES LXIII.

**1.** Find the sum and the product of the roots of the quadratic $3x^2+6x-11=0$.

**2.** Show that the positive root of $x^2-8x-8=0$ is greater than 8.

**3.** If the difference of the roots of $x^2-ax+10=0$ be 3, find $a$.

**4.** Find the sum of the squares of the roots of $2x^2-3x+5=0$; what inference can you draw from your result?

**5.** If $\alpha$ and $\beta$ be the roots of $x^2-6x+13=0$, calculate the value of $\alpha^2\beta^3+\alpha^3\beta^2$, and also of $\alpha^2\beta^4+\alpha^4\beta^2$.

**6.** If $\alpha$ and $\beta$ be the roots of $x^2-px+q=0$, calculate the value of $\alpha^3+\beta^3$ and $\alpha^6+\beta^6$ in terms of $p$ and $q$.

**7.** Find a quadratic equation whose roots are $(b+c)+i(b-c)$ and $(b+c)-i(b-c)$.

Find, without solving them, whether the roots of the following equations are real and distinct, equal or imaginary. When the roots are real, find whether they are both positive, both negative, or of opposite sign. In the last case find whether the positive or negative root is numerically greater:—

---

* If the two quadratics have a root in common, it follows from the remainder theorem that their characteristic functions $x^2+px+q$ and $x^2+p'x+q'$ must have a linear factor in common. If we find the condition for this (see § 140, Ex. 3), we shall arrive at the above result by a third method.

8. $5x^2-6x+4=0$.
9. $5x^2-6x-4=0$.
10. $2x^2-9x+2=0$.
11. $3x^2+7x+3=0$.
12. $4x^2+10x-3=0$.
13. $5x^2-13x-4=0$.
14. $6x^2-7x-3=0$.
15. $4x^2+4x-2=0$.
16. $4x^2-4x+1=0$.
17. $x^2+6x+13=0$.
18. $8x^2+10x-3=0$.
19. $3+10x-8x^2=0$.

20. Discuss the roots of $ax^2+(a+\beta)x+\beta=0$.

21. What is the nature of the roots of $(\lambda+4)x^2+(2\lambda+3)x+(\lambda-1)=0$?

22. For what values of $\lambda$ has the equation $(x^2-2x+1)+\lambda(x^2+3x+5)=0$ equal roots?

23. Determine $k$ so that $x^2-(2k-3)x+2k=0$ may have equal roots.

24. The equation $2x^2+2(p+q)x+p^2+q^2=0$ cannot have real roots unless $p=q$.

25. Find the greatest value of $\lambda$ for which the factors of $(\lambda+1)x^2+\lambda x+(\lambda-1)$ are real.

26. Determine $\lambda$ so that the roots of $2(\lambda x-1)(x-1)-\lambda=0$ may be equal.

27. Show that the roots of $\Sigma(x-a)^2+\Sigma(x-b)(x-c)=0$ are imaginary, provided $\Sigma bc=0$.

28. Show that the roots of $\Sigma(x-a)(x-b-c)=0$ are real, provided $3\Sigma a^2>2\Sigma bc$.

29. Show that for a certain value of $\lambda$ the equation $a/(x+a-\lambda)+b/(x+b-\lambda)=1$ has two equal roots with opposite signs; and find the double roots.

30. If the roots of $x^2-px+q$ be real and differ by less than $f$, then $q$ must be between $\frac{1}{4}p^2$ and $\frac{1}{4}(p^2-f^2)$.

31. Find the square of the difference of the roots of $a/(x-a)+b/(x-b)+c/(x-c)=0$.

32. Show that the roots of $\{(p+q)^2+r^2\}x^2+2(p^2-q^2-r^2)x+\{(p-q)^2+r^2\}=0$ are imaginary, $p$, $q$, $r$ being real, and $r\neq 0$.

33. Solve $(3\lambda-1)x^2-(2\lambda+1)x+\lambda=0$, and discuss the roots when $\lambda$ varies from $-\infty$ to $+\infty$.

34. If $\alpha$ and $\beta$ be the roots of $ax^2+bx+c=0$, show that the roots of $x^2+(a/b+b/c)x+a/c=0$ are $1/\alpha+1/\beta$ and $1/(\alpha+\beta)$.

35. If $\alpha$ and $\beta$ be the roots of $ax^2+bx+c=0$, form the equation whose roots are $\alpha+1/\beta$, $\beta+1/\alpha$.

36. If $\alpha$, $\beta$ be the roots of $ax^2+bx+c=0$, show that the equation whose roots are $1/(\alpha-4\beta)$, $1/(\beta-4\alpha)$ is $(25ac-4b^2)x^2-3abx+a^2=0$.

37. Find the condition that $ax^2+bx+c=0$, and $a'x^2+b'x+c'=0$ have two roots in common.

38. Find the condition that the roots of $a'x^2+b'x+c'=0$ be the roots of $ax^2+bx+c=0$ with the signs changed.

39. Find the value of $c$ in terms of $a$ and $b$ in order that the sum of the roots of the equation $x^2+ax+b=0$ may be equal to the difference of the roots of the equation $x^2+cx+(a+c)b=0$.

40. Find the condition that one of the roots of $ax^2+bx+c=0$ be equal to one of the roots of $a'x^2+b'x+c'=0$ with its sign changed.

41. Find the condition that one root of $ax^2+bx+c=0$ be the reciprocal of one of the roots of $a'x^2+b'x+c'=0$.

**42.** Show that $(3x^2+6x+1)/(4x^2+2x-1)$ can be made equal to any real quantity whatever by giving a suitable real value to $x$.

**43.** If the roots of $x^4+ax^3+bx^2+cx+d=0$ are equal in pairs, then $c^2=a^2d$, $(4b-a^2)^2=64d$.

**44.** The equation $x^4+6x^3+\frac{23}{2}x^2+\frac{15}{2}x+\frac{25}{16}=0$ has its roots equal in pairs; find them.

**45.** Find the equation to a straight line which passes through the point $(0, -3)$ on the $y$-axis and touches the graph of $y=3x^2$.

## Solution of Equations of Higher Degree than the Second by means of Quadratic or Linear Equations

**§ 226.** Since the equation $PQ=0$, where P and Q are integral functions of $x$, is equivalent to the two equations $P=0$, $Q=0$, it follows that, if we can resolve the characteristic of any integral equation into two factors, we can find its roots by means of two equations, each of which is of lower degree than the original equation.

Ex. 1. $x^4-(x+3)^2=0$.

This equation may be written

$$\{x^2+(x+3)\}\{x^2-(x+3)\}=0;$$

and is therefore equivalent to the two

$$x^2+x+3=0, \quad x^2-x-3=0;$$

the roots of which are $x=(-1\pm i\sqrt{11})/2$ and $x=(1\pm\sqrt{13})/2$. We have thus found the four roots of the given biquadratic.

Ex. 2. Find the three cube roots of $+1$.

Let $x$ be a cube root of $+1$, then, by the definition of the cube root, $x^3=+1$. Hence the values of $x$ are the roots of the equation $x^3-1=0$. Now, since $x^3-1\equiv(x-1)(x^2+x+1)$, the equation $x^3-1=0$ is equivalent to $x-1=0$, $x^2+x+1=0$. The former of these two gives $x=1$, the principal value of the root; the latter gives $x=(-1\pm i\sqrt{3})/2$, which are two imaginary cube roots of $+1$. The beginner should verify that in fact $\{(-1\pm i\sqrt{3})/2\}^3=1$.

When one root of an equation, say $x=a$, is known, the remainder theorem furnishes us with a corresponding factor of its characteristic, viz. $x-a$; and the rest of the roots can then be found by means of an equation of lower degree.

Ex. 3. $193x^2-108x-85=0$.

It is obvious at a glance that $x=1$ is one root; the other might be found by factorising the characteristic, but more simply by observing that since the product of the roots is $-85/193$ (by § 218), and one of them is 1, the other is $-85/193$.

Ex. 4. $x^3-3x^2+5x-3=0$.

Obviously one root is $x=1$, hence $x-1$ is a factor of the characteristic $x^3-3x^2+5x-3$. Calculating for the other factor as in § 113, we find that the equation may be written $(x-1)(x^2-2x+3)=0$. The remaining two roots are therefore the roots of $x^2-2x+3=0$, *i.e.* of $x^2-2x+1=-2$, which are $x=1\pm\sqrt{2}i$.

The introduction of an auxiliary variable, either implicitly or explicitly, often simplifies the reduction of an equation.

Ex. 5. $x^4-x^2-2=0$.

Regarding $x^2$ as variable instead of $x$, the characteristic is a quadratic function of $x^2$, viz. $(x^2)^2-(x^2)-2$. Factorising, we get $(x^2+1)(x^2-2)=0$. Hence our equation is equivalent to $x^2+1=0$, $x^2-2=0$, the roots of which are $\pm i$, $\pm\sqrt{2}$.

Ex. 6. $(x^2-5x)^2+4x^2-20x+3=0$.

If we put for a moment $y=x^2-5x$, the equation becomes

$$y^2+4y+3=0,$$

which may be written

$$(y+1)(y+3)=0.$$

This last is equivalent to $y+1=0$, $y+3=0$. Replacing now $y$ by $x^2-5x$, we see that the original equation is equivalent to

$$x^2-5x+1=0, \quad \text{and} \quad x^2-5x+3=0,$$

the roots of which are $(5\pm\sqrt{21})/2$, $(5\pm\sqrt{13})/2$.

Ex. 7. $2x^4+3x^3-9x^2-3x+2=0$.

This biquadratic equation has the peculiarity that the first and last coefficients are equal, and the second and last but one equal with opposite signs; it may be reduced as follows. The equation may be written

$$2(x^4+1)+3x(x^2-1)-9x^2=0,$$

which suggests the form

$$2(x^2-1)^2+3x(x^2-1)-5x^2=0.$$

Now, since $2u^2+3vu-5v^2\equiv(2u+5v)(u-v)$, we have (taking $u=x^2-1$, $v=x$)—

$$2(x^2-1)^2+3x(x^2-1)-5x^2\equiv\{2(x^2-1)+5x\}\{(x^2-1)-x\}.$$

Hence the given equation may be written

$$\{2(x^2-1)+5x\}\{(x^2-1)-x\}=0,$$

and is therefore equivalent to the two

$$2(x^2-1)+5x=0, \quad x^2-1-x=0,$$

whose roots are $(-5\pm\sqrt{41})/4$, $(1\pm\sqrt{5})/2$.

The method of Example 7 will reduce any biquadratic of the form

$$ax^4+bx^3+cx^2\pm bx+a=0,$$

in which the first and last coefficients are equal, and the second and last but one either equal or else equal and of opposite sign. This kind of biquadratic is called a **Reciprocal Biquadratic**; and is of very frequent occurrence.

### EXERCISES LXIV.

**1.** $9(x+3)^2-25(2x-1)^2=0$. **2.** $2(x-1)^2-3(3x+1)^2=0$.
**3.** $2(x-1)^2+3(x+1)^2=0$. **4.** $1821x^2-1872x+51=0$.
**5.** $x^3-1+3(x-1)=0$. **6.** $(x^2-x)^2-3(2x^2-x)^2=0$.
**7.** $4x^2+9=0$. **8.** $(2x+3)^4-16=0$.
**9.** $x^3+8=x+2$.
**10.** $\{(x-3)^2+x^2\}^2-\{(x-3)^2-x^2\}^2=25$.
**11.** $x^4+x^3-x-1=0$. **12.** $x^3-x^2-2x=0$.
**13.** $(2x+1)^3+(x-2)^3=0$.
**14.** $(x^2-4x+4)^2-(x^2-2x+1)^2=0$. **15.** $(x^3+2x^2)^2-x^2=0$.
**16.** $(x^2+x+1)^2-3(x+1)^2=0$. **17.** $x^3-2x+1=0$.
**18.** $x^3-7x^2+11x-2=0$. **19.** $2x^3+3x^2+3x+2=0$.
**20.** $x^3+3x^2+3x+1=0$. **21.** $9x^4-4x^2-4x-1=0$.
**22.** $x^4-50x^2+49=0$. **23.** $6x^4+5x^2-4=0$.
**24.** $x^4-2x^2-21/4=0$. **25.** $x^8+x^4+1=0$.
**26.** $6x^4+5x^3-38x^2+5x+6=0$. **27.** $x^4+5x^3+8x^2+5x+1=0$.
**28.** $2x^4+7x^3-x^2-7x+2=0$.
**29.** Find all the fifth roots of $+1$.
**30.** $6(x^2+4x)^2-7(x^2+4x)-3=0$. **31.** $x+\sqrt{x}=90$.
**32.** $(x^2+3x)^2+4(x^2+3x+1)=0$.
**33.** $(x-3)(x-4)(x-5)(x-6)=24$.
**34.** $(x^2+2x+1)^2+3x^2+6x=105$. **35.** $(x+1)(x+2)(x+3)=6$.
**36.** $(x^2+2x)^2+(3x^2+6x+2)=0$.

### SOLUTION OF RATIONAL FRACTIONAL EQUATIONS BY MEANS OF QUADRATIC OR LINEAR EQUATIONS

**§ 227.** From every rational fractional equation, such as

$$1/(x-1)+1/(x-2)+1=0 \qquad (10),$$

we can derive an integral equation by multiplying both sides by the L.C.M. of all the denominators which occur in the equation. Thus, from (10), we derive, by multiplying by $(x-1)(x-2)$—

$$(x-2)+(x-1)+(x-1)(x-2)=0,$$

that is—

$$x^2-x-1=0 \qquad (11),$$

the roots of which are $(1\pm\sqrt{5})/2$.

It might be thought that we should thus introduce extraneous solutions; but in general this is not the case. The reason of this is that the conditions of § 62 are not in general fulfilled, because the values of $x$ which nullify the integralising factor in general make the characteristic of the fractional equation infinite. For example, $x=1$ makes $1/(x-1)+1/(x-2)+1$, the characteristic of (10), infinite; and it does not, therefore, *necessarily* follow that $x=1$ causes $(x-1)\{1/(x-1)+1/(x-2)+1\}$ to vanish. In point of fact it is obvious in this case that neither $x=1$ nor $x=2$ nullifies $(x-2)+(x-1)+(x-1)(x-2)$. Hence (11) has no roots extraneous to (10); and the roots of (10) are $(1 \pm \sqrt{5})/2$.

It may, however, happen in exceptional cases that extraneous roots are introduced. Example 1 below is a case in point.

Ex. 1. $1/(x-1)+(x^2-x-3)/(x-2)(x-3)-3/(x-3)=0$ (12).

If we treat the equation as it stands and integralise by multiplying by $(x-1)(x-2)(x-3)$, we shall get

$$(x-2)(x-3)+(x-1)(x^2-x-3)-3(x-1)(x-2)=0,$$

that is—

$$x^3-4x^2+2x+3=0,$$

which may be written

$$(x-3)(x^2-x-1)=0 \qquad (13),$$

the roots of which are 3, $(1 \pm \sqrt{5})/2$. Now $x=(1 \pm \sqrt{5})/2$ will be found to satisfy (12); but $x=3$ is not in any sense a root of the equation.

The secret of the exceptional character of this case lies in the fact that $x-3$ is really a superfluous divisor in the characteristic of (12). By decomposing into partial fractions it is readily found that

$$\frac{1}{x-1}+\frac{x^2-x-3}{(x-2)(x-3)}-\frac{3}{x-3} \equiv \frac{1}{x-1}+\frac{1}{x-2}+\frac{x}{x-3}-\frac{3}{x-3}$$

$$\equiv \frac{1}{x-1}+\frac{1}{x-2}+1.$$

Hence the characteristic of (12) remains finite when $x=3$, and the principle of § 62 applies. The equation (12) is in fact merely (10) in disguise.

§ **228.** It is often advisable, before integralising an equation, to effect upon it some transformation in order to simplify the subsequent calculations. This may consist, as in Example 1 above, in resolving certain of the occurring fractions into partial fractions; in reducing occurring fractions to lowest terms, as in Example 2 below; in separating the integral and proper-fractional parts of occurring fractions, as in Example 3 below, and so on.

Ex. 2. $3(x-1)/(x^2-10x+9)-3(x-1)/(x^2-6x+5)=4.$

Since $x^2-10x+9\equiv(x-1)(x-9)$, $x^2-6x+5\equiv(x-1)(x-5)$, the equation may be written

$$3/(x-9)-3/(x-5)=4.$$

Integralising we get

$$3(x-5)-3(x-9)=4(x-9)(x-5),$$

which obviously is not satisfied by $x=9$ or $x=5$, and therefore gives no extraneous solutions. The last equation is

$$12=4(x^2-14x+45)$$

equivalent to $3=x^2-14x+45$, that is, to $x^2-14x+42=0$, the roots of which are $x=7\pm\surd 7$.

Ex. 3. $(x+1)/(x-1)+(x+2)/(x-2)=2(x+3)/(x-3)$.

If we separate each fraction into its integral and proper fractional part (by the rule for division or otherwise), we get

$$1+2/(x-1)+1+4/(x-2)=2\{1+6/(x-3)\}.$$

Hence $$2/(x-1)+4/(x-2)=12/(x-3).$$

Dividing both sides by 2 and integralising, we get

$$x^2-5x+6+2(x^2-4x+3)=6(x^2-3x+2),$$

whence $3x^2-5x=0$, the roots of which are $x=0$, $x=5/3$, both of which satisfy the original equation, as they ought to do, since the only possible extraneous roots that could have been introduced would have been $x=1$, $x=2$, or $x=3$.

Ex. 4. $$\frac{x-a}{x^2+a^2}+\frac{x-b}{x^2+b^2}+\frac{a+b}{a^2+b^2}=\frac{x+b}{x^2+a^2}+\frac{x+a}{x^2+b^2}.$$

If we subtract $(x-a)/(x^2+a^2)+(x-b)/(x^2+b^2)$ from both sides, we get

$$(a+b)/(a^2+b^2)=(a+b)/(x^2+a^2)+(a+b)/(x^2+b^2).$$

If we remove the constant factor, $a+b$ (which we suppose $\neq 0$, otherwise the equation would be an identity and be satisfied by any finite value of $x$), we have

$$1/(a^2+b^2)=1/(x^2+a^2)+1/(x^2+b^2).$$

Integralising, we get

$$x^4+(a^2+b^2)x^2+a^2b^2=(a^2+b^2)\{2x^2+a^2+b^2\},$$

whence $$x^4-(a^2+b^2)x^2-(a^4+a^2b^2+b^4)=0.$$

If we treat this last as a quadratic for $x^2$, we get $x^2=\frac{1}{2}\{a^2+b^2\pm\surd(5a^4+6a^2b^2+5b^4)\}$.

Hence the roots of the given equation are

$$x=\pm\surd[\tfrac{1}{2}\{a^2+b^2\pm\surd(5a^4+6a^2b^2+5b^4)\}].$$

We may take any one of the four different arrangements of sign, so that we have found four different solutions.

Ex. 5. $1/x-1/(x-a-b)=1/a+1/b$.

It is obvious that on integralising we should obtain a quadratic which would give two values for $x$, each of which must (unless for exceptional values of $a$ and $b$, *e.g.* when $a+b=0$) satisfy the original equation. Now, by inspection, we see that $x=a$ and $x=b$ satisfy the original equation. Hence these and no other are the solutions.

§ **229.** Change of variable may also be useful, as in the following example :—

Ex. 6. $(x^2-1)/(x^2+1)+(x^2+1)/(x^2-1)=34/15$.

Let $y=(x^2-1)/(x^2+1)$, then our equation is

$$y+1/y=34/15.$$

Integralised this becomes $15y^2-34y+15=0$, the roots of which, neither extraneous, are 3/5 and 5/3. Hence the original equation is equivalent to the two following :—

$$(x^2-1)/(x^2+1)=3/5, \quad (x^2-1)/(x^2+1)=5/3,$$

which when integralised give

$$5(x^2-1)=3(x^2+1), \quad 3(x^2-1)=5(x^2+1).$$

These last give $x^2=4$ and $x^2=-4$ respectively. Hence the solutions of the original equation are $x=\pm 2$, and $x=\pm 2i$.

## EXERCISES LXV.

1. $x/2+2/x=x/3+3/x$.
2. $3-x(3-3/x)=(3+x)(3+3/x)+3(3-3/x)-3$.
3. $(2x+1)/3=2-1/(2x-1)$.
4. $\dfrac{2x+3}{x}-\dfrac{2x-7}{x-1}=\dfrac{37}{20}$.
5. $(x+6)/(x+1)-(5-x)/2x=15/8$.
6. $(x+2)/(x-2)-5/6=(x-2)/(x+2)$.
7. $2/(2x-3)+1/(x-2)=6/(3x+2)$.
8. $13/(x+2)-6/(x-1)=3/(x-4)$.
9. $x/(x+1)+(x+1)/(x+2)=(x-2)/(x-1)+(x-1)/x$.
10. $(x-1)/(x^2+3x+2)-(x+1)/(x^2-3x+2)=0$.
11. $\dfrac{3x-2}{2x-3}+\dfrac{3-2x}{2-3x}=\dfrac{10}{3}$.
12. $\dfrac{1}{x+2}-\dfrac{1}{x-2}=\dfrac{1}{x-3}-\dfrac{1}{x-7}$.
13. $(2x-1)/(x+1)+(3x-1)/(x+2)=(5x-11)/(x-1)$.
14. $1/7(x-3)(x-2)+(x-4)/(x-1)(x-3)-(x-3)/(x-1)(x-2)=0$.
15. $(x^2-2x+1)/(x^2-4x+3)+(x-5)/(x^2-5x+6)=5/2$.
16. $x^2+x-1/x+1/x^2=2$.
17. $(x^2+3x-4)/(x^2+3x+1)+(x^2+2x+5)/(x^2+2x+1)=2$.
18. $(2x-1)/(x-3)-(x-2)/(x-1)=1/(x-1)$.
19. $(2x-1)/(x-1)+(x-2)/(x-3)=(3x+2)/(x-2)$.
20. $\dfrac{3+2x}{2+5x}-\dfrac{1+x}{5+2x}=1-\dfrac{10(x^2+1)}{10+29x+10x^2}$.
21. $1/(x-1)(x-2)+1/(x-2)(x-3)+1/(x-3)(x-1)=$
$6\{1/(x-1)+1/(x-2)+1/(x-3)\}/11$.
22. $4/(x-2)+9/(x-3)=x^2/(x-2)+13/(x-3)$.
23. $(x+1)/(x-1)+(x+2)/(x-2)=(2x+13)/(x+1)$.
24. $\dfrac{x+7}{x+5}+\dfrac{x+9}{x+7}=\dfrac{x+6}{x+4}+\dfrac{x+10}{x+8}$.

25. $1/(x^2-3x+2)-2/(x^2-4x+3)+1/(x^2-5x+6)=0$.
26. $(x-1)^2/(x+1)+(x-2)^2/(x+2)=2x-13$.
27. $1/(x+1)-15/(x-3)-(2x+2)/(x+1)(x-3)=3-(5x+2)/(x-3)$.
28. $x+\dfrac{x}{x-1}+\dfrac{x}{x^2-1}=\dfrac{x^2}{x+1}+\dfrac{1}{x-1}+\dfrac{5x-4}{x^2-1}$.
29. $(x-3)/(x^2-4x+3)+(x-1)/(x^2-4x+4)=(2x-5)/(2x^2-6x+4)$.
30. $(x^2-5x+4)/(x-1)+(x^2+x+1)/(x+1)+(x^2+2x+1)/(x+2)=3x$.
31. $\dfrac{x^2-3x-1}{x^2-3x+1}+\dfrac{x^2-3x-2}{x^2-3x+2}=\dfrac{14}{15}$.

## EXERCISES LXVI.

1. $\dfrac{1}{x}+\dfrac{1}{a}=\dfrac{1}{x+a}$. 2. $(x+a)/(x+b)=(2x+a+c)^2/(2x+b+c)^2$.
3. $(x+a)(x+b)/(x-a)(x-b)=(x+c)(x+d)/(x-c)(x-d)$.
4. $a/(x-a)+b/(x-b)=2(a+b)/(x-a-b)$.
5. $(p+q)/(x-r)=p/(x-p)+q/(x-q)$.
6. $x/(x+a-b)-x/(x-a+b)=2b/x$.
7. $(x+a)/(x-a)-(x-a)/(x+a)=(x^3+a^3)/(x^3-a^3)-(x^3-a^3)/(x^3+a^3)$.
8. $\Sigma(x+p)/(x+r-p)(x+p-q)=0$.
9. $\underset{abc}{\Sigma}(x+a)/(x-a)=3$.
10. Solve $x^2-x+a^2/(x^2-x)=2a$.
11. $5/(x-a)+5/(x+a)=8/x+1/(x-2a)+1/(x+2a)$.
12. $\Sigma x/(b-c+x)=3$. 13. $\Sigma(x-b)(x-c)/(x+a)=0$.
14. $(x+a)(x+b)/(x-a)(x-b)+(x+c)(x+d)/(x-c)(x-d)=2$.
15. $\Sigma(x-a)/(x+b+c)=3$.
16. $\Sigma(x+b)(x+c)/(x-b)(x-c)=3$.
17. $\dfrac{x-a}{x-3a}-\dfrac{x-b}{x-b-2a}=\dfrac{x-3b+3a}{x-b+a}-\dfrac{x-3b+4a}{x-b+2a}$.
18. $2\Sigma a\Sigma[1/(x-b)(x-c)]=3\Sigma 1/(x-a)$.
19. $\Sigma(1-ax)/(a-x)=a+b+c$.
20. $\Sigma(b+c)/(x-a)=3$.
21. $(ax+b)(cx+d)/(acx+bc+ad)+(ax+e)(cx+f)/(acx+ec+af)=2x$.

## Solution of Irrational Equations by Means of Linear and Quadratic Equations

§ 230. If we multiply by a properly chosen **Rationalising Factor,** we can always derive from any irrational equation a rational equation which shall have all the finite solutions that belong to the original equations. In general, the rationalised equation will have in addition solutions which arise from the nullification of the rationalising factor and are extraneous to the original equation. We shall consider here for the most part quadratic irrationalities only; and in so far as they are concerned, the truth of the statement just made follows from the principles of Chapter XVI. This will be understood from the discussion of the following case:—

Let P, Q, R be any functions of $x$, which may be rational or irrational; and consider the equation

$$\sqrt{P}+\sqrt{Q}-\sqrt{R}=0 \qquad (14).$$

We propose to derive from (14) an equation which shall contain, if any, only such irrationalities as are latent in P, Q, R.

To get rid of the irrationality $\sqrt{P}$, we multiply by $-\sqrt{P}+\sqrt{Q}-\sqrt{R}$, and derive

$$-P+Q+R-2\sqrt{(QR)}=0 \qquad (15),$$

an equation which contains all the solutions of (14), and in addition the solutions of $-\sqrt{P}+\sqrt{Q}-\sqrt{R}=0$.

Finally, to get rid of the irrationality $\sqrt{(QR)}$, we multiply both sides of (15) by $-P+Q+R+2\sqrt{(QR)}$, and derive

$$(-P+Q+R)^2-4QR=0,$$

that is—

$$P^2+Q^2+R^2-2QR-2RP-2PQ=0 \qquad (16).$$

Since $-P+Q+R+2\sqrt{(QR)}\equiv(\sqrt{P}+\sqrt{Q}+\sqrt{R})(-\sqrt{P}+\sqrt{Q}+\sqrt{R})$, (16) has in addition to the solutions of (15) the solutions of $\sqrt{P}+\sqrt{Q}+\sqrt{R}=0$, $-\sqrt{P}+\sqrt{Q}+\sqrt{R}=0$.

We have therefore in (16) found a derivative of (14) which is rational so far as the explicit irrationalities of (14) are concerned, but which has in addition to the solutions of (14) the solutions of

$$\sqrt{P}-\sqrt{Q}+\sqrt{R}=0, \quad \sqrt{P}+\sqrt{Q}+\sqrt{R}=0, \quad \sqrt{P}-\sqrt{Q}-\sqrt{R}=0 \qquad (17).*$$

If (16) contain further square roots, we can arrange it in linear form, use a rationalising factor to get rid of the explicit irrationalities, and so on, until at last all the square roots have been squared.

The special result of (16) occurs so often in practice that it is almost worth while to commit it to memory; and it should be observed that (16) is the rationalised equation not only for (14), but also for any one of the three equations (17).

* It will be noticed that in the equations (17) we have every possible distinct equation that can be derived from (14) by changing the signs of any of the radicals.

§ **231.** The beginner should note that the effect of multiplying both sides of (14) by the factor $-\sqrt{P}+\sqrt{Q}-\sqrt{R}$ is exactly the same as if we subtracted $\sqrt{P}$ from both sides and thereafter squared both sides of the equation.

Thus from (14) we derive

$$\sqrt{Q}-\sqrt{R}=-\sqrt{P};$$

whence

$$(\sqrt{Q}-\sqrt{R})^2=(-\sqrt{P})^2,$$

that is-

$$Q+R-2\sqrt{(QR)}=P,$$

or

$$-P+Q+R-2\sqrt{(QR)}=0,$$

as before.

Conversely, of course, $(\sqrt{Q}-\sqrt{R})^2=P$ is equivalent to $\sqrt{Q}-\sqrt{R}=-\sqrt{P}$, together with $\sqrt{Q}-\sqrt{R}=\sqrt{P}$.

We preferred in § 230 to use the language of rationalising factors in order to make as clear as possible the origin of the extraneous solutions in the rationalised equation. In practice it is usually more convenient to transpose and square.

Ex. 1. $\sqrt{(x+4)}+\sqrt{(x+20)}=8$.

The given equation is equivalent to

$$\sqrt{(x+4)}-8=-\sqrt{(x+20)},$$

from which, by squaring, we derive

$$x+4+64-16\sqrt{(x+4)}=x+20,$$

which is equivalent to

$$\sqrt{(x+4)}=3.$$

From this last, by squaring, we get

$$x+4=9.$$

Hence $x=5$. Inasmuch as the last equation would have resulted equally from $\sqrt{(x+4)}-\sqrt{(x+20)}=8$, $-\sqrt{(x+4)}+\sqrt{(x+20)}=8$, or $-\sqrt{(x+4)}-\sqrt{(x+20)}=8$, it is necessary to verify whether $x=5$ is really a solution of the original equation. Since* $\sqrt{(5+4)}+\sqrt{(5+20)}\equiv 3+5\equiv 8$, $x=5$ is really a solution of the given equation.

Since $x+4=9$, or $x-5=0$, would have resulted equally from the rationalisation of $\sqrt{(x+4)}-\sqrt{(x+20)}=8$, etc., we arrive at the remarkable result that each of these three *irrational* equations possesses no finite solution whatever.†

---

* Of course, as usual in this book, we use $\sqrt{9}$ to denote the principal value of the root, viz. $+3$. It may also be mentioned that we do not at the present stage consider in verifying equations any case where the radicand under the square root is negative, because we have given no definition for distinguishing the two values of the root in such cases.

† It is easy to construct rational *fractional* equations which have no

It is worthy of notice that it is not indifferent, as regards the amount of work to be done, what term we transpose first before squaring. The beginner will readily convince himself of this by working out Example 1, first squaring both sides of the given equation as it stands, then transposing the rational terms to one side and squaring again. The next example also illustrates this point; and it also shows, as does the one that follows, that it is often advantageous to transform an irrational equation before proceeding to rationalise. Such transformations are suggested by the special nature of each equation and no general rule can be given regarding them.

Ex. 2. $\{\surd(x+a^2)+\surd x\}/\{\surd(x+a^2)-\surd x\}=b^2$, where $b>+1$.

If we multiply numerator and denominator on the left-hand side of the equation by $\surd(x+a^2)+\surd x$, and also multiply both sides by $a^2$, we derive the equivalent equation

$$\{\surd(x+a^2)+\surd x\}^2=a^2b^2.$$

This last equation is equivalent to

$$\surd(x+a^2)+\surd x=\pm ab;$$

adding $\mp ab-\surd x$ to both sides and squaring we get

$$x+a^2+a^2b^2\mp 2ab\surd(x+a^2)=x.$$

The last equation is equivalent to

$$\pm\surd(x+a^2)=a(b^2+1)/2b.$$

Squaring the last equation and subtracting $a^2$ from both sides, we get

$$x=a^2\{(b^2+1)^2/4b^2-1\}=a^2(b^2-1)^2/4b^2,$$

which will be found to satisfy the original equation, provided $b>1$. If $b<1$, the principal value of $\surd x$ is $a(1-b^2)/2b$, and the left-hand side of the original equation reduces to $1/b^2$.

Ex. 3. $$\frac{\surd x-\surd(x+1)}{\surd x+\surd(x+1)}+\frac{\surd x+\surd(x-1)}{\surd x-\surd(x-1)}=8 \qquad (\alpha).$$

If we multiply numerator and denominator of the two fractions on the left by $\surd x-\surd(x+1)$ and $\surd x+\surd(x-1)$ respectively, we derive the equivalent equation

$$-\{2x+1-2\surd(x^2+x)\}+\{2x-1+2\surd(x^2-x)\}=8,$$

that is, if we divide by 2 and add 1 to both sides—

$$\surd(x^2+x)+\surd(x^2-x)=5 \qquad (\beta),$$

which is equivalent to ($\alpha$).

---

finite solution, *e.g.* $1/(x-1)-1/(x-2)=0$. There is, in fact, no theory regarding the number of the roots of fractional or irrational equations such as exists (see § 217) for *integral* equations.

From ($\alpha$), by squaring—

$$2x^2+2\sqrt{(x^4-x^2)}=25 \qquad (\gamma),$$

which is equivalent to

$$\sqrt{(x^4-x^2)}=25/2-x^2 \qquad (\delta).$$

From ($\delta$), by squaring—

$$x^4-x^2=25^2/4-25x^2+x^4 \qquad (\epsilon),$$

which is equivalent to

$$x^2=25^2/4\cdot 6 \qquad (\xi);$$

whence

$$x=\pm 25/4\sqrt{6}.$$

Since $x=-25/4\sqrt{6}$ makes $\sqrt{x}$, etc. imaginary, it cannot be considered here. It may be shown that $x=25/4\sqrt{6}$ satisfies ($\alpha$), by the following indirect reasoning. In the first place, $x=25/4\sqrt{6}$ certainly satisfies ($\epsilon$); for ($\epsilon$) is equivalent to ($\xi$). Since $25/2-25^2/96\equiv 25^2/50-25^2/96$ is positive, both sides of ($\delta$) are positive, and ($\delta$) follows from ($\epsilon$); and, since when $x=25/4\sqrt{6}$ both sides of ($\beta$) are positive, ($\beta$) follows from ($\gamma$), which is equivalent to ($\delta$). Since ($\alpha$) is equivalent to ($\beta$), it follows that $x=25/4\sqrt{6}$ satisfies ($\alpha$).

Ex. 4. $2x^2-3x-21=2x\sqrt{(x^2-3x+4)}$.

The given equation is equivalent to

$$x^2-3x+4-2x\sqrt{(x^2-3x+4)}+x^2=25$$

that is—

$$\{\sqrt{(x^2-3x+4)}-x\}^2=25,$$

which is equivalent to the two equations

$$\sqrt{(x^2-3x+4)}-x=\pm 5,$$

or

$$\sqrt{(x^2-3x+4)}=x\pm 5.$$

Now, squaring and reducing, we get

$$-3x\mp 10x=21,$$

which gives $x=3$, and $x=-21/13$, of which only the latter satisfies the given equation.

Occasionally change of variable is useful, as in the following example:—

Ex. 5. $\sqrt{(x^2-3x+1)}+\sqrt{(2x^2-6x+3)}=5$.

Let $y=x^2-3x+1$, then the given equation may be written

$$\sqrt{y}+\sqrt{(2y+1)}=5.$$

If we add $-\sqrt{y}$ to both sides, square, etc., we get

$$y^2-148y+576=0,$$

which gives $y=4$ and $y=144$, the first of which satisfies $\sqrt{y}+\sqrt{(2y+1)}=5$, but not the second. Replacing $y$ by its value in terms of $x$, we get

$$x^2-3x+1=4,$$

which leads at once to $x=(3\pm\sqrt{21})/2$, both satisfying the given equation.

## EXERCISES LXVII.*

1. $\sqrt{(x-7)}+\sqrt{(x+9)}=8$. 2. $\sqrt{(3x+4)}-\sqrt{(x+2)}=2$.

3. $\sqrt{(x+2)}+\sqrt{(x+3)}=\sqrt{(2x+5)}$.

4. $\sqrt{(x-1)}+\sqrt{(x-4)}=\sqrt{(2x-1)}$.

5. $\sqrt{(3x+2)}-\sqrt{(2x-1)}=\sqrt{(x+1)}$.

6. $\sqrt{(x+6)}+\sqrt{(x+9)}=\sqrt{(4-x)}$.

7. $\sqrt{(x-1)}+\sqrt{(3x+1)}=\sqrt{(2x-6)}$.

8. $\sqrt{(14+x)}+\sqrt{(6+x)}=\sqrt{(26+2x)}$.

9. $\sqrt{(6x+16)}+\sqrt{(2x+13)}=\sqrt{(12x+63)}$.

10. $\sqrt{(x+2)}+1/\sqrt{(x+2)}=x+3$.

11. $\dfrac{\sqrt{(x+7)}-\sqrt{(x+1)}}{\sqrt{(x+7)}+\sqrt{(x+1)}}=\dfrac{\sqrt{(2x+32)}-\sqrt{(2x+8)}}{\sqrt{(2x+32)}+\sqrt{(2x+8)}}$.

12. $1/\{1-\sqrt{(x-1)}\}+1/\{1+\sqrt{(x-1)}\}=1/\sqrt{(x^2-1)}$.

13. $\sqrt{(a^2+ax)}=\sqrt{(a^2-x^2)}+\sqrt{(a^2-ax)}$.

14. $\sqrt{(x^2-3x+1)}+\sqrt{(x^2-3x-1)}=\sqrt{\{(x-1)(x-2)\}}$.

15. $\{\sqrt{(x^2+9)}+\sqrt{x}\}/\{\sqrt{(x^2+9)}-\sqrt{x}\}=7/3$.

16. $\sqrt{(x^2-3x)}+\sqrt{(x^2-9)}=12\sqrt{\{(x-3)/(x+3)\}}$.

17. $\sqrt{(3+x)}-\sqrt{(3-x)}=\sqrt{(9-x^2)}$.

18. $x\sqrt{(x^2+1)}+x\sqrt{(x^2-1)}=2$.

19. $\sqrt{(x^2+1)}+3/\sqrt{(x^2+1)}-4=0$.

20. $2x\sqrt{(x^2+a^2)}+2x\sqrt{(x^2+b^2)}=a^2b^2$, where $a>b$.

21. $\dfrac{\sqrt{(ax+p)}-\sqrt{(ax+q)}}{\sqrt{(ax+p)}+\sqrt{(ax+q)}}=\dfrac{\sqrt{(bx+r)}-\sqrt{(bx+s)}}{\sqrt{(bx+r)}+\sqrt{(bx+s)}}$.

22. $1/\{x+\sqrt{(x^2+2)}+\sqrt{(x^2+1)}\}+1/\{x+\sqrt{(x^2+2)}-\sqrt{(x^2+1)}\}=1$.

23. $\sqrt{(x^2+x+4)}+\sqrt{(x^2-x+3)}=2x+1$.

24. $\sqrt{(x^2-3x+1)}+\sqrt{(x^2-x+1)}=2\sqrt{(x^2+1)}$.

25. $\sqrt{(x^3+x+1)}-\sqrt{(x^2-x+1)}=\sqrt{(x^2+1)}-\sqrt{(x^2-1)}$; verify the real solutions either directly or indirectly.

26. $x^2+7x-\sqrt{(2x^2+14x)}-4=0$.

27. $\sqrt{(x^2+3x-1)}+\sqrt{(x^2+5x-1)}=\sqrt{3}+\sqrt{5}$.

28. $2x^2-6x+11=25-2\sqrt{(x^2-3x+5)}$.

29. $\{\sqrt{(2x^2+3x+2)}-\sqrt{(x^2-3x+6)}\}/\{\sqrt{(2x^2+3x+2)}+\sqrt{(x^2-3x+6)}\}=\{\sqrt{(2x+12)}-\sqrt{(x+2)}\}/\{\sqrt{(2x+12)}+\sqrt{(x+2)}\}$.

30. Rationalise $\Sigma\sqrt{\{(b-c)(x-a)\}}=0$.

31. Find the common rational integral equation derivable from $\sqrt{(y-z)}\pm\sqrt{(z-x)}\pm\sqrt{(x-y)}=0$; and show that no one of these has a *real* solution in which the values of $x$, $y$, $z$ are unequal.

---

* No solutions are to be given which do not satisfy the equation; and solutions are to be excluded for which any of the irrational functions have an imaginary value, or which are themselves imaginary.

# CHAPTER XXI

## SYSTEMS OF EQUATIONS WHICH CAN BE SOLVED BY MEANS OF LINEAR OR QUADRATIC EQUATIONS

§ **232.** The rules for the determinateness of systems in general are the same as for a linear system. The solution is in general determinate when the number of equations is equal to the number of variables. It may, however, happen in special cases that the equations of the system are not all independent; and then the solution is indeterminate, *i.e.* there are an infinite number of solutions. It may also happen, as in § 83, Ex. 2, that the system is in part determinate and in part indeterminate.

When the number of equations is less than the number of variables, the system is always indeterminate. When the number of equations exceeds the number of variables, there is in general no solution, and the system is said to be inconsistent. In special cases the equations may not be all independent, and then the system may be determinate or even indeterminate, notwithstanding that the number of equations exceeds the number of variables.

§ **233. Systems** where one or more of the equations are of higher degree than the first have in general more than one solution, but always a finite number of solutions when the system is wholly determinate. The number of solutions may be called the **Order of the System.** Although we cannot here prove it, we give the following simple rule for the order of a system, because it forms a useful guide to the beginner in the solution of systems of higher order:—

*The number of the (finite) solutions of a wholly determinate system of integral equations cannot exceed, and is in general equal to the product of the degrees of the constituent equations.*

Ex. 1. The number of solutions of the system $x^3+y^3=2$, $x^3-y^3=3$ is $3\times3=9$.

Ex. 2. The number of solutions of the system $x+y-z=0$, $x^2+y^2+z^2=1$, $xyz=2$ is $1\times2\times3=6$.

**§ 234.** What may be regarded as the general method of solving a system of two equations in $x$ and $y$ is illustrated in the two following examples :—

Ex. 3. $2x+y-3=0$, $x^2+xy-y^2=1$. Employing the principle of interequational transformation, § 76, we see that the two given equations are equivalent to

$$y=3-2x,\quad x^2+x(3-2x)-(3-2x)^2=1\ ;$$

that is, to

$$y=3-2x,\quad x^2-3x+2=0.$$

The second equation gives $x=1$ and $x=2$ ; corresponding to these the first equation gives $y=1$, $y=-1$ respectively. Hence we have found two solutions, viz. $x=1$, $y=1$ and $x=2$, $y=-1$. Since the order of the system is 2, the number of roots obtained is the maximum possible number assigned by the rule of § 233.

Ex. 4. $x^2+xy-y^2+x+y=0$, $x^2-xy-y^2+x+y-1=0$. We may write the system as follows :—

$$P\equiv y^2-(x+1)y-(x^2+x)=0,$$
$$Q\equiv y^2+(x-1)y-(x^2+x-1)=0.$$

From these we derive the system—

$$Q-P\equiv 2xy+1=0,$$
$$(x^2+x)Q-(x^2+x-1)P\equiv y^2+\{(x^2+x)(x-1)+(x^2+x-1)(x+1)\}y=0\ ;$$

and it is very easy to show that, so far at least as finite solutions are concerned, the new system $Q-P=0$, $(x^2+x)Q-(x^2+x-1)P=0$ is equivalent to $P=0$, $Q=0$, that is, to the original system.

Again, since $y=0$ is obviously no part of a finite solution (for it reduces $2xy+1=0$ to $1=0$), we may replace the new system by

$$2xy+1=0,\quad y+(2x^3+2x^2-x-1)=0.$$

Since $x=0$ is no part of a solution, the last pair may be replaced by the equivalent system

$$2xy+1=0,\quad -1/2x+2x^3+2x^2-x-1=0\ ;$$

that is, by

$$y=-1/2x,\quad 4x^4+4x^3-2x^2-2x-1=0.$$

We have now to find the four roots of the biquadratic $4x^4+4x^3-2x^2-2x-1=0$. Corresponding to each of these the first equation, $y=-1/2x$, will give a value of $y$. We thus get all the four solutions of the system.

It will be observed that in each of these cases we have derived from the given system an equivalent system, one of the equations of which contains $x$ only; while the other equation contains both $x$ and $y$, but $y$ only in the first degree. In general, it is possible to do the like for any given system.* The equation that contains $x$ alone we call the **x-eliminant of the System.** We have therefore in general simply to solve an ordinary integral equation in one of the variables; for each root of this equation the other, which is linear in $y$, will give a single value of $y$; and we obtain a number of solutions equal to the degree of the $x$-eliminant. We might, of course, have "eliminated $x$" and used a $y$-eliminant in the same way as we have used the $x$-eliminant.

*It follows that the degree of the* x*-eliminant of a system is equal to the number of the solutions of the system in which the value of* x *is finite,* i.e. *in general equal to the order of the system.*

In other words, the solution of a system of the $m$th order depends essentially in general on the solution of an integral equation of the $m$th degree in one variable. Any peculiarity in the system will be reflected in its eliminants; in particular, if the system be soluble by means of linear or quadratic equations, each of its eliminants must be so soluble.

§ **235.** The only perfectly general case in which a system of higher order than the first is soluble by quadratic (or linear) equations is that where one equation is of the first degree and the other of the second (a 1-2-system). The eliminant of a 1-3-system would be a cubic, which would not be reducible unless the cubic had a rational root. A 2-2-system has a biquadratic eliminant which is not in general reducible to quadratic or linear equations. Example 4, § 234, is a case where the biquadratic is not reducible; in the following example the biquadratic is reducible:—

Ex. 5. $x^2+xy=4x-2,\quad y^2+xy=4y-1.$

The system is equivalent to

$$y=(-x^2+4x-2)/x,\ (x^2-4x+2)^2-(x^2-4x+2)x^2$$
$$=4x(-x^2+4x-2)-x^2=0;$$

that is, to

$$y=(-x^2+4x-2)/x,\quad 3x^2-8x+4=0.$$

The expected biquadratic reduces to a quadratic whose roots are $x=2$ and $x=2/3$, corresponding to which the other equation gives $y=1$

* There are exceptions, as the beginner will understand by carefully studying the system $x^2+y^2-x-5y+6=0$, $x^2-xy+y^2-4y+5=0$.

and $y=1/3$ respectively. Hence we get two solutions $x=2$, $y=1$ and $x=2/3$, $y=1/3$.*

Frequently the solution of a special system can be facilitated by deriving from it a simpler system wholly or partially equivalent.

Ex. 6. By addition we can derive from the system of Example 5 the following :—

$$(x+y)^2-4(x+y)+3=0, \quad x(x+y)=4x-2,$$

which is obviously equivalent.

Now the first of these gives $x+y=3$ or $x+y=1$. Using these in the second, we see (by § 76) that our new system is equivalent to the two

$$x+y=3, \quad 3x=4x-2,$$
$$x+y=1, \quad x=4x-2,$$

whence $x=2$, $y=1$ and $x=2/3$, $y=1/3$, as before.

Ex. 7. $x^3-y^3=702$, $x-y=6$.

Since $x^3-y^3\equiv(x-y)(x^2+xy+y^2)$, using the second equation to modify the first, we get (by § 76) the equivalent system $x^2+xy+y^2=117$, $x-y=6$, of the second order, which, solved as in Example 3, gives the two solutions $x=9$, $y=3$ ; $x=-3$, $y=-9$.

## EXERCISES LXVIII.

1. $3x-4y^2=14$, $2x+3y^2=32$.
2. $6x+1/y=10=12x-1/2y$.
3. $(x+1)^2+2(y-1)=51$, $2(x+1)^2-3y=92$.
4. $1/(2-x)+1/(1+y)=\frac{4}{3}$, $2/(2-x)+3/(1+y)=3$.
5. $x^2-9y^2=0$, $x+3y-1=0$ ; illustrate graphically.
6. $2x^2y-y=60$, $5x^2-164/y=164$.
7. $7x+7y=22$, $7xy=3$.
8. $x-y=1$, $xy=2$.
9. $x^2+y^2=5$, $x+2y=5$ ; illustrate graphically.
10. $6x^2-2xy+y^2=49$, $3x-y=4$.
11. $2x^2+3xy-20x=36$, $x-4y=1$.
12. $x-3y=16$, $x^2+3y^2-2x+4y=50$.
13. $5x^2+y^2+2x-7y-4=91$, $7x+3y=9$.
14. $x-2y=1$, $2x^2+y^2=54$.
15. $x^2+xy+y^2+x+y=8$, $x+y=1$.
16. $x/a+y/b=1$, $a/x+b/y=1$.
17. $ax+by=1$, $a^2x^2+abxy+b^2y^2=2$.
18. $(x-a)(y-b)-a(y-b)-b(x-a)=0$, $x+y-a-b=0$.
19. $(x+2y-1)^2-(3x-y+1)^2=0$, $3xy-2x+3y-2=0$.
20. $4x+6y-5=0$, $(x+2)/(y+3)-(x+3)/(y+2)=0$.
21. $x^2-3xy+2y^2+x-y=0$, $x^3-y^3+(x-y)(x-1)(y-1)=0$. Remark on a peculiarity.
22. $x^3-y^3=x^2-y^2$, $x^2+3xy+2y^2=0$.

---

* The other two solutions are infinite. See A. XVIII. § 6.

**23.** $x+y=2a$, $(2a-b)x^2+(2a+b)y^2=4a^3$.
**24.** $a(x+y)+b(x-y)=x^2-y^2=a'(x+y)+b'(x-y)$.
**25.** $a/(a-x)+b/(b-y)=c$, $3x-2y=3a-2b$.
**26.** $(x+y)(x+y+1)=56$, $(x-y)(x-y-1)=12$.
**27.** $x+xy=2m$, $y+xy=m+1$.
**28.** $(x+y)/(1+xy)=a$, $(x-y)/(1-xy)=b$.
**29.** $2x+3y=6xy$, $9x-2y=4xy$.
**30.** $x(3y-5)=4$, $y(2x+7)=27$.
**31.** $4x^2+9y^2=34$, $6xy=15$.

§ 236. **Homogeneous Systems.**—When one side of each equation of a system is a homogeneous function of $x$ and $y$, and the other also a homogeneous function of $x$ and $y$, or a constant, the system is spoken of as a Homogeneous System. If we change the variables in such a system from $x$ and $y$ to $x$ and $v=y/x$, it will be found that the $v$-eliminant of the new system can always be readily found, and the solution of the system is thus often more easily obtained. Since $v=y/x$ is in general not finite when $x=0$, solutions for which $x=0$ must be separately obtained.

Ex. 8. $x^2+3xy-y^2=x+2y$, $x^2-3xy+y^2=x+y$.

In the first place, we note that, if $x=0$, the two equations reduce to $y^2+y=0$, $y^2-y=0$, which have the common solution $y=0$, and no other. Hence to $x=0$ corresponds $y=0$, and nothing else. We have therefore the solution $x=0$, $y=0$ occurring once.

Next put $y=vx$, and we get the following system in $x$, $v$:—

$$x^2(1+3v-v^2)=x(1+2v), \quad x^2(1-3v+v^2)=x(1+v).$$

Apart from solutions in which $x=0$, which have already been considered, this system is evidently equivalent to $x=(1+v)/(1-3v+v^2)$, $(1+v)(1+3v-v^2)=(1+2v)(1-3v+v^2)$; that is, to

$$x=(1+v)/(1-3v+v^2), \quad 3v^3-7v^2-5v=0.$$

The last of these gives $v=0$, and $v=(7\pm\sqrt{109})/6$.

When $x$ is finite, $v=0$ gives $x=1$, and $y=vx=0$.

Corresponding to $v=(7\pm\sqrt{109})/6$, the equation $x=(1+v)/(1-3v+v^2)$ gives $x=\frac{1}{4}(11\pm\sqrt{109})$, and $y=vx=\frac{1}{24}(11\pm\sqrt{109})(7\pm\sqrt{109})=\frac{1}{4}(31\pm3\sqrt{109})$.

Hence the four solutions of the given system are—

$$\begin{aligned} x&=0, \quad 1, \quad \tfrac{1}{4}(11+\sqrt{109}), \quad \tfrac{1}{4}(11-\sqrt{109});\\ y&=0, \quad 0, \quad \tfrac{1}{4}(31+3\sqrt{109}), \quad \tfrac{1}{4}(31-3\sqrt{109}).\end{aligned}$$

§ 237. **Symmetric Equations.**—When the characteristics of the equations of a system are symmetric functions * of $x$ and $y$, it is obvious that, if $x=\alpha$, $y=\beta$ be a solution, then $x=\beta$,

* For another kind of symmetric system, and for the reasons underlying the artifice here suggested, see A. XVII. § 13.

$y=a$ is also a solution. In this case the process of solution is always simplified by taking as new variables any two independent symmetric functions of $x$ and $y$, usually the two simplest, viz. $u=x+y$, $v=xy$. The order of the $u$-$v$-system is always lower than the order of the original system. Every solution of the $u$-$v$-system gives a corresponding pair of the original system, viz. we have to solve the system $x+y=u$, $xy=v$.

Ex. 9. $x^5+y^5=33$, $x+y=3$.

Since $x^5+y^5\equiv(x+y)(x^4-x^3y+x^2y^2-xy^3+y^4)$, we see that, so far as finite solutions are concerned, the system is equivalent to

$$x^4-x^3y+x^2y^2-xy^3+y^4=11, \quad x+y=3.$$

Put now $x+y=u$, $xy=v$. Then $x^2+y^2\equiv(x+y)^2-2xy\equiv u^2-2v$; and $x^4+y^4+2x^2y^2\equiv u^4-4u^2v+4v^2$; so that $x^4+y^4\equiv u^4-4u^2v+2v^2$. Hence $x^4-x^3y+x^2y^2-xy^3+y^4\equiv x^4+y^4-xy(x^2+y^2)+x^2y^2\equiv u^4-4u^2v+2v^2-v(u^2-2v)+v^2\equiv u^4-5u^2v+5v^2$. The $u$-$v$-system is therefore

$$u^4-5u^2v+5v^2=11, \quad u=3,$$

which is equivalent to

$$5v^2-45v+70=0, \quad u=3,$$

that is—

$$v^2-9v+14=0, \quad u=3;$$

this last gives $u=3$, $v=2$ and $u=3$, $v=7$.

If $x$ and $y$ be the values corresponding to $u=3$ and $v=2$, we have

$$(z-x)(z-y)=z^2-(x+y)z+xy\equiv z^2-3z+2.$$

Hence $x$ and $y$ are the roots, *taken in either order*, of the quadratic $z^2-3z+2=0$. Hence $x=1$, $y=2$, or $x=2$, $y=1$.

Again, corresponding to $u=3$, $v=7$, we have $z^2-3z+7=0$; whence $x=\frac{1}{2}(3\pm\surd 19i)$, $y=\frac{1}{2}(3\mp\surd 19i)$.

*N.B.*—The values of $x$ and $y$ corresponding to $u=3$, $v=2$ may also be found by solving the system $x+y=3$, $xy=2$, etc., after the manner of Example 3.

We have thus obtained four finite solutions of the given system, viz.—

$$x=1,\ 2,\ \tfrac{1}{2}(3+\surd 19i),\ \tfrac{1}{2}(3-\surd 19i);$$
$$y=2,\ 1,\ \tfrac{1}{2}(3-\surd 19i),\ \tfrac{1}{2}(3+\surd 19i).$$

### EXERCISES LXIX.

1. $\surd(x-y)+\surd(x+y+1)=\surd(x+y)+\surd(x-y+1)=p$, where $p>+1$.

2. $\sqrt{\left(\dfrac{x-y}{x+y}\right)}+\sqrt{\left(\dfrac{x+y}{x-y}\right)}=3$, $ax+by+c=0$.

3. $15/\surd(x-y)+20/\surd(x+y)=9$,
$\surd(x^2-y^2)-3\surd(x+y)+4\surd(x-y)=0$.

4. Find $a$ and $b$ so that $x-y=1$, $2x+y=8$, $ax^3-bx^2y-axy=15$,

$a/x^3 - b/x^2y - a/xy = 15$ shall be a consistent system of equations in $x$ and $y$.

**5.** $x(x-y)=5$, $y(x+y)=36$.
**6.** $x^2+3xy=7$, $xy+3y^2=14$.
**7.** $x^2-xy=2$, $2x^2+y^2=9$.
**8.** $x^2+xy=10$, $y^2+xy=15$.
**9.** $x^2+2xy=85$, $4y^2-3xy=54$.
**10.** $1/x+1/y=5$, $6(x^2-y^2)=5xy$.
**11.** $7(x/y-y/x)=48$, $xy=7$.
**12.** $x^2-2xy-y^2=1$, $3x^2-4xy=35$.
**13.** $x^2+xy+y^2=49$, $x^2-xy+y^2=19$.
**14.** $2x^2-3xy+2y^2=8$, $x^2+xy-2y^2=7$.
**15.** $3x^2+4xy-y^2=14$, $2x^2+5xy+6y^2=9$.
**16.** $x^2+xy+y^2=x^3-y^3=91$.
**17.** $x^2+y^2+3xy=79$, $x+y+2xy=38$.
**18.** $x^2+y^2-x-5y+6=0$, $x^2-xy+y^2-4y+5=0$.
**19.** $x^2+xy+y^2=7(x+y)$, $x^2-xy+y^2=9(x-y)$.
**20.** $x^2-2xy+y^2+2(x-y)=15$, $x^3-y^3=63$.
**21.** $x^2+y^2=34$, $x^3y+xy^3=510$.
**22.** $x^2+4y=y^2+4x=13$.
**23.** $x^2+y^2-5(x+y)+10=0$, $xy-(x+y)=2$.
**24.** $x+y=a$, $x^3+y^3=ab^2$.
**25.** $x^2/y+y^2/x=x+y=1$.
**26.** $(x+1/x)(y+1/y)=28$, $x^2+1/x^2+y^2+1/y^2=61$.
**27.** $x^3+y^3+x^2=3=x^3+y^3+y^2$.
**28.** $x^3y+y^2=a=xy^3+x^2$.
**29.** $x^3y+x^2=a=xy^3+y^2$.
**30.** $3x+2y=3/x+2/y+1$, $2x+3y=2/x+3/y+1$.
**31.** $x^2-y^2=5$, $(x^4+y^4)^2+x^2y^2(x^2-y^2)^2=10309$.
**32.** $(3y+2)/(y+1)=(3x+1)/(x+1)=(2x+y)/(x+y)$.

§ **238.** It would be beyond our present limits to enter into the theory of systems in more than two variables; but we append the following examples of the solution of **Special Systems in Three Variables**:—

Ex. **10.***
$$x+y+z=0 \quad (\alpha);$$
$$x+2y+3z=0 \quad (\beta);$$
$$3x^2+2y^2+z^2=108 \quad (\gamma).$$

This special case can be elegantly treated by means of the principle of § 81, Ex. 3. The equations ($\alpha$) and ($\beta$) are equivalent to $x=\rho$, $y=-2\rho$, $z=\rho$, where $\rho$ is an auxiliary variable whose value we have to find. Substituting in ($\gamma$), we get

$$(3+8+1)\rho^2=108,$$

whence $\rho=\pm 3$. Therefore we get two solutions, viz. $x=3$, $y=-6$, $z=3$; and $x=-3$, $y=6$, $z=-3$, which constitute the complete solution of the given system.

Ex. **11.**
$$2yz+3zx+4xy=14xyz;$$
$$3yz+zx+5xy=30xyz;$$
$$yz-2zx+3xy=26xyz.$$

* This is merely a special example of a system of order 2, which can always be solved by means of a quadratic, no matter how many variables there are; we have only to find the $x$- or $y$- or $z$-eliminant.

It is obvious at sight that $x=0$, $y=0$, $z=0$ is one solution of the system. Apart from this the system is equivalent to

$$2/x+3/y+4/z=14\,;$$
$$3/x+1/y+5/z=30\,;$$
$$1/x-2/y+3/z=26\,;$$

obtained by dividing both sides of each equation of the system by $xyz$. This last system is linear in $1/x$, $1/y$, $1/z$, and gives $1/x=3$, $1/y=-4$, $1/z=5$; whence $x=1/3$, $y=-1/4$, $z=1/5$. We have thus obtained 2 out of the 27 possible solutions of the system.

The system is partly indeterminate, $(\alpha, 0, 0)$, $(0, \beta, 0)$, $(0, 0, \gamma)$ being solutions whatever finite values $\alpha$, $\beta$, $\gamma$ may have.

Ex. 12.
$$(y^2/b^2+1)(z^2/c^2+1)=1\,;$$
$$(z^2/c^2+1)(x^2/a^2+1)=1\,;$$
$$(x^2/a^2+1)(y^2/b^2+1)=1.$$

We employ here the principle that every solution of the system $P=P'$, $Q=Q'$, $R=R'$ is also a solution of the system $QR/P=Q'R'/P'$, $RP/Q=R'P'/Q'$, $PQ/R=P'Q'/R'$.* In the present case this leads us to the derived system

$$(x^2/a^2+1)^2=1, \quad (y^2/b^2+1)^2=1, \quad (z^2/c^2+1)^2=1,$$

which leads to

$$x^2/a^2+1=\pm 1, \quad y^2/b^2+1=\pm 1, \quad z^2/c^2+1=\pm 1.$$

It is easily seen, however, by referring to the original system, that we must take all the upper signs of the ambiguities together, or else all the lower, so that we get only two and not eight systems. These are

$$x^2=0, \quad y^2=0, \quad z^2=0\,;$$
$$x^2=-2a^2, \quad y^2=-2b^2, \quad z^2=-2c^2.$$

Hence we have obtained $x=0$, $y=0$, $z=0$; $x=\sqrt{2}ai$, $y=\sqrt{2}bi$, $z=\sqrt{2}ci$; $x=-\sqrt{2}ai$, $y=-\sqrt{2}bi$, $z=-\sqrt{2}ci$; $x=-\sqrt{2}ai$, $y=\sqrt{2}bi$, $z=\sqrt{2}ci$, etc.—that is, 9 out of the 64 possible solutions. We do not enter into the question as to how often the solution (0, 0, 0) ought to be counted. Even if we count it as 8 identical solutions, we should have only 16 out of the 64 theoretically possible.

## EXERCISES LXX.

1. $x-2y+z=x+y+2z=0$, $\Sigma(x-1)^2=44$.
2. $(y+1)(z+1)=63$, $(x+1)(z+1)=45$, $(x+1)(y+1)=35$.
3. $2x-3y+5z=3x+6y-7z=0$, $x^2+2y^2+3z^2=6$.
4. $yz=40$, $zx=24$, $xy=15$.
5. $x+y+z=3x+4y+5z=0$, $(x-1)^2+(y-2)^2+(z-3)^2=2$.
6. $(x+1)(y+2)=9$, $(y+2)(z+3)=16$, $(z+3)(x+1)=25$.
7. $x^nyz=a^{n+2}$, $xy^nz=b^{n+2}$, $xyz^n=c^{n+2}$.
8. $-yz+zx+xy=a$, $yz-zx+xy=b$, $yz+zx-xy=c$.
9. $ax+by+cz=0$, $a^2x+b^2y+c^2z=0$, $x^3+y^3+z^3=d^3$.

* The beginner should inquire how far the two systems are equivalent.

**10.** $-y^2z^2+z^2x^2+x^2y^2=y^2z^2-z^2x^2+x^2y^2=y^2z^2+z^2x^2-x^2y^2=xyz.$

**11.** $y+z-a^2/x=z+x-a^2/y=x+y-a^2/z=0.$

**12.** $(y+z)(z+x)=210,\ (z+x)(x+y)=182,\ (x+y)(y+z)=195.$

**13.** $xy/(x+y)=1,\ xz/(x+z)=2,\ yz/(y+z)=3.$

**14.** $lx+my+nz=mx+ny+lz=1/l+1/m+1/n,$
$$l^2m^2x^2+m^2n^2y^2+n^2l^2z^2=3.$$

**15.** $2y-yz=4,\ 2z-zx=9,\ 2x-xy=16.$

**16.** $(b-c)x+(c-a)y+(a-b)z=x+y+z=0,\ x^2+y^2+z^2-ax-by-cz=0.$

**17.** $(b-c)x^2+(c-a)y^2+(a-b)z^2=0,\ (c-a)x^2+(a-b)y^2+(b-c)z^2=0,\ ax+\beta y+\gamma z=d.$

**18.** $x^2(x^2+y^2+z^2)=152,\ y^2(x^2+y^2+z^2)=342,\ z^2(x^2+y^2+z^2)=950.$

**19.** $(y+z)/x+(z+x)/y+(x+y)/z=2(y+z)/x+(z+x)/y+3(x+y)/z=0,\ y^2z^2(y+z)^2+z^2x^2(z+x)^2+x^2y^2(x+y)^2=9x^2y^2z^2.$

**20.** $(y+z)^2-x^2=(z+x)^2-y^2=(x+y)^2-z^2=3.$

**21.** $a^2\{(y-b)^2+(z-c)^2\}=(y-b)^2(z-c)^2,\ b^2\{(z-c)^2+(x-a)^2\}$
$$=(z-c)^2(x-a)^2,\ c^2\{(x-a)^2+(y-b)^2\}=(x-a)^2(y-b)^2.$$

**22.** $x^2+y+z=21,\ x^2+yz=22,\ x^2yz=96.$

**23.** $x^2+yz=7,\ y^2+zx=7,\ z^2+xy=11.$

**24.** $x+y+z=10,\ yz+zx+xy=31,\ xyz=30.$

# CHAPTER XXII

## PROBLEMS INVOLVING QUADRATIC EQUATIONS

§ 239. The only new point regarding problems that involve equations of a higher degree than the first is the multiplicity of solutions. Thus, for example, a problem which leads to a quadratic equation has theoretically two solutions. As a matter of fact, however, either both, only one, or neither of the abstract solutions may be solutions of the concrete problem. This point is fully illustrated in the following examples :—

Ex. 1. If $s$ be the height in feet reached after $t$ seconds by a body projected vertically upwards with an initial velocity of $v$ feet per second, then $s=vt-16t^2$. If the initial velocity be 50 feet per second, after how long will the height be 30 feet ?

The formula gives us at once the quadratic equation

$$30=50t-16t^2,$$

to determine $t$. This equation is equivalent to

$$8t^2-25t+15=0,$$

the roots of which are $t=(25\pm\sqrt{145})/16$ — that is approximately $t=\cdot 810$ and $t=2\cdot 315$. Both these solutions are admissible : the first corresponds to the ascent, the second to the subsequent descent of the body.

Ex. 2. A father left a sum of £46,800 to be equally divided among his children ; before the estate was divided two of the children died ; and in consequence each of the survivors got £1950 more than if all had lived to share. How many children were there ?

Let $x$ be the number of children ; then, if all had survived, the share of each would have been $46,800/x$. After two had died, there remained $x-2$ ; and the share of each of these was $46,800/(x-2)$.

Hence $$46,800/(x-2)-46,800/x=1950,$$

which is equivalent to

$$24/(x-2)-24/x=1,$$

that is, since $x=0$, $x=2$ are evidently not solutions, to

$$48=x(x-2),$$

or

$$x^2-2x-48=0.$$

The roots of this equation are $x=8$ and $x=-6$. The first of these alone has any meaning in the concrete problem.

Ex. 3. AB is a line of length $a$; to find a point P in AB such that $AP \cdot PB=b^2$.

Let $AP=x$; then $BP=a-x$, and we have $x(a-x)=b^2$, which is equivalent to

$$x^2-ax+b^2=0.$$

The roots of this equation are $\{a \pm \sqrt{(a^2-4b^2)}\}/2$. If $a<2b$, the roots are imaginary; and the concrete problem admits of no solution. If $a>2b$, the roots are real and (since their product is $+b^2$) obviously both positive; the concrete problem then admits of two solutions. Since the sum of the roots is $a$, if the two roots be $x_1$, $x_2$, we have $x_1=a-x_2$—that is, if $P_1$ and $P_2$ denote the two positions of P, $AP_1=BP_2$, which might have been expected *a priori*. If $a=2b$, the two roots are equal, each being $\frac{1}{2}a$; in this case P is the middle point of AB.

Ex. 4. To find a point P in the line AB such that $AP^2=AB \cdot PB$ (Problem of "Golden Section").

Let $AB=a$, $AP=x$; then the conditional equation for the determination of $x$ is $x^2=a(a-x)$, equivalent to

$$x^2+ax-a^2=0,$$

the roots of which are $(\sqrt{5}-1)a/2$ and $-(\sqrt{5}+1)a/2$. The first of these is positive, and gives a solution of the problem of internal golden section.

The other root, viz. $-(\sqrt{5}+1)a/2$, gives us the solution of the problem of external golden section; for, if we were to take P on the far side of A from B, and take $x=-AP$ to indicate the change of direction, we should have $(-x)^2=a(a+(-x))$—that is, $x^2=a(a-x)$, the same equation of condition as before.

It should be noticed that, if $P_1$ and $P_2$ be the points of internal and external golden section, we have $+AP_1-AP_2=-a$, which gives a simple construction for deriving $P_2$ from $P_1$, or *vice versa*.

It will be noticed that in this problem the root which was not available in the solution of the original problem gives us the solution of a slightly altered problem of the same kind.

Ex. 5. A man invests two sums, each of £7200, one in the 4 per cent consols, the other in the 3 per cents, the income from the first investment exceeding that from the second by £50. If the price of each stock had been £10 higher, the difference of income would have been £48; find the prices of the two stocks.

Let the prices of the stocks be £$x$ and £$y$ respectively; then the equations of condition are

$$\frac{7200}{x} \times 4 - \frac{7200}{y} \times 3 = 50,$$

$$\frac{7200}{x+10}\times 4-\frac{7200}{y+10}\times 3=48.$$

These are obviously equivalent to

$$144(4y-3x)=xy,$$
$$150(4y-3x+10)=(x+10)(y+10).$$

Substituting the value of $xy$ given by the first equation in the second, we get after a little transformation

$$y=2x-100.$$

Using this last value of $y$, we get from $144(4y-3x)=xy$ the quadratic

$$144(5x-400)=x(2x-100),$$

which is equivalent to

$$x^2-410x+90\times 320=0,$$

the roots of which are $x=90$, $x=320$, corresponding to which $y=2x-100$ gives $y=80$, $y=540$.

Strictly speaking, both solutions are available in the concrete problem; but, as consols at £320 and £540 are in the present state of society a financial absurdity, we may conclude that the solution required is $x=90$, $y=80$.

§ 240. **Problems which can be solved by Elementary Geometrical Constructions.**—Examples 3 and 4 of last paragraph naturally suggest to us to consider the connection between linear and quadratic equations and certain geometrical problems. The reader is probably aware that a problem is said to be soluble by *elementary geometrical construction* when each step in the construction that constitutes the solution is either the drawing of a straight line through two given points, or the description of a circle having a given centre and a given radius.

We have already seen (§ 73) that the co-ordinates $(x, y)$ of every point on a straight line satisfy a linear equation, say

$$Ax+By+C=0 \qquad (1).$$

It is easy to see that the co-ordinates of every point on a circle whose centre is the point $(a, b)$, and whose radius is $c$, satisfy the equation

$$(x-a)^2+(y-b)^2-c^2=0 \qquad (2),$$

for this equation simply expresses that the square of the distance of the point $(x, y)$ from the point $(a, b)$ is equal to the square of the radius. The co-ordinates of the intersection of the straight line and the circle are therefore found by solving (1) and (2) as a system. Now this system, being of the second order, depends on a certain quadratic equation (see §§ 233-235). The intersection of two straight lines is (§ 77) found by means of a system of the first order—that is, by means of a linear equation. Hence the following interesting conclusion :—

*The algebraic solution of any problem which can be solved by elementary geometrical construction can be effected by means of linear and quadratic equations.*

*The converse is also true.* We remark, in the first place, that the solution of a linear equation, say $Ax = B$, corresponds to the geometrical problem to find a rectangle of area B, one of whose sides is A, which is a well-known Euclidian problem soluble by elementary construction. We have therefore only to consider the solution of a quadratic equation, and to show that this can always be effected by elementary construction.

We may suppose the quadratic transformed so that the coefficient of $x^2$ is $+1$. We have then four distinct cases to consider, viz.—

$$x^2 - ax + b = 0 \qquad (3);$$
$$x^2 - ax - b = 0 \qquad (4);$$
$$x^2 + ax + b = 0 \qquad (3');$$
$$x^2 + ax - b = 0 \qquad (4');$$

where $a$ and $b$ are real positive quantities.

Of these it is necessary to consider only the first two ; for if we put $x = -\xi$, (3′) and (4′) become

$$\xi^2 - a\xi + b = 0 \qquad (3'');$$
$$\xi^2 - a\xi - b = 0 \qquad (4'');$$

so that the roots of (3′) and (4′) are simply the roots of an equation of the type (3) and an equation of the type (4) respectively with the signs changed.

Now (3) and (4) may be written

$$x(a - x) = b \qquad (3^a);$$
$$x(x - a) = b \qquad (4^a).$$

Let AB be a straight line whose length is $a$ units of ar chosen scale, say $a$ inches ; let us think of $b$ as $b$ square inches and let P be a point in AB on the same side of A as B, who distance from A is $x$ inches. Then we see that to solve ( is to determine an internal point (or points) P in AB, such the area of the rectangle AP·PB is $b$ square inches ;

solve ($4^a$) is to find an external point in AB such that the area of the rectangle AP·PB is $b$ square inches.

We find the number of inches in $\sqrt{b}$ (*Eucl.* II. 14) by finding the side of the square whose area is equal to the rectangle contained by lines whose lengths are 1 inch and $b$ inches respectively.

To solve our first problem, we describe on AB a semicircle; draw BQ perpendicular to AB of length $\sqrt{b}$ inches; through Q

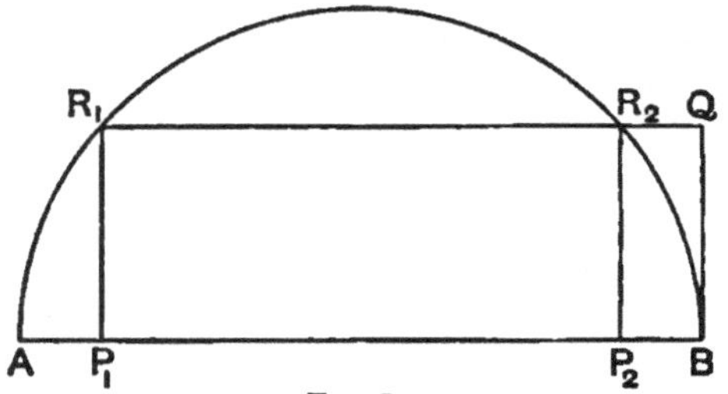

FIG. 5.

draw $R_2R_1$ to meet the semicircle in $R_2$ and $R_1$; draw $R_1P_1$ and $R_2P_2$ perpendicular to AB; then the roots of (3) are $+AP_1$ and $+AP_2$, where $AP_1$ and $AP_2$ denote the number of inches in $AP_1$ and $AP_2$ respectively.

To solve the second problem, describe on AB a semicircle as before, C being its centre; draw any radius CQ; draw QR per-

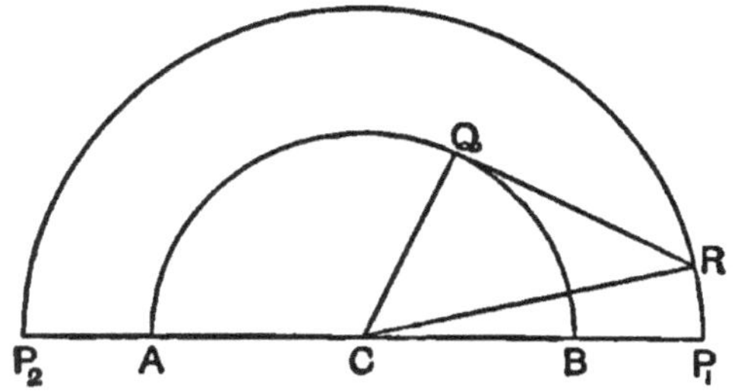

FIG. 6.

pendicular to CQ and of length $\sqrt{b}$ inches; with C as centre and CR as radius describe a circle meeting AB (of course externally) in $P_1$ and $P_2$ respectively; then the roots of (4) are $+AP_1$ and $-AP_2$.

The reader who has mastered the requisite theorems in a systetry will readily demonstrate the correctness of the above constructions: and he will see that every step involved is an *elementary* construction.

The geometric construction shows that the roots of (3) are real and unequal, real and equal, or imaginary, according as $a/2 > = < \sqrt{b}$—that is, according as $a^2 - 4b > = < 0$; also that the roots of (4) in all cases are real and of opposite sign.

It follows from what has been shown that *any geometric problem the algebraic solution of which involves only the operations of addition, subtraction, multiplication, division, and extraction of the square root can be solved by elementary geometric construction; and conversely.*

From this again it follows that, *even when the roots of a quadratic are imaginary, and the above construction fails, they can still be found by elementary geometric construction.*

For the roots of the quadratic $Ax^2 + Bx + C = 0$ when imaginary are $\xi \pm \eta i$, where $\xi = -B/2A$, $\eta = \sqrt{(4AC - B^2)}/2A$ (§ 215). The operations by means of which $\xi$ and $\eta$ are derived from A, B, C are therefore merely addition, subtraction, multiplication, division, and extraction of the square root, all of which, as we have seen in the course of the foregoing discussion, can be represented by elementary geometrical construction.

The above results, besides their purely theoretical interest, are often useful in practice, by enabling us to tell whether a given geometric problem admits of elementary geometric construction or not.

Ex. 6. Show that the five fifth roots of 1 can be found by elementary geometric constructions.

The equation which determines the fifth roots is $x^5 - 1 = 0$, which is equivalent to

$$x - 1 = 0,$$

and

$$x^4 + x^3 + x^2 + x + 1 = 0.$$

The first of these is linear; and the second, being a reciprocal biquadratic (see § 227, Ex. 6), can be solved by means of quadratic equations, viz. it is equivalent to

$$x^2 + \tfrac{1}{2}(1 + \sqrt{5})x + 1 = 0, \quad x^2 + \tfrac{1}{2}(1 - \sqrt{5})x + 1 = 0.$$

Hence the problem is soluble by elementary construction.

*N.B.*—The occurrence of the surd $\sqrt{5}$ shows that this problem is connected with the problem of the "Golden Section." On the other hand, it appears from the theory of the radication of complex numbers* that the problem of finding the fifth roots of 1 is closely connected with

* A. Ch. XII. § 18.

the problem of the quinquesection of the circle. Hence the connection between the problem of the "Golden Section" and the construction of a regular pentagon, the relation between which, although early discovered, is not very obvious from the point of view of the old synthetic geometry.

## EXERCISES LXXI.

**1.** Find a number other than unity such that the difference between it and its cube is 6 times the square of the number which is less by 1 than the given number.

**2.** I have thought of a number; I multiply it by $2\frac{1}{3}$ and add 7 to the product; I then multiply the result by 8 times the number thought of; finally, I divide by 14 and subtract from the quotient 4 times the number thought of; I thus obtain 2352. What number did I think of?

**3.** Find two consecutive odd numbers the sum of whose squares is 394.

**4.** A certain integer of two digits has its tens digit greater by 3 than its units digit, and when the number is multiplied by the units digit the product is 425. What is the integer?

**5.** If a man live $2a^2+a$ years longer, his age will be the square of his age $2a^2+a$ years ago. What is his present age?

**6.** A and B run a race. B, who is slower than A by a mile in 5 hours, gets a start of $2\frac{1}{2}$ minutes, and they reach the fifth milestone together. What are their respective rates?

**7.** A farmer bought some sheep for £72, and found that if he had received 6 more for the same money, he would have paid £1 less for each. How many sheep did he buy?

**8.** On selling goods for £12, I find that the number expressing my profit per cent has been twice that expressing the cost of the goods in pounds. Find the cost price.

**9.** If the number of pence which a dozen apples cost is greater by 2 than twice the number of apples that can be bought for 1s., how many apples can be bought for 9s.?

**10.** A man buys a number of articles for £1, and makes £1 : 1s. by selling all but two at 2d. apiece more than they cost. How many did he buy?

**11.** The bill of a party at a restaurant was £10; but three of the party were unable to pay. The rest divided the bill equally among them, and each had 3s. 4d. more than his proper share to pay. How many were there in the party?

**12.** A grazier bought a certain number of oxen for £240. After losing 3 he sold the remainder for £8 a head more than they cost him, thus gaining £59 by his bargain. How many oxen did he buy?

**13.** A merchant bought a number of barrels of herrings for £15. Twenty were lost in transit, and he sold the remainder for £15, thus gaining 4s. per barrel over the original cost. How many barrels did he buy?

**14.** A party of twenty ladies and gentlemen spent £2 : 8s. in a hotel, the ladies spending altogether as much as the gentlemen. It appeared, however, from the bill that each gentleman spent 1s. more than each lady. How many gentlemen were in the party?

**15.** I bought a number of handkerchiefs for 60s.; if I had got 3 more for the same money they would have cost me 1s. less apiece. How many did I buy?

**16.** Show that it is impossible to find a positive integer such that the sum of its square and its cube is an integral multiple of the square of the next highest integer.

**17.** A vintner draws a certain quantity of wine out of a full vessel containing 256 gallons; he then fills up the vessel with water, draws off the same quantity as before, and so on for four draughts. At the end there are only 81 gallons of pure wine left in the vessel. How much did he draw each time?

**18.** Two men, A and B, bicycle from P to Q. A starts half an hour after B, and overtakes him 24 miles from P. A goes on to Q, and after staying there a quarter of an hour starts back at his previous pace and meets B 2 miles from Q. A then reduces his pace 1 mile an hour, and reaches P 3 hours and 12 minutes afterwards. Find the distance from P to Q.

**19.** Find a positive integer such that the excess of its cube over its square is $(q+q^2)$ times the number itself.

**20.** B can do a certain piece of work in half the time that A can do it. C would require 4 days more than B. Together they can do it in 1 day. In how many days can A do it?

**21.** A walks a quarter of a mile in an hour faster than B, and in consequence takes a quarter of an hour less time to walk 15 miles. Find the rate at which each walks.

**22.** Find the side of a square inscribed in a right-angled triangle whose sides are $a$ and $b$, so that two vertices of the square stand on the hypotenuse.

**23.** If H and H′ be the internal and external points of medial section of AB, so that $AB.BH=AH^2$ and $AB.BH'=AH'^2$, show algebraically that

(1) $AH.BH=(AH+BH).(AH-BH)$; $AH'.BH'=(BH'+AH').(BH'-AH')$.

(2) $AH.(AH-BH)=BH^2$; $AH'.(AH'+BH')=BH'^2$.

(3) $AB^2+BH^2=3AH^2$; $AB^2+BH'^2=3AH'^2$.

(4) $(AB+BH)^2=5AH^2$; $(AB+BH')^2=5AH'^2$.

(For other Exercises of the same kind, see Mackay's *Euclid*, p. 153.)

AB being a given straight line, $l$, $m$, and $c$ positive constants, give algebraical solutions of the problems to find a point P in the line satisfying the following conditions; and discuss in each problem the different cases that arise for different values of the constants:—

**24.** $AP.PB=c^2$. **25.** $lAP^2+mBP^2=c^2$.

**26.** $lAP^2-mBP^2=c^2$. **27.** $(AP-BP)^2=AB.BP$.

**28.** To find a point H internal or external in a given straight line AB such that $AB.BH=nAH^2$. In particular, consider the case where $n=1$. (Problem of "Medial Section.")

**29.** Find the distance from the end of the diagonal of a rectangle of a point on that diagonal the sum of the squares of whose distances from the ends of the other diagonal is double the square on either diagonal.

**30.** Find the radius of the circle inscribed in an isosceles triangle whose base is 4 and whose height is 3 inches.

**31.** Two roads cross at right angles. A and B start at distances of 5 and 6 miles respectively from the crossing to walk towards it at the same rate. How far must they go so that the distance between them is reduced to 2 miles?

**32.** Find a point P in the diagonal of a rectangle whose sides are $a$ and $b$ such that the sum of the areas of the two rectangles "about the diagonal" which have a common vertex at P may be equal to $c^2$, and discuss the possibility of the problem.

**33.** Find the side of a square inscribed in an equilateral triangle whose side is $a$.

**34.** A circle is inscribed in a quadrant of a circle of radius $r$. Find the radius of the inscribed circle.

**35.** The distances of two points A and B from a straight line L are 4 and 8 respectively, and the distance between their projections on L is 9. Find a point in L the sum of whose distances from A and B is 15.

**36.** Two equal spheres are just contained in a cylindrical box of height $h$ and diameter $d$, find the radius of either sphere.

**37.** ABC is a triangle right angled at C. P is a point in AB; L and M the projections of P on BC and CA. It is required to determine P so that PLCM may have a given area.

**38.** The distances of A and B from a given straight line L are $a$ and $b$ respectively, and the distance between their projections on L is $c$. To find a point in L the sum of whose distances from A and B is $d$. Discuss the different cases of the problem.

**39.** ABCD is a rectangle ($AB=a$, $BC=b$); O a point in BA produced such that $OA=c$. OPQR meets AD internally in P, CD internally in Q, and BC externally in R. Find AP so that $AP=CQ$.

**40.** From a point O outside a circle of radius $r$ at a distance $d$ from its centre OPQ is drawn to meet the circle in P and Q, so that $OQ = nOP$; calculate OP.

## EXERCISES LXXII.

**1.** Two circles touch each other, one touches one pair of adjacent sides of a square of side $a$, and the other the other two adjacent sides of the same square. If the sum of the areas of the two circles be equal to the area of a circle of radius $b$, find the radii of the two circles and discuss the possibility of the problem.

**2.** Find the sides of a rectangle of area $c^2$ inscribed in an equilateral triangle whose side is $a$.

**3.** Find the sides of a rectangle of area $c^2$ inscribed in a right-angled triangle whose sides are $a$ and $b$, so that one vertex is at the right angle and each of the remaining three on one side of the triangle.

4. The perimeter of a rectangular field is 306 yards, and the diagonal is 117 yards. What is the area?

5. An integral number consists of two digits. When the number is divided by the sum of its digits the quotient is greater by 2 than the digit in the ten's place; but if the digits be reversed and the resulting number divided by a number greater by 1 than the sum of the digits, the quotient is greater by 2 than the preceding quotient. Required the number.

6. Find two positive numbers whose sum is $a$ such that the difference of their squares is $b$ times their product. Is the problem always possible?

7. If $x$ and $y$ be the distances from B and C respectively of a point on BC, whose distance from any point A is $d$, and $BC=a$, $AB=c$, $AC=b$, show that $b^2x+c^2y=a(d^2+xy)$; and hence show that, if $a=b=c$, then $x$ and $y$ are the roots of the quadratic $\xi^2-a\xi^2+(a^2-d^2)=0$.

8. To construct a right-angled triangle, givên its hypotenuse and its area.

9. The hypotenuse of a right-angled triangle is $c$, and its sides are in continued proportion. Find the sides.

10. ABCD is a square whose side is $a$, Q a point within it, L, M, N, P the projections of Q on AB, BC, CD, DA. If $AQ^2+QC^2=\frac{3}{2}a^2$ and area APQL + area $CMQN=\frac{5}{16}a^2$, find the distances of Q from AB and AD.

11. Find the sides of a rectangle of given area inscribed in a given circle.

12. A sets out from C to D, and at the same time B sets out from D to C. A arrives at D $a$ hours, and B at C $b$ hours after they meet. How long did each take to perform the journey?

13. Find two numbers such that the sum of their squares is $a$ times their product, and the difference of their squares $b$ times their product.

14. The sum of two numbers is 20, and the sum of their cubes is 4940. Find the numbers.

15. A and B distribute £60 each among a certain number of persons. A relieves 40 persons more than B does, and B gives to each 5s. more than A. How many persons did A and B respectively relieve?

16. Two market-women had 88 eggs between them. They sold them all, each realising the same sum. On their way home the one said to the other, "If I had sold your eggs at my price, I would have got 6s. 9d."; and the other replied, "If I had sold yours at my price, I would have got 14s. 1d." How many eggs had each and at what prices per dozen did they sell?

17. Solve the former problem generalised by putting $a$ for 88, $b$ for pence in 6s. 9d., $c$ for pence in 14s. 1d.

18. If $x$ and $y$ be the distances from one vertex of a rectangle whose sides are $a$ and $b$ of the vertices of an inscribed rectangle of area $c^2$, find equations for $x$ and $y$. Show that the equations are soluble by means of quadratics, and work out the solution when $a=b$.

19. To cut a rectangular corner out of a given rectangle whose

sides are $a$ and $b$, so that the remaining part may have a given area $a^2$, and the excised part a given perimeter $2d$.

**20.** Two merchants put together £500 into a temporary business. The one left his capital in it for five months, the other for two. The business was uniformly profitable, and when it was wound up each of them received £450 as returned capital and profit. How much did each put in?

**21.** To cut from a given square of side $a$ a corner whose area is $\frac{9}{25}$ of the area of the square, so that the perimeter of the resulting figure shall be $\frac{13}{20}$ of the perimeter of the square.

**22.** To find five positive numbers such that the products of the first and second, second and third, third and fourth, fourth and fifth, and first and fifth are $a$, $b$, $c$, $d$, $e$, respectively. Show that a similar problem for four, or any even number, of numbers is either impossible or indeterminate.

**23.** Two travellers, A and B, start at the same time from two different places, C and D, to make the journey from C to D and from D to C. When they meet A has done $d$ miles more than B, and it would take A and B $a$ and $b$ days respectively to complete their journeys. Find the distance from C to D.

**24.** Three couriers, A, B, C, start for a certain destination. B rides 3 miles an hour faster than A, and C rides 10 miles an hour. B starts five hours after A, and C two hours after B. If they all arrive together, find the distance and the rates of riding.

**25.** The difference between two positive integers is 8, and the difference between their cubes is 2072. Find the numbers.

# CHAPTER XXIII

## ENUMERATION OF COMBINATIONS AND PERMUTATIONS

§ **241.** By an **r-combination** of a given set of things is meant a group of $r$ of them considered *without reference to order.* By an **r-permutation** is meant a group of $r$ considered as *arranged in a definite order.* It is usual to suppose the things arranged in a straight line, or along an unclosed curve, so that the permutation has a beginning and an end. We might, however, arrange them along a closed curve, *e.g.* a circle; and then the permutation might be regarded as having neither beginning nor end. When the former view is taken we speak of a "**linear permutation,**" or **permutation** simply; in the latter case we distinguish by speaking of a "**circular permutation.**" Unless the contrary is indicated, we distinguish between what is called counter-clock order and cum-clock order circular permutations—that is to say, we regard

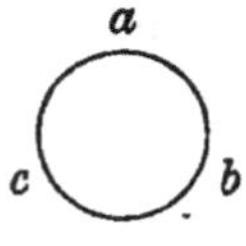


and

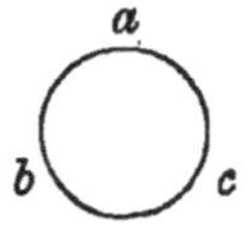


as two distinct circular permutations. If counter-clock and cum-clock arrangements be not distinguished, there are only half as many circular permutations as in the contrary case.

In counting the permutations and combinations of given things we may represent them by letters (or by numbers), all different, or partly alike and partly different, according as the things considered are all different, or partly alike and partly different.

Let us consider, for example, four letters, $a$, $b$, $c$, $d$. The 3-com binations are *bcd, acd, abd, abc.* Since combinations and not permutations are in question, *bdc, cbd,* etc. are not distinct from *bcd*; and the number of distinct 3-combinations of four letters which are regarded as all different is 4. If two of the letters, say *a* and *b*, are regarded as alike, we might write the four 3-combinations set down above *acd, acd, aad, aac,* of which only the last three are distinct. The number of distinct 3-combinations of four letters, two of which are alike, is thus only 3.

Let us next consider the linear 3-permutations that can be formed with four letters. Taking each of the combinations and arranging in all possible orders, we shall find the actual permutations to be

| | | | |
|---|---|---|---|
| *bcd,* | *acd,* | *abd,* | *abc,* |
| *bdc,* | *adc,* | *adb,* | *acb,* |
| *cbd,* | *cad,* | *bad,* | *bac,* |
| *cdb,* | *cda,* | *bda,* | *bca,* |
| *dbc,* | *dac,* | *dab,* | *cab,* |
| *dcb,* | *dca,* | *dba,* | *cba,* |

24 in number.

If we consider circular permutations, only two in each of the four columns is distinct, *e.g.* in the first *bcd* and *bdc*; for if we arrange *bcd* equidistantly round a circle, it will be seen that *dbc* and *cdb* are merely other ways of naming the circular permutation *bcd.* Hence the number of circular 3-permutations of four letters is 8.

Finally, let us consider the 3-permutations of four letters, two of which are alike, say *a, a, c, d.* We now have

| | | |
|---|---|---|
| *acd,* | *aad,* | *aac,* |
| *adc,* | *ada,* | *aca,* |
| *cad,* | *daa,* | *caa,* |
| *cda,* | | |
| *dac,* | | |
| *dca,* | | |

12 in number.

These examples will convey a clear idea of the nature of the problems to be considered in the present chapter; and it will readily be gathered that when the number of letters is large the direct method of enumeration, which consists in setting all the combinations or permutations down and then counting them, would in practice be very tedious, not to speak of the danger of inaccuracy.

§ **242.** The counting of combinations and permutations in the more simple cases requires no particular skill or profound knowledge of Algebra. On the other hand, the course of thought which is involved in these calculations often recurs in Elementary Algebra, and some of the simpler results enter frequently into the elementary formulæ (see, for example, §§ 94,

106). For both reasons, it is desirable that the matter of the present chapter should be mastered early in a course of Algebra.

§ **243.** *In* r *pigeon-holes there are respectively* $n_1, n_2, \ldots, n_r$ *things all different. In how many ways can an* r*-combination of the things be taken with the condition that one, and only one, is to be taken from each pigeon-hole?*

Since there is no question as to the order of the things, we may arrange the things in each combination in the order of the pigeon-holes. The first thing can be selected in $n_1$ different ways; the second thing in $n_2$ different ways. Any one of the ways of selecting the first thing may be combined with any one of the ways of selecting the second. Hence the first two things may be selected in any one of $n_1 \times n_2$ ways. Again, with each one of the $n_1 \times n_2$ ways of selecting, the first two things may be combined any one of the $n_3$ ways of selecting the third thing. There are therefore $n_1 \times n_2 \times n_3$ ways of selecting the first three things. Proceeding in this way we see that *the number of ways of selecting the* r *things is* $n_1 \times n_2 \times n_3 \times \ldots \times n_r$ —*that is to say, it is the product of the numbers of things in each of the* r *pigeon-holes.*

Ex. A man has 5 coats, 6 vests, and 9 pairs of trousers. In how many ways can he make up a complete suit of clothes?—Answer, in $5 \times 6 \times 9 = 270$ ways.

§ **244.** *The number of* r*-permutations of* n *letters all different, when repetition of the letters within the permutation is allowed, is* $n^r$.

We may look upon the formation of the $r$-permutation as the filling of $r$ pigeon-holes placed in order side by side, one thing to be placed in each. We may fill the first hole by putting into it any one of the $n$ things, and the second in like manner. Since repetition to the fullest extent is allowed, the filling of one hole in no way interferes with the filling of any other; in other words, we may combine any way of filling the first with any way of filling the second, and so on. Hence, as in § 243, the whole number of ways of filling the holes is $n \times n \times n \times \ldots \times n$ ($r$ factors)—that is to say, the number of $r$-permutations is $n^r$.

Ex. We may verify the above by writing down the 2-permutations of three letters $a$, $b$, $c$. They are $aa$; $ab$, $ba$; $ac$, $ca$; $bb$; $bc$, $cb$; $cc$, and their number is $9 = 3^2$, which agrees with our general theorem.

§ **245.** *The number of* r*-permutations of* n *things, which are all* *different, is, when no repetition is allowed within the permutation,* n(n − 1)(n − 2) . . . (n − r + 1).*

Let us construct the permutations as before by filling *r* pigeon-holes placed in a row. The first hole can be filled by placing in it any one of the *n* things, *i.e.* in *n* ways. Supposing the first hole filled with any particular thing, this thing is not simultaneously available for filling the second hole; so that for every one of the *n* ways of filling the first hole there are only $n-1$ ways of filling the second. There are thus $n(n-1)$ ways of filling the first two holes. When the first two are filled in any particular way, there are $n-2$ ways of filling the third. Hence the first three holes can be filled in $n(n-1)(n-2)$ different ways. Proceeding in this way, we finally arrive at the result above stated. The number of *r*-permutations of *n* things without repetition is usually denoted by ${}_nP_r$; so that

$${}_nP_r = n(n-1)(n-2) \ . \ . \ . \ (n-r+1) \qquad (1);$$

the number consists of *r* factors, viz. *n* and the next $r-1$ consecutive lower integers.

For example, ${}_6P_3 = 6.5.4 = 120$; ${}_{11}P_7 = 11.10.9.8.7.6.5 = 1,663,200$.

§ **246.** Since every circular *r*-permutation gives rise to *r* different linear *r*-permutations by taking in succession its *r* things as first thing in the linear permutation, whereas every linear permutation by joining its two ends gives one, and only one, circular permutation, it follows that there are *r* times as many linear *r*-permutations as there are circular *r*-permutations. Hence *the number of circular* r-*permutations of* n *things is* ${}_nP_r/r$—*that is to say,* $n(n-1) \ . \ . \ . \ (n-r+1)/r$.

Ex. The number of circular 3-permutations of four things is 4.3.2/3 = 8, in agreement with an enumeration made above in § 241.

§ **247.** The number of linear *n*-permutations of *n* things (usually spoken of simply as the "permutations") is by a special application of the result of § 245.

$${}_nP_n = n(n-1) \ . \ . \ . \ 3.2.1 \qquad (2);$$

that is to say, the product of the first *n* integral numbers. This product is so important that a special name and notation

* To avoid circumlocution, we shall in future speak of permutations and combinations simply, on the understanding that repetition of the same thing is not allowed unless this is expressly stated.

are set apart for it, viz. we call $1 . 2 . 3 \ldots (n-1)n$ "factorial $n$," and denote it by $n\,!$. Thus $3\,! = 1 . 2 . 3 = 6$, $2\,! = 1 . 2 = 2$, $1\,! = 1$. According to the present definition $0\,!$ is meaningless; but, for reasons that will be apparent hereafter, we define $0\,!$ to mean 1.*

Bearing in mind the results of this and the preceding section, we see that *the numbers of linear and circular* n*-permutations of* n *things are* n ! *and* (n − 1) ! *respectively*, it being understood that counter-clock and cum-clock order are distinguished in the circular permutations.

§ **248.** The problem of enumerating **the $r$-permutations of $n$ things when groups of them are regarded as alike** is in its general form very complicated. When $r = n$, the solution is important and very simple, as we shall now show.

Consider first the special case of five letters, $a_1a_2a_3bc$. Let P′ be the number of permutations (5-permutations) when $a_1a_2a_3$ are regarded as alike, and let ${}_5P_5(=5\,!)$ denote, as usual, the number of permutations when all the letters are regarded as unlike. Taking any one of the P′ permutations, say $a_1ba_2a_3c$, we see that from the point of view of the P′ permutations the relative order of $a_1a_2a_3$ is of no consequence, so long as $b$ remains in the second, and $c$ in the fifth place. On the other hand, every alteration of the order of $a_1a_2a_3$ alone, *e.g.* $a_1ba_3a_2c$, $a_2ba_1a_3c$, etc., gives a new permutation from the point of view of the ${}_5P_5$ permutations. Now we can write $a_1a_2a_3$ in $3\,!$ different orders. Hence, from every one of the P′ permutations, by writing the $a_1a_2a_3$ in different orders, but keeping the other letters fixed, we can deduce $P' \times 3\,!$ permutations. Now in the P′ permutations themselves the remaining letters, $bc$, appear in all possible different orders. Hence the $P' \times 3\,!$ permutations, formed by deranging $a_1a_2a_3$, are simply all possible permutations of the five letters when all are considered different. We have thus $P' \times 3\,! = {}_5P_5$, and therefore $P' = {}_5P_5/3\,!$.

Take now the general case of $n$ letters, among which a group of $\alpha$ are considered as alike, a group of $\beta$ as alike but different from the group of $\alpha$, and so on, and let P′ be the number of permutations on this hypothesis. Also let ${}_nP_n(=n\,!)$

* Until lately the universal notation in English books for factorial $n$ was $\underline{|n}$. This, for typographical reasons, is now being discarded. Gauss's symbol $\Pi(n)$ is sometimes used.

denote the number of permutations when all the letters are supposed unlike.

By first deranging the group of $\alpha$ in every possible way, the other letters being fixed, we shall derive from the P′ permutations $P' \times \alpha\,!$, in which the group of $\alpha$ are no longer considered as alike. From these $P' \times \alpha\,!$ permutations, by deranging the group of $\beta$, we deduce $P' \times \alpha\,! \times \beta\,!$ permutations in which the group of $\alpha$ and the group of $\beta$ are no longer considered alike. Proceed in this way until all the groups of like letters have been made unlike. The resulting number of combinations will be finally ${}_nP_n$. Hence $P'\alpha\,! \times \beta\,! \times \ldots \doteq {}_nP_n$, and therefore $P' = {}_nP_n/\alpha\,!\,\beta\,!\,\ldots = n\,!/\alpha\,!\,\beta\,!\,\ldots$

*Hence the number of permutations of* n *things,* α *of which are alike,* β *alike, etc., is* $n\,!/\alpha\,!\,\beta\,!\,\ldots$

Ex. The letters of the word *assessment* are a s s s s e e m n t, 10 in number, of which 4 are *s*'s, and 2 *e*'s. Hence the number of distinct permutations of the letters of the word is $10\,!/4\,!\,2\,! = 10 . 9 . 8 . 7 . 3 . 5 = 75{,}600$.

§ **249.** Since $1\,! = 1$, we may, if we like, write in the denominator of $n\,!/\alpha\,!\,\beta\,!\,\ldots$ a factor $1!$ corresponding to each non-repeated letter, *e.g.* in the case of the permutations of the letters of "assessment" we might write $10\,!/4\,!\,2\,!\,1\,!\,1\,!\,1\,!\,1\,!$. When this is done, the sum of $\alpha$, $\beta$, etc. is $n$. The number $n\,!/\alpha\,!\,\beta\,!\,\ldots$, where $\alpha + \beta + \ldots = n$, is called the **Multinomial Coefficient of the** $n$**th Order of Type** $(\alpha, \beta, \ldots)$ and may be denoted by the symbol ${}_nM_{\alpha,\,\beta}\ldots$

It is an interesting remark, of which we shall make important use hereafter, that ${}_nM_{\alpha,\,\beta,}\ldots$ is the number of different ways in which the single factors of the product $a^\alpha b^\beta \ldots$ can be written out in order.

Ex. Consider the product $ab^2c$. This can be written in ${}_4M_{1,\,2,\,1} = 4\,!/1\,!\,2\,!\,1\,! = 12$ ways. The actual arrangements are *abbc, abcb, acbb*; *babc, bacb, bbac, bbca, bcab, bcba*; *cabb, cbab, cbba.*

### EXERCISES LXXIII.

**1.** Find the number of linear and also of circular 5-permutations of 10 things.

**2.** Find the number of linear permutations of the letters of the words *Uniformity, Penelope, Parallelepiped.*

**3.** Find the number of permutations of the letters of the word *perimeter.*

4. Determine $n$ when the number of 4-permutations of $n$ things is equal to the number of 3-permutations of $n+1$ things.

5. Find the number of the $r$-permutations of $n$ things in which a particular thing occurs.

6. Find the ratio of the number of 5-permutations of 10 things in which none of three given things occur to the number in which one at least of the three occurs.

7. If the number of the $r$-permutations of $n$ things in which a particular thing does not occur be equal to the number of those in which it does occur, find the relation between $n$ and $r$.

8. On a railway there are fifteen stations. Find the number of tickets required in order that it may be possible to book a passenger from every station to every other.

9. One game of lawn tennis is to be arranged out of a party of six ladies and five gentlemen, each side consisting of one lady and one gentleman. In how many ways can this be done?

10. If the number of $r$-permutations of $n$ things in which two particular things occur be equal to the number of those in which neither occurs, find the relation between $n$ and $r$.

11. In how many ways can a company of $2n$ soldiers be arranged, first in one line; second two deep, a given half of the company being in the front rank and the other half in the rear rank?

12. If the number of $r$-permutations of $n$ things be equal to the number of $(r+1)$-permutations of $n-1$ things, show that $r$ must be of the form $s^2-1$, where $s$ is a positive integer not less than 2; and find the value of $n$ in terms of $s$.

13. Find the number of circular permutations of $n$ things, two of which are alike.

14. Taking the vowels to be five in number, viz. $a$, $e$, $i$, $o$, $u$, and all the other letters of the alphabet as consonants, how many words of one syllable can be formed, each containing one vowel and one consonant?

15. How many different integral numbers of three significant digits can be made with the six digits 1 2 3 0 0 4?

16. Show that the number of 7-permutations of 7 things when two of the things are excluded each from a particular position is 3720.

17. The stem of a tree splits into three branches, each of these branches again into three, and so on $n$ times. In how many different ways can an ant climb from the ground to the tip of a branch?

18. In how many different ways can $n$ different coins, in each of which the obverse and reverse are distinct, be arranged in a row; and in how many different ways can they be set in a bracelet?

19. In a *mêlée* of $m$ combatants against an equal number, each combatant is paired with a single adversary. In how many ways can the fight be arranged, if we attend only to the question of who is paired with whom?

20. In how many ways can an equilateral-triangular patch be built up with different equilateral-triangular tiles, nine in number?

21. A word consists of five consonants and four vowels, no two consonants being together. If the alphabet consist of twenty-one consonants and five vowels, how many such words can be formed?

22. Show that five different cards may be placed in four boxes in 1024 different ways.

23. How many words of eight letters can be formed with four vowels and four consonants, the vowels being always in the even places?

24. How many words, each of seven letters, can be formed from three vowels and four consonants, in which no two consonants are next to one another?

25. With the names Jones, Thomson, Wilkinson, how many different groups of three letters may be formed by taking one letter from each name and never two alike in any one group, the order of the letters to be attended to?

26. ABCD is a wire-netting consisting of square meshes, each side of which is an inch long. The side AB is horizontal and $m$ inches long; AD is vertical and $n$ inches long. If AB be the upper side, in how many different ways can an ant crawl from A to C, supposing it never to climb upwards?

27. In how many ways can six boys and six girls be arranged in a ring, so that each girl is between two boys?

28. A typewriter has fifteen distinct letters, and fifteen of each at his disposal. How many distinct words of three letters can he form with them, and how many distinct sets of five such words involving all the fifteen?

29. $n$ swords are laid on the ground crossing each other at the same point. A sword-dancer starts in any one of the angles and springs into another, then into another, and so on, but never goes twice into the same angle until he finally returns to the one he started from. In how many ways can he do this?

30. Given eight points, no three of which are collinear, how many different straight lines can be drawn to connect them; also how many different triangles, quadrilaterals, etc.?

31. At an examination there are eight candidates from one school, and six from another. In how many ways can they be seated at two tables each holding seven, so that candidates from the same school shall not sit next each other, the seats being on one side of a table only?

32. In how many ways can $m$ shillings and $m+n$ florins be given to $m'$ boys and $m'+n'$ girls, one coin to each, where $2m+n=2m'+n'$?

33. In a bicycle race there are $3p$ candidates. They first run in three heats of $p$ each, and then the three winners in a final. If we suppose that there are no dead heats, in how many ways may the race result? First, if we attend only to the winners in each heat; second, if we attend not only to the winner but also to the order of the other competitors in each heat, and suppose none of them to give up?

34. The Morse telegraph signals are a dot and a dash. How many different letters can be telegraphed with five signals, each of which may be either dot or dash; secondly, with five signals, two of which are dots and three dashes?

35. In how many ways can the eight major chessmen of one colour be placed on the board, so that no two pieces are in one row or one column?

§ **250.** *The number of* r-*combinations of* n *things when repetition within the combination is not allowed is*

$$\frac{n(n-1)(n-2)\ .\ .\ .\ (n-r+1)}{1\,.\,2\,.\,3\ .\ .\ .\ r} \qquad (3).$$

Let us denote the number of $r$-combinations by $_nC_r$ and the number of $r$-permutations, as heretofore, by $_nP_r$. We may form the $r$-permutations by first selecting all possible $r$-combinations and then arranging each of these in every possible order. Now each $r$-combination can, by § 246, be arranged in $r!$ different ways; hence the number of $r$-permutations is $_nC_r \times r!$. We must, therefore, have $_nC_r \times r! = {}_nP_r$; and therefore $_nC_r = {}_nP_r/r! = n(n-1)\ .\ .\ .\ (n-r+1)/(1\,.\,2\ .\ .\ .\ r)$, which proves our theorem (compare with § 107*a*).

§ **251.** Since for every $r$-combination we select out of $n$ things we leave a combination of $n-r$ ("**Complementary Combination**") behind, it is clear that the number of $r$-combinations of $n$ things is equal to the number of $(n-r)$-combinations—that is to say, $_nC_r = {}_nC_{n-r}$.

If, therefore, we write out the numbers of combinations of $n$ things taken in every possible way, and take $_nC_0$ (which is, strictly speaking, meaningless) to mean 1, we see that in the series of numbers

$$_nC_0,\quad {}_nC_1,\quad {}_nC_2,\quad .\ .\ .,\quad {}_nC_{n-2},\quad {}_nC_{n-1},\quad {}_nC_n,$$

the first is equal to the last; the second to the last but one; and so on.

The numbers just written are often (for a reason already explained in § 106) called the **Binomial Coefficients.** They are particular cases of multinomial coefficients as defined in § 249. In fact, we have

$$_nC_r = \frac{n(n-1)\ .\ .\ .\ (n-r+1)}{1\,.\,2\ .\ .\ .\ r} \times \frac{(n-r)(n-r-1)\ .\ .\ .\ 2\,.\,1}{1\,.\,2\ .\ .\ .\ (n-r)},$$

$$= \frac{n!}{r!\,(n-r)!} \qquad (4).$$

Now, since $r+(n-r)=n$, $n!/r!\,(n-r)! = {}_nM_{r,\,n-r}$. Therefore $_nC_r$ is a multinomial coefficient of order $n$ and type $(r, n-r)$. Hence we see that *the number of* r-*combinations of* n *things which are all different, is equal to the number of linear permutations of* n *things,* r *of which are alike and the remaining* n − r *also alike.*

§ **252**. The rule, which we have called the "**Addition Theorem for the Binomial Coefficients**" (§ 107), viz.—

$$_nC_r = {}_{n-1}C_r + {}_{n-1}C_{r-1} \qquad (5),$$

can be proved very simply by classifying the $r$-combinations of $n$ things into first, those that do not contain any particular thing $a$; second, those that do contain $a$. The number of $r$-combinations of the first kind is obviously $_{n-1}C_r$. We shall obtain all those that do contain $a$, by first forming all the $(r-1)$-combinations of the remaining $n-1$ letters, the number of which is $_{n-1}C_{r-1}$, and then adding to each of these $a$. Since every $r$-combination either does or does not contain $a$, it follows that $_nC_r = {}_{n-1}C_r + {}_{n-1}C_{r-1}$, a result already deduced in § 107 from the law of distribution.

## EXERCISES LXXIV.

**1.** If $_{2n}C_{n-1}/_{2n-2}C_n = 132/35$, find $n$.

**2.** Find the number of 97-combinations of 100 different things.

**3.** Find the ratio of the number of 5-combinations of 12 things in each of which two given things appear to the number of those in which one but not both of the two given things appear.

**4.** If $_{n-1}C_r : {}_nC_r : {}_{n+1}C_r = 6 : 9 : 13$, find $n$ and $r$.

**5.** Find the number of $r$-permutations of $n$ things in which $s$ particular things occur $(s<r)$.

**6.** Find the number of 6-combinations of 10 things, two of which are alike.

**7.** If the number of $r$-combinations of $n$ different things in which a particular thing does not occur be $p$ times the number of those in which it does occur, show that $n=(p+1)r$.

**8.** If the number of $r$-combinations of $n$ different things in which two particular things occur be equal to the number in which neither occurs, find the relation between $n$ and $r$.

**9.** If a guard of $r$ men be formed out of a company of $m$ men, and guard duty be equally distributed, show that two particular men will be together on guard $r(r-1)$ times out of $m(m-1)$.

**10.** In how many ways can a committee of three be chosen from four married couples, provided that in no case can a husband and his wife be both chosen?

**11.** How many trios can be formed by taking one senior and two juniors from $n$ seniors and $2n$ juniors?

**12.** In how many ways can four red (all alike) and six black balls (all alike) be placed in a row so that no two of the red balls are next to one another?

**13.** How many different words each consisting of one vowel and two consonants can be formed from ten given consonants and five given vowels?

**14.** In how many different orders may a cricket eleven be sent in to bat, the first two players being regarded as sent in together?

**15.** How many words of seven different letters, viz. three vowels and four consonants, can be constructed with the condition that no two vowels shall ever be adjacent?

**16.** A polygon is formed by joining $n$ points in a plane. Find the number of straight lines, not sides of the polygon, which can be drawn joining any two angular points.

**17.** From a company of $n$ soldiers a guard of $r$ men is formed every night. If the guard duty were equally distributed, and every possible arrangement of the guard taken, show that the ratio of the number of days on which a soldier is on guard duty to that on which he is off would be $r : n-r$.

**18.** In a company of twenty men how long would it be before exactly the same guard of ten men would recur, supposing guard duty equally distributed?

**19.** A guard of $r$ men is formed from a company of $m$ in every possible way. It is found that two particular men are three times as often together on guard as three particular men are. Find $m$, $r$ being given.

**20.** How many numbers of five digits can be formed in each of which every digit is greater than the one which follows it on the right?

**21.** A storming party is to be made by selecting $p$ men from a regiment of $l$ men, $q$ from another of $m$, and $r$ from a third of $n$; find an expression for the number of ways in which this may be done.

**22.** A cricket club consists of fourteen members, of whom only four can bowl; how many elevens can be made up so that there shall be in each of them two at least of those who can bowl?

**23.** There are ten things of which two are alike; find the number of permutations of the things taken five at a time.

**24.** If $r$ straight lines in a plane pass through a point A, and $s$ through a point B, how many points of intersection are there, including A and B?

**25.** If $l$ straight lines pass through a point A, $m$ through B, and $n$ through C, and no one of the straight lines contains more than one of the points A, B, C, and no three meet in any point except A, B, or C, find how many triangles are formed by the lines.

**26.** In how many ways can two numbers of three figures each be formed from six given digits of which two are alike?

**27.** A party of three men, six ladies, and nine boys is to cross a river at a ferry. The boat carries each time a man, two ladies, and three boys. In how many ways may the boatman take in his passengers the first time; and in how many ways may the whole party be ferried across?

**28.** Show that ten men, of whom three are brothers, may be arranged in a row so that no two brothers are together in $42 \times 8!$ different ways.

**29.** In how many ways can a dozen things be divided into three sets of four each?

**30.** In how many ways can $m$ different things be put into $n$ pigeon-holes, so that no pigeon-hole contains more than one thing $(m<n)$:

first, when the order of the things in the holes is attended to; second, when the order is not attended to?

**31.** A regiment of $nr$ soldiers is to be distributed in detachments of $r$ each among $n$ different towns; in how many ways can this be done?

**32.** In a college of 100 men there are four coxswains; show that there are $4!\,96!/64!(8!)^4$ ways in which four crews each consisting of a coxswain and eight men may be chosen from the members of the college.

**33.** I have four black balls (exactly alike), and also one red, one white, one green, and one blue ball. In how many ways can I make up a row of four balls, no two rows being alike?

**34.** How many different sums of money can be made with the following coins:—A penny, a sixpence, a shilling, a half-crown, a crown, and a sovereign.

**35.** There are $4n$ things, of which $n$ are alike and all the rest are different. Show that the number of permutations of the $4n$ things taken $2n$ together, each permutation containing the $n$ alike things, is $(3n)!/(n!)^2$.

**36.** There are twelve golfers at the first tee. In how many ways may they play off in couples? In how many ways, if the first is a foursome and the rest couples; (2) if two foursomes and the rest couples; (3) if 1 foursome, 1 couple, 4 singles, and 1 couple?

**37.** A man puts his hand into a bag containing $n$ things, all different; and he may draw 0, 1, 2, . . ., or any number up to $n$; show that he can make $2^n$ different drawings.

**38.** Prove that the whole number of ways in which a selection of one or more things can be made out of $n$ different things is $2^n-1$. If $p$ of the things are alike, prove that the whole number of ways in which such a selection can be made is $(p+1)2^{n-p}-1$.

**39.** In how many ways may a man vote at an election where every voter gives six votes, which he may distribute as he pleases amongst three candidates?

**40.** With $n$ letters $a_1, a_2, \ldots, a_n$, how many different products of the form $a_1^2a_2^2a_3^2a_4^3a_5^4$ can be formed?

**41.** If $r$ pigeon-holes contain respectively $n_1, n_2, \ldots, n_r$ things, all different, how many $s$-combinations can be formed by taking $s$ things from the pigeon-holes, but never more than one thing from each?

## Multinomial Theorem

§ **253.** We can now establish the **Multinomial Theorem for any Positive Integral Exponent,** which is a rule for writing out in standard form the expansion of $(a_1+a_2+\ldots+a_m)^n$; it may be stated symbolically as follows:—

$$(a_1+a_2+\ldots+a_m)^n = \sum\nolimits_n \mathrm{M}_{\alpha_1\alpha_2\ldots\alpha_m}\Sigma a_1^{\alpha_1}a_2^{\alpha_2}\ldots a_m^{\alpha_m} \quad (6),$$

where the smaller sigma denotes summation of all the terms of the type $(\alpha_1, \alpha_2 \ldots, \alpha_m)$ that can be formed with the $m$

letters $a_1, a_2, \ldots, a_m$; ${}_n\mathrm{M}_{a_1 a_2 \ldots a_n}$ denotes the multinomial coefficient of $n$th order and type $(a_1, a_2, \ldots, a_m)$, viz. $n!/a_1!\, a_2! \ldots a_m!$ (where $a_1 + a_2 + \ldots + a_m = n$), (see § 249). The larger sigma denotes summation with respect to all possible types of products of the $n$th degree that can be formed with the $m$ letters.

In the demonstration we shall confine our attention to the case $(a+b+c+d)^4$; clearness will thereby be gained and nothing essential lost in generality.

Since $(a+b+c+d)^4$ is a symmetric function of $a, b, c, d$, the coefficients of all the terms of any given type are equal (see § 100).

The possible types are $a^4$, $a^3b$, $a^2b^2$, $a^2bc$, $abcd$. Hence

$$(a+b+c+d)^4 = \mathrm{A}\Sigma a^4 + \mathrm{B}\Sigma a^3b + \mathrm{C}\Sigma a^2b^2 + \mathrm{D}\Sigma a^2bc + \mathrm{E}\Sigma abcd \qquad (7),$$

where A, B, C, D, E are evidently integers denoting the numbers of times that each term of the respective types $a^4$, $a^3b$, $a^2b^2$, $a^2bc$, $abcd$ occurs in the distribution of the product

$$(a+b+c+d)(a+b+c+d)(a+b+c+d)(a+b+c+d).$$

Consider the partial product $a^2b^2$. We may write $a^2b^2 = aabb$, which may be understood to indicate that $a^2b^2$ may be obtained by taking $a$ from the first bracket, $a$ from the second, $b$ from the third, and $b$ from the fourth. We may also write $a^2b^2 = abab$, which corresponds to taking $a$ from the first, $b$ from the second, $a$ from the third, and $b$ from the fourth bracket; in short, it is obvious that for every distinct way of obtaining the partial product $a^2b^2$ there is a distinct permutation of the letters $aabb$. Now as we have seen in § 249 the number of these permutations is ${}_4\mathrm{M}_{2,2} = 4!/2!\,2!$.

The same reasoning applies to the other four types. We have therefore

$$\begin{aligned}(a+b+c+d)^4 &\\ &= {}_4\mathrm{M}_4\Sigma a^4 + {}_4\mathrm{M}_{3,1}\Sigma a^3b + {}_4\mathrm{M}_{2,2}\Sigma a^2b^2 \\ &\qquad + {}_4\mathrm{M}_{2,1,1}\Sigma a^2bc + {}_4\mathrm{M}_{1,1,1,1}\Sigma abcd, \\ &= \frac{4!}{4!}\Sigma a^4 + \frac{4!}{3!\,1!}\Sigma a^3b + \frac{4!}{2!\,2!}\Sigma a^2b^2 + \frac{4!}{2!\,1!\,1!}\Sigma a^2bc \\ &\qquad + \frac{4!}{1!\,1!\,1!\,1!}\Sigma abcd, \\ &= \Sigma a^4 + 4\Sigma a^3b + 6\Sigma a^2b^2 + 12\Sigma a^2bc + 24\Sigma abcd \qquad (8).\end{aligned}$$

It will be observed that the formation of the coefficients has nothing whatever to do with the number of the letters *abcd*, so long as there are a certain minimum number present. Each coefficient depends merely on the exponent $n$ and on the type indices $\alpha_1, \alpha_2, \ldots, \alpha_m$. Take, for example $(a+b+c+d+e)^4$, the types are the same as before, viz. $a^4$, $a^3b$, $a^2b^2$, $a^2bc$, $abcd$. Hence the coefficients are the same as before. We have, in fact—

$$(a+b+c+d+e)^4 = \Sigma a^4 + 4\Sigma a^3b + 6\Sigma a^2b^2 + 12\Sigma a^2bc + 24\Sigma abcd \qquad (9),$$

The difference between (8) and (9) consists solely in the difference of the meaning of the sigmas. Thus in (8) $\Sigma a^4 = a^4 + b^4 + c^4 + d^4$; in (9) $\Sigma a^4 = a^4 + b^4 + c^4 + d^4 + e^4$; in (8) $\Sigma abcd$ is simply $abcd$; in (9) $\Sigma abcd = bcde + acde + abde + abce + abcd$; and so on.

It will now be evident that (8) holds for any number of variables not less than 4.

For three letters we have simply to put $d=0$, we thus get

$$(a+b+c)^4 = \Sigma a^4 + 4\Sigma a^3b + 6\Sigma a^2b^2 + 12\Sigma a^2bc \qquad (10);$$

and so on.

The truth of the general theorem (6) will now be obvious; after what has been said, there is nothing new in it except the generality of the notation.

It will be observed that, if $m=n$ or $>n$, the terms of type $a_1a_2\ldots a_n$ can occur; and therefore terms of the required degree of all possible types will occur on the right of (6).

If $m<n$ certain of the types will be wanting; in particular no term of the type $a_1a_2\ldots a_n$ can occur.

If $m=2$, the only possible types are $a_1^n, a_1^{n-1}a_2, a_1^{n-2}a_2^2, \ldots, a_1^{n-r}a_2^r, \ldots, a_2^n$, each type containing only a single term. The equation (6) then becomes

$$(a_1+a_2)^n = a_1^n + \frac{n!}{(n-1)!\,1!}a_1^{n-1}a_2 + \ldots + \frac{n!}{(n-r)!\,r!}a_1^{n-r}a_2^r + \ldots + a_2^n \qquad (11),$$

which is simply the binomial theorem.

Ex. 1. If ${}_nC_r$ have its usual meaning, and $r \not> n-1$, prove that

$$1 - {}_nC_1 + {}_nC_2 - \ldots + (-1)^r{}_nC_r = (-1)^r{}_{n-1}C_r \qquad (12).$$

Since $1-x^n \equiv (1-x)(1+x+x^2+\ldots+x^{n-1})$, we have the identity

$$(1-x^n)(1-x)^{n-1} \equiv (1-x)^n(1+x+x^2+\ldots+x^{n-1});$$

that is to say—

$$(1-x^n)(1-{}_{n-1}C_1x+{}_{n-1}C_2x^2+\ldots+(-1)^{n-1}{}_{n-1}C_{n-1}x^{n-1})$$
$$\equiv(1-{}_nC_1x+{}_nC_2x^2-\ldots+(-1)_n{}^n C_n x^n)(1+x+x^2+\ldots+x^{n-1}) \quad (13).$$

If we now compare the coefficients of $x^r$ $(r \not> n-1)$ on both sides of (13), we see at once that $(-1)^r{}_{n-1}C_r=1-{}_nC_1+{}_nC_2-\ldots+(-1)^r{}_nC_r$.

Ex. 2. Find the coefficient of $a^2b^2c^2d^2$ in the expansion of $(a+b+c+d)^8$.

The coefficient is $8!/2!2!2!2!=2520$.

Ex. 3. Find the coefficient of $x^6$ in the expansion of $(1+3x+2x^2)^{10}$.

We have, by (6)—

$$(1+3x+2x^2)^{10}=\Sigma\frac{10!}{\alpha!\beta!\gamma!}1^\alpha(3x)^\beta(2x^2)^\gamma$$
$$=\Sigma\frac{3^\beta 2^\gamma 10!}{\alpha!\beta!\gamma!}x^{\beta+2\gamma} \quad (14),$$

where $\alpha+\beta+\gamma=10$ (15).

We have therefore to select those terms on the right of (14) for which

$$\beta+2\gamma=6 \quad (16),$$

and add the coefficients.

Since $\alpha$, $\beta$, $\gamma$ are all either zero or positive integers, it is easy to see that the admissible values are given by the following table:—

| $\alpha$ | $\beta$ | $\gamma$ |
|---|---|---|
| 4 | 6 | 0 |
| 5 | 4 | 1 |
| 6 | 2 | 2 |
| 7 | 0 | 3 |

therefore the required coefficient is

$$\frac{3^6 . 10!}{4!6!}+\frac{3^4 . 2 . 10!}{5!4!1!}+\frac{3^2 . 2^2 . 10!}{6!2!2!}+\frac{2^3 . 10!}{7!3!}=403,530.$$

## EXERCISES LXXV.

1. Write down the eighteenth term of $(2x-y)^{19}$.
2. Find the ninth term of the expansion of $(4y-x)^{11}$.
3. What is the coefficient of $x^6$ in $(3x-\frac{1}{3})^9$; and what is the sum of all the coefficients in $(2x-1)^n$?
4. Find the coefficient of $x$ in the expansion of $(2x-3/x)^9$.
5. Find the coefficient of $x^r$ in $(1-x)^n(1+x)^2$.
6. Show that, if $m<n$, $r<n-m$, then

$${}_{n+m}C_{r+m}={}_nC_{r+m}+m\,{}_nC_{r+m-1}+\frac{m(m-1)}{1.2}{}_nC_{r+m-2}+\ldots+{}_nC_r.$$

**7.** The three middle terms in the expansion of $(x+1)^{n^2-1}$, where $n$ is an odd integer, are in A. P.; show that $x=(n+1)/(n-1)$ or $(n-1)/(n+1)$.

**8.** If the coefficients of $x^n$ in $(ax+b)^{2n}$ and $(bx+a)^{2n+1}$ be equal, show that $a=(n+1)/(2n+1)$.

**9.** The product of three expressions each of the form $a+b+c+\ldots$ is to be formed. There are seven letters in the first, six in the second, and five in the third. How many distinct terms are there if four of the letters in the first are common to the other two?

**10.** There are ten letters $a, b, c, d, \ldots$; how many different terms are there of the type (1) $a^3bd$; (2) $a^2b^2c^2$; (3) $a^2b^2c^3$; (4) $a^4bc^2$; (5) $abcd^2e$? How many different types are there of the fourth and fifth degrees respectively with these letters?

**11.** How many different types exist in the expansion of $(a+b+c+d+e)^5$?

**12.** Show that the expansion of $(a_1+a_2+\ldots+a_m)^r$ can be made to depend on the expansion of $(a_1+a_2+\ldots+a_r)^r$, $m>r$.

**13.** Prove that if $2s \ngtr n$, $(-1)^s {}_nC_{2s} = {}_nC_{2s} - {}_nC_1\,{}_nC_{2s-1} + {}_nC_2\,{}_nC_{2s-2} - \ldots$

**14.** Expand $(a+b+c)^7$.

**15.** Find the coefficient of $a^5b^5$ in the expansion of $(a+b+c+d)^{10}$.

**16.** Expand $(a+b+c+d)^6$.

**17.** Find the coefficient of $ab^2cd$ in the expansion of $(a-2b+2c-d)^5$.

**18.** Find the coefficient of $ab^2c^2d$ in the expansion of $(a+2b+3c+4d)^6$.

**19.** Find the coefficient of $x^{15}$ in the expansion of $(1+x^3+x^5)^{11}$.

**20.** Find the coefficient of $x^9$ in the expansion of $(1-2x+x^4)^{12}$.

**21.** Find the coefficient of $x^{13}$ in the expansion of $(1-x+2x^2)^{11}$.

**22.** Find the coefficient of $x^{10}$ in the expansion of $(1-x+x^2-x^3)^{10}$.

# CHAPTER XXIV

## DERIVATION OF CONDITIONAL EQUATIONS

§ 254. By the **Derivation of Conditional Equations** we understand the process of deriving from a system of equations, A, another system B which has all the solutions of A. As this kind of work is an art in which almost every algebraic resource can be utilised, we have reserved it for the end of this little work. The learner will find here an algebraic palæstra in which he can practise every kind of gymnastic that he knows.

The systems A and B spoken of may each contain one or more equations. If B contain fewer equations than A, there can of course, in general, be no question of equivalence; because (supposing all the equations of A independent) B will have infinitely more solutions than A.

If the number of equations in B be the same as the number of independent equations in A, there will naturally arise two questions: first, whether all the equations in B are independent; second, supposing them all independent, is B equivalent to A—that is, has it exactly the same solutions as A, or has it more? When the given system A contains only one equation, the derived system B can of course contain only one independent equation; but the question of equivalence still remains.

It is easy enough as a rule to make derivations; but not so easy to settle the questions just alluded to, *i.e.* to make it clear what is the exact meaning of the derivation. We point this out because an unfortunate practice has arisen of setting loosely worded exercises on this subject in examination papers, which not unfrequently amount to asking the examinee to establish conclusions which are not rigorously true.

§ 255. The **Use of the Principle of the Minimum Number**

**of Variables** is of course the theoretically fundamental method in the derivation of equations. By means of the given system A we express all the variables in terms of the smallest number possible; then, if B be a true derivative of A, every equation in B will become an identity.

Ex. 1. If $$yz+zx+xy=0 \qquad (1),$$
and none of the six $x$, $y$, $z$, $y+z$, $z+x$, $x+y$ vanish, show that
$$\Sigma 1/(y+z) = -\Sigma x^2/xyz \qquad (2).$$

Since $y+z \neq 0$, we have, by the given equation, $x=-yz/(y+z)$; and obviously we cannot use fewer than the two variables $y$ and $z$. We have therefore, since
$$z+x=z-yz/(y+z)=z^2/(y+z),$$
$$x+y=y-yz/(y+z)=y^2/(y+z),$$
to establish the identity
$$\frac{1}{y+z}+\frac{y+z}{z^2}+\frac{y+z}{y^2} \equiv \left\{\frac{y^2z^2}{(y+z)^2}+y^2+z^2\right\}\Big/\frac{y^2z^2}{y+z},$$
$$\equiv \left\{\frac{y^2z^2}{(y+z)^2}+y^2+z^2\right\}\frac{y+z}{y^2z^2},$$
which is immediately obvious, if we distribute the crooked bracket.

We shall give a more elegant solution of this example presently; meantime the learner should try to solve the more difficult question—whether (2) is equivalent to (1) under the restrictions imposed upon $x$, $y$, $z$, etc.

§ 256. The **Use of the Elementary Symmetric Functions** in many cases simplifies the application of the principle of the minimum number of variables. If we denote $\Sigma x$, $\Sigma xy$, $\Sigma xyz$, . . ., where there may be any number of variables, by $p_1$, $p_2$, $p_3$, . . ., it is a fundamental theorem in the calculus of identities (the general proof of which the beginner will meet with later on),* that any integral symmetric function can be expressed as an integral function of $p_1$, $p_2$, $p_3$, . . ., which we call the elementary symmetric functions.

A convenient method for calculating the monotypic symmetric functions in terms of $p_1$, $p_2$, $p_3$, . . ., was given by Cayley. It runs as follows :—

$$p_1=\Sigma x;$$
$$\left.\begin{array}{l} p_2=\Sigma xy, \\ p_1^2=\Sigma x^2+2\Sigma xy; \end{array}\right\}$$

* See A. XVIII. § 4.

$$\left.\begin{aligned} p_3 &= \Sigma xyz, \\ p_1p_2 &= \Sigma x^2y + 3\Sigma xyz, \\ p_1^3 &= \Sigma x^3 + 3\Sigma x^2y + 6\Sigma xyz\,; \end{aligned}\right\}$$

$$\left.\begin{aligned} p_4 &= \Sigma xyzu, \\ p_1p_3 &= \Sigma x^2yz + 4\Sigma xyzu, \\ p_2^2 &= \Sigma x^2y^2 + 2\Sigma x^2yz + 6\Sigma xyzu, \\ p_1^2p_2 &= \Sigma x^3y + 2\Sigma x^2y^2 + 5\Sigma x^2yz + 12\Sigma xyzu, \\ p_1^4 &= \Sigma x^4 + 4\Sigma x^3y + 6\Sigma x^2y^2 + 12\Sigma x^2yz + 24\Sigma xyzu\,; \\ &\text{etc.} \end{aligned}\right\}$$

From these the reader will calculate without difficulty

$$\Sigma x^2 = p_1^2 - 2p_2 \qquad (3),$$

$$\left.\begin{aligned} \Sigma x^3 &= p_1^3 - 3p_1p_2 + 3p_3 \\ \Sigma x^2y &= p_1p_2 - 3p_3 \end{aligned}\right\} \qquad (4)\,;$$

$$\left.\begin{aligned} \Sigma x^4 &= p_1^4 - 4p_1^2p_2 + 2p_2^2 + 4p_1p_3 - 4p_4 \\ \Sigma x^3y &= p_1^2p_2 - 2p_2^2 - p_1p_3 + 4p_4 \\ \Sigma x^2y^2 &= p_2^2 - 2p_1p_3 + 2p_4 \\ \Sigma x^2yz &= p_1p_3 - 4p_4 \end{aligned}\right\} \qquad (5)\,;$$

etc.

These formulæ hold for any number of variables, provided we notice that, when there are only two variables, we must put $p_3 = 0$, $p_4 = 0$, etc.; if there are only three, $p_4 = 0$, $p_5 = 0$, etc., and so on.

Knowing that expressions of the kind exist in all cases, we can employ a variety of special methods for obtaining them quickly in particular cases (see A. Chap. XVIII. §§ 2, 3, 4).

Ex. 2. Given that $x \neq 0$, $y \neq 0$, $z \neq 0$, $y+z \neq 0$, $z+x \neq 0$, $x+y \neq 0$, is the equation (2) of Ex. 1, § 255, equivalent to (1), from which it is derived?

Since (2) is a derivative of (1), if we write it in the form

$$xyz\Sigma(x+y)(x+z) + (y+z)(z+x)(x+y)\Sigma x^2 = 0 \qquad (6),$$

which is equivalent to (2), since $x \neq 0$, etc., we should expect that $\Sigma xy$, the characteristic of (1), will be a factor in the characteristic of (6).

Introducing the elementary symmetric functions as variables, by means of (4), we find that (6) becomes

$$p_3(p_1^2 + p_2) + (p_1^2 - 2p_2)(p_1p_2 - p_3) = 0,$$

that is— $$p_2(p_1^3 - 2p_1p_2 + 3p_3) = 0.$$

Hence (6) is equivalent to

$$p_2=0 \quad \text{and} \quad p_1^3-2p_1p_2+3p_3=0\ ;$$

that is to say, to (1) together with

$$(\Sigma x)^3-2\Sigma x\Sigma xy+3xyz=0,$$

which may also be written

$$\Sigma x^3+\Sigma x^2y+3xyz=0.$$

Ex. 3. Eliminate $x$ and $y$ from the equations

$$x+y=a, \quad x^3+y^3=b^3, \quad x^4+y^4=c^4,$$

where $a \neq 0$, $b \neq 0$, $c \neq 0$.

If we introduce as variables the elementary symmetric functions $p_1=x+y$, $p_2=xy$, the three equations become

$$p_1=a,$$
$$p_1^3-3p_1p_2=b^3,$$
$$p_1^4-4p_1^2p_2+2p_2^2=c^4.$$

The first of these gives $p_1=a$, consequently the second becomes $3ap_2=a^3-b^3$, which reduces the third to

$$9a^6-12a^3(a^3-b^3)+2(a^3-b^3)^2=9a^2c^4,$$

which is equivalent to

$$a^6-8a^3b^3-2b^6+9a^2c^4=0 \qquad (7)\ ;$$

and this is the required eliminant, viz. it follows that $a$, $b$, $c$ must satisfy this last equation if the three given equations in $x$ and $y$ be consistent.

Since $a \neq 0$, it can readily be shown by reversing the analysis that any common solution of $x+y=a$, and $x^3+y^3=b^3$ satisfies $x^4+y^4=c^4$, provided $a$, $b$, $c$ are connected by the relation (7).

§ **257. Indirect Methods of Derivation.**—It often happens that the direct application of the principle of the minimum number of variables is impossible or inconvenient, by reason of the complication of the calculations required. In most cases, for example, irrational operations would be required ; and these it is a matter of convenience and also to some extent a point of honour for the algebraist to avoid, unless some obvious special advantage or elegance can be gained by their use.

In such cases indirect methods of derivation are resorted to, such as **Interequational Transformation**, which has already been explained in § 76.

If this process be used among the equations of a given system A, it has the advantage that the derived system B, if it contains a sufficient number of independent equations, is equivalent to A.

We may also use **Irreversible Processes of Derivation,**

such as squaring both sides of an equation or multiplying or dividing by functions of the variables; in such cases the question of equivalence, if it arise, requires special examination.

Ex. 4. We may solve Example 1 indirectly in the following, more elegant way :—

Since $x \neq 0$, $y \neq 0$, $z \neq 0$, we have

$$\Sigma 1/(y+z) = \Sigma x/(xy+xz).$$

Now, since $\Sigma xy = 0$, we have $xy + xz = -yz$. Therefore

$$\begin{aligned}\Sigma x/(xy+xz) &= -\Sigma x/yz, \\ &= -\Sigma x^2/xyz,\end{aligned}$$

which establishes the derivation.

Ex. 5. Show that we can always derive from the system

$$lx + my + n = 0 \qquad (8),$$

$$ax^2 + 2hxy + by^2 + 2gx + 2fy + c = 0 \qquad (9),$$

where $l$ and $m$ do not both vanish, an equivalent system of the form

$$\left.\begin{aligned} lx + my + n = 0 \\ a'(x^2+y^2) + 2g'x + 2f'y + c' = 0 \end{aligned}\right\} \qquad (10).$$

The system

$$lx + my + n = 0 \qquad (11),$$

$$ax^2 + 2hxy + by^2 + 2gx + 2fy + c + (\lambda x + \mu y + \nu)(lx + my + n) = 0 \qquad (12),$$

is equivalent to (8) and (9); for, by using (11) in (12), we reduce (12) at once to (9).

Now $\lambda$, $\mu$, $\nu$ are constants absolutely at our disposal. Let us choose them so that the coefficients of $x^2$ and $y^2$ in (12) are equal, and so that the coefficient of $xy$ shall vanish.

The necessary and sufficient conditions are

$$l\lambda - m\mu + a - b = 0, \quad m\lambda + l\mu + 2h = 0,$$

whence

$$\begin{aligned}\lambda &= \{(b-a)l - 2hm\}/(l^2+m^2), \\ \mu &= \{(a-b)m - 2hl\}/(l^2+m^2).\end{aligned}$$

The values of $\lambda$ and $\mu$ will therefore be finite and determinate, so long as $l$ and $m$ do not both vanish. It will be observed that $\nu$ has not been conditioned in any way. The required derivation can therefore be accomplished in a one-fold infinity of ways.

The reader will afterwards learn that the above analysis amounts to proving that the intersections of a conic and a straight line can be found in an infinity of ways by elementary geometric construction—a result which follows also from the considerations of § 240.

Ex. 6. If $(y^2+z^2+yz)/(y+z) = (z^2+x^2+zx)/(z+x)$, and if $x \neq y$, $y+z \neq 0$, $z+x \neq 0$, $x+y \neq 0$, then each of these fractions is equal to $(x^2+y^2+xy)/(x+y)$.

Since $y+z \neq 0$, $z+x \neq 0$, the given equation is equivalent to

$$(z+x)(y^2+z^2+yz) - (y+z)(z^2+x^2+zx) = 0;$$

that is— $z^2(x-y)+y^2(z+x)-x^2(y+z)+z\{y(z+x)-x(y+z)\}=0$;

that is— $z^2(x-y)-z(x^2-y^2)-xy(x-y)-z^2(x-y)=0$;

that is— $-(x-y)(yz+zx+xy)=0.$

Therefore, since $x\neq y$, the given equation is equivalent to

$$yz+zx+xy=0 \qquad (13).$$

From the symmetry of this equation, we might at once infer the required conclusion; but it will probably be more convincing if we actually *derive* the result from (13).

Multiplying both sides of (13) by $-(y-z)$, we derive

$$-(y-z)(yz+zx+xy)=0;$$

from which in succession—

$$x^2(y-z)-x(y^2-z^2)-yz(y-z)-x^2(y-z)=0;$$
$$x^2(y-z)+z^2(x+y)-y^2(z+x)+x\{z(x+y)-y(z+x)\}=0;$$
$$(x+y)(z^2+x^2+zx)-(z+x)(x^2+y^2+xy)=0.$$

Hence, since $x+y\neq 0$, $z+x\neq 0$ we derive finally

$$(z^2+x^2+zx)/(z+x)=(x^2+y^2+xy)/(x+y).$$

It will be observed that the steps in the latter half of the calculation might have been derived from the first half by reversing the order of the equations, and executing in each the circular permutation of $x, y, z$ into $y, z, x$.

Ex. 7. If

$$1/(1+x+xy)+1/(1+y+yz)+1/(1+z+zx)=1 \qquad (14),$$

and $1+x+xy\neq 0$, $1+y+yz\neq 0$, $1+z+zx\neq 0$, then either $xyz=1$, or else $\Pi(1+x)=-1$.

Let us put, for the moment, A for $x+xy$, B for $y+yz$, and C for $z+zx$. Then, since, by our conditions, $1+A\neq 0$, $1+B\neq 0$, $1+C\neq 0$, (14) is equivalent to

$$\Sigma(1+B)(1+C)=\Pi(1+A);$$

that is, to $3+2\Sigma A+\Sigma AB=1+\Sigma A+\Sigma AB+ABC$;

that is, to $2+\Sigma A-ABC=0$;

or, if we replace A, B, C by their values, to

$$2+\Sigma x+\Sigma xy-xyz\Pi(1+x)=0.$$

Now this last equation may be written

$$1+\Pi(1+x)-xyz-xyz\Pi(1+x)=0,$$

or $$(1-xyz)\{1+\Pi(1+x)\}=0.$$

Hence either $1-xyz=0$, or $1+\Pi(1+x)=0$, which is virtually the required conclusion.

§ **258.** In the applications of Algebra, to co-ordinate geometry, for example, which is one of the most beautiful, indirect derivation of conditional equations is one of the commonest operations. The student of mathematics who has mastered the bare elements should therefore practise this exercise assiduously. But we repeat our warning that, important as is dexterity of the fingers, he who cultivates this at the expense of accurate thought may become a successful worker of certain kinds of superficial examination problems, but a mathematician, or a successful applier of mathematics to things practical, never!

## EXERCISES LXXVI.

**1.** If $a+b+c=0=bc-ca-ab$, and $a$, $b$, $c$ be all real, show that the product $bc$ is negative.

**2.** If $(y/x+b/a)(z/x+c/a)=bc/a^2$, then $\Sigma a/x=0$, provided $a$, $x$, $y$, $z$ be all finite both ways.

**3.** If $x^2=x+1$, $x^3=2x+1$.

**4.** If $a+b+c+d=0$, prove that $a^3+b^3+c^3+d^3+3(a+b)(b+c)(c+a)=0$, and $(a+b)(a+c)(a+d)=(b+c)(b+d)(b+a)=(c+d)(c+a)(c+b)=(d+a)(d+b)(d+c)$.

**5.** Eliminate $x$ and $y$ from $x+y=4$, $3x+5y=14$, $ax^2+bxy+cy^2=d$.

**6.** If $a$, $b$, $c$, $d$ be all positive and the sum of any three greater than the fourth, and if $\{(c+d)^2-(a-b)^2\}/\{(a+c+d)^2-b^2\}+\{(d+a)^2-(b-c)^2\}/\{(a+b+d)^2-c^2\}=1$, prove that $a+b=c+d$.

**7.** Show that if $a^2=bc$, then $(b^2-ca)^3+(c^2-ab)^3$ can be expressed as the square of a rational function of certain of the letters $a$, $b$, $c$.

**8.** If $(x+1)^2=x$, find the value of $11x^3+8x^2+8x-2$.

**9.** Show that, if $x \neq y \neq z$, the equations $(x-z)/(x-y)=(y-x)/(y-z)=(z-y)/(z-x)$ are equivalent to a single equation only.

**10.** Eliminate $x$ and $y$ from the system $x+y=a+b$, $(a+\lambda)x+(b+\lambda)y=ab$, $(b+2\lambda)x+(a+2\lambda)y=a^2+b^2$.

**11.** If $x=pq$, $y=qr$, $z=rs$ and $x+y+z=ps$, then $(x+y)^2+(y+z)^2=(x+z)^2$.

**12.** If $x$, $y$, $z$, $u$ are real and $x^2+u^2=2(xy+yz+zu-y^2-z^2)$, prove that $x=y=z=u$.

**13.** If $x-\alpha : y-\beta : z-\gamma = l : m : n$, show that $(m\gamma-n\beta)x+(n\alpha-l\gamma)y+(l\beta-m\alpha)z=0$.

**14.** If $a:b=c:d$, then $a^2+b^2 : c^2+d^2=ab:cd$. Is the converse necessarily true?

**15.** If $y+z+u=ax$, $z+u+x=by$, $u+x+y=cz$, $x+y+z=du$, then $1=\Sigma 1/(a+1)$ or $\Sigma x=0$.

**16.** If $a:b=c:d$, $a$, $b$, $c$, $d$ being each finite both ways, prove that $a^2+b^2+c^2+d^2 : a^2-b^2+c^2-d^2 = a^2+b^2-c^2-d^2 : a^2-b^2-c^2+d^2$. Is the converse true?

**17.** If $x$ and $y$ be real and $1+x^3=y^3-3xy$, then either $1+x=y$ or $x=-y=1$.

**18.** If $ax^2+bxy+cy^2=1$, $cx^2+bxy+ay^2=1$, $x+y=1$, and $c \neq a$, show that $a+b+c=4$.

**19.** If $x^2+y^2=1$ and $x+y=2$, show that $(x^2+x+1)^2+(y^2+y+1)^2=7/2$.

**20.** If $yz+ay+bz=zx+az+bx=xy+ax+by=0$, $yz+zx+xy=xyz$, and $x\neq 0$, $y\neq 0$, $z\neq 0$, then $a+b=-3$.

**21.** Given $(bz+cy)/ayz=(cx+az)/bzx=(ay+bx)/cxy$, $x\neq 0$, $y\neq 0$, $z\neq 0$, show that $x:y:z=a/(b+c-a):b/(c+a-b):c/(a+b-c)$, provided $a$, $b$, $c$ satisfy certain restrictions.

**22.** If $\Sigma x=0$, $x\neq 0$, $y\neq 0$, $z\neq 0$, then $\Sigma 1/(y^2+z^2-x^2)=0$.

**23.** If $A=b-c$, $B=c-a$, $C=a-b$, then $\Sigma A^4=2\Sigma A^2B^2$.

**24.** If $x=a(y+z)$, $y=b(z+x)$, $z=c(x+y)$, and $x+y+z\neq 0$, $x\neq 0$, $y\neq 0$, $z\neq 0$, find a relation between $a$, $b$, $c$ independent of $x$, $y$, $z$; and show that $x^2/a(1-bc)=y^2/b(1-ca)=z^2/c(1-ab)$.

**25.** Find systems of the forms $lx+my+nz=0$, $Ax^2+By^2+Cz^2=0$ ($\alpha$), $lx+my+nz=0$, $Dyz+Ezx+Fxy=0$ ($\beta$), equivalent to $lx+my+nz=0$, $ax^2+by^2+cz^2+2fyz+2gzx+2hxy=0$.

**26.** If $a+b+c=0$, and $ax+by+cz=0$, then $\Sigma(cy+bz)^3=3\Pi(cy+bz)$.

**27.** If $bz+cy=cx+az=ay+bx$ and $\Sigma x^2=2\Sigma yz$, then $a\pm b\pm c=0$.

**28.** If $2s=a+b+c$, $S=a^2+b^2+c^2$, prove that $\Sigma(S-2a^2)(S-2b^2)=16s(s-a)(s-b)(s-c)$.

**29.** If $\Sigma x=0$, then $\Sigma x^4/(y^3+z^3-3xyz)=0$.

**30.** If $\Sigma a=0$, and $a(by+cz-ax)=b(cz+ax-by)=c(ax+by-cz)$, show that $\Sigma x=0$, provided $a\neq 0$, $b\neq 0$, $c\neq 0$. What follows if $a=0$, $b\neq 0$?

**31.** Eliminate $x$ and $y$ from $x/a+y/b=1$, $x^2+y^2=c^2$, $xy=d^2$.

**32.** If $y_1{}^2-4ax_1=y_2{}^2-4ax_2=0$, $(y-y_1)(x_1-x_2)+(x-x_1)(y_2-y_1)=0$, and $y_1\neq y_2$, $a\neq 0$, then $y(y_1+y_2)-y_1y_2-4ax=0$.

**33.** If $\Sigma ab=0$, prove $(\Sigma a)^2=\Sigma a^2$; $(\Sigma a)^3=\Sigma a^3-3abc$; $(\Sigma a)^4=\Sigma a^4-4abc\Sigma a$.

**34.** Eliminate $x$ and $y$ from $x+y=a$, $x^2+y^2=b^2$, $x^3-x^2y-xy^2+y^3=c^3$.

**35.** If $3\Sigma a^2b-2\Sigma a^3-12abc=0$, one of the three $a$, $b$, $c$ must be an arithmetic mean between the other two.

**36.** If $\Sigma a\neq 0$, $\Sigma ab\neq 0$, and $\Sigma a^3/\Sigma a^2=\Sigma a^2/\Sigma a$, then each of these $=\Sigma a(b^2+c^2)/2\Sigma ab$.

**37.** Show that the equations $\Sigma x=0$, $\Pi(y+z)=a$, $\Sigma x^3=b$ are inconsistent unless $3a+b=0$. If this condition be satisfied, there are an infinity of common solutions.

**38.** If $a\neq b\neq c$, $x+y+z=0$, and $x^2/(b-c)+y^2/(c-a)+z^2/(a-b)=0$, then $x/(b-c)=y/(c-a)=z/(a-b)$.

**39.** If $x/(yz-x^2)=a$, $y/(zx-y^2)=b$, $z/(xy-z^2)=c$, then under certain restrictions $a/(bc-a^2):b/(ca-b^2):c/(ab-c^2)=x:y:z$; and if $yz+zx+xy=0$, then $a=b=c$, or $\Sigma a=0$.

**40.** If $(yz-x^2)/(y+z)=(zx-y^2)/(z+x)$, $x\neq y$, $y+z\neq 0$, $z+x\neq 0$, show that each of them $=x+y+z$.

**41.** If $\Sigma x=a$, $\Sigma 1/x=1/a$, then $\Sigma x^{2n+1}=a^{2n+1}$, $\Sigma 1/x^{2n+1}=1/a^{2n+1}$.

**42.** If $x\neq y\neq z$, $y+z\neq 0$, $z+x\neq 0$, $x+y\neq 0$, and $(yz-x^2)/(y+z)=(zx-y^2)/(z+x)$, prove that each of these fractions is equal to $x+y+z$, and also to $(xy-z^2)/(x+y)$. Show that $x$, $y$, $z$ cannot be all real.

**43.** If $x$, $y$, $z$ be all finite both ways, and all unequal, then $(yz-x^2)/x=(zx-y^2)/y=(xy-z^2)/z$ are equivalent to one equation only, and $\Sigma y^4z^4=2x^2y^2z^2\Sigma x^2$.

**44.** If $x^2-yz-a^2=y^2-zx-b^2=z^2-xy-c^2=0$, and $a^4-b^2c^2\neq 0$, $b^4-c^2a^2\neq 0$, $c^4-a^2b^2\neq 0$, show that $x/(a^4-b^2c^2)=y/(b^4-c^2a^2)=z/(c^4-a^2b^2)$; and find $x$, $y$, $z$ in terms of $a$, $b$, $c$.

**45.** If $l^2+m^2+n^2=1$, $al^2+bm^2+cn^2=0$, $a\neq b\neq c$, then $\Sigma a^2l^2=\Sigma m^2n^2(b-c)=\sqrt{\{-\Sigma m^2n^2bc(b-c)^2\}}=S$ say. If $b=c$, $a\neq b$, then $S=-ca$.

**46.** If $(x-a)/yz=(y-b)/zx=(z-c)/xy$, and $x\neq 0$, $y\neq 0$, $z\neq 0$, show that $\Sigma ax(y^2-z^2)=0$, $\Sigma a^2x^2(y^2-z^2)=-(y^2-z^2)(z^2-x^2)(x^2-y^2)$.

**47.** If $x$, $y$, $z$ are each finite both ways, $y+z\neq 0$, $z+x\neq 0$, $x+y\neq 0$, and $(y^2+z^2-x^2)/(y+z)=(z^2+x^2-y^2)/(z+x)=(x^2+y^2-z^2)/(x+y)$, show that at least two of the three $x$, $y$, $z$ must be equal; and find all the admissible values of the ratios $x:y:z$.

# CHAPTER XXV

## GRAPHICAL DISCUSSION OF RATIONAL FUNCTIONS

### Simple Typical Cases

§ **259.** The general character of the graph of a rational function is largely determined by the position and nature of the **Zeros** and **Infinities** of the function—that is to say, by the values of $x$ for which the values of the function become zero or cease to be finite, and by the way in which the value of the function varies in such neighbourhoods. We shall, therefore, gain simplicity as well as clearness and brevity in the following discussion if we begin by working out carefully a few simple typical cases to illustrate the behaviour of a rational function and the corresponding peculiarities of its graph in the neighbourhood of a zero or an infinity of the function.

In working through this chapter, the student is supposed to trace out all the fundamental curves for himself, using either plotting paper or square and scale. In no other way can he arrive at that kind of "lively conviction" which will enable him to discuss new cases for himself without hesitation or liability to error.

§ **260. Case I.** $\qquad y = ax \qquad$ (1).

We have already seen (§ 58) that the graph for this equation is a straight line passing through the origin. The graph will run to an infinite distance in the first and third quadrants, or in the second and fourth quadrants, according as $a$ is positive or negative. It should be observed that in passing through its zero, which corresponds to $x = 0$, the value of the function changes its sign, and the graph crosses the $x$-axis at the origin, making with the $x$-axis an angle which is not infinitely small, supposing, as we tacitly do, that $a \neq 0$.

**§ 261. Case II.** $$y = ax^2 \qquad (2).$$

The nature of the graph for this equation (see §§ 55, 58) has already been fully indicated in the case where $a$ is positive. To get the graph when $a$ has any negative value, say $a = -3$, we

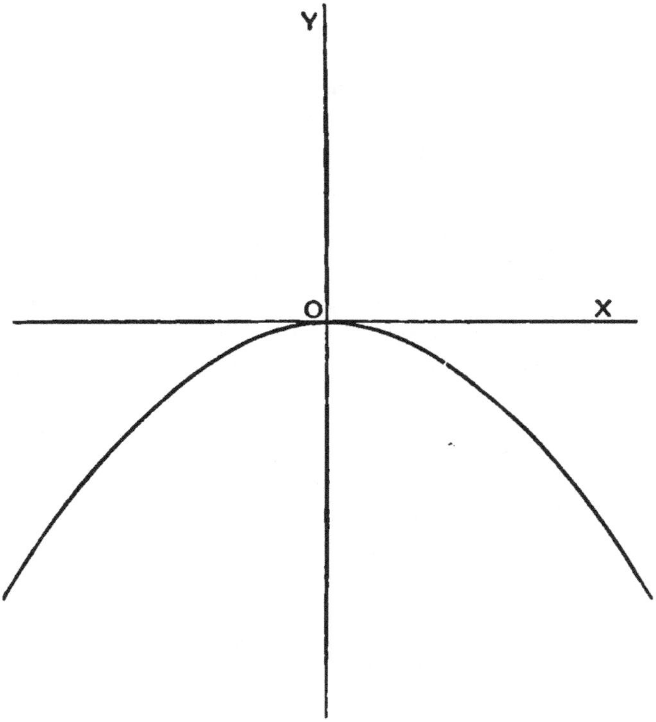


FIG. 7.

have obviously merely to draw the graph for the corresponding positive value of $a$, viz., in the case supposed, the graph of $y = 3x^2$, and take its image in a plane mirror whose plane is perpendicular to the co-ordinate plane, and whose reflecting surface meets that plane in the $x$-axis.*

We thus get a graph like Fig. 7. We may speak of these

* This process we shall hereafter briefly describe as reflection in the $x$-axis.

two graphs as a **Festoon** and an **Inverted Festoon** respectively.

The graph of $y=ax^2$ is typical of $y=ax^{2m}$, where $2m$ is any even positive integer. The only difference observable when the graphs are drawn accurately to scale is that the curves are flatter at the origin and run more steeply to infinity as $2m$ is taken greater and greater (see § 55).

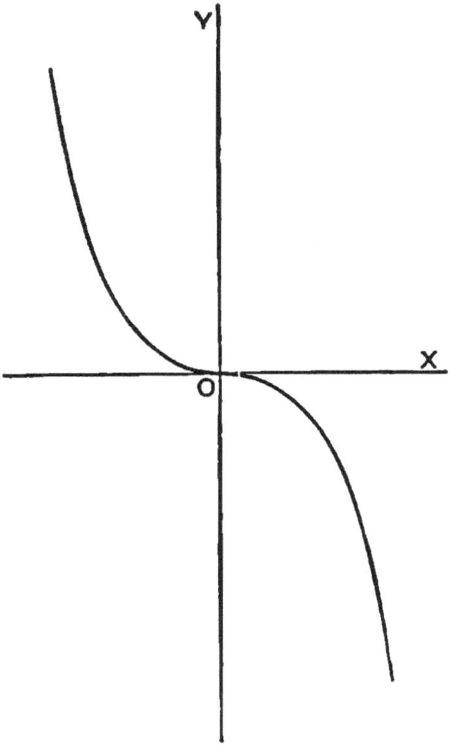


FIG. 8.

It will be observed in the present case that the function $ax^{2m}$ passes through its zero (corresponding to $x=0$) without changing its sign; and that the graph touches, but does not cross, the $x$-axis at the origin; also that the graph passes to infinity in the first and second quadrants, or else in the third and fourth quadrants, according as $a$ is positive or negative; this is expressed analytically by saying that $ax^{2m}$ becomes infinite only when $x$ is infinite, and that the sign of its infinity is the same as the sign of $a$. We may also say that as $x$ passes through infinity (*i.e.* passes from $+\infty$ to $-\infty$, or *vice versa*), $ax^{2m}$ passes through infinity without change of sign.

**§ 262. Case III.** $y=ax^3$ (3).

A graph, typical of this case, when $a$ is positive, viz. the graph of $y=x^3$, has already been drawn (§ 55). To get a type for the case where $a$ is negative, say the graph for $y=-x^3$, we have merely to take the image of the graph of $y=x^3$ in the $x$-axis, we thus get Fig. 8. As has been pointed out in § 58, to get the graph when $a$ has any value differing from unity, we have merely to alter the scale of the ordinates ($y$'s) in the graph for $y=x^3$ or $y=-x^3$.

Like $ax$, the function $ax^3$ changes sign as it passes through its zero; but, unlike $ax$, the graph touches the $x$-axis at the origin which is the point on the graph corresponding to the zero value. A point, like the origin in the present case, where a curve crosses its tangent and changes the direction of bending, is called a **Point of Inflexion.**

The graph of $ax^3$ runs to infinity in two opposite quadrants —the first and third if $a$ be positive, the second and fourth if $a$ be negative. Analytically this amounts to saying that $ax^3$ passes through infinity and changes its sign as $x$ passes from $+\infty$ to $-\infty$, or *vice versa.*

**§ 263. Case IV.** $y=a/x$ (4).

For simplicity, let us consider the case $y=1/x$, which is typical of all cases where $a$ is positive, all other cases being derivable therefrom by merely altering the scale of the ordinates.

When $x=1$, $y=1$. For any positive value of $x$, which is less than 1, $1/x>1$; and the smaller we take $x$ the larger $1/x$ becomes. For example, we have the following table of corresponding values of $x$ and $y$:—

| | | | | | | |
|---|---|---|---|---|---|---|
| $x$ | 1, | ·1/10, | 1/100, | 1/1000, | 1/10,000, | . . . |
| $y$ | 1, | 10, | 100, | 1000, | 10,000, | . . . |

In short, by making $x$ a sufficiently small positive quantity, we can make $y$ as large a positive quantity as we please. Graphically the result is that, as we approach the origin from the point $x=1$ on the $x$-axis, the graph runs away towards the positive end of the $y$-axis, getting nearer and nearer thereto, as

$x$ is made smaller and smaller, but never quite reaching it, although we can get a point on the graph as near the $y$-axis as we choose by taking $x$ sufficiently small and positive. In such a case as this the graph is said to approach the $y$-axis **asymptotically**; and the $y$-axis is spoken of as an **asymptotic straight line**, or simply as an **asymptote**.*

Consider next values of $x$ greater than 1; for such values, $1/x < 1$, and by making $x$ sufficiently large, we can make $1/x$ as

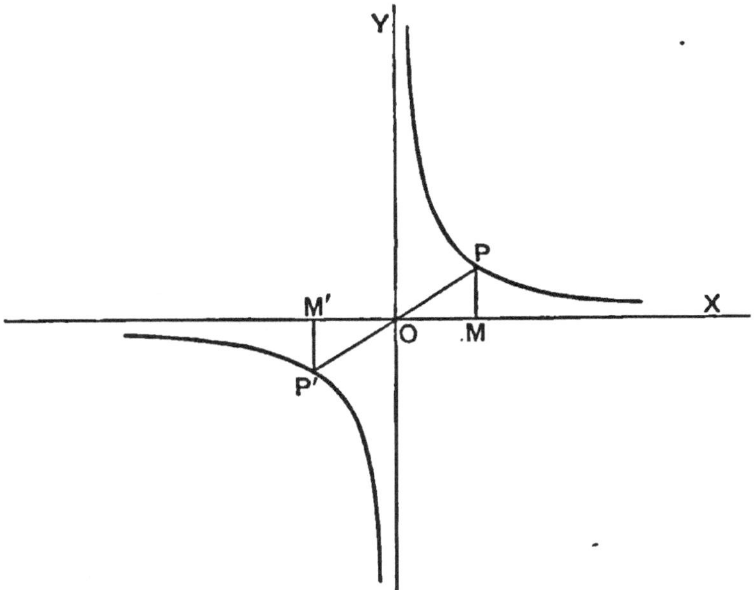


FIG. 9.

small a positive quantity as we please, but never quite zero so long as $x$ is finite and definite. For example, we have the following corresponding pairs of values :—

| | | | | | |
|---|---|---|---|---|---|
| $x$ | 1, | 10, | 100, | 1000, | 10,000, . . . |
| $y$ | 1, | 1/10, | 1/000, | 1/1000, | 1/10,000, . . . |

It appears therefore that, as we proceed from $x = 1$ towards

* From the Greek vocables α-negative σύν = "together with" and πτωτός "falling." "Curvilinear asymptotes" are sometimes used in mathematics; but when no adjective is used, "rectilinear" is understood.

the positive end of the $x$-axis, the graph runs down more and more nearly to the $x$-axis, never quite reaching it, but approaching it in such a way that by taking a positive value of $x$ sufficiently large, we can find a point on the graph above the $x$-axis, but as near it as we please. In short, the $x$-axis is an asymptote to the graph.

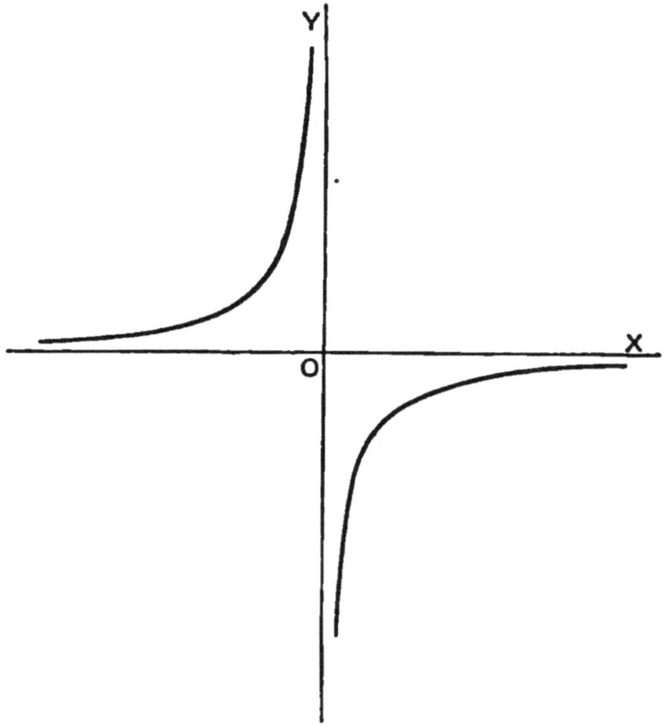


FIG. 10.

So far as the positive side of the $y$-axis is concerned, the graph must therefore be as in Fig. 9.

Since, if we change the sign of $x$ without altering its absolute value, we merely change the sign of $1/x$, it appears at once that the part of the graph corresponding to negative values of $x$ will be found by reflecting the part already drawn first in the $y$-axis, and then again in the $x$-axis. In other words, if P and P′ be the points on the graph, whose abscissæ are $+h$ and $-h$ respectively, P and P′ are collinear with the origin, and $OP = OP'$.

To get the graph of $y = -1/x$, which is typical of the case of $y = a/x$ where $a$ is negative, we have merely to reflect the graph of Fig. 9 in the $x$-axis. We thus get Fig. 10.

The case $y = a/x$ is evidently, so far as the general form of the graph is concerned, typical of the general case $y = a/x^{2m+1}$, where $2m + 1$ is any odd positive integer.

The analytical peculiarities of the function $a/x^{2m+1}$, which are clearly brought out by the above graphical discussion, are that it has no zero for any finite value of $x$, but passes through zero and changes its sign as $x$ passes from $+\infty$ to $-\infty$; and that its infinity corresponds to a finite value of $x$, viz. $x = 0$, the function changing its value suddenly from $-\infty$ to $+\infty$ as $x$ passes through the value 0.

To indicate that $y$ is very large and positive when $x$ is very small and positive, but very large and negative when $x$ is very small and negative, it is usual and convenient to say that, "when $x = +0$, $y = +\infty$, and when $x = -0$, $y = -\infty$." This is, of course, merely a piece of mathematical shorthand, and must not be magnified by the beginner into a mystery of any kind. In the same way we shall sometimes use $a - 0$ to mean "a quantity less than $a$ by very little;" and $a + 0$ to mean "a quantity greater than $a$ by very little."

## § 264. Case V. $y = a/x^2$ (5).

We may consider first the special case $y = 1/x^2$. It is obvious, as in Case IV., that (1, 1) is a point on the graph; and also that the graph will run asymptotically towards the positive ends of the axes of $x$ and $y$, just as in Case IV., the only difference being that $1/x^2$ decreases more rapidly as $x$ increases from 1, and increases more rapidly as $x$ decreases from 1, than does $1/x$. The graph of $y = 1/x^2$ will therefore approach the asymptote OX more rapidly, and the asymptote OY less rapidly than the graph of $y = 1/x$.

When we consider negative values of $x$, there is a fundamental difference between Cases IV. and V.; for, since $1/(-x)^2 = 1/x^2$, it follows that the values of $y$ for positive and negative values of $x$ of equal absolute value are equal. Hence the part of the graph to the left of the $y$-axis is the image in that axis of the part to the right. The graph is therefore like the continuously drawn curve in Fig. 11.

The graph of $y = -1/x^2$ is obviously the image in the $x$-axis of the graph of $y = 1/x^2$—that is, like the dotted curve in Fig. 11.

From one or other of these curves the graph of (5), when the absolute value of $a$ is not unity, can be obtained by merely altering the scale of the ordinates.

The graph of $y=a/x^{2m}$, where $2m$ is any even positive integer, is evidently of the same general form as the graph of $a/x^2$. The analytical peculiarities of the function $a/x^{2m}$

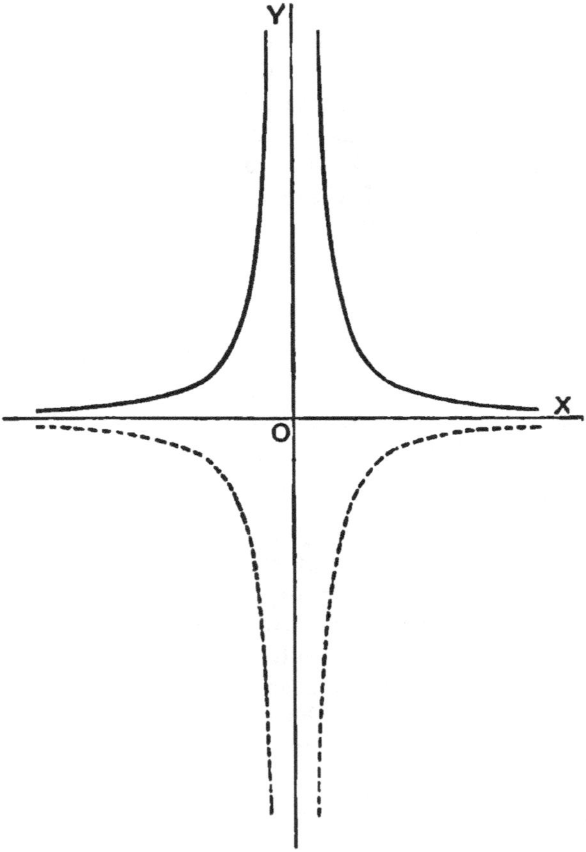


FIG. 11.

corresponding to the peculiarities of the graph just indicated are that it has no zero for any finite value of $x$, but passes through zero without change of sign as $x$ passes from $+\infty$ to $-\infty$; and that it passes through $\infty$ without change of sign as $x$ passes through 0, the sign of the infinity being the same as the sign of $a$.

**§ 265. Case VI.** $y = ax + b + c/x$ (6),

for very large values of $x$.

We first draw the graph of

$$y_1 = ax + b \qquad (7),$$

where we have attached a suffix to the $y$ to distinguish it from the $y$ of equation (6). The graph of (7), as we have already seen in § 58, is a straight line AB, as in Fig. 12, if, to fix our

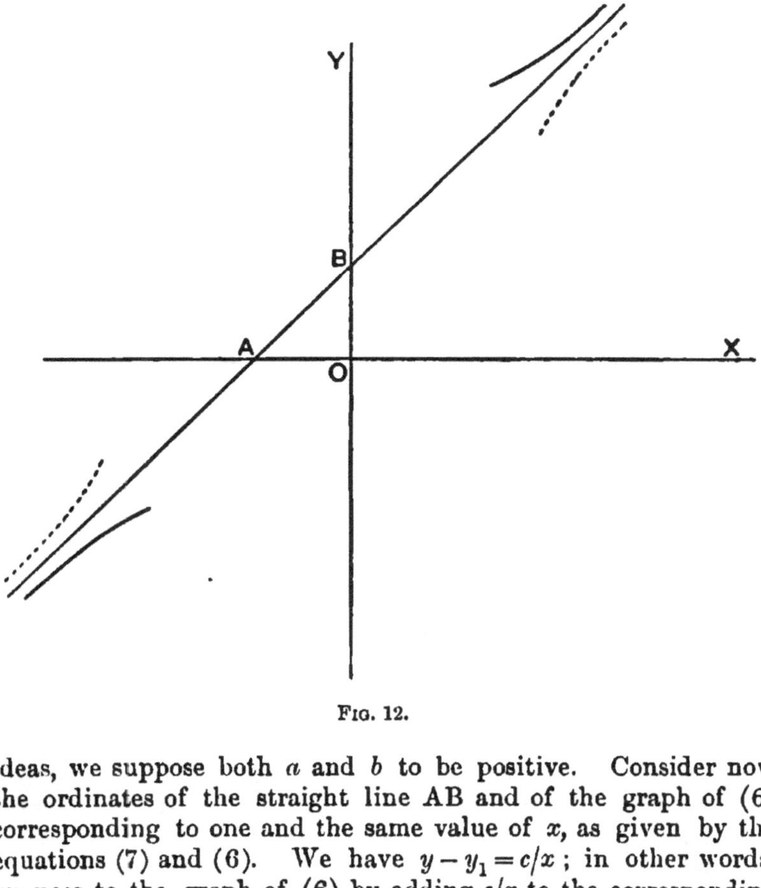


Fig. 12.

ideas, we suppose both $a$ and $b$ to be positive. Consider now the ordinates of the straight line AB and of the graph of (6) corresponding to one and the same value of $x$, as given by the equations (7) and (6). We have $y - y_1 = c/x$; in other words, we pass to the graph of (6) by adding $c/x$ to the corresponding ordinate of AB. The point Q on the graph of (6) will therefore be above or below the corresponding point P on AB, according as $c/x$ is positive or negative. Suppose for a moment that $c$ is positive, then $c/x$ is positive when $x$ is positive, and negative when $x$ is negative. Hence the graph of (6) is above AB towards the far right and below towards the far left of the co-ordinate diagram. Furthermore, $c$ being constant, as $x$ is made larger and larger, $c/x$ becomes smaller and smaller; and $c/x$ can be made as small a positive or negative quantity as we please by making $x$ a sufficiently large positive or negative quantity. Hence by going sufficiently far to the right or left we shall find

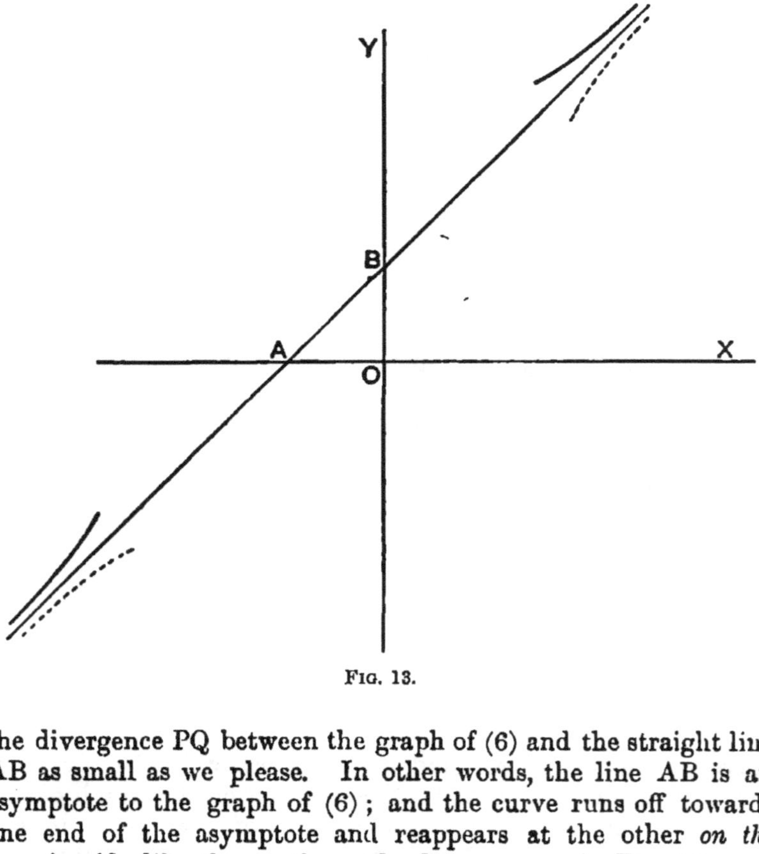


FIG. 13.

the divergence PQ between the graph of (6) and the straight line AB as small as we please. In other words, the line AB is an asymptote to the graph of (6); and the curve runs off towards one end of the asymptote and reappears at the other *on the opposite side*, like the continuously drawn curve in Fig. 12.

If $c$ had been negative instead of positive, the only difference would have been that the graph would have lain below the asymptote on the far right and above on the far left, as is indicated by the dotted curve in Fig. 12.

§ **266. Case VII.** $y = ax + b + c/x^2$ (8),

for very large values of $x$.

Here, as in Case VI., we first draw the graph of $y = ax + b$ and compare it with the graph of (8). The conclusions are the same as before, except that now $c/x^2$ has the same sign for very large values of $x$ whether positive or negative; hence the curve is above the straight line AB both on the far right and on the far left, or else below in both cases, according as $c$ is positive or negative. The straight line is an asymptote as before, and the curve runs away towards one end of the asymptote and returns from the other end on the same side as shown by the continuously drawn curve in Fig. 13, in which $a$, $b$, $c$ are supposed to be all positive. When $c$ is negative, the graph runs like the dotted curve in Fig. 13.

## EXERCISES LXXVII.

**1.** Plot the graph of $y=3x^5$. **2.** Plot the graph of $y=-4x^4$.

**3.** Show that the graphs of $y=ax^m$ and $y=bx^n$, where $a$ and $b$ are both positive, and $m$ and $n$ positive integers, have only one real point in common besides the origin, unless $m$ and $n$ be either both even or both odd, in which case they have two.

**4.** Indicate the ultimate forms towards which the graph of $y=ax^m$ tends when $m$ is increased more and more.

**5.** Plot the graph of $y=-3/x$. **6.** Plot the graph of $y=2/x^2$.

**7.** Discuss the finite intersections of the graphs of $y=1/x^m$, $y=1/x^n$.

**8.** To what ultimate form does the graph of $y=a/x^m$ tend when $m$ is increased more and more.

Make out as much as you can regarding the forms of the following graphs by using the typical cases, and afterwards verify your conclusions by plotting a number of points from the equations:—

**9.** $y=2x-1-1/x$. **10.** $y=-x-1+2/x^2$.

**11.** $y=-x+4/x$. **12.** $y=x-16/x^3$.

**13.** $y=1-1/x$. **14.** $y=2-8/x^2$.

**15.** $y=x^2+1/x$. **16.** $y=x^2-1/x^2$.

**17.** $y=1/x+1/x^2$. **18.** $y=1/x-1/x^3$.

**19.** $y=1/x^2-1/x^3$. **20.** $y=1/x^2+1/x^4$.

**21.** $y=x-x^3+1/x$. **22.** $y=x-1/x+1/x^3$.

§ **267.** By the artifice of **Changing the Origin of Co-ordinates** we can at once considerably extend the scope of the standard Cases I.-V.

Consider, for example—

$$y = a(x-a)^3 \qquad (9),$$

where, to get a definite figure, we shall suppose $a$ and $\alpha$ both positive. Let $\omega$ (Fig. 14) be the point whose co-ordinates are $(\alpha, 0)$; and let P be any point whose co-ordinates with reference to the original axes OX, OY are $(x, y)$. Take a new pair of axes, viz. $\omega$X and $\omega$Y′, $\omega$Y′ being parallel to OY. If $(\xi, \eta)$ be the co-ordinates of P relative to $\omega$X, $\omega$Y′, we have obviously $\xi = x - \alpha$,

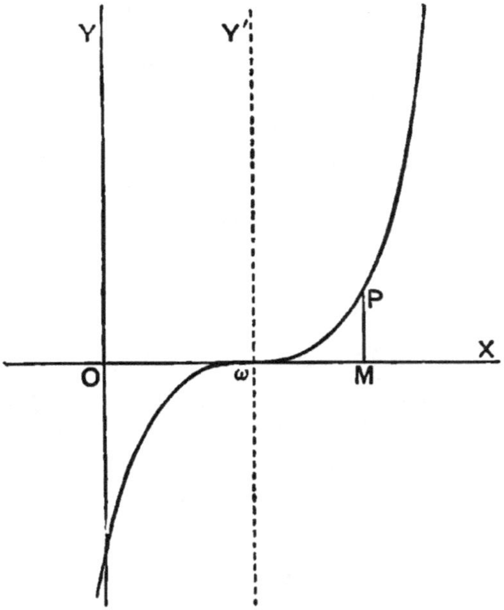


FIG. 14.

$\eta = y$. Hence if we refer the graph of (9) to the new axes, its equation becomes

$$\eta = a\xi^3.$$

Now this is Case III., the plotting being of course carried out with reference to the new axes. Hence the graph is as drawn in Fig. 14. In short, the graph of (9) is simply the graph of (3) shifted through a distance $\alpha$ to the right. If $\alpha$ had been negative, the shift would have been to the left.

The reader will have no difficulty in deducing the graphs of

$$y = a(x - \alpha) \qquad (10),$$

and

$$y = a(x - \alpha)^2 \qquad (11),$$

from the graphs of (1) and (2) respectively.

Consider next

$$y = \beta + c/x \qquad (12);$$

and suppose, for the present, that $\beta$ is positive

Let $\omega$ (Fig. 15) be the point $(0, \beta)$, and let $\omega X'$ and $\omega Y$ be a new pair of rectangular axes parallel to the former pair, having $\omega$ for origin. If $(\xi, \eta)$ be the co-ordinates with reference to $\omega X'$, $\omega Y$ of a point P, whose co-ordinates with reference to OX, OY are $(x, y)$, we have obviously $\xi = x$, $\eta = y - \beta$. Hence

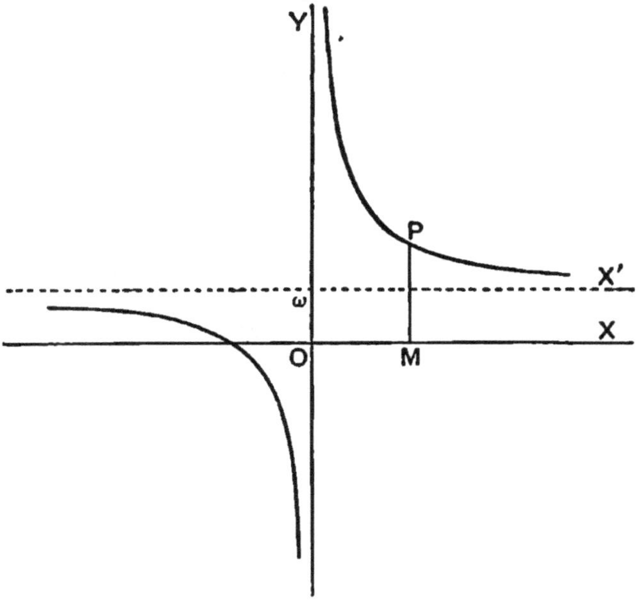


FIG. 15.

the equation with reference to the axes $\omega X'$, $\omega Y$ which represents the graph of (12) is

$$\eta = c/\xi.$$

But this is Case IV. Hence the graph of (12) is simply the graph of (4) shifted a distance $\beta$ parallel to the $y$-axis, as in Fig. 15. The shift is upwards or downwards according as $\beta$ is positive or negative.

In exactly the same way we see that the graph of

$$y = \beta + c/x^2 \qquad (13)$$

can be derived by shifting the graph of $y = c/x^2$ through a distance $\beta$ parallel to the $y$-axis, up or down according as $\beta$ is positive or negative.

Lastly, consider

$$y = x^2 - 4x + 5 \qquad (14),$$

which might be written

$$y - 1 = (x - 2)^2.$$

If we put $x = 2$, we get $y = 1$; therefore (2, 1) is a point on

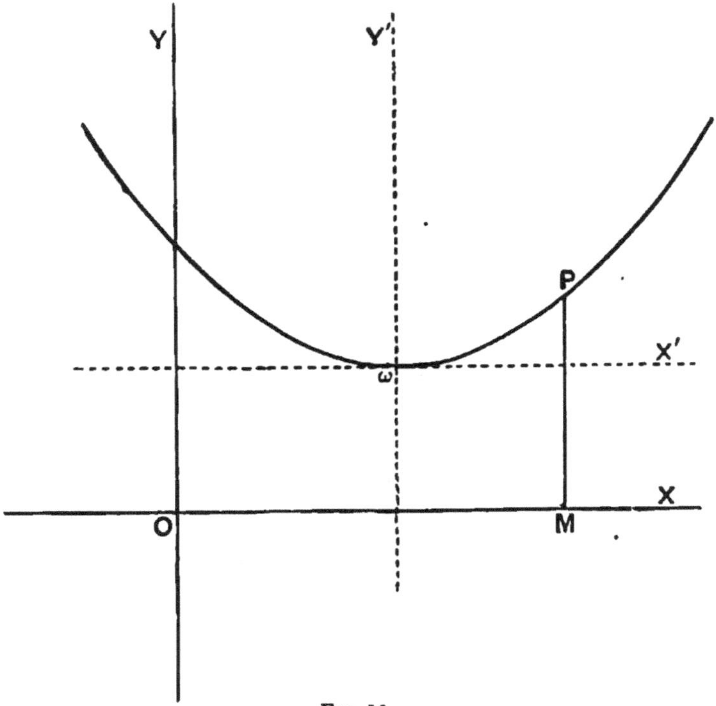


Fig. 16.

the curve. Let us examine the shape of the curve by taking the point (2, 1), ω say (Fig. 16), as origin, the new axes ω'X, ωY' being parallel to the original axes OX, OY respectively. If $(\xi, \eta)$ be the co-ordinates with reference to ωX', ωY' of the point P, whose co-ordinates with reference to the axes OX, OY are $(x, y)$, we have obviously $x = 2 + \xi$, $y = 1 + \eta$. Hence the equation to the graph of (14) referred to ωX', ωY' is

$$1 + \eta = (\xi + 2)^2 - 4(\xi + 2) + 5$$

which reduces to

$$\eta = \xi^2 ;$$

but this is Case II. Hence the graph of (14) is obtained by shifting the graph of $y=x^2$ a distance 2 to the right, parallel to the $x$-axis, and then a distance 1 upwards, parallel to the $y$-axis, so that the apex of the festoon comes to the point (2, 1). The graph for (14) is therefore given by Fig. 16.

It is obvious that in this way the examination of the nature of a curve at any point P can be reduced to the discussion of the nature of a curve at the origin of co-ordinates shifted for the purpose to the particular point P.

§ **268.** It will be convenient at this stage to define precisely what is meant by **Maxima, Minima,** and **Turning Values** of a function.

*When any value,* b, *of a function is greater than the immediately*

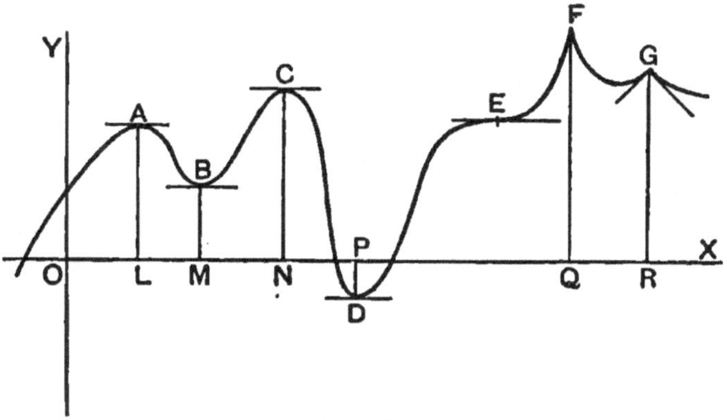


FIG. 17.

*neighbouring values, both before and after, it is called a maximum value of the function.*

*When any value,* b, *of a function is less than the immediately neighbouring values both before and after, it is called a minimum value of the function.*

Observe that here greater and less are to be taken in the algebraic and not in the absolute sense; and that a maximum value is not necessarily the greatest of all the values that a function may have, or a minimum necessarily the least of all, see Fig. 17, where A and C are maxima and B and D minima values.

*Maxima and minima values are spoken of collectively as turning values.*

It is obvious that at a maximum value the function ceases to increase and begins to decrease, and at a minimum value ceases to decrease and begins to increase, if we imagine $x$ increased so that the function passes through the turning value.

The points on the graph of a function which correspond to its maxima and minima are evidently points of maximum and minimum distance from the $x$-axis—that is to say, tops of hills and bottoms of valleys on the curve. It is also obvious that *in general at turning-points the tangents to the graph are parallel to the* x-*axis*. Thus, for example, in Fig. 17, A, B, C, D are turning-points, at which the tangents are parallel to the $x$-axis.

It should, however, be noted that points, such as E in Fig. 17, may occur at which the tangent is parallel to the $x$-axis and which are not turning-points.

Again, turning-points may occur at which the graph has a sharp point or a nick, such as F and G in Fig. 17, and at which the tangents are not parallel to the axis of $x$. It will not be found, however, that such points occur on the graphs of rational functions.

It follows at once from the most obvious notions of geometric continuity that *so long as a function is continuous, maxima and minima values must follow each other alternately, i.e.* every hill on the graph must be followed by a valley before another hill can occur.

## The Graph of the General Quadratic Function

$$ax^2 + bx + c$$

§ **269.** In discussing the graph of the general quadratic function, whose equation is

$$y = ax^2 + bx + c \qquad (15),$$

it is convenient to transform the function as was done in factorising it (see § 130). We thus reduce (15) to the equivalent form

$$y = a(x + b/2a)^2 + (4ac - b^2)/4a \qquad (16).$$

If we now change the origin (as in § 267) to the point $(-b/2a,\ (4ac - b^2)/4a)$, by putting $x = -b/2a + \xi$, $y = (4ac - b^2)/4a + \eta$, the equation (16) becomes simply

$$\eta = a\xi^2,$$

which is Case II. (§ 261).

It appears therefore *that the graph of the quadratic function is a festoon, or an inverted festoon, according as* a *is positive or negative.*

The co-ordinates (with reference to the original axes) of the vertex of the festoon are $x = -b/2a$ and $y = (4ac - b^2)/4a$; and this vertex is obviously a minimum or a maximum turning-point according as $a$ is positive or negative.

The reader should draw for himself the figures corresponding to the twelve cases indicated in the following table, in which $\Delta$ stands for the discriminant of the quadratic function, viz. $b^2 - 4ac$:—

| $a$ | $\Delta$ | $b$ | $a$ | $\Delta$ | $b$ |
|---|---|---|---|---|---|
| + | + | + | − | + | + |
| + | + | − | − | + | − |
| + | 0 | + | − | 0 | + |
| + | 0 | − | − | 0 | − |
| + | − | + | − | − | + |
| + | − | − | − | | − |

When the graph is a festoon, the condition that it shall cut the $x$-axis in two real distinct points is that the $y$-co-ordinate of the minimum point be negative—that is, since in this case $a$ is positive, that $b^2 - 4ac$ shall be positive. When the graph is an inverted festoon, the corresponding condition is that the $y$-co-ordinate of the maximum point shall be positive—that is, since in this case $a$ is negative, that $b^2 - 4ac$ shall be positive. Now the abscissæ of the points where the graph cuts the $x$-axis—that is of the points for which $y = 0$—are the roots of the equation

$$ax^2 + bx + c = 0 \qquad (17).$$

Hence we have arrived by graphical considerations at the result of § 215, viz. that the necessary and sufficient condition that the roots of (17) be real and distinct, is that $b^2 - 4ac > 0$. It is easy to see by similar considerations that the condition for equal roots is $b^2 - 4ac = 0$, and for imaginary roots $b^2 - 4ac < 0$.

The reader should here observe the association of the occurrence of equal roots with tangency by the $x$-axis.

Ex. Trace the graph of $-3x^2+2x+1$, and find its turning value, and the value of $x$ corresponding thereto.

The equation to the graph may be written $y=-3x^2+2x+1$. If we transform this as above indicated, we have $y=-3(x-1/3)^2+4/3$.

The co-ordinates of the turning-point are $x=1/3$, $y=4/3$. The equation to the graph referred to $(1/3, 4/3)$ as origin is $\eta=-3\xi^2$. The graph is therefore an inverted festoon, the vertex of which, $(1/3, 4/3)$, is a maximum point. The function $-3x^2+2x+1$ has therefore a maximum value $4/3$ corresponding to $x=1/3$.

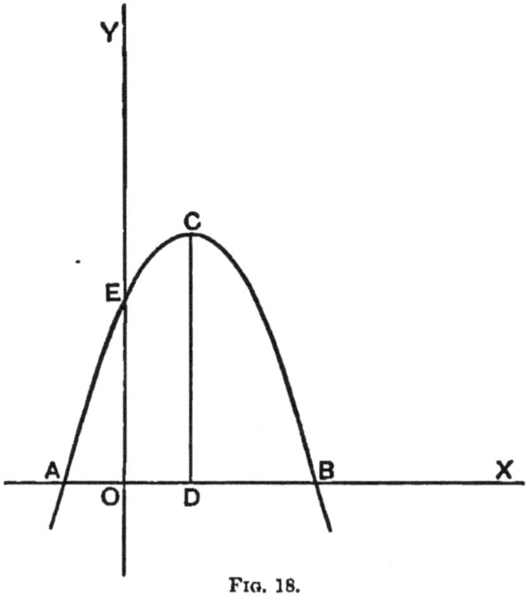


Fig. 18.

The graph is roughly indicated in Fig. 18, where $OD=1/3$, $CD=1\frac{1}{3}$; and $OA=1/3$, $OB=1$, $OE=1$.

## EXERCISES LXXVIII.

Draw graphs for the following. Use the artifice of changing the origin where necessary to get the general form of the graph; and finally check your result by plotting a sufficient number of test points:—

1. $y=x(x-1)$.
2. $y=x^2+2x$.
3. $y=(x-1)(x-2)$.
4. $y=(1-x)(x-2)$.
5. $y=x^2-4x+3$.
6. $y=x^2-4x+4$.
7. $y=x^2-4x+5$.
8. $y=2+x-x^2$.
9. $y=x^2+x+1$.
10. $y=2x^2-2x+2$.
11. $y=3x^2-10x+4$.
12. $y=a^2-x^2$.
13. $y=-2x^2+9x-4$.
14. $y=-4x^2-12x-9$.

**15.** $y = -1/(x-3)^2$.

**16.** $y = 1/(x+1)^2$.

**17.** $y = 2/(x+2)^2$.

**18.** $y = x^3 + 3x^2 + 3x$.

**19.** $y = 3 - 2/(x-1)$.

**20.** $y = 3 + 1/(x+2)^2$.

**21.** If the graph of a quadratic function cut the $x$-axis in two distinct points equidistant from the origin, its turning-point is on the $y$-axis, and conversely.

**22.** Find a quadratic function whose graph has its turning-point at (3, 9), and which cuts the $y$-axis at the point (0, 1).

**23.** The graph of a quadratic function meets the $x$-axis in the points $(-1, 0)$, $(1/2, 0)$, and the $y$-axis in the point $(0, -3)$; find the turning value of the function.

## Method of Approximation

**§ 270.** In practice every arithmetical calculation is carried out to a finite number of digits, say to a given decimal place, *e.g.* to hundreds only, or to tens, to units, to tenths, to hundredths, etc. Likewise every measurement has in practice only a limited degree of accuracy. A land measurer may measure within an inch or two; a joiner to 1/32 of an inch; and so on. Alike, therefore, by arithmetical calculation and by practical measurements we have directly suggested to us the notion of successive approximations of greater and greater accuracy to a given quantity. Consider, for example, the diagonal of a square whose side is 1 inch. A rough approximation to the length of the diagonal is 1·4 in.; we might call this a **First Approximation.** A closer approximation is 1·41 in., which we might call a **Second Approximation.** A nearer approximation still is 1·414 in., which we might call a **Third Approximation**, and so on.

The application of this order of ideas to the calculation of the values of functions and to the plotting of their graphs leads us to a method of the highest importance both in theory and in practice. In particular, we shall find that the method of approximation enables us greatly to extend the usefulness of the seven typical cases studied at the beginning of this chapter.

Since every magnitude that occurs in practice may be supposed to be represented by the length of a straight line, we shall suppose that the practical purpose that we have in view in calculating the values of a function is the plotting of its graph to various scales. The larger the scale the greater of course is the accuracy with which we can represent the value of the abscissa, and the greater the accuracy required in the calculation of the ordinate of the graph.

§ 271. **Approximations when $x$ and $y$ are both Small.**—Consider the graph of

$$y = x - x^3 \qquad (18);$$

and let us suppose that the scale of our diagram is such that we cannot lay down $y$ to nearer than 1/1000th of the ordinate-scale-unit. Then, if we confine ourselves to values of $x$ which lie between $-\cdot01$ and $+\cdot01$, we need not consider the term $x^3$ at all, since* $|x^3| \not> \cdot000001$ if $|x| \not> \cdot01$. For the part of the graph considered, $y = x$ is therefore a sufficient approximation. Moreover, since $x^3/x = x^2$ we see that, as $x$ is made smaller and smaller, the importance of $x^3$ relatively to $x$ becomes smaller and smaller. In other words, the straight line represented by $y = x$ becomes a better and better approximation to the graph of (18) the nearer we approach the origin; and is a sufficient practical approximation for all purposes that do not require the ordinate to be calculated nearer than to $\cdot001$, so long as $|x| \not> \cdot01$. We therefore call this straight line a **First Approximation to the Graph** of (18) at the origin.

Next, consider the graph of

$$y = x - x^3 + x^4 - x^5 \qquad (19).$$

If we suppose $-\cdot1 \not> x \not> +\cdot1$, and accuracy in the value of $y$ up to $\cdot001$ be required, the first approximation $y = x$ would no longer be sufficient for the whole interval considered. We must, therefore, consider the curve

$$y = x - x^3,$$

which is said to give a **Second Approximation to the Graph** of (19).

If accuracy were required up to $\cdot0001$, we should have to consider $x^4$; and we might replace the graph of (19) within the interval $-\cdot1 \not> x \not> \cdot1$ by the graph of

$$y = x - x^3 + x^4,$$

which we call a **Third Approximation** to the graph.

It should be noticed that, besides giving us arithmetical accuracy up to the third place of decimals in the interval $-\cdot1 \not> x \not> \cdot1$, the second approximation gives us the important information that near the origin on the right the graph of (19) falls below the straight line $y = x$, which is the first approxima-

* The reader will recollect that $|x^3|$ means the absolute or mere arithmetical value of $x^3$.

tion just at the origin, and rises above the same straight line to the immediate left of the origin.

By looking at Fig. 2, p. 67, where the graphs of $y=x$ and $y=x-x^3$ are drawn to scale, the reader will see how closely the line $y=x$ coincides with the graph of $y=x-x^3$ very near the origin. He will also note that it is only near the origin that the approximation holds good; as $x$ is increased or decreased more and more, the two graphs separate more and more, and for large values of $x$ they diverge utterly.

In the above examples we have supposed that the graph passes through the origin—that is that $x$ and $y$ vanish together.

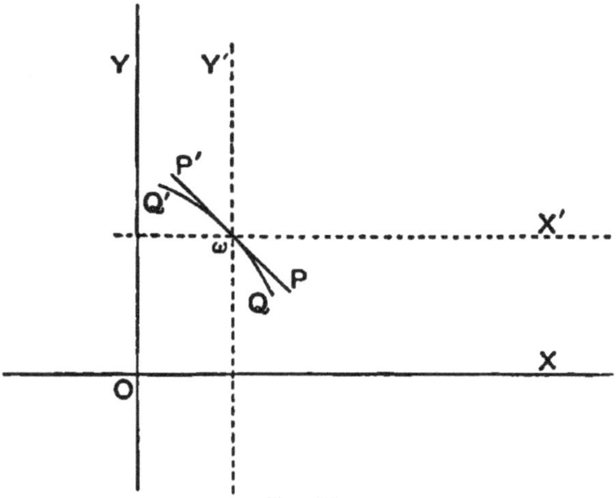


FIG. 19.

We can, however, find an approximation to the form of the graph at any point $(\alpha, \beta)$ on itself by merely shifting the origin, as in § 267.

Ex. Find the shape of the graph

$$y=3/2+(x-1)(x-2)(2x-1) \qquad (20)$$

near the point on it whose abscissa is 1.

When $x=1$, $y=3/2$. Shift the origin to $(1, 3/2)$, $\omega$ say, and let the co-ordinates of any point P with reference to the new origin be $(\xi, \eta)$. Then, if $(x, y)$ be the co-ordinates of P with reference to the old origin, we have $x=1+\xi$, $y=3/2+\eta$. Hence the equation to the graph with reference to the new origin is

$$\eta=\xi(\xi-1)(2\xi+1);$$

that is—

$$\eta=-\xi-\xi^2+2\xi^3$$

A first approximation to the graph is given by

$$\eta = -\xi \qquad (21),$$

which represents a straight line P$\omega$P′ (Fig. 19).

A second approximation is given by

$$\eta = -\xi - \xi^2 \qquad (22).$$

Since, for one and the same value of $\xi$, we derive the ordinate of (22) from the ordinate of (21) by subtracting $\xi^2$, which has the same sign whether $\xi$ is positive or negative, it follows that (22) represents a curve like Q$\omega$Q′, lying below P$\omega$P′, and convex towards it.

§ 272. **Approximations when $x$ and $y$ are both very Large.**—Consider again the graph whose equation is (18). We may write the equation in the equivalent form

$$y = -x^3(1 - 1/x^2) \qquad (23).$$

If now we suppose $x$ to be very large, $1/x^2$ will be very small compared with 1, and for a first approximation to the graph we may replace $1 - 1/x^2$ by 1. We thus get

$$y = -x^3$$

which is Case III. § 262.* Hence for very large values of $x$ the graph of $y = x - x^3$ has nearly the same shape as the graph of $y = -x^3$ (which is drawn to scale in Fig. 8, p. 367). This is sometimes expressed by saying that the graph of $y \doteq -x^3$ is a **First Approximation at Infinity** to the graph of $y = x - x^3$.

In the same way, since the equation

$$y = x^4 - x^3 + x^2 - 1 \qquad (24)$$

may be written

$$y = x^4(1 - 1/x + 1/x^2 - 1/x^4) \qquad (25),$$

a first approximation at infinity to (24) is given by

$$y = x^4 \qquad (26);$$

a second approximation would be given by

$$y = x^4(1 - 1/x),$$

that is—

$$y = x^4 - x^3 \qquad (27);$$

a third approximation by

$$y = x^4(1 - 1/x + 1/x^2),$$

that is—

$$y = x^4 - x^3 + x^2 \qquad (28);$$

and so on.

* For example, if $x = 1000$, the difference between the ordinates of (22) and (23) is 1000, a quantity which is in itself large, no doubt, but which is small compared with 1,000,000,000.

Another good example of the present kind of approximation may be given in connection with

$$y = x + 1 + 1/x - 1/x^2 \qquad (29).$$

When $x$ is very large the order of importance of the terms is evidently $x$, 1, $1/x$, $1/x^2$.* Hence we get for first, second, and third approximations

$$y = x \qquad (30);$$
$$y = x + 1 \qquad (31);$$
$$y = x + 1 + 1/x \qquad (32).$$

The first and second approximations are straight lines. The third approximation shows (see § 265) that the second approximation is an asymptote, and that the first is a parallel to an asymptote. The graph, in so far as points to the far right and the far left are concerned, is like the full-drawn curve in Fig. 12.

§ 273. **Approximations when $y$ is very Great and $x$ very Small.**—The nature of cases of this kind will be understood by considering the nature of (29) when $x$ is very small. The order of importance of the terms is now $1/x^2$, $1/x$, 1, $x$.† Hence first and second approximations are given by

$$y = -1/x^2 \qquad (33),$$

and

$$y = -1/x^2 + 1/x \qquad (34).$$

Hence to a first approximation the curve for very small values of $x$ is like the dotted curve in Fig. 11.

The second approximation shows (always of course when $x$ is small) that on the right of the $y$-axis the graph of (29) is above the dotted curve just mentioned; and on the left below, as is roughly indicated in Fig. 20, where the continuously drawn curve is part of the graph of (29).

The case where $y$ is very large when $x$ has a finite value [*e.g.* $y = x + 1 + 1/(x-1)$] can of course be brought under the present by change of origin.

§ 274. **Approximation when $y$ is Finite for a very large Value of $x$.**—As an example of this case we may take

$$y = 1 + 1/x - 1/x^2 + 1/x^4 \qquad (35).$$

* *e.g.* if $x = 1000$, the values of these terms are 1000, 1, ·001, ·000,001 respectively.

† *e.g.* if $x = ·01$, the values of these terms are 10,000, 100, 1, ·01 respectively.

The order of importance of the terms when $x$ is very large is $1$, $1/x$, $1/x^2$, $1/x^4$. Hence first, second, and third approximations are

$$y=1 \qquad (36);$$
$$y=1+1/x \qquad (37);$$
$$y=1+1/x-1/x^2 \qquad (38).$$

The first approximation is a straight line parallel to the axis of $x$. The second approximation shows that this line is an asymptote to the graph of (35), and that the graph approaches

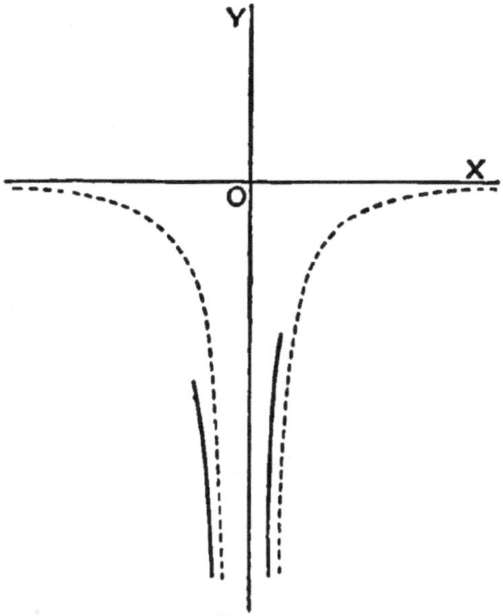

FIG. 20.

the asymptote from above on the far right, and from below on the far left. The reader should examine the geometrical meaning of the third approximation.

§ **275.** Since the calculation of successive approximations to a rational function of $x$, when $x$ is very small or very large, is an operation of frequent occurrence in practice, it will be well to point out how this can be systematically accomplished by quite elementary means. We shall at the same time show how the limits of error in such approximations can be found when limits are assigned for $x$. All that is required is the process used in calculating the integral quotient and remainder for a quotient of integral functions of $x$, with some slight modifications and extensions which will be fully understood from the following examples :—

Ex. 1. To find a third approximation to $(2+3x-x^2)/(1+x+x^2)$ when $x$ is very small: and to find an upper limit for the error of this approximation when $-\cdot 01 \not> x \not> +\cdot 01$.

We arrange the terms of the dividend and divisor in the order of their importance, which in the present case is $1, x, x^2, \ldots$, since $x$ is small, and then proceed to subtract from the dividend multiple after multiple of the divisor, so as to destroy in succession the terms of the dividend, exactly as in the process of long division (§ 110), until we reach the term containing $x^2$ in what now takes the place of the integral quotient. Thus

$$\begin{array}{rl|l}
2+3x-x^2 & & 1+x+x^2 \\
\underline{2+2x+2x^2} & & \overline{2+x-4x^2} \\
x-3x^2 & & \\
\underline{x+x^2+x^3} & & \\
-4x^2-x^3 & & \\
\underline{-4x^2-4x^3-4x^4} & & \\
+3x^3+4x^4 & . &
\end{array}$$

The meaning of this work is that

$$\frac{2+3x-x^2}{1+x+x^2} \equiv 2+x-4x^2+\mathrm{R},$$

where

$$\begin{aligned}
\mathrm{R} &\equiv (3x^3+4x^4)/(1+x+x^2), \\
&\equiv x^3(3+4x)/(1+x+x^2).
\end{aligned}$$

From this we see that $2+x-4x^2$ is the third approximation required (2 being the first, and $2+x$ the second approximation).

For the ratio of R to the last term retained (viz. $-4x^2$) is $-\frac{1}{4}x(3+4x)/(1+x+x^2)$, which can be made as small as we choose by sufficiently diminishing $x$.

If a fourth or higher approximation were required, we have only to continue the above calculation until the proper term in the continued quotient is reached. The process is commonly spoken of as **Ascending Continued Division.**

To estimate the error, let us first suppose that $x$ is positive, and $|x| \not> 1/100$, then obviously R is positive and

$$|\mathrm{R}| < 3\cdot 04/10^6, \text{ say } < 4/10^6.$$

Next suppose R negative, and $|x| \not> 1/100$. Then, since $4x$ and $x$ are now negative, we have

$$\begin{aligned}
|\mathrm{R}| &< 3/(1-\cdot 01)10^6 < 3/\cdot 99\,.\,10^6, \\
&< 4/10^6.
\end{aligned}$$

The utmost error of the third approximation is therefore numerically less than $\cdot 000{,}004$, when $|x| \not> \cdot 01$.

Ex. 2. To find a third approximation to $(2+3x-x^2)/(1+x+x^2)$ when $x$ is large: and to estimate its accuracy when $x > 100$.

The order of importance of terms is now . . ., $x^2$, $x$, 1, $1/x$, $1/x^2$. . . . We therefore arrange both dividend and divisor according to descending powers of $x$. Otherwise we proceed as before

$$\begin{array}{r|l} -x^2+3x+2 & x^2+x+1 \\ \underline{-x^2-x-1} & \overline{-1+4/x-1/x^2} \\ 4x+3 & \\ \underline{4x+4+4/x} & \\ -1-4/x & \\ \underline{-1-1/x-1/x^2} & \\ -3/x+1/x^2. & \end{array}$$

Hence

$$\frac{2+3x-x^2}{1+x+x^2} \equiv -1+\frac{4}{x}-\frac{1}{x^2}+\mathrm{R},$$

where

$$\mathrm{R} \equiv \frac{-3/x+1/x^2}{x^2+x+1},$$
$$\equiv \frac{1}{x^3} \cdot \frac{-3+1/x}{1+1/x+1/x^2}.$$

This process is called **Descending Continued Division.**

It is at once seen that the ratio of R to the last term of the quotient, viz. $-1/x^2$, can be made as small as we please by sufficiently increasing $x$. Hence $-1+4/x-1/x^2$ is the required third approximation (the first and second approximations being $-1$ and $4/x$ respectively).

Finally, if $x$ be positive, since $|x|>100$—

$$|\mathrm{R}|<3/10^6;$$

and, if $x$ be negative—

$$|\mathrm{R}|<3{\cdot}01/(1-{\cdot}01)10^6<3{\cdot}01/{\cdot}99\,.\,10^6,$$
$$<4/10^6.$$

Therefore the utmost numerical error of the third approximation is less than ·000,004, when $x>100$.

## EXERCISES LXXIX.

1. Show that $1+x+x^2+\ldots+x^n$ is an $(n+1)$th approximation to $1/(1-x)$, in the sense that $\{1/(1-x)-(1+x+x^2\ldots+x^n)\}/x^n$ can be made as small as we please by sufficiently decreasing $x$. Show also that, if $0 \ngtr x \ngtr 1/10^p$, then the error of this $n$th approximation does not exceed $1/10^{pn+p-1}$.* Establish a similar theorem for $1/(1+x)$; and, in particular, show that the error of the $n$th approximation in this case does not exceed $1/10^{pn+p}$, provided $0 \ngtr x \ngtr 1/10^p$.

Find second approximations to the following rational functions when $x$ is small; and assign upper limits to the error of the approximation supposing $0 \ngtr x \ngtr {\cdot}01$:—

2. $(1-x)^2(1+x^2)$.
3. $(1-x)(1-3x)(1+x)^3$.
4. $1/(1+2x)$.
5. $1/(1+x+2x^2)$.

---

* Use the identity $1-x^{n+1}=(1-x)(1+x+x^2+\ldots+x^n)$.

6. $1/(2+x-x^2)$.
7. $(1+x)/(1-x)$.
8. $(1+x+x^2)/(1-x)^2$.
9. $(1+x)^3/(1-x)^3$.

Discuss by approximation, combined with change of origin where necessary, the forms of the graphs of the following at the points indicated :—

10. $y=x(x-2)^3$ at $(0, 0)$.
11. $y=x^3(x-1)(x-2)$ at $(0, 0)$.
12. $y=x/(x-2)^2$ at $(0, 0)$.
13. $y=x^2/(x-1)$ at $(0, 0)$.
14. $y=x^3/(1+x+x^2)$ at $(0, 0)$.
15. $y=(x^2-6x+8)/(x^2-2x+1)$ at $(2, 0)$.
16. $y=(2x^2+3x)/(x^2+x+1)$ at $(2, 2)$.
17. $y=(x-1)^2/(x^2-x+1)$ at $(1, 0)$.
18. $y=(x-2)^3/(x^2+3x+1)$ at $(2, 0)$.
19. $y=(x^2+5x+6)/(x-1)(x-2)$ at $(1, \infty)$.
20. $y=(x^2+5x+6)/(x-1)^2$ at $(1, \infty)$.
21. $y=(x^3+3x^2+3x+1)/x^3$ at $(-1, 0)$.
22. $y=(x^2+1)/(x-2)^3$ at $(2, \infty)$.
23. Show that $y=x(2-x)/(2-x+x^2)$ has an inflexion at $(0, 0)$.
24. Show that the equation to the graph of an integral function which passes through the point $(\alpha, \beta)$ can be put into the form $y-\beta=a_1(x-\alpha)+a_2(x-\alpha)^2+ \ . \ . \ .$, where one or more of the coefficients $a_1, a_2, \ . \ . \ .$ may be zero. Hence show how to find the condition that the graph have an inflexion at $(\alpha, \beta)$.
25. Show that the graph of a rational function which passes through the point $(\alpha, \beta)$ may be represented by an equation of the form

$$y-\beta=\{a_1(x-\alpha)+a_2(x-\alpha)^2+ \ . \ . \ . \ \}/\{b_0+b_1(x-\alpha)+b_2(x-\alpha)^2+ \ . \ . \ .\}$$

where one or more of the coefficients $a_1, a_2, \ . \ . \ ., b_1, b_2, \ . \ . \ .$ may vanish, but $b_0 \neq 0$. Show that the graph will have an inflexion at $(\alpha, \beta)$, provided $a_2b_0-a_1b_1=0$, $a_1b_2-a_3b_0 \neq 0$.
26. Show that $y=(x-\alpha)(x-\beta)/(A+Bx+Cx^2)$ will have an inflexion at $(\alpha, 0)$, provided $A+B\beta+C(2\alpha\beta-\alpha^2)=0$.

Discuss the shape of the graphs of the following at the points indicated :—

27. $y=(3x^2-2x+3)/(x^2+1)$ at $(1, 2)$.
28. $y=(-3x^2+3x-2)/(x^3+1)$ at $(1, -1)$.
29. $y=(-2x+4x^3-x^5)/(x^2-1)$ at $(0, 0)$.
30. $y=(2-x-x^2-2x^3)/(1-x)$ at $(0, 2)$.
31. $y=(1-x+4x^2-x^3)/x$ at $(1, 3)$.

## Examples of Graphs of Rational Functions

**§ 276.** We shall now give some examples of the application of the foregoing principles to the plotting of graphs of rational functions.

The following remarks regarding the **Quadratic Rational Function**—that is, *the quotient of two integral functions, one at least* *of which is of the second degree*—will be helpful in practice ; and most of them apply to rational fractional functions in general :—

Let the rational quadratic function be $(ax^2+bx+c)/(Ax^2+Bx+C)$, so that the equation to the graph is

$$y=(ax^2+bx+c)/(Ax^2+Bx+C) \qquad (39).$$

We suppose that $ax^2+bx+c$ and $Ax^2+Bx+C$ have no common factor ; it follows by the remainder theorem that they cannot both vanish for the same value of $x$. Also, since each of them contains only a finite number of terms, and there is no division by $x$, neither can cease to be finite for a finite value of $x$. It follows (see §§ 62, 207) that the only finite values of $x$ for which $y$ can vanish are the roots of the quadratic

$$ax^2+bx+c=0 \qquad (40) ;$$

and the only finite values of $x$ for which $y$ can become infinite are the roots of the quadratic

$$Ax^2+Bx+C=0 \qquad (41).$$

The first step in plotting the graph consists in discussing the roots of (40) and (41), and laying down the points corresponding to them (so far as they are real) on the $x$-axis.

The next step is to discover by the method of approximations, by shift of origin, or simply by examining the sign of $y$ for values of $x$ immediately before and after the real roots of (40) and (41), how the graph passes through the zeros and infinities that correspond to finite values of $x$.

It will have been observed from what precedes that, if the factor in the numerator of the rational function which corresponds to a zero be not a repeated factor, the graph crosses the $x$-axis at a non-evanescent angle. If the factor be repeated an odd number of times, the graph crosses, but also touches the $x$-axis. If the factor be repeated an even number of times, the graph touches, but does not cross the $x$-axis.

Again, if the factor in the denominator which corresponds to the infinity be not repeated, or repeated an odd number of times, the graph is related to its asymptote, as in Figs. 9 or 10. If the factor be repeated an even number of times, the relation of graph to asymptote is as in Fig. 11.

Next the behaviour of the graph as it crosses the $y$-axis should be examined.

Then the nature of the grapn when $x$ is very large should

be determined, and the asymptotes, if any, laid down. The method of approximation, or the calculation of properly selected single values of $y$, will settle how the graph approaches its asymptotes.

Finally, the various pieces should be joined in the simplest manner that will embody all the peculiarities established in the discussion. Unless a large number of individual points throughout the graph are accurately plotted, there is, of course, a considerable element of guessing in this last proceeding; but not so much as the beginner would at first be inclined to suppose. The proper way to acquire skill and conviction on this head is to draw a considerable variety of graphs to scale by means of plotting paper, after having made out the general form by the process just described.

In settling the general form of a graph much depends on the position of its turning points, if it have any; and these should always be accurately plotted, wherever it is possible to determine them from the equation. There is a uniform elementary method for doing this in the case of the quadratic rational function, which will be understood from Example 4 below.

Ex. 1. $y=x(x-1)^2$ (42).

The finite zeros of $y$ correspond to $x=0$, $x=1$.

In the neighbourhood of $x=0$, first and second approximations are $y=x$ and $y=x-2x^2$. The first gives a straight line bisecting the first and third quadrants; the second shows that near the origin the graph falls below this line.

If we shift the origin to the point (1, 0), (A, Fig. 21), the equation (42) becomes $y=(1+\xi)\xi^2$; a first approximation to the graph is therefore $y=\xi^2$, a festoon having its vertex at (1, 0). We may write (42) in the equivalent form $y=x^3(1-1/x)^2$. From which we see that a first approximation when $x$ and $y$ are both very great is given by $y=x^3$, which is Case III., § 262. A short table of corresponding values of $x$ and $y$ confirms these results.

| $x$ | $y$ | |
|---|---|---|
| $-\infty$ | $-\infty$ | |
| $-0$ | $-0$ | |
| $+0$ | $+0$ | |
| $1/2$ | $1/8$ | Max. |
| $1-0$ | $+0$ | |
| $1+0$ | $+0$ | Min. |
| $+\infty$ | $+\infty$ | |

The general form of the graph is indicated in Fig. 21. A is obviously a minimum turning point; and (since $y$ cannot become infinite for any finite value of $x$) there must also be a maximum turn-

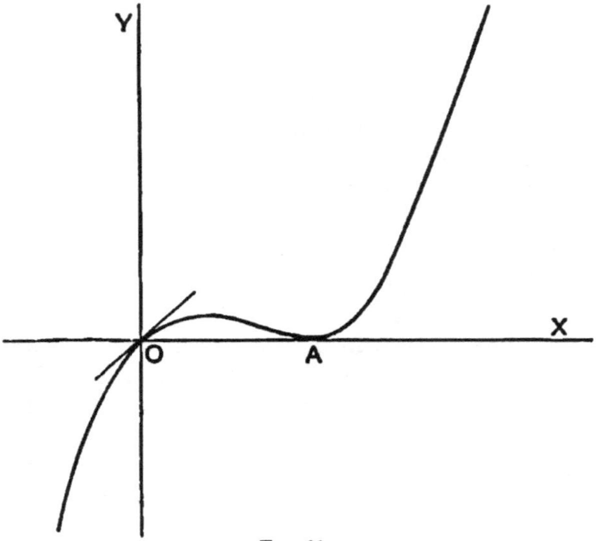


FIG. 21.

ing point between OA, the exact determination of which is, however, beyond our present methods.

Ex. 2. $y=x^2(x-1)^3$ (43).

Zero, $x=0$, O. First approximation, $y=-x^2$.
Zero, $x=1$, A. First approximation, $y=\xi^3$.
First approximation when $x$ and $y$ are both very great, $y=x^5$.

| $x$ | $y$ | |
|---|---|---|
| $-\infty$ | $-\infty$ | |
| $-0$ | $-0$ | Max. |
| $+0$ | $-0$ | |
| $1/2$ | $-1/32$ | |
| $1-0$ | $-0$ | |
| $1+0$ | $+0$ | |
| $+\infty$ | $+\infty$ | |

The general form is indicated in Fig. 22.

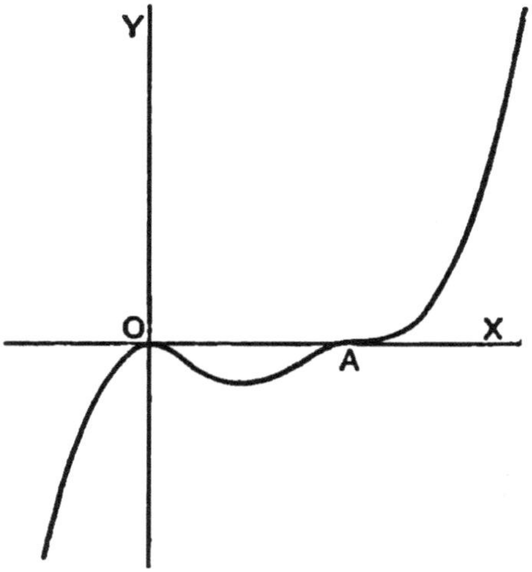


Fig. 22.

Ex. 3. $y=(x-1)^2(x-2)^2$ (44).

Zero, $x=1$, A. First approximation, $y=\xi^2$.
Zero, $x=2$, B. First approximation, $y=\xi^2$.
Approximation at $(\infty, \infty)$, $y=x^4$.

| $x$ | $y$ | |
|---|---|---|
| $-\infty$ | $+\infty$ | |
| 0 | 4 | |
| $1-0$ | $+0$ | Min. |
| $1+0$ | $+0$ | |
| 3/2 | 1/16 | Max. |
| $2-0$ | $+0$ | |
| $2+0$ | $+0$ | Min. |
| $+\infty$ | $+\infty$ | |

The general form of the graph is given in Fig. 23. Since $y$ is always positive, the maximum positive value of $(x-1)(x-2)$, which corresponds to $x=3/2$, gives a maximum value of $\{(x-1)(x-2)\}^2$. Hence the maximum turning point is (3/2, 1/16).

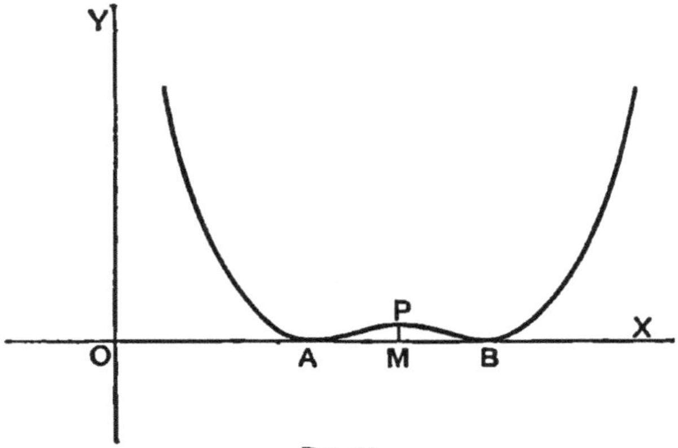


FIG. 23.

Ex. 4. $y=x+1+1/x=(x^2+x+1)/x$ (45)

No real zeros when $x$ is finite.
First approximation at $(0, \infty)$, $y=1/x$.
Second approximation at $(\infty, \infty)$, $y=x+1$.
Third approximation at $(\infty, \infty)$, $y=x+1+1/x$.

| $x$ | $y$ | |
|---|---|---|
| $-\infty$ | $-\infty$ | |
| $-1$ | $-1$ | Max. |
| $-0$ | $-\infty$ | |
| $+0$ | $+\infty$ | |
| $1$ | $3$ | Min. |
| $+\infty$ | $+\infty$ | |

These results suggest the curve drawn in Fig. 24.

If this be correct, there ought to be a minimum turning point to the right, and a maximum turning point to the left of the $y$-axis.

These points may be found as follows:—Let us consider the points on the graph of (45) which have a given ordinate $y$. In general there are two such, whose abscissæ are the roots of the quadratic equation

$$x^2+(1-y)x+1=0 \quad (46),$$

regarded as an equation to determine $x$, when $y$ is given.

If $y$ have a value somewhat greater than the turning value $+\text{PL}$ (Fig. 24), say $y = +\text{ON}$, we shall get two points, viz. Q and Q′, where NQQ′ is parallel to the $x$-axis, each of which has the given ordinate, and which have different abscissæ, viz. $+\text{OM}$ and $+\text{OM}'$. As we diminish the given value of $y$, the points Q and Q′ come nearer and nearer; and, when $y = +\text{PL}$, the two points coincide, and the two corresponding abscissæ become equal. If we make $y$ a little less than $+\text{PL}$, the parallel to the $x$-axis will no longer meet the graph at all. We shall therefore find the ordinate, $y$, of a turning value by so determining $y$ that the roots of (46) shall be equal; and this value will be a minimum turning value, if on making $y$ a *little less* the roots of (46) cease to be real; on the contrary, a maximum turning

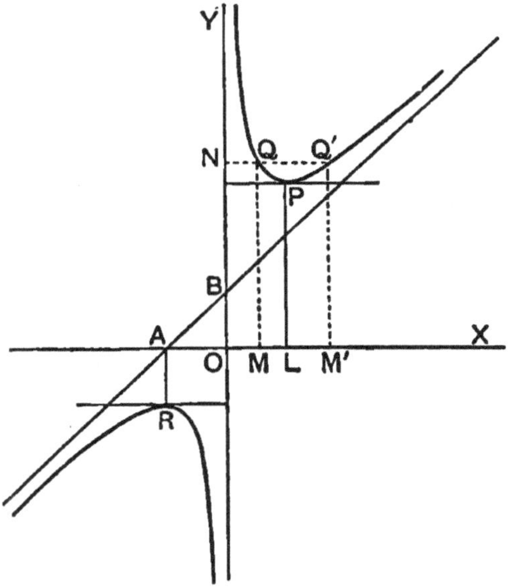


FIG. 24.

value, *e.g.* $y = -\text{AR}$, if on making $y$ a *little greater* the roots cease to be real.

Now the discriminant of (46) is

$$\Delta \equiv (1-y)^2 - 4 \equiv (y-3)(y+1).$$

$\Delta$ vanishes when $y=3$; and becomes negative if $y$ is a little less than 3. Hence $y=3$ is a minimum turning value.

$\Delta$ vanishes when $y=-1$, and becomes negative if $y$ is (algebraically) a little greater than $-1$. Hence $y=-1$ is a maximum turning value.

To determine the value of $x$ corresponding to $y=3$, we observe that, when $y=3$, the roots of (46) are equal. Now the sum of the roots (see § 218) is $-(1-y)$; hence in the present case each of them is $-(1-y)/2$, that is, $x=+1$.

Similarly, when $y=-1$, $x=-(1-y)/2=-1$. The maximum turning point is therefore $(-1, -1)$; and the minimum $(1, 3)$, which verifies our previous results.

This method for finding turning values is evidently applicable to any rational quadratic function. $\Delta$ is in general a quadratic function of $y$; and its factors may be real and distinct, coincident, or imaginary according to circumstances. In any particular case, therefore, we may have two real and distinct turning points, or none at all. When there are two, one is always a maximum and the other a minimum.*

In the following examples where turning points are given, they have been found by the present method. The details are left to the reader :—

Ex. 5. $y=(x^2-x+1)/(x^2+x+1)$ (47).

Since the roots of $x^2-x+1=0$ and of $x^2+x+1=0$ are all imaginary, there are no zeros or infinities of $y$ for finite values of $x$.

When $x=0$, $y=1$. To get an approximation to the form of the graph at (0, 1), B, put $y=1+\eta$. We thus get

$$\eta=-2x/(1+x+x^2).$$

Neglecting $x$ and $x^2$ in comparison with 1, we may replace $1+x+x^2$ by 1. We thus have for a first approximation at (0, 1)

$$\eta=-2x,$$

viz. a straight line in the second and fourth quadrants, whose ordinate is double its abscissa.

To get a second approximation we retain $x$, but neglect $x^2$ in the denominator. We thus get

$$\eta=-2x/(1+x)=-2x(1-x)/(1-x^2),\dagger$$

wherein we may neglect $x^2$. Thus, finally—

$$\eta=-2x+2x^2,$$

is a second approximation, which shows that the curve lies above the first approximation at (0, 1).

To get an approximation when $x$ is infinite, we write (47) in the equivalent form

$$\begin{aligned} y&=1-2x/(x^2+x+1),\\ &=1-2/x(1+1/x+1/x^2). \end{aligned}$$

For a second approximation, when $x$ is very large, we may neglect $1/x$ and $1/x^2$ in comparison with 1 in the denominator. We thus get

$$y=1-2/x;$$

hence we see that $y=1$ is an asymptote; and that the curve is below this line on the far right, and above on the far left. Hence the graph is like the curve in Fig. 25.

---

* It may also happen that $\Delta$ is only a linear function of $y$, or does not contain $y$ at all. This means geometrically that one or both of the turning points has passed to infinity. See A. Ch. XVIII. § 11, regarding this and other special points regarding the turning values of quadratic rational functions.

† We might also use ascending continued division as in § 275.

| $x$ | $y$ | |
|---|---|---|
| $-\infty$ | $1+0$ | |
| $-1$ | $3$ | Max. |
| $0$ | $1$ | |
| $1$ | $1/3$ | Min. |
| $+\infty$ | $1-0$ | |

It will be found that the minimum and maximum turning points are (1, 1/3) and (−1, 3) respectively.

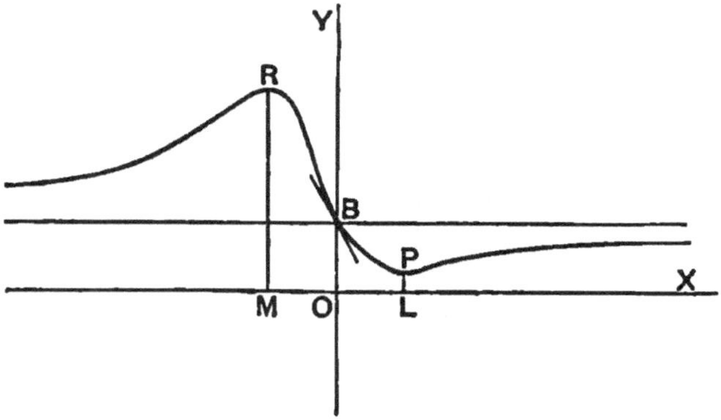


Fig. 25.

Ex. 6. $y=(x-1)(x-2)/(x^2+x+1)$ (48).

Zero, $x=1$, A. First approximation, $y=-\frac{1}{3}\xi$.
Second approximation, $y=-\frac{1}{3}\xi+\frac{2}{3}\xi^2$.
Zero, $x=2$, B. First approximation, $y=\frac{1}{7}\xi$.
Second approximation, $y=\frac{1}{7}\xi+\frac{2}{49}\xi^2$.
Point (0, 2), C. First approximation, $\eta=-5x$.
Second approximation, $\eta=-5x+4x^2$.
Second approximation at $(\infty, 1)$, $y=1-4/x$.

It is obviously suggested that the graph crosses the asymptote $y=1$ at a finite point. Putting $y=1$ in (48), we get $x^2+x+1=x^2-3x+2$ which gives $x=1/4$.* The point in question is therefore (1/4, 1).

It will be found that there are maximum and minimum turning points whose co-ordinates are approximately (−·9, 6·1) and (1·4, −·06) respectively (see Fig. 26).

---

* The other root of the quadratic that we should ordinarily get is infinite, as it ought to be (see A. Ch. XVIII. § 5).

| $x$ | $y$ | |
|---|---|---|
| $-\infty$ | $1+0$ | |
| $-\cdot 9$ | $6\cdot 1$ | Max |
| $0$ | $2$ | |
| $1/4$ | $1$ | |
| $1-0$ | $+0$ | |
| $1+0$ | $-0$ | |
| $2-0$ | $-0$ | |
| $2+0$ | $+0$ | |
| $1\cdot 4$ | $-\ \cdot 06$ | Min. |
| $+\infty$ | $1-0$ | |

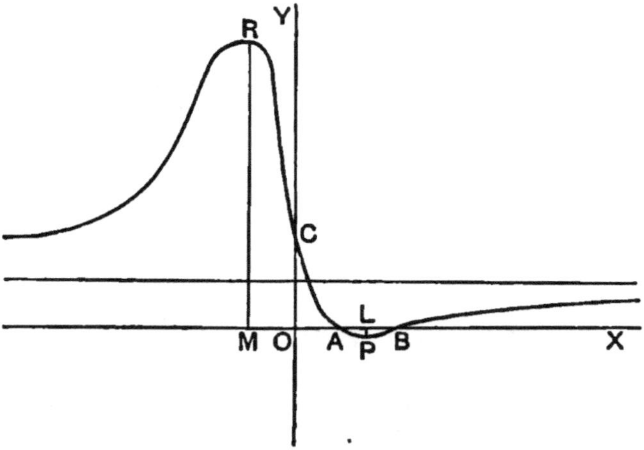


FIG. 26.

Ex. 7. $y=(x^2+x+1)/(x-1)(x-2)$ (49).
There are no zeros for finite values of $x$.

Infinity, $x=1$, A. First approximation, $y=-3/\xi$.
Infinity, $x=2$, B. First approximation, $y=\ \ 7/\xi$.
When $x=\infty$, $y=1$. First approximation, $y=1+4/x$.
Minimum point $(-\cdot 9,\ \cdot 16)$, approximately.
Maximum point $(1\cdot 40,\ -18\cdot 2)$, approximately.

| $x$ | $y$ | |
|---|---|---|
| $-\infty$ | $1-0$ | |
| $-\cdot 9$ | $\cdot 16$ | Min. |
| $0$ | $1/2$ | |
| $1/4$ | $1$ | |
| $1-0$ | $+\infty$ | |
| $1+0$ | $-\infty$ | |
| $1\cdot 40$ | $-18\cdot 2$ | Max. |
| $2-0$ | $-\infty$ | |
| $2+0$ | $+\infty$ | |
| $+\infty$ | $1+0$ | |

Fig. 27 indicates the nature of the graph.

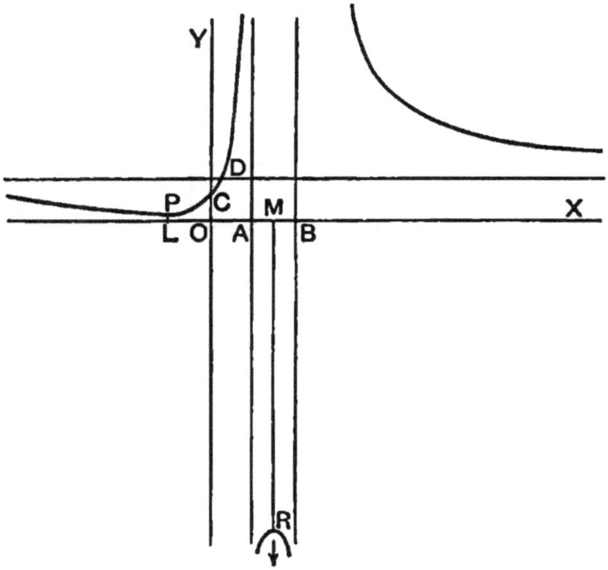


FIG. 27.

Ex. 8. $y=(x-1)(x-3)/(x-2)(x-4)$ (50).

The zeros of $y$ correspond to $x=1$ and $x=3$ ; the infinities to $x=2$, $x=4$. When $x=\infty$, $y=1$.

There are no real turning-points (see Fig. 28).

| $x$ | $y$ |
|---|---|
| $-\infty$ | $1-0$ |
| $0$ | $3/8$ |
| $1-0$ | $+0$ |
| $1+0$ | $-0$ |
| $2-0$ | $-\infty$ |
| $2+0$ | $+\infty$ |
| $5/2$ | $1$ |
| $3-0$ | $+0$ |
| $3+0$ | $-0$ |
| $4-0$ | $-\infty$ |
| $4+0$ | $+\infty$ |
| $+\infty$ | $1+0$ |

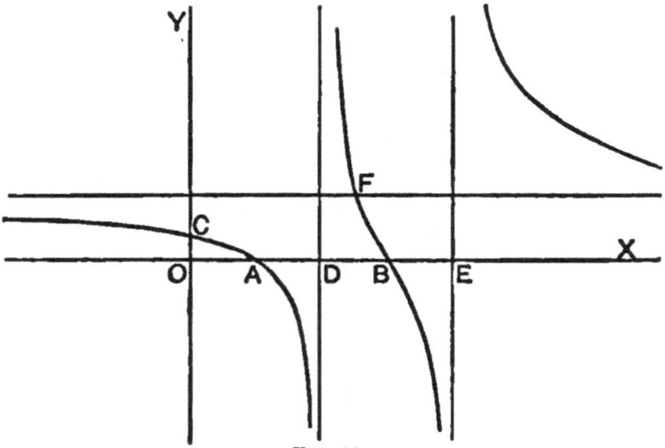


FIG. 28

Ex. 9. $y=(x-1)(x-2)/(x-4)$ (51).

$$y=0, \text{ when } x=1, \text{ or } x=2.$$
$$y=\infty, \text{ when } x=4.$$

To get the form of the graph when $x$ is very large, we put (51) into the equivalent form

$$y=x+1+6/(x-4).$$

Whence $$y=x+1+6/x(1-4/x).$$

A third approximation at $(\infty, \infty)$ is therefore given by

$$y=x+1+6/x.$$

There are maximum and minimum turning-points whose co-ordinates are approximately (1·5, ·1) and (6·45, 9·9) respectively.

| $x$ | $y$ | |
|---|---|---|
| $-\infty$ | $-\infty$ | |
| 0 | $-1/2$ | |
| $1-0$ | $-0$ | |
| $1+0$ | $+0$ | |
| 1·5 | ·1 | Max. |
| $2-0$ | $+0$ | |
| $2+0$ | $-0$ | |
| $4-0$ | $-\infty$ | |
| $4+0$ | $+\infty$ | |
| 6·45 | 9·9 | Min. |
| $+\infty$ | $+\infty$ | |

The graph is roughly indicated in Fig. 29.

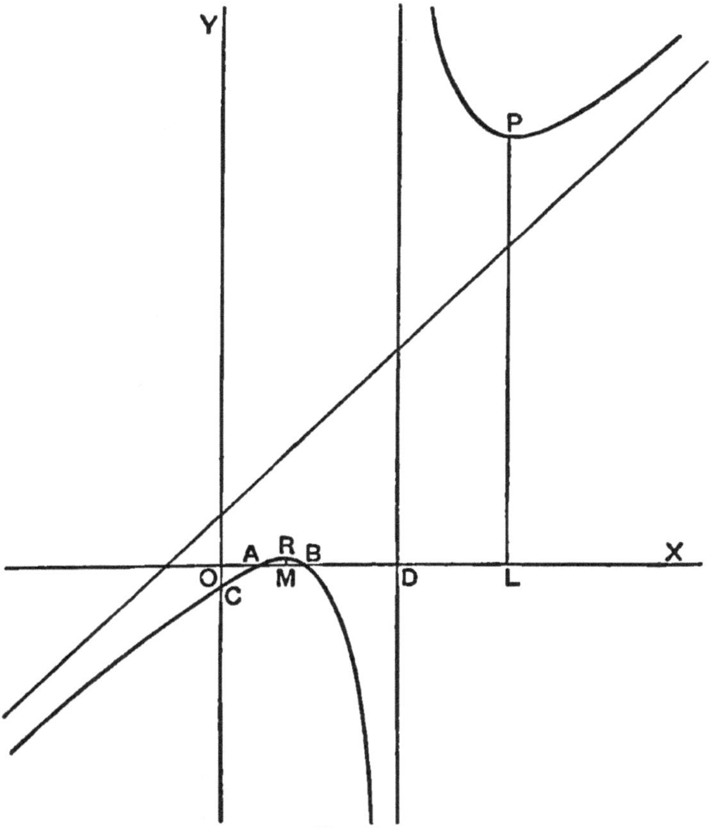


FIG. 29.

Ex. 10. $y=(x^3+1)/x^2$ (52).

$y=0$, when $x=-1$; there is no other finite zero.

When $x=0$, $y=\infty$, the first approximation being given by $y=1/x^2$.

The second approximation at $(\infty, \infty)$ is given by $y=x+1/x^2$. For the general form of the graph see Fig. 30.

| $x$ | $y$ |
|---|---|
| $-\infty$ | $-\infty$ |
| $-1-0$ | $-0$ |
| $-1+0$ | $+0$ |
| $-0$ | $+\infty$ |
| $+0$ | $+\infty$ |
| $+\infty$ | $+\infty$ |

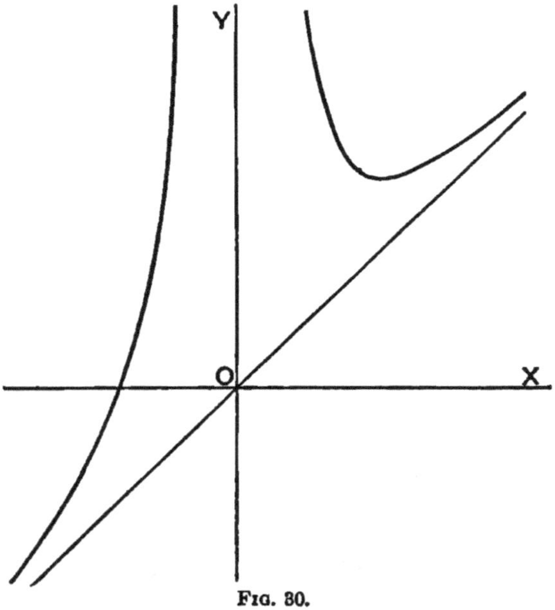


FIG. 30.

## EXERCISES LXXX.

Determine the general form of the graphs of the following; and verify your results by plotting test points. A number of the graphs should be fully plotted to scale :—

1. $y=x^2(x-1)$.
2. $y=x(x-2)^2$.
3. $y=x(x-2)^3$.
4. $y=x^3(x-1)$.

5. $y=x^3(x-1)^3$.
6. $y=x^3(1-x)^3$.
7. $y=x(x-1)(x-2)$.
8. $y=x(x-1)^2(x-2)$.
9. $y=x^2(x-1)(x-2)^2$.
10. $y=x^3(x-1)^3(x-2)^3$.
11. $y=(x-1)/(x-2)$.
12. $y=(x-1)^2/(x-2)$.
13. $y=(x-1)/(x+3)^2$.
14. $y=x/(x-1)^3$.
15. $y=3x+1-2/x$.
16. $y=3x-1/x$.
17. $y=-3x+1-1/x^2$.
18. $y=-2x-1/x^2$.
19. $y=(x-1)(x-2)/(x-3)(x-4)$.
20. $y=(x-1)(x-3)/(x-2)(x-4)$.
21. $y=(x^2+x+1)/(x-1)$.
22. $y=(x^2-x+1)/(x+2)$.
23. $y=(x^2+2x+1)/(x-1)$.
24. $y=(x^2-5x+6)/(x-1)$.
25. $y=(x-1)^2/(x^2-x+1)$.
26. $y=(x-1)^2/(x+1)(x-3)$.
27. $y=(x-1)^2/(x+1)^2$.
28. $y=(x^2+4x+3)/(x^2-x-6)$.

29. Discuss the gradual change in the graph of $y=(x-a)(x-b)(x-c)$ first, as $b$ is made more and more nearly equal to $a$; second, as $b$ and $c$ are both made more and more nearly equal to $a$.

30. Find the condition that $y=(ax^2+bx+c)/(Ax^2+Bx+C)$ may have an asymptote which is not parallel to either axis, and state the conditions for a rational function generally.

31. Show that the graph of a rational function of $x$ cannot cross the $x$-axis at right angles.

## Delimitation of the Roots of Equations

**277.** We shall next point out some applications of the graphical method to the problem of finding the number and approximate value of the real roots of an equation in which the coefficients are real.

In the first place, we lay down the following important principle:—

*If the function* f(x) *be finite and continuous for all values of* x *in the interval* $\alpha \ngtr x \ngtr \beta$, *and if* $f(\alpha)$ *and* $f(\beta)$ *have opposite signs, then one root at least, and, if more than one, an odd number of roots, of the equation* $f(x)=0$ *lie in the interval* $\alpha \ngtr x \ngtr \beta$.

This is at once obvious, if we consider the graph of the function $f(x)$, the equation of which is $y=f(x)$. If $f(\alpha)$ and $f(\beta)$ have opposite signs, then the two points P and Q on the graph, whose co-ordinates are $\{\alpha, f(\alpha)\}$ and $\{\beta, f(\beta)\}$ respectively, lie on opposite sides of the $x$-axis. Since, as $x$ passes continuously from $\alpha$ to $\beta$, the graphic point travels continuously from P to Q, beginning on one side of the $x$-axis, and ending on the other, it must have crossed the $x$-axis at least once, and if more than once, an odd number of times. This proves the theorem, because $f(x)$ vanishes whenever the graphic point crosses the $x$-axis. It should be remarked that there is no restriction on $f(x)$, except that it shall be continuous. Thus $f(x)$ may, if we choose, be a rational fractional function of $x$, provided it do not become infinite between $x=\alpha$ and $x=\beta$. The necessity for the restriction of continuity will be seen by considering Fig. 24 above, where $(x^2+x+1)/x$ is negative when $x=-\text{OA}$, and positive when $x=+\text{OL}$, and yet there is no root of $(x^2+x+1)/x$ in the interval $-\text{OA} \ngtr x \ngtr \text{OL}$.

If $f(x)$ be an integral function of $x$, say $f(x) \equiv a_n x^n + a_{n-1}x^{n-1} + \ldots + a_1 x + a_0$, since, as we have seen, the highest term is the most important and dominates the sign of the value of the function when $x$ is very large, it follows that $f(\infty)$ and $f(-\infty)$ have always opposite signs when $n$ is odd, and always the same sign when $n$ is even. Hence

*Every integral equation of odd degree (whose coefficients are real) has an odd number of real roots, and has at least one real root.*

*Every integral equation of even degree (whose coefficients are real) has an even number of real roots, if it has any.*

*Every integral equation (whose coefficients are real) has an even number of imaginary roots, if it has any.*

Ex. $f(x) \equiv x^3 - 6x^2 + x - 5 = 0$.

The equation being of odd degree must have at least one real root. It can have no negative root; for, if we put $x=-\xi$, where $\xi$ is any positive quantity, $f(-\xi) = -(\xi^3+6\xi^2+\xi+5)$, which obviously cannot vanish, since all the terms within the bracket have the same sign. On the other hand, we find $f(5)=-25$ and $f(6)=+1$; hence there is at least one real root between $+5$ and $+6$.

**§ 278.** The most obvious method of finding the roots of the equation $f(x)=0$, as has already been seen, is to draw the graph of $y=f(x)$, taking care to get it very accurate (by plotting the graphic points close together) when the values of $y$ are small. We have then merely to measure the abscissæ of the points where it crosses the $x$-axis, attach the proper sign, and these are the roots of $f(x)=0$.

The process just described amounts to finding the intersection of the two graphs $y=0$ and $y=f(x)$. The following generalisation of the method, which consists in **using the intersection of any two properly chosen Graphs,** is often very useful in delimiting roots. We can arrange any equation in an infinite variety of ways in the form

$$\phi(x) = \psi(x).$$

Let the functions $\phi(x)$ and $\psi(x)$ be chosen so that the general forms at least of their graphs $y=\phi(x)$ and $y_1=\psi(x)$ are easily found.

The problem of solving the equation $\phi(x)=\psi(x)$ is then reduced to finding the abscissæ of the points for which $y=y_1$, that is to say, the abscissæ of the points of intersection of the two graphs.

Ex. Consider the equation $x^3-6x^2+x-5=0$, already partly dis-

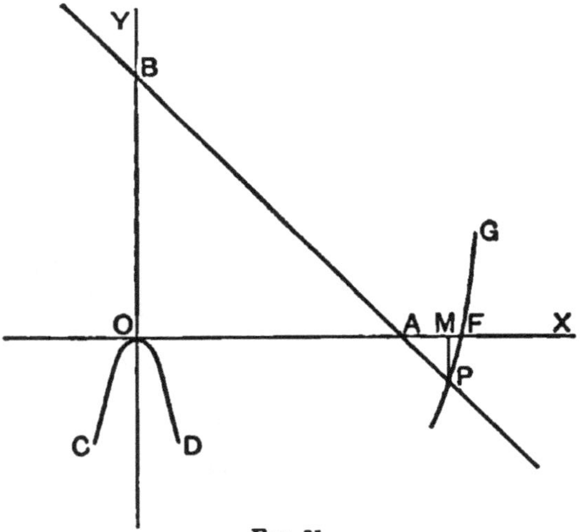


Fig. 31.

cussed in last paragraph. We may write the equation in the equivalent form

$$x^3-6x^2=-x+5.$$

Now draw roughly the graphs of $y=-x+5$ and $y_1=x^3-6x^2$. The result (Fig. 31) is the straight line AB, where OA=OB=5; and the curve CODPFG, where OF=6. Only part of the curve is drawn, as it descends very steeply to a large minimum between $x=0$ and $x=6$. The part with which we are concerned is PFG. That this part is very steep may be seen by observing that the value of $x^3-6x^2$ is $-15\cdot1\ldots$ when $x=5\cdot5$.*

The abscissa of the intersection P is therefore very little less than 6; and it is obvious that there is one real root a little less than $+6$; and no other real root positive or negative.

* Also, of course, by shifting the origin to F, and considering the first approximation, which is $y=36\xi$.

§ 279. All methods for the **Calculation of the Roots of Equations by successive Approximation** rest ultimately on the simple principle of § 277. For integral equations there exists a systematic process, called after its inventor Horner's method,* which is practically perfect for calculating digit by digit the value of any real root which has been isolated, either tentatively or by means of graphical methods, so far that we know the highest significant digit or a sufficient number of the highest digits to distinguish the root from others nearly equal

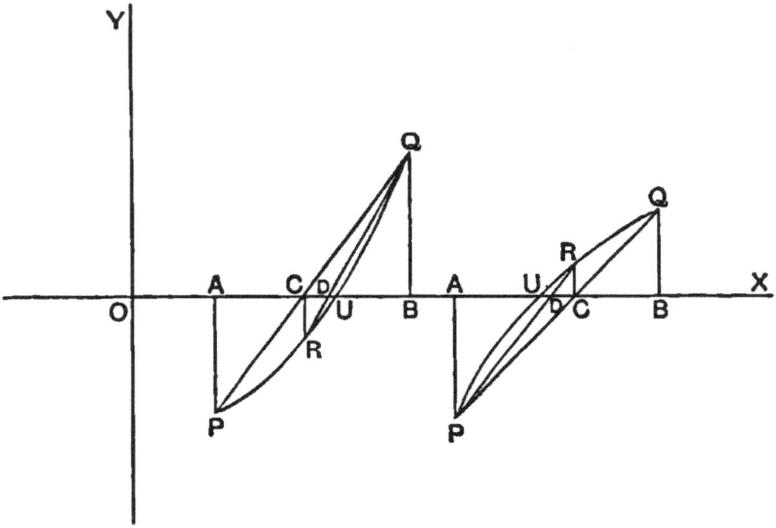


FIG. 32.

to it. The ordinary rules for extracting the square and cube root are special cases of this method.

We shall give here an elementary method which is applicable to any kind of equation, but which cannot compete in rapidity and convenience with Horner's method when integral equations are in question. It is a case of what the old mathematicians † used to call **the Rule of Falsehood**, or **the Rule of False Position**; and the student will meet with it again under the names of **Interpolation by First Differences**, or **the Rule of Proportional Parts**.

* See A. Ch. XV.

† See, for example, Recorde's *Arithmeticke* (1540).

Suppose that the equation in question is

$$f(x) = 0,$$

regarding which we suppose that $f(x)$ is continuous in the neighbourhood of the roots of the equation. To avoid useless discussion, we shall suppose that the graph of $y = f(x)$ has no inflexion and is smooth in the parts which we have occasion to use. Let now $a$ and $\beta$ be found tentatively such that $f(a) = -h$ and $f(\beta) = k$, where $h$ and $k$ are positive; we shall suppose $a < \beta$, but it will be seen that this does not affect our final result.

Let P and Q be the points $(a, -h)$ and $(\beta, k)$, obviously points on the graph, since $-h = f(a)$ and $k = f(\beta)$. Since $f(a)$ and $f(\beta)$ have opposite signs, the part of the graph between P and Q must meet the $x$-axis at least once, in U say; let us suppose that it meets it only once. Then there is only one root of $f(x) = 0$ between $x = a$ and $x = \beta$, viz. $x = +\mathrm{OU}$. We are not supposed to know the exact course of the graph between P and Q; but we can get an approximation by replacing it by the straight line PQ; and this will be nearer to the truth the shorter the portion PQ is—that is to say, the smaller the values of both $h$ and $k$. The approximation to the root thus obtained is $\xi = +\mathrm{OC}$; and it is obvious that it is nearer to the real value, $+\mathrm{OU}$, than either $+\mathrm{OA}$ or $+\mathrm{OB}$, *i.e.* than $a$ or $\beta$.

It is easy to calculate $\xi$ in terms of $h$, $k$, $a$, $\beta$. We have, in fact, from the similar triangles ACP and BCQ, $(\xi - a)/h = (\beta - \xi)/k$, whence

$$\left.\begin{aligned} &\xi = (\beta h + ak)/(h + k) \\ \text{or} \quad &\xi = a + h(\beta - a)/(h + k) \end{aligned}\right\} \qquad (53),$$

which is the "Rule of Falsehood." In practice, the second form is usually most convenient.

If it is desired to come still nearer to the root, we next calculate $f(\xi)$. If $f(\xi)$ turns out to be negative $= -h'$, say, we use $(\xi, -h')$ and $(\beta, k)$ as before and calculate $\xi_1 = +\mathrm{OD}$ by means of (53), viz. we get

$$\xi_1 = (\beta h' + \xi k)/(h' + k) \qquad (54).$$

If $f(\xi)$ is positive, $= +h'$, say, we use $(\xi_1, h)$ and $(a, -h)$, and get

$$\xi_1 = (\xi h + ah')/(h + h').$$

It is often better in the second step, instead of using either P or Q along with R, to find tentatively some new value of $x$,

$\gamma$ say, between $\alpha$ and $\xi$, or between $\xi$ and $\beta$, as the case may be, such that $f(\gamma)$ is of opposite sign to $f(\xi)$, and use the corresponding point along with R.

Ex. Consider once more the equation

$$f(x) \equiv x^3 - 6x^2 + x - 5 = 0.$$

We have seen that it has one real root very near to $+6$. Let us try $5 \cdot 9$, and calculate $f(5 \cdot 9)$ and $f(6)$—

$$\begin{array}{r|rrrr} & 1 & -6 & +1 & -5 \\ 5\cdot 9 & 0 & +5\cdot 9 & -\cdot 59 & +2\cdot 42 \\ \hline & 1 & -\cdot 1 & +\cdot 41 & | -2\cdot 58 \end{array}$$

$$\begin{array}{r|rrrr} & 1 & -6 & +1 & -5 \\ 6 & 0 & +6 & +0 & +6 \\ \hline & 1 & +0 & +1 & | +1 \end{array};$$

that is, $f(5 \cdot 9) = -2 \cdot 58$, $f(6) = +1$.

Therefore, by the "Rule of Falsehood" in its second form—

$$\xi = 5 \cdot 9 + 2 \cdot 58 \times \cdot 1/3 \cdot 58$$
$$= 5 \cdot 97.$$

Since it will be found that $f(5 \cdot 97) = -\cdot 099227$, $f(5 \cdot 98) = +\cdot 264792$, it follows that $5 \cdot 97$ is correct to the second place of decimals.

A second approximation, founded on the values of $f(5 \cdot 97)$ and $f(5 \cdot 98)$, gives $\xi_1 = 5 \cdot 972725$, which agrees with the calculation by Horner's method to the sixth place of decimals.

§ 280. Many problems regarding maxima and minima can be solved by the methods of this chapter; and in practice when difficulty is experienced in solving a problem of this kind which depends on a single variable, the use of a graph will often help to make the matter clear.

Ex. 1. Show that if the area of a rectangle is given, its perimeter is a minimum when it is a square.

Let $a$ be the side of a square having the given area; then, if $x$ be one side of the rectangle, the other adjacent side is $a^2/x$; and the perimeter is given by

$$y = 2x + 2a^2/x.$$

We have to find the minimum value of $y$, which, as also $x$, must from the nature of the problem be positive. We may use the method of § 276, Example 4. The equation for $x$, $y$ being supposed given, is

$$2x^2 - yx + 2a^2 = 0;$$

the discriminant of which is $\Delta = y^2 - 16a^2 = (y - 4a)(y + 4a)$. Negative values of $y$ being out of the question, we have merely to examine the sign of $\Delta$ for values of $y$ near $4a$. When $y$ is a little less than $4a$, $\Delta$ is negative; and when $y = 4a$, $\Delta = 0$. Hence $y = 4a$ is a minimum value of $y$; to this corresponds $x = a$ and $a^2/x = a$, so that the minimum value of the perimeter occurs when the rectangle is a square.

Ex. 2. ABC is a triangle, right angled at C, whose sides satisfy the condition $l\text{AC}^2+m\text{BC}^2=d^2$, where $l$, $m$, $d^2$ are given positive quantities. Find the lengths of the sides when the area is a maximum.

Let the sides be $x$ and $y$, so that $x$ and $y$ in the present case are both necessarily positive.

We have to find the maximum value of $\frac{1}{2}xy$ subject to the condition

$$lx^2+my^2=d^2 \qquad (\alpha).$$

Since $l$ and $m$ are positive, we may put $\xi=lx^2$, $\eta=my^2$, where $\xi$ and $\eta$ are positive quantities.

Since $x$ and $y$ are both positive, $\frac{1}{2}xy$ will be a maximum or minimum when $\frac{1}{4}x^2y^2$ is a maximum or minimum; and therefore also when $(lx^2)(my^2)/4lm$—that is, $\xi\eta/4lm$ is a maximum or minimum. Now, since $4lm$ is constant, $\xi\eta/4lm$ will be a maximum or minimum when $\xi\eta$ is a maximum or minimum.

We have from ($\alpha$)

$$\xi+\eta=d^2 \qquad (\beta).$$

Hence

$$\begin{aligned}\xi\eta&=\xi(d^2-\xi),\\&=\tfrac{1}{4}d^4-(\tfrac{1}{2}d^2-\xi)^2.\end{aligned}$$

Therefore $\xi\eta$ is a maximum when the essentially positive quantity $(\frac{1}{2}d^2-\xi)^2$ has its least value—that is to say, when $\xi=\frac{1}{2}d^2$. Using ($\beta$), we see that the maximum value of the area occurs when $\xi=\eta=\frac{1}{2}d^2$. Reverting to the meanings given to $\xi$ and $\eta$, we have $lx^2=my^2=\frac{1}{2}d^2$, which lead to $x=d/\surd(2l)$, $y=d/\surd(2m)$.

## EXERCISES LXXXI.

1. Show that the equation $x^3-x^2-1=0$ has only one real root; and calculate it to two places of decimals.

2. Show that $x^3+2x^2-5x-7=0$ has three real roots; and calculate each to one place of decimals.

3. Show that $x^3-2x+2=0$ has a real negative root; and calculate it to two places of decimals.

4. Calculate approximately the real positive root of $x^3+2x^2-4=0$.

5. Show that $x^4-x^3+1=0$ has no real roots.

6. Show that $x^4-2x-1=0$ has two real roots, and no more.

7. Show that $x^5-x^2-2x-1=0$ has only one real root; and calculate it to one place of decimals.

8. Show that $x^4+2x^3+x^2+x-1=0$ has two real roots; and delimit them.

9. Discuss the roots of $x^4+2x^3-x^2-1=0$.

10. If $xy=4$, discuss the turning values of $x+y$.

11. If $x+y=10$, find the minimum value of $3x^2+2y^2$, and the values of $x$ and $y$ corresponding thereto.

12. The triangle ACB is right angled at C, and AB is given. If D be the projection of C on AB, find the greatest and least values of $\text{AC}^2+\text{BD}^2$.

**13.** Show that the area of the maximum rectangle inscribed (within) a given triangle is half the area of the triangle.

**14.** Find the minimum square that can be inscribed in a given square.

**15.** The sum of the squares of the parallel sides of a trapezium is given, and also the distance between them. When is the area a maximum?

**16.** ACB is right angled at C. P is a point in AB whose projections on BC and CA are L and M respectively. For what position of P is the area PLCM a maximum?

**17.** A right angle PAQ turns about a fixed point A, and meets a fixed straight line in P and Q. When is the distance PQ a maximum?

**18.** If $x^2+y^2=50$, discuss the turning values of $xy$.

**19.** If $x^2-y^2=50$, discuss the turning values of $xy$.

**20.** If $2x^2+5y^2=50$, discuss the turning values of $xy$.

**21.** If the hypotenuse of a right-angled triangle is given, when is its area a maximum?

**22.** If the area of a right-angled triangle is given, when is its hypotenuse a minimum?

**23.** Find the maximum rectangle which can be inscribed in a given semicircle.

**24.** The corner of a rectangular printed page is turned down so that the corner always falls on a particular printed line parallel to the top of the page. What is the distance of the corner from the crease when the ratio of this distance to the length of the crease is a maximum?

**25.** ACB is right angled at C. O is the middle point of AB; and D the projection of C on AB. If AB be given ($=c$, say), what is the distance of D from A when the area OCD is a maximum?

**26.** ACB is a given triangle, right angled at C. L and M are the projections on BC and AC respectively of a movable point on AB. When is $CL^2+CM^2$ a minimum?

# ANSWERS TO EXERCISES

## I.

**1.** 132. **2.** 104. **3.** 8 ; 45 ; 0. **4.** $48\frac{1}{7}$. **5.** $71\frac{1}{15}$. **6.** $4\frac{1}{7}$. **7.** $1\frac{3}{4}$. **8.** 18,496. **9.** $21\frac{3}{17}$. **10.** $1\frac{1}{8}$. **11.** $\frac{14}{19}$. **12.** 28/33. **13.** 144. **14.** 2304. **15.** 9/16. **16.** 262,144. **17.** $1\frac{7}{9}$. **18.** $1\frac{7}{9}$. **19.** ·378. **20.** 37·795. **21.** 1. **22.** ·378. **23.** 1. **24.** 14·697. **25.** 4.

## II.

**1.** $3\times 240+6\times 12+8$. **2.** $a\times 240+b\times 12+c$. **3.** $x+10y+100z$. **4.** $2n+7$. **5.** $2n-3,\ 2n-1,\ 2n+1,\ 2n+3,\ 2n+5$. **6.** $5(a+b+c)+6(a'+b'+c')+8(a''+b''+c'')$ ; 158. **7.** $30a+24b+12c+6d$. **8.** $5a+b+c/4+d/8+e/20$. **9.** $6/a+7/3$ hrs. **10.** $Pnr/100$ ; $P(1+nr/100)$, where P is the principal in pounds, $r$ the rate per cent in pounds, and $n$ the number of years. **11.** $P/(1+nr/1200)$, P and $r$ being as in last exercise, and $n$ the number of months. **12.** $A\div 100\times p-A\div 100\times P\div Q\times q$. **13.** $c\{(a\alpha+b\beta)/(a+b)-(\alpha+\beta)/2\}$. **14.** $\{10P+10(P+Q)+10(P+2Q)\}/30$. **15.** $(ap+bq)/(p+q)$ ; 8·31 approx. **16.** $ap/100,\ bq/100,\ cr/100$ ; $ap/(a+b+c),\ bq/(a+b+c),\ cr/(a+b+c)$. **17.** $100\{ap+bq+cr-(a+b+c)s\}/(ap+bq+cr)$. **18.** $(pq/100)^3$. **19.** $P(1+r/100)^n$, P being the principal in pounds, $r$ the rate per cent in pounds, and $n$ the number of years. **20.** $p_0+p_1r+p_2r^2+\ldots+p_nr^n$. **21.** $p_1/10+p_2/10^2+p_3/10^3+p_4/10^4$. **22.** 15, 852. **23.** ·7104. **24.** $a(1-b/a)^n$.

## III.

**1.** $3+5-6,\ 5+3-6,\ 5-6+3$. **2.** $-6$. **3.** $-9$. **4.** $+8$. **5.** $-4$. **6.** $+2$. **7.** 0. **8.** $a+a+a+a+b+b+c+c$. **9.** $-12$. **10.** $b+b+b+b+c+c$. **11.** $a-b$. **12.** $+1$, if $n$ be of the form $4m+2$ or $4m+3$ ; $-1$, if $n$ be of the form $4m$ or $4m+1$, where $m$ is an integer. **13.** 0, if $n$ is odd ; $+a$, if $n$ is even. **14.** $3-(5-6)$ is greater. **15.** The second is the greater. **18.** 0.

## IV.

**1.** 4. **2.** 27/10. **3.** 1/4. **4.** 3/4. **5.** $2aa$. **6.** 9/4. **7.** $aa/dd$. **8.** 1/20. **9.** 144/5. **10.** 1/2. **11.** 729/64. **12.** 1.

## V.

**1.** ·0081. **2.** 548·8. **3.** 1/43,046,721. **4.** 9/3125. **5.** $a^{13}b^{10}c^{10}$. **6.** $(1/3a)z^2/y^4$. **7.** $35abc^2d^4$. **8.** $a^{40}$. **9.** $\frac{2}{3}a^{12}$. **10.** $a^{30}$. **11.** $a^{1\cdot 2\cdot 3\cdot \ldots n}$. **12.** $a^3b^3c^3x^4y^4z^4$. **13.** $1/x^4y^4z^4$. **14.** $x^{4l}y^{4m}z^{4n}$. **15.** $x^{l(n-m)}y^{m(l-n)}z^{n(m-l)}$, which cannot be integral, since $n-m$, $l-n$, $m-l$ cannot be all positive; when $x=y=z$, the value is 1. **16.** $2^4 \cdot 3^4 \cdot 5^3 \cdot 7$. **17.** $5^2 \cdot 7^2 \cdot 11^3 \cdot 73^5/2^{15} \cdot 3^8$. **20.** $\frac{1}{2}\{(\alpha+1)(\beta+1)(\gamma+1)+1\}$, if $\alpha$, $\beta$, $\gamma$ be all even; otherwise $\frac{1}{2}(\alpha+1)(\beta+1)(\gamma+1)$ : 4. The factorisation $1 \times a^\alpha b^\beta c^\gamma$ is included in this enumeration. **21.** $2^7 \cdot 5=640$. **22.** $Byz$, where B is an undetermined constant. **23.** $Axy^2z^2$, where A is an undetermined constant. **25.** It can be deduced from the data that $x^{xyz}=x$, whence $xyz=1$, which requires that $x=y=z=1$, since $x$, $y$, $z$ are integral. **26.** 4, 4, 4, 12; $4l$, $4m$, $4n$, $4(l+m+n)$. **27.** $Ax^2y^3z^5$, where A is an undetermined constant. **28.** $\frac{1}{3}x^3y^2z$. **29.** $4x^3y^4$, $2xy^2$ is one solution.

## VI.

**1.** $-a^5$. **2.** $-a^{4n+1}$. **3.** $a^{40}$. **4.** $a^{15}$. **5.** $a^3b^3c^3x^4y^4z^4$. **6.** $a^{30}$. **7.** $-5^9 \cdot 2^{30} \cdot 3^{12}=-1,114,512,556,032 \times 10^9$. **8.** $a^{1\cdot 2\cdot 3 \cdots 4n}$. **9.** $-a-2b+2c$. **10.** $-x^2+\frac{13}{10}x$. **11.** $2a$. **12.** $2b$. **13.** 0. **14.** $2a$. **15.** $-a+b-c$. **16.** $-3y-4z$. **17.** $-5x+3y$. **18.** $2x^2+2y^2+2z^2-yz-zx-xy$. **19.** $3x^3+x^2+x+3$. **20.** $2a-2b+x$. **21.** $-2a+x$. **22.** $\frac{3}{8}a-\frac{1}{4}b$. **23.** $a_1-a_n$. **24.** 2. **25.** $ab+bc+cd+de-ac-bd-ce-da$. **26.** $4x$. **27.** $2bx+2ay$. **28.** $a-ab+abc-abcd+abcde$. **29.** $a_1+a_2+\ldots+a_{n-1}-(n-1)a_n$. **30.** $-26x+14y$. **31.** $3ab$. **32.** $\Sigma a^2$. **33.** $x^3-2x^2+2x-1$. **34.** $3xyz+yz^2-y^2z+zx^2-z^2x+xy^2-x^2y$. **35.** $x^2+2x-48$. **36.** $-x^2-14x-48$. **37.** $x^2-\frac{1}{6}x-\frac{1}{6}$. **38.** $\frac{1}{2}x^2+\frac{3}{4}x-\frac{1}{2}$. **39.** $\frac{1}{3}x^4+\frac{1}{12}x^3+\frac{3}{8}x^2+\frac{3}{4}x-\frac{1}{2}$. **40.** $abx^2+(a+b)x+1$. **41.** $acx^2-(ad+bc)x+bd$. **42.** $xy/ab+x/a+y/b+1$. **43.** $-\frac{3}{4}x^2+x-\frac{1}{2}$. **44.** $-5xy+x+2y+2z-2$. **47.** $x=1$, $x=2$. **48.** $x=1$, $x=-4$. **49.** $x=-2$, $x=-2$. **50.** $x=0$, $x=1$. **51.** $x=a^2$, $y=b^2$. **52.** 1° $y^3z$, $yz^3$, $z^3x$, $zx^3$, $x^3y$, $xy^3$; 2° $xy^2z^2$, $yz^2x^2$, $zx^2y^2$; 3° $yz/x$, $zx/y$, $xy/z$. **53.** 1° $xy^2$, $x^2y$, $xz^2$, $x^2z$, $xu^2$, $x^2u$, $yz^2$, $y^2z$, $yu^2$, $y^2u$, $zu^2$, $z^2u$; 2° $xy^3$, $x^3y$, $xz^3$, $x^3z$, $xu^3$, $x^3u$, $yz^3$, $y^3z$, $yu^3$, $y^3u$, $zu^3$, $z^3u$; 3° $yz/x$, $yu/x$, $zu/x$, $xz/y$, $xu/y$, $zu/y$, $xy/z$, $xu/z$, $yu/z$, $xy/u$, $xz/u$, $yz/u$; 4° $x^2y/z$, $x^2z/y$, $x^2y/u$, $x^2u/y$, $x^2z/u$, $x^2u/z$, $y^2x/z$, $y^2z/x$, $y^2x/u$, $y^2u/x$, $y^2u/z$, $y^2z/u$, $z^2x/y$, $z^2y/x$, $z^2x/u$, $z^2u/x$, $z^2y/u$, $z^2u/y$, $u^2x/y$, $u^2y/x$, $u^2x/z$, $u^2z/x$, $u^2y/z$, $u^2z/y$; 5° $y^2z^2/x^2$, etc., as in 3°, only $x$, $y$, $z$ are each squared. **54.** $x^3$, $y^3$, $z^3$, $u^3$; $xy^2$, $x^2y$, $xz^2$, $x^2z$, $xu^2$, $x^2u$, $yz^2$, $y^2z$, $yu^2$, $y^2u$, $zu^2$, $z^2u$; $yzu$, $xzu$, $xyu$, $xyz$. **55.** 1° $ax^2+by^2+cz^2+du^2+exy+fxz+gxu+hyz+kyu+lzu$; 2° $ax^2y+bxy^2+cx^2z+dxz^2+exu^2+fx^2u+gy^2z+hyz^2+ky^2u+lyu^2+mz^2u+nzu^2$. **56.** $a+bx+cy+dx^2+exy+fy^2+gx^3+hx^2y+kxy^2+ly^3$. **57.** $a+b\Sigma x+c\Sigma x^2+d\Sigma xy$. **58.** 81/50. **59.** $-2$. **60.** $\frac{10}{11}\Sigma yz$. **61.** From the data we can deduce $(B-C)(xy/z-zx/y)=0$, whence, since $xy/z-zx/y$ is not necessarily 0, $B-C=0$.

## VII.

**1.** $\frac{1}{3}x^4+\frac{2}{3}x^3-\frac{1}{2}x^2$. **2.** $3x^3-6x^2+4x-1$. **3.** $3x^3-2x+1$. **4.** $x^4-\frac{34}{15}x^3+4x^2-\frac{10}{3}x+\frac{8}{5}$. **5.** $2x^3+\frac{13}{2}x^2+\frac{11}{2}x+1$. **6.** $x^3-\frac{8}{15}x^2-\frac{1}{60}x$

$+\frac{1}{60}$. **7.** $x^4-y^4$. **8.** $(a-b)x^3+(a-c)x^2+(b-d)x+(c-d)$. **9.** $x^2-\{(a^4-b^4)/a^2b^2\}xy-y^2$. **10.** $4x^2$. **11.** $x^4-2x^3y-x^2y^2-2xy^3+y^4$. **12.** $x^4-y^4$. **13.** $(a+b)x^4-(a+b)y^4$. **14.** $9x^4-y^4$. **15.** $x^{2n}-a^{2n}$. **16.** $x^{2n}-2x^n+1$. **17.** $x^{2n}-2+1/x^{2n}$. **18.** $2^nx^3+5.2^{n-1}x^2+(2^n+1)x+2$. **19.** $3x^{n+1}-x^n-2x^{n-1}+x-1$. **20.** $x^{3n}+1$. **21.** $-x^{2p+q}+x^{2p-q}+x^{p+q}+x^{p-q}+1$. **22.** $3x^{2p}+x^{2p+b-c}+x^{2p-b+c}+x^{2p+c-a}+x^{2p-c+a}+x^{2p+a-b}+x^{2p-a+b}$. **23.** $x^2+xy+y^2$. **24.** $x^2-2x-2$. **25.** $3(x-1)^4+14(x-1)^3+21(x-1)^2+13(x-1)+5$. **26.** $c=2$. **27.** $A=-1$. **28.** $A=1$, $B=-4$, $C=1$. **29.** $16/x-1$. **30.** If $a=-1$, the function is $x^2+x+1$.

## VIII.

**1.** 2. **2.** $a_1a_3/a_6a_8$. **3.** 0. **4.** $(2x^2+2x)/(x-1)$. **6.** $4u$. **7.** $a-b+3$. **8.** $2x^4-4x^2+2$. **10.** $(2x^3+2x)/(x^3+1)$. **11.** $3-z/y-x/z-y/x$. **12.** 4. **14.** $(2+2y)/(1-y)$. **15.** $x+y$. **16.** 1. **17.** 1. **18.** $x^2/(a^2-b^2)+2(a^2+b^2)x/(a^2-b^2)^2+1/(a^2-b^2)$. **19.** 0. **20.** 3. **21.** $-(6x^2+2)/(x-1)^3$. **22.** $(-\Sigma x^3+\Sigma x^2y-3xyz)/(y-z)(z-x)(x-y)$. **23.** $(x^3-4x^2+3x)/(x^2-3x+1)$. **24.** $(x^4-x^3y-3x^2y+2xy^2+y^2)/(x^3-x^2y-2xy+y^2)$.

## IX.

**5.** $2x^3y^3$. **6.** $\Sigma a^2b^2c^2+\Sigma a^3bcd$. **7.** $a^2-d^2$. **8.** $x^2+y^2-z^2-u^2-2xy+2zu$. **11.** $x^4-y^4$; $x^4-y^4-z^4-4y^3z-6y^2z^2-4yz^3$. **14.** $-3$. **15.** 2. **16.** $-2$. **17.** $c^2-b^2+ab-ac$. **18.** $b/c+c/b$. **19.** $x^4-y^4-z^4+2y^2z^2$. **20.** $x^3+8y^3-z^3-12y^2z+6yz^2+3z^2x-3zx^2+6x^2y+12xy^2-12xyz$. **21.** $x^4+y^4+z^4-4y^3z-4yz^3-4z^3x-4zx^3+4x^3y+4xy^3+6y^2z^2+6z^2x^2+6x^2y^2-12x^2yz-12xy^2z+12xyz^2$. **22.** $\Sigma x^3-y^2z-yz^2-z^2x-zx^2+3x^2y+3xy^2-2xyz$. **23.** $\Sigma a^3x^3-3abcxyz$. **24.** $2\Sigma ab-\Sigma a^2$. **25.** $b^2c^4-b^4c^2+c^2a^4-c^4a^2+a^2b^4-a^4b^2$. **26.** $x^6-x^4y+x^4z-x^4u-x^2yz+x^2yu-x^2zu+yzu$. **27.** $x^4-10x^3+35x^2-50x+24$. **28.** $x^6-14x^4+49x^2-36$. **29.** 0. **30.** $b^3c-bc^3+c^3a-ca^3+a^3b-ab^3$. **31.** $a^3(b^2-c^2)+b^3(c^2-a^2)+c^3(a^2-b^2)+a(b^4-c^4)+b(c^4-a^4)+c(a^4-b^4)$. **32.** $(x+a-b)(x-a+b)$. **33.** $(x^2+a^2+b^2)(x^2+a^2-b^2)$. **34.** $x^4(x-1)^2$. **35.** $(x^2+2\sqrt{2}x+4)(x^2-2\sqrt{2}x+4)$. **36.** $-(x+y+1)(-x+y+1)(x-y+1)(x+y-1)$. **37.** $(b+c)(-c-a)(a-b)$. **38.** $(x^2-a)(x^2+b)$. **39.** $(x^2-a^2)(x^2-b^2)(x^2-c^2)\equiv(x+a)(x+b)(x+c)(x-a)(x-b)(x-c)$.

## XI.

**26.** 3·414, ·586. **27.** ·618, $-1$·618.

## XII.

**1.** 1. **2.** 3. **3.** $-1$. **4.** $-7/3$. **5.** $-1$. **6.** 1. **7.** 1. **8.** 0. **9.** $-11/7$. **10.** 254/35. **11.** $-1$. **12.** 2. **13.** 4/7. **14.** 1. **15.** 0. **16.** 3. **17.** 11. **18.** 0. **19.** 37/14. **20.** $-3$. **21.** $-1/2$. **22.** $-1/5$. **23.** 1, $-13/16$. **24.** $-2$. **25.** 0. **26.** 0. **27.** No finite solution. **28.**

4195/884 = 4·745. . . . **29.** 118/119. **30.** 307/142. **31.** ·905. **32.** $-2·294$. **33.** 1·479795. **34.** 8·225637. **35.** $\pm 1/\sqrt{3}$. **36.** 2, 2/3. **37.** 3, $-2$. **38.** 1, $-4$. **39.** 1, 1, 2. **40.** 0, 2/5. **41.** 1, 1, $\pm 2$. **42.** 3, $\pm 1$. **43.** $-1$, $-1$, $-1/3$, $-7$. **44.** 1, $\pm\sqrt{3}$. **45.** 1, $\pm\sqrt{(5/2)}$. **46.** 1, 12/5. **48.** No finite solution. **49.** An identity. **50.** No finite solution. **51.** $x = \pm 3/\sqrt{2}$. **52.** 0, $\pm\sqrt{15}$. **53.** 0, 1, 14. **54.** $x^2 + 2x - 15 = 0$. **55.** $16x^3 - 12x^2 - 16x - 3 = 0$. **56.** $x^2 - 6x + 4 = 0$. **57.** $x^5 - 6x^4 + 2x^3 + 28x^2 - 23x - 14 = 0$.

## XIII.

**1.** $a$. **2.** $61a/156$. **3.** 1. **4.** 0. **5.** $(ap - bq)/(a + b)$, $(ap + bq)/(a - b)$. **6.** $-1$. **7.** $(a^2 + ab + b^2)/(a + b)$. **8.** $(a^2 + ab + b^2)/(a + b)$. **9.** $(2a^2 - ab + 2b^2)/(a + b)$. **10.** $b$. **11.** $\pm a$. **12.** $p$. **13.** $a + b + c$. **14.** $(a^2 - b^2)/2ab$. **15.** $\{qm^2 + (2p + q)n^2\}/\{(p + 2q)m^2 + pn^2\}$. **16.** $a + b$. **17.** $a$. **18.** 0. **19.** $1/4a(a + b)$. **20.** $\frac{1}{4}\Sigma a$. **21.** 0, $-\frac{2}{3}\Sigma a$. **22.** $\{a^3(b - c) + b^3(c - a) + c^3(a - b)\}/3\{a^2(b - c) + b^2(c - a) + c^2(a - b)\} = \frac{1}{3}\Sigma a$. **23.** $\frac{1}{3}\Sigma a$. **24.** $h = \frac{1}{2}(a + b)$; if $a = b$, the equations are consistent for all values of $h$. **26.** $\pm 6a$. **27.** $\pm(a - b)$. **28.** $\pm\sqrt{(a^2 + b^2)}$. **29.** $\pm\sqrt{[\{a^3(b^2 - c^2) + b^3(c^2 - a^2) + c^3(a^2 - b^2)\}/3\{a(b^2 - c^2) + b(c^2 - a^2) + c(a^2 - b^2)\}]} = \pm\sqrt{(\frac{1}{3}\Sigma bc)}$.

## XIV.

**1.** 6. **2.** 48 + 16. **3.** 36. **4.** 360. **5.** 44 and 45. **6.** 48 and 39. **7.** 70 and 50. **8.** 42 and 12. **9.** £250. **10.** £6480. **11.** 40 half-crowns, 20 shillings, 80 sixpences. **12.** 11 half-crowns, 13 florins, 24 shillings. **13.** 10s. **14.** 1s. 4d. **15.** 6. **16.** 16. **17.** Capital £1,250,000; receipts £128,048 : 15 : 7. **18.** 2s. 4d. **19.** 60 sixpences, 30 shillings, 12 half-crowns. **20.** 16 ft. by 48 ft. **21.** $bc/(b - ca)$. **22.** $(42 - 3b)/(a + b)$. **23.** 98. **24.** £1 : 8s. **25.** 1s. 3d.; 4s. 1d. **26.** A, £$3\frac{1}{9}$; B, £$15\frac{1}{9}$. **27.** £$498\frac{71}{78}$; £$512\frac{33}{38}$. **28.** 8 to 3. **29.** £50,000, £200,000. **30.** £750, £250. **31.** $16\frac{2}{3}$ per cent. **32.** 143. **33.** £240. **34.** $(107a/100 - 21b/20a)/(a - b)$. **35.** 4500. **36.** After 181·6 days. **37.** $\frac{6}{7}$ hr. **38.** Miner 42 days, partner $23\frac{1}{3}$ days. **39.** After $4\frac{4}{29}$ hours; $14\frac{14}{29}$ miles from the starting-point of the first pedestrian. **40.** After A has gone $abd/(bd + ce)$ yards. **41.** 11 h. 47 m. $46\frac{2}{103}$ s. **42.** 25 miles. **43.** $l(h - k)/(h + k)$ miles an hour. **44.** 2 miles an hour. **45.** $1/176(m - m')$ hrs.; $51/176(m - m')$ hrs. **46.** $27\frac{3}{11}$ minutes past two. **47.** 7 miles. **48.** Loses $1\frac{383}{737}$ min. per hr. **49.** $a + \sqrt{\{(x_1 - a)(x_2 - a)\}}$, $a - \sqrt{\{(x_1 - a)(x_2 - a)\}}$. **50.** $p^2 + pc + c^2$. **51.** Ratio of gold to silver is $\{\sigma w - (\sigma - 1)\mathrm{W}\}/\{(\rho - 1)\mathrm{W} - \rho w\}$. **52.** Sum invested $a(1 + \mathrm{R} + \mathrm{R}^2 + \ldots + \mathrm{R}^{n-1})/(\mathrm{R}^n - q)$, where $\mathrm{R} = 1 + p/100$.

## XV.

**1.** $(-7/6, -3)$. **2.** (0, 0), (1, 1/2), $(-1, -1/2)$, (0, 1), $(1, -2)$, $(-1, 4)$. **3.** (1/13, 2/13). **4.** $(-1, 1)$. **5.** $a(x + 2) + b(y - 3) = 0$. **6.** $ax + by = 0$. **17.** (8, 0), (6, 3), (4, 6), (2, 9), (0, 12). **18.** (17, 3), (10, 9), (3, 15). **19.** (14, 1), (1, 8). **20.** (3, 6). **21.** Two pairs, viz. 2/3, 7/5 and 5/3, 2/5. **22.** Three subscriptions, one allowance. **23.** $x=4z+3$, $y=3z+1$, where $z$ is any integer; (19, 13), (23, 16), (27, 19). **24.** In one way only: A gives B 8 florins and B gives A 2 half-crowns. **25.** $x=5z+3$, $y=4z-2$, where $z$ is any integer. **26.** The general integral solution is $x=18-15u$, $y=-3+13u$, from which it appears that there is only one positive integral solution, viz. (3, 10). **27.** The equation leads to $x+3y=13/2$, an arithmetical absurdity if $x$ and $y$ be integers.

## XVI.

**1.** (5, 3). **2.** (3, 5). **3.** (1, 1). **4.** (10, 10). **5.** (7, 9). **6.** (10, 7). **7.** (3, 7). **8.** (10, −7). **9.** (2, 3). **10.** (3, 1). **11.** (5, 2). **12.** (0, 0). **13.** (3, 4). **14.** (7, 11). **15.** (23/7, 47/7). **16.** (2, 5). **17.** (5, 3). **18.** (385/34, 495/34). **19.** (1, 3). **20.** (10, 9). **21.** (−6268/11, 2968/11). **22.** (6·309, ·089). **23.** (16·68564, −·88254). **24.** (165/13, 196/13). **25.** (15/2, 9).

## XVII.

**1.** $(5a,\ 3b)$. **2.** $(3a^2,\ 5b^2)$. **3.** $(a+b,\ 0)$. **4.** $(0,\ 0)$. **5.** $\{ab(ad-bc)/(a^2-b^2),\ ab(ac-bd)/(a^2-b^2)\}$. **6.** $(b+a,\ b-a)$. **7.** $\{-b^2/(a-b),\ a^2/(a-b)\}$. **8.** $(a,\ b)$. **9.** $\{b^4(a-1)/(a-b),\ -a^4(b-1)/(a-b)\}$. **10.** $(b^n,\ a^n)$. **11.** $\{\frac{1}{2}(a+b),\ \frac{1}{2}(a-b)\}$. **12.** $(2a,\ 2b)$. **13.** $\{(c^2+b^2-a^2)/2a,\ (c^2+a^2-b^2)/2b\}$. **14.** $(a+b,\ a-b)$. **15.** $(l+m,\ l-m)$. **16.** $\{(p-a)(q-a)/(b-a),\ (p-b)(q-b)/(a-b)\}$. **17.** $\{(a+b)a^2,\ (a+b)b^2\}$. **18.** $\{(2b-a)(c-a)(c+a-b)/(b-a)(3c-a-b),\ (2a-b)(c-b)(c+b-a)/(a-b)(3c-a-b)\}$. **19.** $[(a^2-\lambda)\{b^2(a+b)-ab^3-(a+b)\lambda\}/\lambda b(b^2-a^2),\ (b^2-\lambda)\{a^2(a+b)-a^3b-(a+b)\lambda\}/\lambda a(a^2-b^2)]$. **20.** $(1,\ 1)$. **21.** $(a/2,\ b/2)$. **22.** $\{(1-b)(a^2-c^2)/a(a-b),\ (1-a)(b^2-c^2)/b(b-a)\}$.

## XVIII.

**7.** $3x+y=0$. **8.** $8x+7y-51=0$. **9.** The other equation is $a(x-2)+b(y-5)=0$, where $a$ and $b$ are any constants such that $4a+6b\neq 0$; the problem has a onefold infinity of solutions. **10.** The solution will be finite and determinate unless $a^2+b^2=0$, which requires that $a=0$, $b=0$, supposing $a$ and $b$ to be real. **11.** $\lambda=1$. **13.** Inconsistent. **14.** $k=-31/7$. **15.** $p=-33/2$. **16.** If $c=0$, the system has the common solution $x=0$, $y=0$; if $b=c$, the solution $x=0$, $y=-1$; if $c=a$, the solution $x=-1$, $y=0$; these are the only finite and determinate common solutions. The student should also discuss graphically the cases where $a=0$, $b=0$, $a=b$. **17.** $\lambda=0$. **19.** If $ab'-a'b\neq 0$, the system has the determinate solution $x=0$, $y=0$; if $ab'-a'b=0$, the two equations are equivalent, and the system has a onefold infinity of solutions.

## XIX.

**1.** $(4/5,\ 17/5,\ -8/5)$. **2.** $(-2,\ 3,\ 1)$. **3.** $(0,\ 1,\ 2)$. **4.** $(6/11,\ -2/11, 7/11)$. **5.** $(33/89,\ -2/89,\ -11/178)$. **6.** $(-9/7,\ -26/7,\ -25/7)$. **7.** $(11,\ 5,\ 7)$. **8.** $(3,\ 0,\ -2)$. **9.** $(3,\ 1,\ -1)$. **10.** $(-9,\ 20,\ 25)$. **11.** The solution is partly indeterminate, viz. $x=1$ together with any values of $y$ and $z$ that satisfy $3y+2z=5$. **12.** $(1,\ 2,\ 3)$; $(20-19u,\ 4u-2,\ 11u-8)$, where $u$ is any integer. **13.** $(-a/2,\ -a/2,\ -a/2)$. **14.** $\{12c/(12a+b),\ c/(12a+b),\ -13c/(12a+b)\}$. **15.** $(a,\ a,\ a)$. **16.** $x=\{a^3+2a^2(b+c)+abc+a^2bc\}/(a-b)(a-c)$, $y=\{b^3+2b^2(c+a)+abc+ab^2c\}/(b-c)(b-a)$, $z=\{c^3+2c^2(a+b)+abc+abc^2\}/(c-a)(c-b)$. **17.** $\{(m+n)/2,\ (n+l)/2,\ (l+m)/2\}$. **18.** $(a,\ b,\ c)$. **19.** $\{(a+b/2+c/2)\Sigma a,\ (b+c/2+a/2)\Sigma a,\ (c+a/2+b/2)\Sigma a\}$. **20.** $\Sigma a^3+abc=0$.

## XX.

**1.** $(1/2,\ -3/2)$. **2.** $(1/2,\ 1/2)$. **3.** $\{(a^2-b^2)/2a,\ 0\}$. **4.** $(l+m,\ l+m)$. **5.** $(3/10,\ -1/5,\ 0)$; $(1/5,\ -2/15,\ -1/15)$. **6.** $(7/11,\ 19/11)$; $(5/33,\ 15/11)$; $(19/13,\ 1/13)$; $(1/32,\ 9/32)$. **7.** $(-4/7,\ 1/14)$; $(-2,\ -1)$; $(5,\ 1)$; $(-5/7,\ -1/7)$. **8.** $(5,\ 3)$; $(5,\ -3)$; $(-5,\ 3)$; $(-5,\ -3)$. **9.** $\{\sqrt{(21/4)},\ \sqrt{(83/2)}\}$; $\{\sqrt{(21/4)},\ -\sqrt{(83/2)}\}$; $\{-\sqrt{(21/4)},\ \sqrt{(83/2)}\}$; $\{-\sqrt{(21/4)},\ -\sqrt{(83/2)}\}$. **10.** $(1/9,\ 1/3)$; $(3/11,\ 1/11)$. **11.** $(b,\ a)$. **12.** $\{\sqrt{(a^2-b^2)},\ \sqrt{(a^2-b^2)}\}$; $\{\sqrt{(a^2-b^2)},\ -\sqrt{(a^2-b^2)}\}$; $\{-\sqrt{(a^2-b^2)},\ \sqrt{(a^2-b^2)}\}$; $\{-\sqrt{(a^2-b^2)},\ -\sqrt{(a^2-b^2)}\}$. **13.** $(1, 1, 1)$; $(1,\ 1,\ -1)$; $(1,\ -1,\ 1)$; $(-1,\ 1,\ 1)$; $(1,\ -1,\ -1)$; $(-1,\ 1,\ -1)$; $(-1,\ -1,\ 1)$; $(-1,\ -1,\ -1)$.

## XXI.

**1.** 30 half-crowns; 40 florins. **2.** A, £500; B, £700. **3.** A, £900; B, £2400. **4.** 84, 63, 42, 21; the given conditions are not independent. **5.** 42/63. **6.** 17. **7.** $3\frac{3}{4}$ miles per hour. **8.** A, 3s. 6d.; B, 4s. 2d. **9.** 3/5. **10.** 43. **11.** 7/12. **12.** 193. **13.** $13+27+60$. **14.** $13+26+4+80$. **15.** A, $\frac{1}{2}(b+c-a)$; B, $\frac{1}{2}(c+a-b)$; C, $\frac{1}{2}(a+b-c)$; $b+c>a$, $c+a>b$, $a+b>c$. **16.** A, 80; B, 60. **17.** £105. **18.** 11 ft. by 7 ft. **19.** A, £12; B, £4. **20.** Breakdown after 15 miles; speed 10 miles an hour. **21.** A, 15%; B, 35%. **22.** Estate £150; owed to creditors £200 and £100. **23.** $30\frac{59}{67}$ yds. **24.** $(ja-ib)/(j-i-mij/100)$. **25.** $7x-25$. **26.** $x$. **27.** $\frac{1}{3}(-4x-2y+23)$. **28.** $\frac{1}{4}(x-y)^2$. **29.** $0x^2+x+2$. **30.** $(x-1)(-\frac{1}{6}x+\frac{4}{3})$. **31.** 9. **33.** $8x-y-19=0$. **35.** £1200 + £900 = £2100. **36.** 900 gals. **37.** A, 4 miles; B, 1 mile an hour. **38.** 120 lbs. **39.** A, 350″; B, 400″. **40.** Franc $=\frac{6149}{25252}$ th.; pound $=6\frac{377}{8313}$ th. **41.** 74 lbs. tin; 46 lbs. lead. **42.** 0, 3; 1, 2; 2, 1; 3, 0. **44.** 42, 44, 41, 43; 10, 12, 5, 7. **46.** Rise $(bp-aq)/(a-b)$; acreage $\{a(r-q)+b(p-r)\}/(p-q)$. **47.** $2abc/(bc+ca+ab)$. **48.** Distance $ab(a-b)hk(h-k)/(ak-bh)^2$; speed $ab(h-k)/(ak-bh)$. **49.** Distance $ab(h-k)/(a-b)$; speed $ab(h-k)/(ah-bk)$. **50.** $y=x+1$; when $x=7$, $y=8$. **51.** $\sqrt{3}(b-a)/2$. **52.** $AP=bc/(a+2c)$; $BQ=b(c+a)/(a+2c)$. **53.** $AP=bc/a$ or $bc/(a+2c)$. **54.** $AP=bc/a$ or $bc/(a+2c)$.

## XXII.

**1.** $\frac{1}{12}$ sq. ft. **2.** 28 sq. ft. **3.** $\frac{1}{2}\{x_1(y_2-y_3)+x_2(y_3-y_1)+x_3(y_1-y_2)\}$. **6.** $cp=ab$; $cx=a^2$; $cy=b^2$; $a^2+b^2=c^2$. From these any of the required expressions can be readily found. **9.** 45, 53. **10.** Area $\frac{1}{2}$; perpendicular $\frac{1}{3}\sqrt{6}$. **11.** $2(1+\sqrt{2})a^2$. **12.** $(\rho x_1 \pm x_2)/(\rho \pm 1)$, where $\pm a(\rho^2-1)=2(x_1-x_2)\rho$. **13.** $2\frac{2}{5}$. **15.** If O be the middle point of AB$(=a)$, OP$=c^2/2a$. **16.** $20\frac{5}{7}$. **17.** 34. **19.** $\frac{3}{2}\sqrt{3}a^2$, $a$ being the radius. **23.** $(c^2-x_1{}^2+x_2{}^2)/2(x_2-x_1)$. **27.** 168. **30.** $5(\sqrt{3}-1)$. **31.** $\frac{72}{13}\sqrt{15}$; $\frac{1}{2}\sqrt{865}$. **40.** $\Delta(1-\rho+\rho^2)/(1+\rho)^2$. **45.** $(\rho+1)(\rho-\sigma)/(\rho-1)(\rho+\sigma)$. **47.** $2abc/(b^2 \sim c^2)$. **48.** 12,291,460 sq. miles. **50.** 17·29 feet. **51.** $\frac{1}{8}$.

## XXIII.

**1.** $adg-aeg-afg+bdg-beg-bfg+cdg-ceg-cfg-adh+aeh+afh-bdh+beh+bfh-cdh+ceh+cfh$. **2.** $adg+aeg-afg-bdg-beg+bfg+cdg+ceg-cfg+adh+aeh-afh-bdh-beh+bfh+cdh+ceh-cfh$. **3.** $a^2-4b^2-c^2+4bc$. **4.** $a^2+4b^2+c^2-4bc-2ca+4ab$. **5.** $a^2+4b^2+4c^2+d^2+4ab+4ac+2ad+8bc+4bd+4cd$. **6.** $a^2+4b^2+4c^2+d^2+4ab-4ac-2ad-8bc-4bd+4cd$. **7.** $1+2x+3x^2+4x^3+3x^4+2x^5+x^6$. **8.** $1-2x+3x^2-4x^3+3x^4-2x^5+x^6$. **9.** $1+4x+10x^2+12x^3+9x^4$. **10.** $a^2-4b^2+9c^2-16d^2+6ac-16bd$. **11.** 60; 480.

## XXIV.

**1.** $(\alpha)$ 2; 2; 3 : 4. $(\beta)$ 2; 2; 3 : 7. $(\gamma)$ 2; 2; 2 : 3. **2.** 4; 9. **3.** 6; heterogeneous; symmetrical; neither. **4.** $(x^2+y^2)(x-y)xy \equiv x^4y-x^3y^2+x^2y^3-xy^4$. **5.** $2x^2-xy+2y^2$; $2x^2+xy$. **6.** $17x^2-xy+2y^2$. **8.** (i) $a+bx+cy+dz+eu+fx^2+gy^2+hz^2+ku^2+lxy+mxz+nxu+pyz+qyu+rzu$; (ii) $a+b\Sigma x+f\Sigma x^2+l\Sigma xy$; (iii) $fx^2+gy^2+hz^2+ku^2+lxy+mxz+nxu+pyz+qyu+rzu$; (iv) $f\Sigma x^2+l\Sigma xy$. **9.** $a+b(x^2+y^2+z^2)$. **10.** $b=c$. **11.** $a+bx+cy+dxy$; $b=c$. **12.** $a+bx+cy+du+ev+fxu+gxv+hyu+kyv$. **13.** $ax^2y^2+bx^2y+cxy^2+dx^2+exy+fy^2+gx+hy+k$; $b=c$, $d=f$, $g=h$. **14.** $a\Sigma x^4+b\Sigma y^2z^2$. **15.** $a\Sigma x^2yz$. **16.** Even, $a\Sigma x^4$, $a\Sigma x^2y^2$; odd, $axyzu$; neither even nor odd, $a\Sigma x^3y$, $a\Sigma x^2yz$. **17.** In the case of an even function, the two parts of the graph on opposite sides of the $y$-axis are the images of each other in that axis. In the case of an odd function, the parts of the graph in opposite quadrants are derivable from each other by successive reflections in the axes of $x$ and $y$. **18.** $x^3y^2z+y^3z^2x+z^3x^2y$; $x^2-yz+y^2-zx+z^2-xy$; $x^2-y^2z+y^2-z^2x+z^2-x^2y$; $x+y-z+y+z-x+z+x-y$; the second and fourth are absolutely symmetrical.

## XXV.

**1.** $a^2+b^2-c^2-d^2+2ab+2cd$. **2.** $\Sigma a^4-4(b^3d+bd^3+a^3c+ac^3)+6b^2d^2+6a^2c^2-2(a^2b^2+a^2d^2+b^2c^2+c^2d^2)+4(a^2bd+ab^2c+bc^2d+acd^2)-8abcd$.

**3.** $(b+c)x^2 - by^2 - cz^2 - (b+c)yz + bzx + cxy$. **4.** $4\Sigma x^2$. **6.** 64. **10.** $4\Sigma ax$. **11.** $\Sigma a^4(b-c) + \Sigma b^2c^2(b-c)$. **12.** $8a^3$. **13.** $-[\Sigma a(b-c)^3x^2 + (b-c)(c-a)(a-b)\Sigma(a-b-c)yz]/\Pi a(b-c)$. **14.** $3x^3 + 3\Sigma ax^2 + 2\Sigma abx$. **15.** $3x^4 + 2\Sigma ax^3 + \Sigma a^2x^2 + 2\Sigma a(b^2-c^2)x$. **16.** $\Sigma x^3y^3 + \Sigma x^4yz + 3x^2y^2z^2$. **17.** $\Sigma x^2y - 6xyz$. **18.** $3xyz - \Sigma x^3$. **19.** $\Sigma a^3b + 5\Sigma a^2bc + 2\Sigma a^2b^2$. **20.** $2\Sigma a^3 - 2\Sigma a^2b + 6abc$. **21.** 0. **23.** $\Sigma a^4$. **24.** $2\Sigma x^4 + 2\Sigma y^3z + 3\Sigma y^2z^2$. **27.** $2\Sigma(b^3c - bc^3)$. **28.** $\Sigma x^3 + \Sigma x^2y$. **30.** $\Sigma x^4yz - \Sigma y^3z^3$. **31.** $\Sigma a^3b^2c - \Sigma a^4bc - 2a^2b^2c^2$.

## XXVI.

**1.** $x^8 + x^7 - 7x^6 - 2x^5 + 14x^4 - 2x^3 - 7x^2 + x + 1$. **2.** $2x^5 - 7x^4y + 11x^3y^2 - 11x^2y^3 + 7xy^4 - 2y^5$. **3.** $\frac{1}{8}x^6 - \frac{1}{4}x^5 - \frac{1}{4}x^4 + x^3 - \frac{1}{2}x^2 - x + 1$. **4.** $x^8 + 2x^7 + x^6 + 5x^5 - 6x^4 - 15x^3 + 12x^2 + 20x + 4$. **5.** 131/7. **6.** $x^4 - (a + 1/a)x^3 + (a^2 + 1 - 1/a^2)x^2 - (a - 1/a)x - 1$. **8.** $x^4 + 1 + 1/x^4$. **9.** $1 - (a^2 + 1/a^2)x^2 + x^4$. **10.** $(b+c-a)x^4 + 2(a-b)x^3 + (-a+3b-c)x^2 - 2(b-c)x + (a+b-c)$. **11.** $x^6 - a^6$. **12.** $x^8 + a^4x^4 + a^8$. **13.** $4x^{15} - 12x^{12} + 14x^9 - 36x^6 + 6x^3$. **14.** 6. **15.** $2x^6 + 12x^4 + 12x^2 + 2$. **16.** $-p^3x^6 + p^2qx^5 + (pq^2 + p^2r)x^4 - (q^3 + 2pqr)x^3 + (pr^2 + q^2r)x^2 + qr^2x - r^3$. **17.** $(a^2-b^2)x^4 + 2ab^2x^3y + (2a^2 + 2b^2 - a^2b^2)x^2y^2 - 2ab^2xy^3 + (a^2-b^2)y^4$. **18.** $1 - 2x^4 + x^8$. **19.** $2 + 12x^2 + 24x^4 + 20x^6 + 6x^8$. **20.** 85. **21.** $-\Sigma a^2 + 2\Sigma ab$. **22.** $-12ab$; $6b^2$. **23.** $12b^2$. **24.** $a^2 + 4ab + b^2 + 2a + 2b$. **28.** $A = \frac{1}{8}$, $B = -\frac{1}{8}$, $C = \frac{1}{2}$. **29.** $L = 29/2$; $M = 7/2$; $N = -16$. **30.** $l = (pa^2 + qa + r)/(a-b)(a-c)$; $m = (pb^2 + qb + r)/(b-c)(b-a)$; $n = (pc^2 + qc + r)/(c-a)(c-b)$. **31.** $A = 3$; $B = -3$; $C = 1$; $D = 0$. **32.** $A = 2$; $B = 3$; $C = 1$; $D = 11$. **33.** $a = 2$; $b = 9$; $c = -13$; $d = 3$. **34.** $(x+1)^6 - 3(x+1)^5 + 13(x+1)^4 - 21(x+1)^3 + 44(x+1)^2 - 34(x+1) + 40$.

## XXVII.

**1.** $x^5 + 10x^4 + 40x^3 + 80x^2 + 80x + 32$. **2.** 70. **3.** 1120. **4.** $-252$. **5.** $2x^{11} + 440x^9 + 10560x^7 + 59136x^5 + 84480x^3 + 22528x$. **6.** 105. **7.** $59049x^{10} - 131220x^8 + 116640x^6 - 51840x^4 + 11520x^2 - 1024$. **8.** $x^{10} - 5x^9 + 15x^8 - 30x^7 + 45x^6 - 51x^5 + 45x^4 - 30x^3 + 15x^2 - 5x + 1$. **9.** $a^{20} + a^{18}b^2 + 3a^{14}b^6 + 3a^{12}b^8 + 3a^8b^{12} + 3a^6b^{14} + a^2b^{18} + b^{20}$. **10.** $1 + x - 5x^2 - 5x^3 + 10x^4 + 10x^5 - 10x^6 - 10x^7 + 5x^8 + 5x^9 - x^{10} - x^{11}$. **11.** $x^{30} - 5x^{24} + 10x^{18} - 10x^{12} + 5x^6 - 1$. **12.** 10. **13.** $x^{12} + 4x^{11} + 2x^{10} - 12x^9 - 17x^8 + 8x^7 + 28x^6 + 8x^5 - 17x^4 - 12x^3 + 2x^2 + 4x + 1$. **14.** $x^{12} - 2x^6y^6 + y^{12}$. **15.** 21. **16.** $-120$. **17.** 98. **18.** $-20$; $-62$. **19.** $p = -4$; $q = 10$. **20.** $1 - 3x + 6x^2 - 10x^3$. **21.** $-40$. **22.** 5922. **23.** $4x^5y - 8x^3y^3 + 4xy^5$. **26.** $a^5 - b^5 \equiv (a-b)^5 + 5(a-b)^3ab + 5(a-b)a^2b^2$; $(x-y)^5 + (y-z)^5 \equiv (x-z)^5 - 5(x-z)^3(x-y)(y-z) + 5(x-z)(x-y)^2(y-z)^2$; $(x-y)^5 - (y-z)^5 \equiv (x-2y+z)^5 + 5(x-2y+z)^3(x-y)(y-z) + 5(x-2y+z)(x-y)^2(y-z)^2$. **27.** $x^n - y^n$. **28.** $x^n + y^n$. **29.** $x^n - y^n$. **30.** $x^{n+2} - x^{n+1}y - xy^{n+1} + y^{n+2}$. **32.** $n = 99,999,000$. **33.** Follows from the identity $(2x-3)(2x-1)(2x+1)(2x+3) - 1 \equiv 8(2x^4 - 5x^2 + 1)$. **34.** $n^2 - n + 1 \equiv n(n-1) + 1$. Now, if $n$ be an integer, one of the two $n$ or $n-1$ is even; therefore $n(n-1) + 1$ is of the form $2m+1$. **35.** Follows from the identity $(2x-1)^2 + (2x+1)^2 + (2x+3)^2 + 1 \equiv 12(x^2 + x + 1)$. **36.** An upper limit for the increase of the product is 5,500,015.

## XXVIII.

**1.** $3x^3+5x^2+17x+51+150/(x-3)$. **2.** $x^2-2x+1$. **3.** $x^5+x^4-x^2-x+x/(x^2-x+1)$. **4.** $x^3+3x^2-2x+3$. **5.** $x+8+(18x-6)/(x-1)^2$. **6.** $x^7-x^5-x-1+(2x-3)/(x-1)^2$. **7.** $3x^4+2x^3+5x^2-x+6$. **8.** $x^4-x^2+1$. **9.** $x^3-3x^2+6x-8$. **10.** $x^3+x^2-x-1$. **11.** $(x^2+1)+(2x+3)/(x^2+x+1)+(x^2+1)+(-2x+3)/(x^2-x+1)\equiv 2x^2+2+(2x^2+6)/(x^4+x^2+1)$. **12.** $x^4-\frac{1}{12}x^3-\frac{247}{180}x^2+\frac{337}{540}x+\frac{4109}{1620}+(-\frac{10175}{4860}x-\frac{3299}{810})/(x^2+\frac{1}{3}x+2)$. **13.** $x^3-2x^2+\frac{5}{2}x-8+(\frac{101}{4}x^2-3x+49)/(x^3+2x^2+\frac{3}{2}x+6)$. **14.** $x^8-x^4+1$. **15.** $(x-a)^8-(x-a)^4+1$. **16.** $55y^4$; $\frac{55}{16}x^4$. **17.** $20xy^4-8y^5$; $-20yx^4+8x^5$. **18.** $x^4+10x^3+65x^2+350x+1701+7770/(x-4)$. **19.** $x^4+x^3-8x^2-6x+8$. **20.** $x^3-ax^2+(a^2+b-3)x+(a^3+ab-6a)+\{(+a^4+a^2b-9a^2-3b+15)x+(+3a^3+3ab-18a+1)\}/(x^2+ax+3)$. **21.** $(y+z)x+(z^2+2yz)-y^2z/(x+y)$. **22.** $x+y-z$. **23.** $2x^2+6y^2+2z^2+4xz$. **24.** $x-1$. **25.** $(a+b)x^3+(a-b)x^2+(a+b)x+(a-b)$. **26.** $px^3-qx^2-rx+2s$. **27.** $p=8$; $q=-3$. **28.** $p=1$; $q=-1$. **29.** $a=b$. **30.** $a=\frac{19}{15}$; $b=-\frac{1}{15}$. **31.** $p^2-p-q+2=0$; $pq-q+2=0$. **32.** $a=1$; $c=2$. **33.** $\lambda=-3$. **34.** $p=-3$, $q=-2$.

## XXIX.

**1.** $-166$. **2.** $3x^3-12x^2+45x-180$; $721$. **4.** $0$. **5.** $(x^2+3x+4)(2x^2+3x+2)\equiv 2x^4+9x^3+19x^2+18x+8$. **6.** $-55/9$. **7.** $l=-7$, $m=3$. **8.** $\lambda=-8$, $\mu=-17$. **9.** $a=-10$, $b=-10$. **10.** $x^{n-1}+x^{n-2}+\ldots+x-1$. **11.** $\Sigma xy$. **13.** $\Sigma a^2-\Sigma ab$. **14.** $ax^2+2bx+c$. **15.** $3\Sigma x^2-9\Sigma xy$. **16.** $(x-2)^4+8(x-2)^3+25(x-2)^2+36(x-2)+21$. **17.** $A=1$, $B=7$, $C=6$, $D=1$. **18.** $-11x-12+(28x-32)Q_1+(x+7)Q_1Q_2$. **19.** $a=-1$, $b=5$, $c=2$, $d=7$, $e=3$; $a_0=9$, $a_1=25$, $a_2=30$, $a_3=16$, $a_4=3$. **20.** $1+5(x-1)+10(x-1)^2+10(x-1)^3+5(x-1)^4+(x-1)^5$; $90y+422-(45y+249)z+(9y+42)z^2+z^3$. **21.** $\{a(x-3)+3\}(x-1)(x+2)$.

## XXX.

**1.** $(a+b-c-d)(a-b+c-d)$. **2.** $(4x+2)(-2x-4)$. **3.** $\{(a+b)x-(a-b)y\}\{(a-b)x+(a+b)y\}$. **4.** $4x^2(x+1)$. **5.** $(1+x)(1-x)(1+y)(1-y)$. **6.** $(x-p+q)(x-p-q)$. **7.** $(x+1+p+q)(x+1+p-q)$. **8.** $(3x-y)(3y-x)$. **9.** $(x+2)^2(x-2)^2$. **10.** $\{(2+\sqrt{3})x+1-2\sqrt{3}\}\{(2-\sqrt{3})x+1+2\sqrt{3}\}$. **11.** $(x+1)^2$. **12.** $(x+a)(x-b)$. **13.** $4x^2$. **14.** $(x-\sqrt[3]{a})\{x^2+\sqrt[3]{a}x+(\sqrt[3]{a})^2\}(x+\sqrt[3]{b})\{x^2-\sqrt[3]{b}x+(\sqrt[3]{b})^2\}$. **15.** $\{(1+\sqrt{2})x+(1-\sqrt{2})y+1-\sqrt{2}\}\{(1-\sqrt{2})x+(1+\sqrt{2})y+1+\sqrt{2}\}$. **16.** $(x^m+a)(x^m+b)$. **17.** $(x-y-a)(x-y-b)$. **18.** $(xy-a^2)(xy-b^2)$. **19.** $(x+y+a+b)(x+y-a-b)(x-y+a-b)(-x+y+a-b)$. **20.** $(2x+y+1)^2$. **21.** $(3x-1)^3$. **22.** $(x+y)^4(x-y)^4$. **23.** $(x+a)(x-a)(x+b)(x-b)(x+c)(x-c)$. **24.** $(2x+a)(2x-b)(2x+c)$. **25.** $(x^4+y^4)(x^2+y^2)(x+y)(x-y)$. **26.** $(x-y)(x^6+x^5y+x^4y^2+x^3y^3+x^2y^4+xy^5+y^6)$. **27.** $(x+y)(x^6-x^5y+x^4y^2-x^3y^3+x^2y^4-xy^5+y^6)$. **28.** $(x^2+3x+9)(x^2-3x+9)$. **29.** $(x+y-z)(x^2+y^2+z^2+yz+zx-xy)$. **30.** $-(x-1)^2(x^2+x+1)^2$. **31.** $x^2+y^2+z^2-yz-zx-xy$.

## XXXI.

**1.** $(x-1)(x-121)$. **2.** $(x-3a)(5x-7)$. **3.** $(x-y)(-2x+8y)$. **4.** $(x-z)(x-6y+z)$. **5.** $(x-1)^2(x+1)$. **6.** $(x+1)(x-1)(x+3)$. **7.** $(x+y)(x^2+y^2)$. **8.** $(x-3)(x^2+9)$. **9.** $(x^2+y^2)(x-y)^2(x+y)^2$. **10.** $(x^2+1)(2x-15)$. **11.** $(x^2+p)(x-a)$. **12.** $-(x-1)(2x-3)$. **13.** $(x+2)(3x+1)$. **14.** $(1+x)(1-x)(1+ax)$. **15.** $(x+p-1)(x^2+x+1)$. **16.** $(x-2)(x^2+25)$. **17.** $xy(x-y)(x^2+4xy+y^2)$. **18.** $(x+y)(x^4-x^2y^2+y^4)$. **19.** $(x+y)(z-u)$. **20.** $(x-y)(x^2+4xy+2y^2)$. **21.** $(x+1)(x^3+2x^2+1)$. **22.** $(x-1)^2(x^2+2x+3)$. **23.** $ab(b-a)(b+a+3c)$. **24.** $(x+1)(x^2+p^2-1)$. **25.** $(x+p)(x^2+x+1)(x^2-x+1)$. **26.** $(x-3)(2x+3)(x^2+3x+3)$. **27.** $(x+1)(x+2)\{(l+m)x-m-n\}$. **28.** $2x(x-2)(2x-5)$. **29.** $(lx+m)(px+q)(x^2+x+1)$. **30.** $2(z-u)(2y+z+u)$.

## XXXII.

**1.** $(x+2)(x+6)$. **2.** $(x+11)(x+13)$. **3.** $(x-2)(x-9)$. **4.** $(x-13)(x-15)$. **5.** $(x-2)(x+7)$. **6.** $(x+5)(x-10)$. **7.** $(x+29)(x-100)$. **8.** $(x-33)(x+111)$. **9.** $(x+1)(x-8)$. **10.** $(x^3+1)(x^3+8)\equiv(x+1)(x^2-x+1)(x+2)(x^2-2x+4)$. **11.** $\frac{1}{4}(2x+1)^2$. **12.** $(x-\frac{1}{3})(x-\frac{1}{5})$. **13.** $x(x+2)(x+4)(x+6)$. **14.** $(x-1)(x+1)(x-3)(x+3)$ **15.** $(1-x)(1-2a+x)$. **16.** $(4x+y)(2x+y)$. **17.** $(x-1)(x-2)(x-4)$. **18.** $(x-1)(2x+3)$. **19.** $2(3x+5y)(x+y)$. **20.** $(10x-9)(x+1)$. **21.** $(x+7)(7x-5)$. **22.** $(4x-5)(5x+6)$. **23.** $3(x+1)(2x+3)$. **24.** $(3x+7)(2x+15)$.

## XXXIII.

**1.** $58+6i$. **2.** $-2+2i$. **3.** $9-46i$. **4.** $10$. **5.** $-8i$. **6.** $(ac+bd)+(bc-ad)i$. **7.** $(ace-adf-bcf-bde)+(bce+ade+acf-bdf)i$. **8.** $-16$. **9.** $8i$. **10.** $-2+2i$. **11.** $2\Sigma mn(m-n)+(-\Sigma l^3+\Sigma l^2(m+n)-4lmn)i$. **12.** $(12a^2-18)+(18-12a^2)i$. **13.** $-8a^5-40a^4b+80a^3b^2+80a^2b^3-40ab^4-8b^5$. **14.** $-96p^5q+320p^3q^3-96pq^5$. **15.** $(ac+bd)/(c^2+d^2)+i(bc-ad)/(c^2+d^2)$. **16.** $\frac{5}{13}+\frac{25}{13}i$. **17.** $5i$. **18.** $-i$. **19.** $2/5$. **20.** $3\cdot7+2\cdot9i$. **21.** $\frac{2}{17}-\frac{9}{17}i$. **22.** $i$. **23.** $\frac{4}{5}+\frac{7}{5}i$. **24.** $\frac{8302}{2373}+\frac{6355}{5746}i$. **25.** $\frac{1273}{200}+\frac{1061}{200}i$. **26.** $-\frac{1}{2}+\frac{1}{2}\sqrt{3}i$. **27.** $2x(x^2-y^2)$. **30.** $x=-8/3$, $y=8/3$. **31.** $x^2-6x+10=0$. **32.** $x^4-4x^2+16=0$. **33.** $x^4-4x^3+4x^2+8=0$. **34.** $x^6-2x^5+8x^3-32x+64=0$.

## XXXIV.

**1.** $(x+4)(x-13)$. **2.** $(29x+30)(x-1)$. **3.** $(x+3+\sqrt{15})(x+3-\sqrt{15})$. **4.** $(2x+1+\sqrt{3})(2x+1-\sqrt{3})$. **5.** $2(3x+11)(3x-7)$. **6.** $(2x+3+\sqrt{5})(2x+3-\sqrt{5})$. **7.** $(2x+3+\sqrt{3})(2x+3-\sqrt{3})$. **8.** $(6x-11)(13x-1)$. **9.** $\frac{1}{4}(4x+3+i\sqrt{19})(4x+3-i\sqrt{19})$. **10.** $(3x-4+3i)(3x-4-3i)$. **11.** $(111x+113)(101x-102)$. **12.** $(2x+4+i\sqrt{3})(2x+4-i\sqrt{3})$. **13.** $(x+3+2i)(x+3-2i)$. **14.** $4(x-\omega)(x-\omega^2)$, where $\omega=(-1+i\sqrt{3})/2$, an imaginary cube root of $+1$. **15.** $(4x+1+2i)(4x+1-2i)$. **16.** $(\frac{5}{6}x+\frac{3}{5})(\frac{2}{7}x-\frac{7}{3})$. **17.** $(31x-23)(9x-13)$. **18.** $(2x^2+1+\sqrt{17})(2x^2+1-\sqrt{17})$. **19.** $(x+1)(x-1)(x-a+1)(x+a-1)$. **20.** $(x+1)(px+q)$. **21.** $\{x-m$

$-n+(m-n)i\}\,\{x-m-n-(m-n)i\}$. **22.** $x(x+a-b)(x-a-b)$. **23.** $\{(a+b)x+a^2+b^2+(a-b)\sqrt{(-ab)}\}\,\{(a+b)x+a^2+b^2-(a-b)\sqrt{(-ab)}\}/(a+b)$. **24.** $a=-29, -67, -103, -209$, etc., corresponding to $(x+1)(6x-35)$, $(2x+1)(3x-35)$, $(3x+1)(2x-35)$, $(6x+1)(x-35)$, etc. There are sixteen different cases.

## XXXV.

**1.** $\frac{1}{23}(23x+\{12+i\sqrt{431}\}y)(23x+\{12-i\sqrt{431}\}y)$. **2.** $2(x-y)^2(x+y)^2(x^2+y^2i)(x^2-y^2i)$. **3.** $(\sqrt{3}x+2\sqrt{2})(\sqrt{3}x-2\sqrt{2})(x+i\sqrt{2})(x-i\sqrt{2})$. **4.** $\frac{1}{4}(2x^2+1+\sqrt{5})(2x^2+1-\sqrt{5})$. **5.** $(xy-2+i\sqrt{21})(xy-2-i\sqrt{21})$. **6.** $(3x+5)\{3(x+1)-2\omega\}\,\{3(x+1)-2\omega^2\}$. **7.** $5(x+1)\{(2x+3)-\omega(3x+2)\}\,\{2x+3)-\omega^2(3x+2)\}$. **8.** $\frac{1}{16}x^6(x-y)(x+y)\{2x+(-1+\sqrt{3}i)y\}\,\{2x+(-1-\sqrt{3}i)y\}\,\{2x+(1+\sqrt{3}i)y\}\,\{2x+(1-\sqrt{3}i)y\}$. **9.** $(x+2iy)(x-2iy)\{x+(\sqrt{3}+i)y\}\,\{x+(\sqrt{3}-i)y\}\,\{x-(\sqrt{3}+i)y\}\,\{x-(\sqrt{3}-i)y\}$. **10.** $(x-y-1)(x-z-1)$. **11.** $(x-1)(x+2)\{x+\frac{1}{2}(1+i\sqrt{23})\}\,\{x+\frac{1}{2}(1-i\sqrt{23})\}$. **12.** $(x+1)^3$. **13.** $(x-1)(2x-1)(x-2)$. **14.** $(2x+1)(x+2)(3x+1)(x+3)$. **15.** $\frac{1}{256}[4x-3+\sqrt{13}+\sqrt{(6+6\sqrt{13})}][4x-3+\sqrt{13}-\sqrt{(6+6\sqrt{13})}][4x-3-\sqrt{13}+\sqrt{(6-6\sqrt{13})}][4x-3-\sqrt{13}-\sqrt{(6-6\sqrt{13})}]$. **16.** $\frac{1}{256}[4x+\{1+\sqrt{5}+\sqrt{(2+2\sqrt{5})}\}y]\times$ etc. **17.** $\frac{1}{256}[4x-\{1+\sqrt{5}+\sqrt{(2+2\sqrt{5})}\}y]\times$ etc. **18.** $(x-y)(x^4+x^3y+x^2y^2+xy^3+y^4)\equiv$ etc. **19.** $(x+y)(x^4-x^3y+x^2y^2-xy^3+y^4)\equiv$ etc. **21.** $\frac{1}{8}(4x-1-\sqrt{17})(4x-1+\sqrt{17})(x+1+\sqrt{2})(x+1-\sqrt{2})$. **22.** $(x-1)(x+1)\{x^2-\frac{1}{2}(3+\sqrt{21})x+1\}\,\{x^2-\frac{1}{2}(3-\sqrt{21})x+1\}\equiv$ etc. **23.** $\frac{1}{16}(x+1)\{2x+1+\sqrt{7}+\sqrt{(4+2\sqrt{7})}\}\times$ etc. **24.** $(x+y)(x^2+xy+y^2)(x^2-xy+y^2)\equiv$ etc. **25.** $(x-y)(x^2+xy+y^2)^2\equiv$ etc. **26.** $(a+b)^2\{a^2-(4-2\sqrt{3}i)ab+b^2\}\,\{a^2-(4+2\sqrt{3}i)ab+b^2\}$.

## XXXVI.

**1.** $(x+2)(x+3)(x-2)^2$. **2.** $(x+4)(x+5)(x+6)$. **3.** $(x-1)(x-2)(x-3)(x+3)$. **4.** $p=3$, $p=\frac{1}{2}$. **5.** $(x-1)(x+1)(x-2)$. **6.** $(x-1)(x-1+\sqrt{3})(x-1-\sqrt{3})$. **7.** $\frac{1}{8}(x-1)(4x+5+i\sqrt{31})(4x+5-i\sqrt{31})$. **8.** $(x-2)(x+1+i\sqrt{2})(x+1-i\sqrt{2})$. **9.** $(x-3)(2x+1)(2x-3)$. **10.** $(x-2)^2(x-2+\sqrt{3})(x-2-\sqrt{3})$. **11.** $\alpha=0$, $\beta=3$; the other factor is $x^2+2xy+3y^2$. **12.** $x^2(x-1)^2(x+1+i)(x+1-i)$. **13.** $(x-1)^2\{(a-1)x+(a-2)\}$. **14.** $2(x-y)(1-xy)$. **15.** $3(y-z)(z-x)(x-y)$. **17.** $(x+1)(x+a+1)(x+a-1)$. **18.** $p=-3a^2$, $q=2a^3$. **19.** $(a+b+c)(b+c)(c+a)(a+b)$. **20.** $-(b-c)(c-a)(a-b)(bc+ca+ab)$.

## XXXVII.

**1.** $(x-3)(y+5)$. **2.** $(x+3)(2y+7)$. **3.** $(x+6)(x+3y+2)$. **4.** $(x+y-1)(x-y+1)$. **5.** $(x+3)(x+y-1)$. **6.** $(x+y+2)(x-y-1)$. **7.** Indecomposable. **8.** $(x-1)(x-y+2)$. **9.** $\{\sqrt{3}(x-2y)+1\}\,\{\sqrt{3}(x-2y)-1\}$. **10.** $(x-2y+3)(x-2y-4)$. **11.** $(2x-3y+4)(3x-2y+5)$. **12.** $(3x-y-1)(x-y+1)$. **13.** $(3x-y+2)(x+y-1)$. **14.** $(2x+3y+1)(3x+2y-1)$. **15.** $r=pq$; the factors are $(x+q)(y+p)$. **17.** $\lambda=-6$;

factors $(2x-5y-2)(3x+2y+3)$. **18.** $(ax+by+a-b)(bx-ay-a-b)$. **20.** $(x+y-1)(x^2+y^2+x+y+1)$. **21.** $(x+y)(x^2+2xy+y^2+x+2y)$.

## XXXVIII.

**1.** $3(x-1)(x-2)(3-2x)$. **2.** $(x^2-x+1)(x+1)(x^2+1)$. **3.** $\{a+b+(1+i)(c+d)/\sqrt{2}\}\{a+b-(1+i)(c+d)/\sqrt{2}\}\{a+b+(1-i)(c+d)/\sqrt{2}\}\{a+b-(1-i)(c+d)/\sqrt{2}\}$. **4.** $2(x+1)^2\{x^2+(2+i\sqrt{3})x+1\}\{x^2+(2-i\sqrt{3})x+1\}\equiv$ etc. **5.** $(x+a)\{x-a+\sqrt{(a^2+b^2)}\}\{x-a-\sqrt{(a^2+b^2)}\}$. **6.** $\Sigma ax.\Sigma x/a$. **7.** $\Sigma a\{2\Sigma a^2+ab-2ac+ad+bc-2bd+cd\}$. **8.** $(bx/c+cy/a+az/b)(cx/b+ay/c+bz/a)$. **9.** $4x(x^2+3y^2+3z^2)$. **10.** $(x-2)(x+p+q)(x+p-q)$. **11.** $3(x-a)\{x-a+\sqrt{2(b-c)i}\}\{x-a-\sqrt{2(b-c)i}\}$. **12.** $-(y-z)(z-x)(x-y)(x+y+z)$. **13.** $(a+b)(a^m+b^n+a^2-ab+b^2)$. **15.** $(\Sigma x^2+\Sigma yz)(\Sigma x^2-\Sigma yz)$. **16.** $x\Sigma x(\Sigma x^2-\Sigma xy)$. **17.** $\Sigma yz(\Sigma yz-\Sigma x^2)$. **18.** $\frac{5}{2}(y-z)(z-x)(x-y)\Sigma(y-z)^2$. **19.** If $a=2m+1$, $b=2n+1$, where $m$ and $n$ are integers, we have $a^4-b^4\equiv(a-b)(a+b)(a^2+b^2)\equiv 8(m-n)(m+n+1)(2m^2+2n^2+2m+2n+1)$.

## XXXIX.

*Only one of the square roots and only the real cube root is given.*

**1.** $2x-3a-2$. **2.** $x^2+2x-1$. **3.** $x^3+x^2-3x+4$. **4.** $11x^3+2x^2-x+1$. **5.** $1-2x+3x^2-4x^3+5x^4$. **6.** $1-x+2x^2-3x^3+4x^4$. **7.** $xy-x-y$. **8.** $x-2y+3$. **9.** $x^3+y^3+x^2+y^2+x+y$. **10.** $3x^2-3x+1$. **11.** $x^2+x+1$. **12.** $x^2+2x+1$. **13.** $4x^2-4x+1$. **14.** $x/2+2/3x^2$. **15.** $-(x+1)/x-1/2+x/(x+1)$. **16.** $ax^2+bx+b+b/x+a/x^2$. **17.** $x-2-1/x$. **18.** $\lambda=6$, $\mu=\pm 6$, $\nu=\pm 4$, square root $3x+2y\pm 2$; or $\lambda=-6$, $\mu=\pm 6$, $\nu=\mp 4$, square root $3x-2y\pm 2$. **19.** $c=-168$, $d=196$, square root $x^2+6x-14$.

## XL.

**1.** $x^3y$. **2.** $x^2yz$. **3.** $xyzu$. **4.** $x^2yz$; or $(a+b)x^2yz$, if $a$ and $b$ be included as variables. **5.** $xy(x^2-y^2)$. **6.** $x^2(x-1)^2$. **7.** $(x-1)^2$. **8.** $(x-1)^2$. **9.** $x^2-y^2$. **10.** $x^2-2y^2$. **11.** $x^2-y^2$. **12.** $x+2y$. **13.** $2x-1$. **14.** $(x+y)^2-(x+y)+1$. **15.** $x^4+x^2y^2+y^4$. **16.** $x+1$. **17.** $x-1$. **18.** $x^3+x^2y+xy^2+y^3$. **19.** $xy-z^2$. **20.** $(x-2)(x-3)$. **21.** $(x-1)^2$. **22.** $x-1$. **23.** $(x-1)^2$. **24.** $x^2+x+1$. **25.** $(x-1)^2$.

## XLI.

**1.** $x^3+x^2+x+1$. **2.** $x^2+2x+3$. **3.** $x^2+1$. **4.** $(x+2)(x+3)$. **5.** $x^2+2x+5$. **6.** $(x^2+1)^2$. **7.** $x^2+x+1$. **8.** $x^4+x^2+1$. **9.** $(x-1)(x+3)$. **10.** $x^2+x+1$. **11.** $2x^2-2x-3$. **12.** $2x^3-1$. **13.** $x-3$. **14.** $x^2+x+2$. **15.** $x^2+3x+1$. **16.** $x^2+2x+3$. **17.** $2x-1$. **18.** $x+1$. **19.** $(x+1)(x+2)$. **20.** $x^2-x+7$. **21.** $x^2+px+p^2$. **22.** $x-1$. **23.**

$px^2+qx+r$. **24.** $x^2+ax+1$. **27.** $l$ must satisfy the equation $5l^2-21l+27=0$. **28.** $p^3-2p^2+2p=0$. **29.** $p+r=0$, $q=-1$. **30.** $2x+3$; if the condition be satisfied, $(2x+3)(2x+1)$. **31.** $x=5$, $x=7$.

## XLII.

**1.** $x^6y^4$. **2.** $x^3y^2z^3$. **3.** $x^3y^3z^3u^3$. **4.** $x(x-1)^2(x+1)^2(x^2+x+1)^2$. **5.** $(x^2-1)^2(x-2)(x-3)^2$. **6.** $(x^2+1)(x^2+3)(x^2-1)^2$. **7.** $(x^4+16x^2+1)(x^6-64)$. **8.** $x^4+x^2+1$. **9.** $(x^2-1)^2(x^6-1)^2$. **10.** $(x^2+y^2)(x^6-y^6)(x^4+x^3y+x^2y^2+xy^3+y^4)$. **11.** $(x^6-y^6)(x^6-x^3y^3+y^6)(x^4-x^2y^2+y^4)$. **12.** $x^{12}-64$. **13.** $(x-1)(x^2-2)(x^2+x+2)$. **14.** $x^8+x^4+1$. **15.** $(x+1)(x-3)(x^2-4)$. **16.** $(x-1)(x-3)(x+2)(x+4)(2x-3)$. **17.** $(x^6-1)(3x^2+3x+1)$. **18.** $7x-5y$; $(2x+5y)(4x-3y)(3x+2y)(7x-5y)$. **19.** $(x^4+x^2+1)(2x^2-3x+2)$. **20.** $(x^4+x^2+1)(11x^2+10x+9)$. **21.** $(x-1)^2(x^4+x^2+1)$. **22.** $(2x^2-3x+1)(3x^4+2x^3-2x^2+3x-2)$.

## XLIII.

**1.** $9x^2/4y^2z$. **2.** $4/3xyu$. **3.** $(a+b)/(a-b)yz$. **4.** $1/(x+y)$. **5.** $(x-y)/(x+y)$. **6.** $2(x-1)/(x+1)$. **7.** $4/(x+3)$. **8.** $x^4/(x^2-2)$. **9.** $1/(x^4+a^4)$. **10.** $x^2-a^2$. **11.** $1/(x^2y^2+9)$. **12.** $(x-2)/(x+1)$. **13.** $-(x+2)/(x^2+x+1)$. **14.** $-(4x^2+4x-3)/(3x^2+8x+5)$. **15.** $-2(x^2+x+1)/(x^3+x^2+x+1)$. **16.** $(2x-7)/(4x^2+8x+7)$. **17.** $(5x-3)/(5x+3)$. **18.** $x/2(x-1)$. **19.** $1/xy(x^{m-1}+y^{m-1})$. **20.** $(x^2+y^2)/(x^2-xy+y^2)$. **21.** $(x-y)/(x^4+x^2y^2+y^4)$. **22.** $(x^3+3x^2y+3xy^2+y^3)/(x^3-y^3)$. **23.** $(x-3)/(2x+1)$. **24.** $(x^3-4x)/(x^2+7)$. **25.** $(x^2+3y^2)/(x^4+10x^2y^2+5y^2)$. **26.** $(y-1)/(x+2y+1)$. **27.** $(x+p)/(x-p)$. **28.** $(x-a+1)/(x^2-xa+a^2)$. **29.** $(x-a)/(x-b)$. **30.** $x^4+x^3+x+1$. **31.** $(x+y-2)/(3x+2y-2)$. **32.** $(x^2-1)/(x+3)$. **33.** $1+a+b-ab$. **34.** $a^4-b^4$. **35.** $\frac{1}{2}\Sigma x$. **36.** $x^2-\sqrt{2}x+1$. **37.** $(a-b)/(b-c)$. **38.** $(1+3x)/(1-2x-4x^2)$. **39.** $(x^2-2x-3)/(x^2+x+7)$. **40.** $x+1$. **41.** $(2x+1)/(3x+1)$. **42.** $7(x^2+x+1)/5$. **43.** $(x^2-2x+3)/(x^3+3x^2+2x+1)$. **44.** $\Sigma bc/(\Sigma a^2+\Sigma bc)$. **45.** Multiply numerator and denominator by $a$; put $ad=bc$; and the fraction reduces to $c(a+b)/(a+c)$.

## XLIV.

**1.** $(x^2-x+1)/(x-3)$. **2.** $-2(x+3)/(x-3)$. **3.** $-4(x-1)/(x+2)$. **4.** $(x^2-16)/(x^3-6x^2+3x+10)$. **5.** 1. **6.** $(x+3)/(x+2)$. **7.** $(x^2+x+1)^2/(x^2-x+1)^2$. **8.** $x+1$. **9.** $2/(x^2+y^2)$. **10.** $(x+1)(x-3)(2x-1)(x^3-x^2+x-5)/(x-1)(x-2)(x+2)(x-5)$. **11.** $(x+1)^2/(x^2+1)$. **12.** $(x+1)(x+3)(x-3)/(x-1)(x^2+6x-9)$. **13.** $2x+2y+6z$. **14.** $3(xy-1)/2(xy+1)$. **15.** 1. **16.** $\{(x+y)^2+(x+y)+1\}/\{(x-y)^2-(x-y)+1\}$. **17.** $2\Sigma a$. **18.** $(x^2-4)/(x-1)$. **19.** $7x+1$. **20.** $\Sigma a^2/(\Sigma a^3-3abc)$.

## XLV.

**1.** $1/ax$. **2.** $-2/(x^2-4x+3)$. **3.** $2x/(x^2+3x+2)$. **4.** $(8x^2+6)/(4x^2-9)$. **5.** $(x-6)/(x^2-9)$. **6.** $(b-a)(a^2+b^2)/ab(a+b)$. **7.** $15/(x-1)(x+2)(2x+3)$. **8.** $ab(a+b)^2/(a-b)(a^2+b^2)$. **9.** $x/(x+1)(x^2+1)$. **10.** $3/(a+c)$. **11.** $(18-6x)/(x+3)^2$. **12.** $4/(1-x^2)^2$. **13.** $2x(x+1)/(x-1)(x^2+1)$. **14.** $1/(5x-4y)$. **15.** $2x^3/(x-1)(x^2-x+1)$. **16.** $-32(x^3+16x)/(x^2-16)^2$. **17.** 0. **18.** $4a^2/(a^2-b^2)$. **19.** 3. **20.** $2/x$. **21.** $-x^2/y^2$. **22.** $1/(x+1)$. **23.** $(x^2+8x+14)/(x+2)(x+4)(x+6)$. **24.** $x/(x^6-1)$. **25.** $4x^2y^2/(x^6-y^6)$. **26.** $6x^2/(x^{12}-1)$. **27.** $xy(3y-x)/(x^6-y^6)$. **28.** 2. **29.** $(3x^3+15x^2-71x-315)/(x+3)(x+5)(x+7)$. **30.** $(x+4)/(x-1)(x-2)(x+3)$. **31.** $(3x-4)/(x-1)(x-2)(x-3)$. **32.** $-2(2x^2+x-4)/(4x^2-1)(x^2-4)$. **33.** $1/(x-2)^2$. **34.** $(1+x-x^2)/(1+x+x^2)$. **35.** $1/(x+1)^2$. **36.** $-4x/(x^2-1)$. **37.** 0. **38.** $(x^2+8xy-6y^2)/(x-2y)(x+3y)(x+4y)$. **39.** 0. **40.** $4x/(x^2-1)^2$. **41.** $-8x/(x^2-9)$. **42.** $-4qx/\{(x-p)^2-q^2\}$. **43.** $2(x^3-1)/\{(x-1)^2-y^2\}$. **44.** $(x^5+x^4+3x^3+x^2+3x-1)/(x^4+x^2+1)(x^3+x^2+x+1)$. **45.** $2(1+xy)$. **46.** $(x^6+15x^4y^2+15x^2y^4+y^6)/xy(3x^4+10x^2y^2+3y^4)$. **47.** $2a^2(3x^4+a^4)/(x^4-a^4)^2$. **48.** $2x^2y^2(x^2+y^2)^2/(x^6-y^6)(x^4+y^4)$.

## XLVI.

**1.** 1. **2.** $1/\Pi(y+z)$. **3.** 1. **4.** $\Sigma x/\Pi(y+z-x)$. **5.** 0. **6.** $-3$. **7.** $\{(p\Sigma a+3q)x^2-2(p\Sigma ab+q\Sigma a)x+3pabc+q\Sigma bc\}/\Pi(x-a)$. **8.** 0. **9.** $(2abc\Sigma a-2\Sigma a^2b^2)/\Pi(b-c)$. **10.** 1. **11.** $3x-\Sigma a$. **12.** $(\Sigma ax^2-2\Sigma abx+3abc)/\Pi(x-a)$. **13.** $-\Sigma a^2/x\Pi(x-a)$. **14.** 3. **15.** $1/\Pi(x-a)$. **16.** $\Sigma a$. **17.** $2(\Sigma ax-\Sigma ab)/\Pi(x-a)$. **18.** $-\Pi(b-c)(x^2+\Sigma bc)/\Pi(b+c)\Pi(x-a)$. **19.** $(x^2+\Sigma ab)/\Pi(x-a)$. **20.** $(3-\Sigma a^2)/\Pi(1-a^2)$. **21.** 1. **22.** 0. **23.** $-9\Pi(y-z)/\Pi(y+z-2x)$. **24.** $(x^2+x+1)/\Pi(x-a)$. **25.** $\Sigma x-\Sigma a$. **26.** $x^2/\Pi(1+lx)$. **27.** 0. **28.** $(3-2\Sigma a^2+\Sigma a^4)/\Pi(1-a^2)$. **29.** $\{\Sigma ax^2-\Pi(b-c)x+\Sigma a^3\}/\Pi\{(b-c)x+a\}$.

## XLVII.

**1.** $8ab(a^2+b^2)/(a^2-b^2)^2$. **2.** $2/c$. **3.** $1/(8x^2-1)$. **4.** $(x^2+ax+a^2)/a(2x+a)$. **5.** $x^4y^4/2(x^2+y^2)^2$. **6.** 2. **7.** $x^2y^2$. **8.** $(2x-3)/(2x-5)$. **9.** $-(x^4+1)/(x^2+1)^2$. **11.** $-(p-q)(x^2-pq)/(p+q)(x^2+pq)$. **12.** $(a-c)/(1+ac)$. **13.** $(b^2c+c^2a+a^2b)/abc$. **14.** $1/(x^4+x^2y^2+y^4)$. **15.** $(x^4+2x^3+6x^2+2x+1)/(x+1)^2(x^2+1)$. **16.** $a^3/(a^3-b^3)$. **17.** $(a^2+c^2)/ac$. **18.** $1/xy$. **19.** 2. **20.** $-4x^3/(x^4-1)$. **21.** $-2x^2/(2x+1)$. **22.** $-(3a^4+5a^3b+6a^2b^2+5ab^3+3b^4)/(a^2-b^2)(a^2+ab+b^2)$. **23.** $-4a^3b^3/(a^4-b^4)$. **24.** 1. **25.** $x^2(x-1)/(x+1)$. **26.** $(x^4+x^2+x)/(x^2+1)$. **27.** $x^2/(x^2+1)$. **28.** $2x^5y^4/(x^8+x^4y^4+y^8)$.

## XLVIII.

**1.** $15/2(x-9)+1/2(x-11)$. **2.** $1+a^2/(a-b)(x-a)+b^2/(b-a)(x-b)$. **3.** $\Sigma(pa+q)/(a-b)(a-c)(x-a)$. **4.** $-5/(x-5)+6/(x-6)$. **5.** $1/(x-1)-5/(x-2)+5/(x-3)$. **6.** $2/3(x-1)-5/4(x-2)-5/12(x+2)$. **7.** $2/(x$

$-1)+4/(x-1)^2+3/(x-2)$. **8.** $14/(x-3)-13/(x-2)-7/(x-2)^2$. **9.** $2/(x-2)-1/(x-1)$. **10.** $-1/2(x-1)+1/2(x-1)^2+x/2(x^2+1)$. **11.** $1/(x-1)+2/(x-1)^2-(x-1)/(x^2+1)$. **12.** $x+1/3(x-1)+(2x+4)/3(x^2+x+1)$. **13.** $1+2/3(x-1)-(2x+4)/3(x^2+x+1)$. **14.** $1/(x-1)+2/(x-1)^2-(x-1)/(x^2-x+1)$. **15.** $(x+1)/(x^2+2x+3)-(x-1)/(x^2+x+2)$. **16.** $-1/4(x-1)+2/7(x-2)-(x+9)/28(x^2+3)$. **17.** $1/6(x-1)-(x+2)/6(x^2+x+1)-1/6(x+1)+(x-2)/6(x^2-x+1)$. **18.** $1/(x^2-x+1)-1/(x^2+x+1)$. **19.** $1/2(x-1)-18/13(2x+3)+5(x+5)/26(x^2+1)$. **20.** $4/9(x-1)-1/18(x+2)-1/2(x+2)^2-(7x+1)/18(x^2+2)$. **21.** $1/6(x-1)-(x-1)/6(x^2+x+1)-1/6(x+1)+(x+1)/6(x^2-x+1)$. **22.** $1/(\omega-\omega^2)(x-\omega)-1/(\omega-\omega^2)(x-\omega^2)$, where $\omega=(-1+i\sqrt{3})/2$, $\omega^2=(-1-i\sqrt{3})/2$. **23.** $(2x^3+4x^2+5x+2)/4(x^2+x+1)^2-(2x^3-4x^2+5x-2)/4(x^2-x+1)^2$.

## XLIX.

**1.** $\sqrt{y}$. **2.** $(a^2/b)\sqrt{a}$ ; $\sqrt{(a^5/b^2)}$. **3.** $(x^2/y^2)\sqrt[6]{(x/y)}$ ; $\sqrt[6]{(x/y)^{13}}$. **4.** 1. **5.** $6x$. **6.** $x^2y^2\sqrt[3]{(xy)^2}$ ; $\sqrt[3]{(xy)^8}$. **7.** $x$. **8.** $\sqrt[mn]{x^{m^2+n^2+1}}$. **9.** $\sqrt[16]{x^{15}}$. **10.** $\sqrt[qs]{(x^{ps}/y^{qr})}$. **11.** $x/(x-y)$. **12.** $x^2/(x-y)$. **13.** $\sqrt{(x+y-1)}$. **14.** $\sqrt[6]{(x^2-y^2)^5}$. **15.** $x^4-y^4$. **16.** $(4x^3+x)\sqrt{x}$ ; $\sqrt{(16x^7+8x^5+x^3)}$. **17.** $(x+a+1)\sqrt{(x+a)}$ ; $\sqrt{\{(x+a)(x+a+1)^2\}}$. **18.** $\sqrt{x}$. **19.** $2\sqrt{(x^2-y^2)}$ ; $\sqrt{(4x^2-4y^2)}$. **20.** $-\{2ax/(a-x)\}\sqrt{(a^2-x^2)}$ ; $-\sqrt{\{4a^2x^2(a+x)/(a-x)\}}$. **21.** $\sqrt[6]{(1/x)}$. **22.** $(2x^2+3x+6)\sqrt[3]{x}$ ; $\sqrt[3]{\{x(2x^2+3x+6)^3\}}$. **23.** 0. **24.** $x^2+x+1$. **25.** $a^2+2x\sqrt{(a^2-x^2)}$. **26.** $x+1+1/x$. **27.** $3+2\sqrt{(x+1)}$. **28.** $x+1+\sqrt{x}$. **29.** $(x^2+6xy+y^2)/xy+4(x+y)\sqrt{(xy)}/xy$. **30.** $(x-1)$. **31.** $1+x\sqrt[3]{x}+\sqrt[3]{x^2}$. **32.** $\sqrt[3]{3}=\sqrt[12]{81}$, $\sqrt[4]{7}=\sqrt[12]{343}$, $\sqrt{5}=\sqrt[12]{15,625}$.

## L.

**1.** 16. **2.** 1/64. **3.** 1/5. **4.** 3. **5.** $x^3\sqrt[12]{x^7}$. **6.** $x^7\sqrt{x}$. **7.** $(1/xyz)\sqrt[12]{(xy^2z^9)}$. **8.** $(xyz)^{1/5}$. **9.** $2^{7/12}/2$. **10.** $x^{5/6}/x^4$. **12.** $(b/a)\sqrt[24]{(a^{11}b^5)}$. **13.** $xyz(x^4yz)^{1/8}$. **14.** $(y^3/x^6)x^{1/3}$. **15.** 1·22.
**16.** $a^{\alpha^2+2\beta^2+2\gamma^2+3\alpha\beta-\beta\gamma+\gamma\alpha}\,b^{\alpha\beta+3\beta\gamma+5\gamma\alpha-\gamma^2}$ ; $2^{118}$. **17.** $(xy)^{(m^2-n^2)/mn}$.
**19.** 1. **20.** 1. **22.** 1. **23.** $(x/y)^{ab}$. **24.** $x^{2(a+b+c)}$. **26.** $(1/2)^{1/2}<(2/3)^{1/3}$.

## LI.

**1.** $-(6x+20+6/x)+(x+15-15/x+1/x^2)\sqrt{x}$. **2.** $(x+y)^3+(x+y)^{-3}-2$. **3.** $x^{\frac{7}{2}}-3x^{\frac{5}{2}}+5x^{\frac{3}{2}}-7x^{\frac{1}{2}}+7x^{-\frac{1}{2}}-5x^{-\frac{3}{2}}+3x^{-\frac{5}{2}}-x^{-\frac{7}{2}}$. **4.** $x+1+x^{-1}$. **5.** $x^6+2x^2-7x^{-2}-16x^{-6}$. **6.** $7.3^{\frac{2}{3}}-3.3^{\frac{1}{3}}-17$. **7.** $(x^{\frac{3}{4}}+x^{\frac{1}{2}}y^{\frac{1}{4}}+x^{\frac{1}{4}}y^{\frac{1}{2}}+y^{\frac{3}{4}}+x^{\frac{1}{4}}+y^{\frac{1}{4}}+1)/(x-y)$. **8.** $a^{\frac{2}{3}}+a^{\frac{1}{3}}b^{\frac{1}{3}}+b^{\frac{2}{3}}$. **9.** $x^{-\frac{2}{3}}+x^{-\frac{1}{3}}y^{-\frac{1}{3}}+y^{-\frac{2}{3}}$. **10.** $x^{\frac{4}{3}}+x^{\frac{2}{3}}y^{\frac{2}{3}}+y^{\frac{4}{3}}$. **11.** $x-x^{\frac{3}{4}}y^{\frac{1}{2}}+x^{\frac{1}{2}}y-x^{\frac{1}{4}}y^{\frac{3}{2}}+y^2$. **12.** $2(x^{\frac{2}{3}}-y^{\frac{2}{3}})$. **13.** $2(x^{\frac{4}{3}}-y^{\frac{4}{3}})$. **14.** 0. **15.** $x^{\frac{1}{3}}+2+x^{-\frac{1}{3}}$. **16.** $(3.3^{\frac{2}{3}}+3^{\frac{1}{3}}+3)/4$. **17.** $(1-x^2)^{\frac{1}{2}}\{x^2-4-2(x+2)(1-x)^{\frac{1}{2}}+2(x-2)(1+x)^{\frac{1}{2}}-4(1-x^2)^{\frac{1}{2}}\}/x^2$. **18.** $x^{(n-1)/n}+x^{(n-2)/n}y^{1/n}+\ldots+y^{(n-1)/n}$. **19.** $x^n+1+x^{-n}$. **20.** 1

$+5x^{\frac{1}{2}}+5x^{\frac{3}{4}}+9x+12x^{\frac{5}{4}}+13x^{\frac{3}{2}}+7x^{\frac{7}{4}}+2x^2$. **21.** $-(x^{\frac{2}{3}}+3x^{\frac{1}{3}}+1)/(x^{\frac{2}{3}}-3x^{\frac{1}{3}}+1)$. **22.** $x-x^{-1}-2$. **23.** $2a^{\frac{5}{2}}-3b^{\frac{4}{3}}+4c^{\frac{3}{2}}$. **24.** $\sqrt{(x/y)}+\sqrt{(y/x)}+1/2$. **25.** $3x^2-1+4/\sqrt{x}$. **27.** $\xi^{14}+\xi^{13}\eta+\ldots+\eta^{14}$, where $\xi=\sqrt[3]{x}$, $\eta=\sqrt[5]{y}$. **28.** $\xi^{14}-\xi^{13}\eta+\ldots+\eta^{14}$, where $\xi=x^{\frac{2}{3}}$, $\eta=y^{\frac{3}{2}}$.

## LII.

**1.** $x-y$. **2.** $a^4$. **3.** $x-2\sqrt{x}$. **4.** $6x^2y-4y^3$. **5.** $4\Sigma x$. **6.** $(x^3+15x^2+15x+1)+(6x^2+20x+6)\sqrt{x}$. **7.** $(x^2-20x+1)+(15x-6)x^{\frac{1}{3}}-(6x-15)x^{\frac{2}{3}}$. **9.** $\{a^4+a^2+1+(3-2a^2)b+b^2+2(b+1-a^2)\sqrt{b}\}/\{(a^2+a+1)^2-(2a^2+2a-1)b+b^2\}$. **10.** $b/c$. **11.** $-\{x^2+xy+y^2+(x+y)\sqrt{(x^2+y^2)}\}/xy$. **12.** $1/(a-x)$. **13.** $\{x^2-2+2\sqrt{(1-x^2)}\}/x^2$. **14.** $\Sigma(b-c)^2\sqrt{\{(a+b)(a+c)\}}/\Pi(b-c)$. **15.** $\{4x+2\sqrt{(x^2+ax)}-2\sqrt{(x^2-ax)}\}/a^2$. **16.** $-2(p+q+\sqrt{p}+\sqrt{q}+2\sqrt{p}\sqrt{q})/(1-p)(1-q)$. **17.** $\{2\sqrt{(1-x^2)}+(x-2)\sqrt{(1+x)}-(x+2)\sqrt{(1-x)}+2\}/2x^2$. **18.** $\{(7-4x^2)+(8x^2+8x-2)\sqrt{(1-x)}+(8x^2-8x-2)\sqrt{(1+x)}-(8x^2+2)\sqrt{(1-x^2)}\}/(4x^2-3)^2$. **19.** $\{(p-q)\sqrt{(p+q)}+(p+q)\sqrt{(p-q)}+\sqrt{(2p^3-2pq^2)}\}/2(p^2-q^2)$. **20.** $\{4(2-x)\sqrt{(1+x)}-8\sqrt{(1-x^2)}\}/x^2$. **21.** $\sqrt{2}$. **22.** $\sqrt{\{2(a^2-b^2)\}}/(a^2-b^2)$. **23.** $-8\Sigma(b-c)\sqrt{a}$.

## LIII.

**1.** $\sqrt{2}$. **2.** $\frac{1}{3}\sqrt{3}$. **3.** $\sqrt{5}$. **4.** $32\sqrt{2}$. **5.** 16. **6.** $\frac{1}{2}\sqrt{3}$. **7.** $\frac{3}{4}\sqrt[3]{4}$. **8.** 3. **9.** $2\sqrt{6}$. **10.** 30. **11.** $\sqrt[30]{(3^{15}2^{6}5^{10})}$. **12.** $2\sqrt[5]{8}$. **13.** $-5\sqrt{2}$. **14.** 0. **15.** $3\sqrt{6}$. **16.** $-\frac{2}{3}\sqrt{3}$. **17.** $9\sqrt{2}$. **18.** $-\sqrt{5}$. **19.** $\frac{5}{4}\sqrt{2}$. **20.** $\frac{7}{10}\sqrt[3]{25}$. **21.** 11·314. **22.** 1·568. **23.** ·684. **24.** 8·196. **25.** 1·366. **26.** ·303. **27.** ·241. **28.** 3. **29.** $(x^4-7x^2+12)+(2x^2-8)\sqrt{2}+(6-2x^2)\sqrt{3}-4\sqrt{6}$. **30.** 4. **31.** 17/7. **32.** 2. **33.** $-3$. **34.** $24-8\sqrt{6}$. **35.** $42+16\sqrt{6}$. **36.** 2·679492. **37.** $-\frac{1}{10}+\frac{1}{4}\sqrt{2}+\frac{3}{20}\sqrt{6}$. **38.** $(31\sqrt{10}-39\sqrt{6}-19\sqrt{35}+20\sqrt{21})/121$. **39.** $1+\frac{3}{2}\sqrt{2}+\sqrt{3}+\frac{1}{2}\sqrt{6}$. **40.** $1+\frac{5}{6}\sqrt{6}-\frac{1}{2}\sqrt{10}-\frac{1}{3}\sqrt{15}$. **41.** $\frac{3}{2}+\frac{1}{4}\sqrt{2}+\frac{3}{5}\sqrt{5}-\frac{1}{4}\sqrt{6}+\frac{1}{20}\sqrt{10}-\frac{1}{10}\sqrt{15}-\frac{7}{20}\sqrt{30}$. **42.** $-(1+\frac{2}{3}\sqrt{6}+\frac{2}{5}\sqrt{10}+\frac{8}{15}\sqrt{15})$. **43.** $2\sqrt{2}+\frac{4}{5}\sqrt{5}-\sqrt{3}-\frac{1}{2}\sqrt{30}-\frac{1}{2}\sqrt{42}+\frac{2}{5}\sqrt{70}$. **44.** $\frac{208}{103}\sqrt{2}+\frac{22}{103}\sqrt{5}$. **45.** $x^4-\sqrt{6}x^3+5x^2-\sqrt{6}x+1$. **46.** $(x^2+1)^2-x(x^2+1)\sqrt{2}$. **47.** $\frac{1}{13}+\frac{3}{13}\sqrt{3}+\frac{60}{143}\sqrt[4]{3}+\frac{89}{143}\sqrt[4]{27}$. **49.** $2\sqrt{3}+\sqrt{5}>\sqrt{2}+\sqrt{5}$. **50.** $\sqrt{3}+\sqrt{7}<\sqrt{6}+2$. **51.** $\sqrt{10}+\sqrt{7}<\sqrt{19}+\sqrt{3}$. **52.** $\sqrt[3]{5}+1<2\sqrt{2}$.

## LIV.

**2.** $\frac{1}{2}\sqrt{6}+\frac{1}{2}\sqrt{10}$. **3.** $1+\sqrt{6}$. **4.** Cannot be expressed in linear form. **5.** $\frac{1}{2}\sqrt{10}-\frac{1}{6}\sqrt{15}$. **6.** $2\sqrt{6}+\sqrt{13}$. **7.** $\sqrt{7}+\sqrt{14}$. **8.** $\sqrt{3}$. **9.** $-\frac{90}{61}-\frac{20}{61}\sqrt{3}-\frac{22}{61}\sqrt{5}-\frac{18}{61}\sqrt{15}$. **10.** $2\sqrt{2}$. **11.** $\sqrt{7}+\sqrt{8}$. **12.** 1. **13.** $\frac{1}{2}+\frac{1}{6}\sqrt{6}+\frac{1}{6}\sqrt{15}$. **14.** $3\sqrt{2}$. **15.** $10\sqrt{2}$. **16.** 144. **17.** $\sqrt{\{(a+b-c)/2\}}+\sqrt{\{(a-b+c)/2\}}$. **18.** $\sqrt{p}+\sqrt{(p-1)}$. **19.** $1+\sqrt{2}-\sqrt{3}$. **20.** $6+\sqrt{5}+\sqrt{7}$. **21.** Cannot be reduced to linear form. **22.** $-\sqrt{6}+\sqrt{7}+\sqrt{8}$. **23.** $\frac{1}{2}+\frac{1}{2}\sqrt{2}+\frac{1}{3}\sqrt{6}$. **24.** $1+\sqrt{3}$. **26.** $\sqrt{7}-\sqrt{3}$. **27.** $\frac{1}{2}\sqrt[4]{2}(\sqrt{2}+\sqrt{6})$. **28.** $\frac{1}{3}\sqrt[4]{27}(\sqrt{2}-1)$. **29.** $\sqrt[4]{3}+\sqrt[4]{75}$. **30.** $\sqrt[8]{2}(\sqrt{2}+\sqrt{7})$.

## LV.

**1.** 1601 : 1041. **2.** 8277 : 10,000. **3.** 1 : 2. **4.** $bc/a$. **7.** 10. **8.** $4\frac{3}{7}$. **9.** 48, 64. **10.** 8. **11.** $ab$. **12.** $44\frac{4}{5}$, $67\frac{1}{5}$. **13.** $1+x:1+2x>1:1+x>1-x:1$. **14.** Greater if $3x>y$, less if $3x<y$. **15.** 1 : 8. **16.** $c(a+b)=ab$. **17.** 1 : 3 or 3 : 1. **18.** 44 : 325. **20.** 1800 : 1. **22.** 2000, 300. **23.** $A/(1+a)+B/(1+b)+C/(1+c) : aA/(1+a)+bB/(1+b)+cC/(1+c)$.

## LVI.

**1.** $5:6=45:54$. **2.** 4. **3.** 9/4. **4.** 16. **5.** $\pm 12$. **6.** 3 : 5 or 1 : 4. **7.** $3\pm\sqrt{17}:4$. **8.** $6a^3b$. **9.** $(a-b)^3(a+b)^{\frac{3}{2}}$. **10.** $3\sqrt{2}$. **11.** $3+\sqrt{2}$. **12.** 12, 24, 48, 96. **13.** If $a$ be the third and $b$ the fifth, the first is $a^2/b$, and the fourth $\sqrt{(ab)}$. **14.** 12, 3. **15.** 637, 91, 13. **16.** 9, 12, 16. **32.** 1, or $-bd/ac$; provided $ad-bc\neq 0$. **38.** $a(a-c)y^2+2ac(b-c)y+c^2(b^2-ac)=0$. **41.** $\{a(c+d)x+c(a+b)y\}/(a+b)(c+d)(x+y)$, $\{b(c+d)x+d(a+b)y\}/(a+b)(c+d)(x+y)$; $x:y=(a+b)(d-c):(a-b)(d+c)$. **43.** 156 lbs. **45.** £21, £24, £27.

## LVII.

**1.** 20/3. **2.** 640/81. **3.** 78. **4.** 135/4. **8.** 4·71. **9.** 14222·22 lbs. **10.** 22·5 fr. **11.** $20\frac{1}{4}$ ft. **12.** 1650 ft. **13.** Four days. **14.** 7 ft. **15.** 21 in.; 9702 cub. in. **16.** £833 : 6 : 8. **17.** $\frac{1}{2}t+\frac{1}{6}\sqrt{\{21t^2-36(a+b)t+36(a^2+b^2)\}}$. **18.** Radius of earth 4000·0053 miles; distance of horizon 200·062 miles.

## LVIII.

**1.** 45. **2.** 1; $(6n-1)/5$. **3.** 275. **4.** $-169/2$. **5.** 1878·75. **6.** $n\{(2n-1)a+(4n-7)b\}$. **7.** 781; 2 or 80. **8.** 221,761. **9.** 28, 25, 22, 19. **10.** $(n-1)u_2-(n-2)u_1$. **11.** $-20/7$. **12.** $5n(3n-7)/8$. **15.** 10,625. **18.** 6. **19.** 1/2. **20.** 17. **21.** 4. **22.** 297. **23.** 4, 84, 164, 244, 324. **24.** 110. **26.** $20(2x-y)$. **29.** 3, 7, 11, 15. **30.** 4, 7, 10; or 10, 7, 4. **31.** First term is 2; common difference 2. **33.** 2. **34.** 10, 7, 4; or 4, 7, 10. **35.** 111 days; 637 men.

## LIX.

**1.** 781. **2.** $5\{1-(-2)^n\}/3$. **3.** $5\frac{473}{729}$. **4.** $72\frac{139}{256}$; 64. **5.** 81/40. **6.** $3(2+\sqrt{2})\{(\sqrt{2})^n-1\}$. **7.** $15\frac{63}{64}$. **8.** $(12+13\sqrt{3})/3$. **9.** 14/99. **10.** 52/165. **11.** $a(b+a)^2\{a^n(b+a)^n-1\}/\{a(b+a)-1\}$. **12.** $\{r^{n+1}-r-1+1/r^n\}/(r-1)$. **13.** $n(3n-1)/2$; $(4^n-1)/3$. **14.** 3, 9, 27, 81, 243. **15.** 5, $5.3^{\frac{3}{5}}$, $5.3^{\frac{6}{5}}$, $5.3^{\frac{9}{5}}$, $5.3^{\frac{12}{5}}$, 135. **16.** 3, $3.5^{\frac{2}{3}}$, $3.5^{\frac{4}{3}}$, $3.5^{\frac{6}{3}}$, $3.5^{\frac{8}{3}}$, $3.5^{\frac{10}{3}}$, 1875. **17.** $-72·9$. **18.** $-3$. **19.** 3367/9. **20.** 2. **21.** $a\{(m+2)r^n$

$+mr^{n-1}-mr-(m+2)\}/2(r-1)$. **23.** $28-21+$etc. **25.** $\{P^{p+q-1}/Q^{p-q-1}\}^{1/2q}$, $\{Q/P\}^{1/2q}$. **34.** $32/3+16/3+$etc.; or $32-16+$etc. **36.** $\frac{1}{3}ABC$. **38.** $(a_1^{3n-3}-a_2^{3n-3})/a_2^{3n-6}(a_1^3-a_2^3)^2$. **40.** $a\{r^{n+1}-(n+1)r+n\}/(r-1)^2$. **42.** $x=1$, $y=1$; or $x=\frac{1}{4}$, $y=-\frac{1}{2}$. **43.** 2, 8, 32, 128; or 135/8, 225/8, 375/8, 625/8. **44.** A. P. 16, 26, 36, 46; G. P. 16, 24, 36, 54.

## LX.

**1.** $21/(11-4n)$. **2.** 7, 63/8, 9. **3.** 32/16, 32/13, 32/10, 32/7, 34/4. **4.** 1/2, 1/6, 1/10, 1/14, 1/18, 1/22. **5.** $-3$. **15.** 6, 54. **17.** 6, 18, 54.

## LXI.

**1.** £14 : 9 : 2; 20. **2.** $A=\frac{1}{3}a$, $B=\frac{1}{2}a+\frac{1}{2}b$, $C=\frac{1}{6}a+\frac{1}{2}b+c$; $\frac{1}{6}n\{2an^2+3(a+b)n+a+3b+6c\}$. **4.** $n(n+1)(4n+5)/6$. **5.** $n(2n+1)(2n-1)/3$. **6.** $a^2n+abn(n-1)+\frac{1}{6}b^2n(n-1)(2n-1)$. **7.** $n(n+1)(n+2)(n+3)/4$. **8.** 22,969; 103. **9.** $-n^2(4n+3)$. **10.** $\frac{1}{6}n(n+1)(2n+1)-\Sigma an(n+1)+n\Sigma ab$. **11.** $na\alpha+\frac{1}{2}(a\beta+b\alpha)n(n+1)+\frac{1}{6}b\beta n(n+1)(2n+1)$. **12.** $C=-1$. **13.** $n/(n+1)$. **14.** $1/15-1/5(5n+3)$. **15.** $n+3/2-1/(n+1)-1/(n+2)$. **16.** $n-19/3-14/(n+2)+40/(n+3)$. **17.** $11/4-3/n^2-2/(n-1)^2$. **18.** $1+1/\sqrt{2}-1/\sqrt{n}-1/\sqrt{(n+1)}$, $1+1/\sqrt{2}$. **19.** $1/2-1/(1+\sqrt{n})$.

## LXII.

**1.** 1, 2. **2.** 1, $-2$. **3.** 3, 10. **4.** 7, 13. **5.** 11, $-12$. **6.** 3, $-4\frac{1}{3}$. **7.** $1\frac{1}{4}$, $-2\frac{1}{3}$. **8.** 3/2, 5/3. **9.** 3/5, $-5/2$. **10.** 1, $3\frac{1}{3}$. **11.** $1\pm\sqrt{3}$. **12.** $3\pm\sqrt{7}$. **13.** $(-11\pm\sqrt{13})/10$. **14.** $(-5\pm\sqrt{2})/3$. **15.** $(25\pm5\sqrt{19})/3$. **16.** $2\pm i$. **17.** $3\pm5i$. **18.** $(-1\pm4i)/2$. **19.** $7\pm i\sqrt{3}$. **20.** $(-5\pm i\sqrt{7})/2$. **21.** $(9\pm2i)/5$. **22.** 11/7, $-7/11$. **23.** 171/2, 11/3. **24.** $(118\pm5\sqrt{517})/3$. **25.** 1, 3. **26.** 7, $-1/3$. **27.** 1, $-b/a$. **28.** $a$, $-1/a$. **29.** $-a+b$, $-b$. **30.** $p+q\pm\sqrt{(p^2+q^2)}$. **31.** $(-a-c\pm b)/3$. **32.** $-ab/(a+b)$, $-(a^2+b^2)/(a+b)$. **33.** $-p/q$, $-q/p$. **34.** $\{-c\pm\sqrt{(2c^2-b^2)}\}/a$. **35.** $\pm m\sqrt{q}/p\sqrt{l}$; if $l=q$, the equation is an identity.

## LXIII.

**1.** $-2$, $-11/3$. **3.** $\pm7$. **4.** $-11/4$; the roots are imaginary. **5.** 1014; 1690. **6.** $p^3-3pq$; $p^6-6p^4q+9p^2q^2-2q^3$. **7.** $x^2-2(b+c)x+2(b^2+c^2)=0$. **8.** Imaginary. **9.** Real; one positive, one negative; the former greater. **10.** Real; both positive. **11.** Real; both negative. **12.** Real; one positive, one negative; the latter greater. **13.** Real; one positive, one negative; the former greater. **14.** Real; one positive, one negative; the former greater. **15.** Real; one positive, one negative; the latter greater. **16.** Equal. **17.** Imaginary. **18.** Real; one positive, one negative; the latter greater. **19.** Real; one positive, one negative; the former greater. **20.** The roots are real; and are rational functions of $\alpha$ and $\beta$. **21.** The roots are rational functions of $\lambda$, with arithmetically rational coefficients. **22.** 0, or $-36/11$. **23.** $1/2$, or $9/2$. **25.** $2/\sqrt{3}$. **26.** $\lambda=(1\pm i\sqrt{2})/3$. **29.** $\lambda=0$; $x=\pm\sqrt{(ab)}$. **31.** $4\{\Sigma a^2b^2-abc\Sigma a\}/(\Sigma a)^2$. **33.** The roots are real when $(2-\sqrt{6})/4 \not> \lambda \not> (2+\sqrt{6})/4$. **35.** $acx^2+b(a+c)x+(a+c)^2=0$. **37.** $a/a'=b/b'=c/c'$. **38.** $b'/a'=-b/a$, $c'/a'=c/a$. **39.** $a+c=0$, or $a-c+4b=0$. **40.** $(ca'-c'a)^2+(bc'+b'c)(ab'+a'b)=0$. **41.** $(cc'-aa')^2=(ba'-b'c)(ab'-c'b)$. **44.** $-1/2$, $-1/2$, $-5/2$, $-5/2$. **45.** $y=\pm 6x-3$.

## LXIV.

**1.** 2, $-4/13$. **2.** $(-11\pm4\sqrt{6})/25$. **3.** $(-1\pm2\sqrt{6}i)/5$. **4.** 1, $17/607$. **5.** 1, $(-1\pm\sqrt{15}i)/2$. **6.** 0, 0, $(5\pm\sqrt{3})/11$. **7.** $\pm3i/2$. **8.** $-1/2$, $-5/2$, $(-3\pm2i)/2$. **9.** $-2$, $1\pm\sqrt{2}i$. **10.** $(3\pm i)/2$, $(3\pm\sqrt{19})/2$. **11.** 1, $-1$, $(-1\pm\sqrt{3}i)/2$. **12.** 0, $-1$, 2. **13.** $1/3$, $(-3\pm5\sqrt{3}i)/6$. **14.** $3/2$, $(3\pm i)/2$. **15.** 0, 0, $-1$, $-1$, $-1\pm\sqrt{2}$. **16.** $(-1+\sqrt{3}\pm\sqrt{2}\sqrt[4]{3})/2$, $(-1-\sqrt{3}\pm\sqrt{2}\sqrt[4]{3}i)/2$. **17.** 1, $(-1\pm\sqrt{5})/2$. **18.** 2, $(5\pm\sqrt{21})/2$. **19.** $-1$, $(-1\pm\sqrt{15}i)/4$. **20.** $-1$, $-1$, $-1$. **21.** 1, $-1/3$, $(-2\pm2\sqrt{2}i)/6$. **22.** $\pm1$, $\pm7$. **23.** $\pm\sqrt{2}/2$, $\pm2\sqrt{3}i/3$. **24.** $\pm\sqrt{14}/2$, $\pm\sqrt{6}i/2$. **25.** $\pm\sqrt{\{(\pm1\pm\sqrt{3}i)/2\}}$, eight solutions. **26.** 2, $1/2$, $-3$, $-1/3$. **27.** $-1$, $-1$, $(-3\pm\sqrt{5})/2$. **28.** $(-3\pm\sqrt{13})/2$, $(-1\pm\sqrt{17})/4$. **29.** 1, $\{-1-\sqrt{5}\pm\sqrt{(10-2\sqrt{5})}i\}/4$, $\{-1+\sqrt{5}\pm\sqrt{(10+2\sqrt{5})}\}/4$. **30.** $(-4\pm\sqrt{22})/2$, $-2\pm\sqrt{33}/3$. **31.** 81. **32.** $-1$, $-1$, $-2$, $-2$. **33.** 2, 7, $(9\pm\sqrt{15}i)/2$. **34.** 2, $-4$, $-1\pm2\sqrt{3}i$. **35.** 0, $-3\pm\sqrt{2}i$. **36.** $-1$, $-1$, $-1\pm i$.

## LXV.

**1.** $\pm\sqrt{6}$. **2.** $-2$. **3.** 1, 2. **4.** 5, $12/37$. **5.** 4, $5/3$. **6.** 10, $-2/5$. **7.** $50/29$. **8.** $(14\pm3\sqrt{10})/2$. **9.** $(-1\pm\sqrt{3})/2$. **10.** $\pm\sqrt{2}i/2$. **11.** $3/7$ $7/3$. **12.** 2·5. **13.** 5, $-5/4$. **14.** 8. **15.** 4. **16.** $\pm1$, $(-1\pm\sqrt{5})/2$. **17.** 1. **18.** $-2$. **19.** $2\pm2\sqrt{3}/3$. **20.** $-10\pm\sqrt{113}$. **21.** 4, $11/6$. **22.** 2, $-1$. **23.** 5, $6/5$. **24.** $-6$. **25.** An identity. **26.** $-4\pm2\sqrt{2}$. **27.** $-3/2$. **28.** $3/2$. **29.** 0, $3/2$. **30.** $(-5\pm\sqrt{5})/4$. **31.** 4, $-1$, $(6\pm\sqrt{14})/4$.

## LXVI.

**1.** $(-1\pm\sqrt{3}i)a/2$. **2.** $(c^2-ab)/(a+b-2c)$; if $a=b$ the equation is an identity. **3.** 0, $\pm\sqrt{[\{ab(c+d)-cd(a+b)\}/\{a+b-c-d\}]}$. **4.** 0, $(a^2+b^2)/(a+b)$. **5.** $pq(p+q-2r)/\{p^2+q^2-r(p+q)\}$. **6.** $\pm(a-b)\sqrt{(b/a)}$. **7.** 0, $\pm\sqrt{(\pm i)}a$. **8.** 0, $-\Sigma p/3$. **9.** $[\Sigma bc\pm\sqrt{\{\Sigma b^2c^2-abc\Sigma a\}}]/\Sigma a$. **10.** $\{1\pm\sqrt{(1+4a)}\}/2$. **11.** $\pm4a$. **12.** $3(b-c)(c-a)(a-b)/2(\Sigma a^2-\Sigma bc)$. **13.** $\pm\frac{1}{3}\sqrt{3}\sqrt{\{\Sigma a^2\pm\sqrt{(\Sigma a^4-\Sigma a^2b^2)}\}}$. **14.** 0 and the roots of $(a+b+c+d)x^2-2(a+b)(c+d)x+(a+b)cd+(c+d)ab=0$. **15.** $\{-2\Sigma a\pm\sqrt{(\Sigma a^2-\Sigma bc)}\}/3$. **16.** $\Sigma ab/\Sigma a$, 0. **17.** $(4a^2+6ab-b^2)/(8a-b)$, assuming $a \neq b \neq 0$. **18.** $\{2\Sigma a \pm \sqrt{(2\Sigma a^2 + \Sigma ab)}\}/3$. **19.** The roots of $(3 - \Sigma a^2)x^2 + (\Sigma a^2 b - 2\Sigma a)x + \Sigma bc - abc\Sigma a = 0$. **20.** One root is obviously $x = a + b + c$; to solve completely, put $y = x - a - b - c$, and solve for $y$. **21.** $-\{bd(ce + af) + cf(bc + ad)\}/ac(bd + cf)$.

## LXVII.

*N.B.*—No extraneous solutions are given, and no solution which is imaginary or renders any of the irrational functions in the original equation imaginary.

**1.** 16. **2.** 7. **3.** $-2$. **4.** 5. **5.** $(\sqrt{17} - 1)/4$. **6.** $-5$. **7.** No solution of the kind required. **8.** $-5$. **9.** 3/2. **10.** $-1$. **11.** 2. **12.** $2(\sqrt{7} - 1)/3$. **13.** $\sqrt{(2\sqrt{3} - 3)a}$. **14.** $\{9 \pm \sqrt{(57 + 24\sqrt{7})}\}/6$. **15.** 4, 9/4. **16.** 3, 27/7. **17.** $\sqrt{(1 + 2\sqrt{7})}$. **18.** $\sqrt[4]{(4/3)}$. **19.** 0, $\pm 2\sqrt{2}$. **20.** $(a^2 - b^2)/2\sqrt{(2a^2 + 2b^2)}$. **21.** $(ps - qr)/\{a(r - s) + b(q - p)\}$. **22.** 1. **23.** 3. **24.** 0. **25.** $\sqrt{(5/3)}$. **26.** 1, $-8$. **27.** 1. **28.** 4, $-1$. **29.** 2. **30.** $(\Sigma a^2 - \Sigma ab)x^2 + (6abc - \Sigma a^2 b)x + \Sigma a^2 b^2 - abc\Sigma a = 0$. **31.** $(y - z)^2 + (z - x)^2 + (x - y)^2 = 0$.

## LXVIII.

**1.** (10, 2), (10, $-2$). **2.** (1, 1/4). **3.** (6, 2), ($-8$, 2). **4.** (1, 2). **5.** (1/2, 1/6); the other solution is not finite. **6.** $\{\sqrt{(2419/7)}i, -28/323\}$, $\{-\sqrt{(2419/7)}i, -28/323\}$. **7.** (3, 1/7), (1/7, 3). **8.** (2, 1), ($-1$, $-2$). **9.** (1, 2), (1, 2). **10.** (3, 5), ($-11/9$, $-23/3$). **11.** (9, 2), ($-16/11$, $-27/44$). **12.** (7, $-3$), (3/2, $-29/6$). **13.** (3, $-4$), ($-321/94$, 1031/94). **14.** (5, 2), ($-43/9$, $-26/9$). **15.** (3, $-2$), ($-2$, 3). **16.** $\{(1 + \sqrt{3}i)a/2, (1 - \sqrt{3}i)b/2\}$, $\{(1 - \sqrt{3}i)a/2, (1 + \sqrt{3}i)b/2\}$. **17.** $\{(1 + \sqrt{5})/2a, (1 - \sqrt{5})/2b\}$, $\{(1 - \sqrt{5})/2a, (1 + \sqrt{5})/2b\}$. **18.** $(a, b)$, $(2a - b, 2b - a)$. **19.** ($-1$, 4), ($-1/6$, 2/3), (0, 2/3), ($-1$, 0). **20.** ($-35/2$, 25/2). **21.** $\{\sqrt{3}i/3, (3 + \sqrt{3}i)/6\}$, $\{-\sqrt{3}i/3, (3 - \sqrt{3}i)/6\}$; the solution is partly indeterminate, viz. it contains all the pairs of values of $x$ and $y$ that satisfy $x - y = 0$. **22.** (0, 0) five times over, and (2/3, $-1/3$). **23.** $(a, a)$, $(a + b, a - b)$. **24.** $[(ab' - a'b)\{1/2(a - a') - 1/2(b - b')\}, (ab' - a'b)\{1/2(a - a') + 1/2(b - b')\}]$, (0, 0). **25.** $\{a - (3a + 2b)/3c, b - (3a + 2b)/2c\}$. **26.** (11/2, 3/2), (2, 5), ($-2$, $-6$), ($-11/2$, $-5/2$). **27.** $(m, 1)$, $(-2, -1 - m)$. **28.** $[(1 + ab + \sqrt{\{(1 - a^2)(1 - b^2)\}})/(a + b), (1 - ab + \sqrt{\{(1 - a^2)(1 - b^2)\}})/(a - b)]$, $[(1 + ab - \sqrt{\{(1 - a^2)(1 - b^2)\}})/(a + b), (1 - ab - \sqrt{\{(1 - a^2)(1 - b^2)\}})/(a - b)]$. **29.** (0, 0), (31/46, 31/24). **30.** (1, 3), (14/5, 15/7). **31.** (5/2, 1), ($-5/2$, $-1$), (3/2, 5/3), ($-3/2$, $-5/3$).

## LXIX.

*N.B.*—For the sake of brevity, "ambiguous signs" ($\pm$) are used in giving the solutions in LXIX. and LXX. In general it is to be understood that the upper signs of the ambiguities are to be taken together, and the lower together; when the contrary is the case, the number of solutions is stated. Thus "$(\pm 1, \pm 1)$" means $(+1, +1)$ and $(-1, -1)$; but "$(\pm 1, \pm 1)$, four solutions," means $(+1, +1)$, $(+1, -1)$, $(-1, +1)$, $(-1, -1)$.

**1.** $\{(p^2-1)^2/4p^2, 0\}$. **2.** $\{-3c/(3a\pm\sqrt{5b}), \mp\sqrt{5c}/(3a\pm\sqrt{5b})$; $x$ must be positive. **3.** $(5125/98, 4675/98)$. **4.** $a=-1615/21$, $b=-815/9$. **5.** $(\pm 5, \pm 4)$, $(\pm\sqrt{2}/2, \mp 9\sqrt{2}/2)$. **6.** $(\pm 1, \pm 2)$; the other two solutions are not finite. **7.** $(\pm 2, \pm 1)$, $(\pm\sqrt{3}/3, \mp 5\sqrt{3}/3)$. **8.** $(\pm 2, \pm 3)$; the other two solutions are not finite. **9.** $(\pm 5, \pm 6)$, $(\pm 17\sqrt{10}/5, \mp 9\sqrt{10}/20)$. **10.** $(1/2, 1/3)$, $(1/15, -1/10)$. **11.** $(\pm 7, \pm 1)$, $(\pm i \mp 7i)$. **12.** $(\pm 5, \pm 2)$, $(\pm 7/\sqrt{17}, \mp 16/\sqrt{17})$. **13.** $(\pm 5, \pm 3)$, $(\pm 3, \pm 5)$. **14.** $(\pm 3, \pm 2)$, $(\pm 5/2, \pm 3/4)$. **15.** $(\pm 3, \mp 1)$, $(\pm 31/\sqrt{197}, \mp 1/\sqrt{197})$. **16.** $(6, 5)$, $(-5, -6)$. **17.** $(3, 5)$, $(5, 3)$, $\{(-15\pm\sqrt{139}i)/4, (-15\mp\sqrt{139}i)/4\}$. **18.** $(1, 2)$, $(1, 3)$, $\{(1\pm\sqrt{7}i)/2, 1\}$. **19.** $(0, 0)$, $(6, 3)$, $(9\pm\sqrt{3}i, -3\pm 2\sqrt{3}i)$. **20.** $(4, 1)$, $(-1, -4)$, $\{\pm\sqrt{(377/60)}i-5/2, \pm\sqrt{(377/60)}i+5/2\}$. **21.** $(\pm 5, \pm 3)$, $(\pm 3, \pm 5)$. **22.** $(1, 3)$, $(3, 1)$, $(-2\pm\sqrt{17}, -2\pm\sqrt{17})$. **23.** $(4, 2)$, $(2, 4)$, $\{(1\pm\sqrt{11}i)/2, (1\mp\sqrt{11}i)/2\}$. **24.** $[\{3a\pm\sqrt{(12b^2-3a^2)}\}/6, \{3a\mp\sqrt{(12b^2-3a^2)}\}/6]$. **25.** $(1/2, 1/2)$, $(1/2, 1/2)$. **26.** $\{2\pm\sqrt{3}, (7\pm 3\sqrt{5})/2\}$, $\{(7\pm 3\sqrt{5})/2, 2\pm\sqrt{3}\}$, $\{-2\pm\sqrt{3}, (-7\pm 3\sqrt{5})/2\}$, $\{(-7\pm 3\sqrt{5})/2, -2\pm\sqrt{3}\}$. **27.** $(1, 1)$, $\{(-3\pm\sqrt{15}i)/4, (-3\pm\sqrt{15}i)/4\}$, $(\pm\sqrt{3}, \mp\sqrt{3})$. **28.** $x=y=\pm\sqrt{[\{-1\pm\sqrt{(1+4a)}\}/2]}$, four solutions; $x=-y=\pm\sqrt{[\{1\pm\sqrt{(1-4a)}\}/2]}$, four solutions, $[\{\sqrt{(a+2)}\pm\sqrt{(a-2)}\}/2, \{\sqrt{(a+2)}\mp\sqrt{(a-2)}\}/2]$, $[\{-\sqrt{(a+2)}\pm\sqrt{(a-2)}\}/2, \{-\sqrt{(a+2)}\mp\sqrt{(a-2)}\}/2]$. **29.** The system is equivalent to $x^2-y^2=0$, $x^3y+x^2=a$, together with $xy+1=0$, $x^3y+x^2=a$; the former gives $x=y=\pm\sqrt{[\{-1\pm\sqrt{(1+4a)}\}/2]}$, four solutions; and $x=-y=\pm\sqrt{[\{1\pm\sqrt{(1-4a)}\}/2]}$; the latter has no finite solutions. **30.** $x=y=(1\pm\sqrt{101})/10$; $x=-1/y=(1\pm\sqrt{101})/10$. **31.** $(\pm 3, \pm 2)$, four solutions; $(\pm 2i, \pm 3i)$, four solutions; $[\pm\sqrt{\{(5+2\sqrt{61}i)/2\}}, \pm\sqrt{\{(-5+2\sqrt{61}i)/2}]$, four solutions; $[\pm\sqrt{\{(5-2\sqrt{61}i)/2\}}, \pm\sqrt{\{(-5-2\sqrt{61}i)/2}]$, four solutions. **32.** $(1, 0)$, $(0, -1/2)$.

## LXX.

**1.** $(-5, -1, 3)$, $(41/7, 41/35, -123/35)$. **2.** $(4, 6, 8)$, $(-6, -8, -10)$. **3.** $(\pm 27/\sqrt{4629}, \mp 87(\sqrt{4629}, \mp 63/\sqrt{4629})$. **4.** $(\pm 3, \pm 5, \pm 8)$. **5.** $(\pm\sqrt{2i}, \mp 2\sqrt{2i}, \pm\sqrt{2i})$. **6.** $(11/4, 2/5, 11/3)$, $(-19/4, -22/5, -29/3)$. **7.** The real positive solution, $a$, $b$, $c$ being supposed all real and positive, is $a(a^2/bc)^{1/(n-1)}$, $b(b^2/ca)^{1/(n-1)}$, $c(c^2/ab)^{1/(n-1)}$. **8.** $[\pm\sqrt{\{(a+b)(a+c)/2(b+c)\}}, \pm\sqrt{\{(b+c)(b+a)/2(c+a)\}}, \pm\{(c+a)(c+b)/2(a+b)\}]$. **9.** $x=\rho bc(b-c)d/\sqrt[3]{\{\Sigma b^3c^3(b-c)^3\}}$, $y=\rho ca(c-a)d/\sqrt[3]{\{\Sigma c^3a^3(c-a)^3\}}$, $z=\rho ab(a-b)d/\sqrt[3]{\{\Sigma b^3c^3(b-c)^3\}}$, where $\rho=1$, or $(-1+\sqrt{3}i)/2$, or $(-1-\sqrt{3}i)/2$. **10.** $(0, 0, 0)$, $(1, 1, 1)$, $(1, -1, -1)$, $(-1, 1, -1)$, $(-1, -1, 1)$; the system is partly indeterminate, being satisfied by $x=$any finite quantity, $y=0$, $z=0$, etc. **11.** $(\pm a/\sqrt{2}, \pm a/\sqrt{2}, \pm a/\sqrt{2})$. **12.** $(\pm 6, \pm 7, \pm 8)$. **13.** $(12/7, 12/5, -12)$. **14.** One solution is obvious by inspection, viz. $x=1/lm$, $y=1/mn$, $z=1/nl$; put $x=1/lm+\xi$, $y=1/mn+\eta$, $z=1/nl+\zeta$, and we find $\xi=(lm-n^2)\rho$,

$\eta=(mn-l^2)\rho$, $\zeta=(nl-m^2)\rho$, where $\rho=2(lmn\Sigma l-\Sigma l^2m^2)/(\Sigma l^4m^4-2l^2m^2n^2\Sigma lm+l^2m^2n^2\Sigma l^2)$. **15.** $(-40/7, 24/5, 7/6)$. **16.** $(0, 0, 0)$, $\{(2a-b-c)/3$ $(2b-c-a)/3$, $(2c-a-b)/3\}$. **17.** $x=\pm\delta/(\pm\alpha\pm\beta\pm\gamma)$, $y=\pm\delta/(\pm\alpha\pm\beta\pm\gamma)$, $z=\pm\delta/(\pm\alpha\pm\beta\pm\gamma)$; where the ambiguities are to be taken, so that the sign of $\alpha$ is the same as the sign of the numerator of $x$, the sign of $\beta$ the same as the sign of the numerator of $y$, and the sign of $\gamma$ the same as the sign of the numerator of $z$, but are otherwise unrestricted. **18.** $(\pm 2, \pm 3, \pm 5)$, eight solutions; $(\pm 2i, \pm 3i, \pm 5i)$, eight solutions. **19.** $(\pm\sqrt{6}, \mp\sqrt{6}/2, \mp\sqrt{6}/2)$. **20.** $(\pm 1, \pm 1, \pm 1)$. **21.** $\{a\pm 2abc/\sqrt{(2a^2b^2+2a^2c^2-2b^2c^2)}$, $b\pm 2abc/\sqrt{(2b^2c^2+2b^2a^2-2c^2a^2)}$, $c\pm 2abc/\sqrt{(2c^2a^2+2c^2b^2-2a^2b^2)}\}$, eight solutions. **22.** $(\pm 4, 2, 3)$, $(\pm 4, 3, 2)$, $\{\pm\sqrt{6}, (15+\sqrt{161})/2, (15-\sqrt{161})/2\}$, $\{\pm\sqrt{6}, (15-\sqrt{161})/2, (15+\sqrt{161})/2\}$. **23.** $x=y=(-11\mp\sqrt{233})\sqrt{(19\pm\sqrt{233})/16}$, $z=\sqrt{(19\pm\sqrt{233})/2}$; $x=y=-(-11\mp\sqrt{233})\sqrt{(19\pm\sqrt{233})/16}$, $z=-\sqrt{(19\pm\sqrt{233})/2}$; $(2, 1, 3)$, $(1, 2, 3)$, $(-2, -1, -3)$, $(-1, -2, -3)$. **24.** The solutions are evidently the roots of the equation $u^3-10u^2+31u-30=0$ taken in any order, viz. $(2, 3, 5)$, $(2, 5, 3)$, $(3, 2, 5)$, $(3, 5, 2)$, $(5, 2, 3)$, $(5, 3, 2)$.

## LXXI.

**1.** 2; 3. **2.** 42. **3.** 13, 15. **4.** 85. **5.** $2a^2+3a+1$. **6.** 5 and $4\frac{1}{6}$ miles an hour. **7.** 18. **8.** £10. **9.** 72. **10.** 20. **11.** 15. **12.** 16. **13.** 50. **14.** 8. **15.** 12. **17.** 64 g. **18.** 50 miles. **19.** $q+1$. **20.** $(\sqrt{105}-3)/2=3{\cdot}624\ldots$ **21.** A 4, and B $3\frac{3}{4}$ miles an hour. **22.** $ab\sqrt{(a^2+b^2)}/(a^2+ab+b^2)$.—In (24)–(27), if $x$ be the distance of P from the middle point of AB, and $2a$ the length of AB, the values of $x$ are as follows, positive values corresponding to positions of P in the right half of AB, negative values to positions in the left half. **24.** $\pm\sqrt{(a^2-c^2)}$. **25.** $[-a(l-m)\pm\sqrt{\{(l+m)c^2-4lma^2\}}]/(l+m)$. **26.** $[-a(l+m)\pm\sqrt{\{(l-m)c^2+4lma^2\}}]/(l-m)$. **27.** $a/2$; $-a$. **28.** If $x=\overline{AH}$, then $x=a\{-1\pm\sqrt{(4n+1)}\}/2n$. **29.** $(\sqrt{3}\pm 1)d/2$, $d$ being the length of the diagonal. **30.** $2(\sqrt{13}-2)/3=1{\cdot}07$ inches. **31.** $(11\pm\sqrt{7})/2=4{\cdot}18$ or $6{\cdot}82$ miles; in the second case they pass the crossing. **32.** If $m=c^2/ab$, and the distance of P from one end of the diagonal be $x$ times the diagonal, then $x=\{1\pm\sqrt{(2m-1)}\}/2$. **33.** $(2\sqrt{3}-3)a$. **34.** $(\sqrt{2}-1)r$. **35.** The distances of the point from the projections of A and B are 3 and 6 respectively. **36.** $\{h+d-\sqrt{(2hd)}\}/2$. **37.** The distance of P from the middle point of AB is $\pm\sqrt{(1-4m)}c/2$, where $m$ is the ratio of the given area to twice the area of the triangle. **38.** If $x$ be the distance of the required point from the point midway between the projections, $x$ is given by $4(c^2-d^2)x^2+4c(a^2-b^2)x+d^4-(2a^2+2b^2+c^2)d^2+(a^2-b^2)^2=0$. **39.** $AP=[a+c-\sqrt{\{(a+c)^2-4bc\}}]/2$. **40.** $OP=\sqrt{\{(d^2-r^2)/n\}}$.

## LXXII.

**1.** The radii of the circles are $[(2-\sqrt{2})a\pm\sqrt{\{2b^2-(6-4\sqrt{2})a^2\}}]/2$, and $[(2-\sqrt{2})a\mp\sqrt{\{2b^2-(6-4\sqrt{2})a^2\}}]/2$. **2.** The sides are $\sqrt{3}\{1\pm\sqrt{(1-4m)}\}a/4$ and $\{1\mp\sqrt{(1-4m)}\}a/2$, where $m=2c^2/\sqrt{3}a^2$. **3.** The sides are $\{1 \pm \sqrt{(1-4m)}\}\ a/2$, $\{1 \mp \sqrt{(1-4m)}\} b/2$, where $m = c^2/ab$. **4.** 4860 sq. yds. **5.** 24. **6.** $a\{b+2-\sqrt{(b^2+4)}\}/2b$ and $a\{b-2+\sqrt{(b^2+4)}\}/2b$. **8.** If $c$ be the hypotenuse, and $d^2$ four times the area, the sides are $\{\sqrt{(c^2+d^2)}+\sqrt{(c^2-d^2)}\}/2$, $\{\sqrt{(c^2+d^2)}-\sqrt{(c^2-d^2)}\}/2$. **9.** $(\sqrt{5}-1)c/2$, $\sqrt{(2\sqrt{5}-2)}c/2$. **10.** $(5 \pm \sqrt{7})a/8$, $(5 \mp \sqrt{7})a/8$; or $(3 \pm \sqrt{7})a/8$, $(3 \mp \sqrt{7})a/8$. **11.** If $d$ be the diameter of the circle, and $c^2$ the given area, the sides are $\{\sqrt{(d^2+2c^2)}+\sqrt{(d^2-2c^2)}\}/2$, $\{\sqrt{(d^2+2c^2)}-\sqrt{(d^2-2c^2)}\}/2$. **12.** A, $\sqrt{a}(\sqrt{a}+\sqrt{b})$; and B, $\sqrt{b}(\sqrt{a}+\sqrt{b})$ hours. **13.** The problem is impossible, unless $a^2-b^2=4$; and then there are an infinity of solutions; we have merely to take $x/y=(a \pm b)/2$. **14.** 3, 17. **15.** 120, 80. **16.** 36 eggs, 52 eggs; 3s. 3d. and 2s. 3d. per doz. **17.** Eggs $a\sqrt{b}/(\sqrt{b}+\sqrt{c})$, $a\sqrt{c}/(\sqrt{b}+\sqrt{c})$; prices $12\sqrt{c}(\sqrt{b}+\sqrt{c})/a$, $12\sqrt{(b}\sqrt{b}+\sqrt{c})/a$. **18.** The equations for $x$ and $y$ are $x^2-y^2-ax+by=0$, $2xy-ay-bx+c^2=0$; the $x$-resultant of this system is $4(x^2-ax)^2+(a^2+b^2)(x^2-ax)+c^2(ab-c^2)=0$, which is obviously soluble by means of quadratics. When $a=b$, the values of $x$ and $y$ are $\{a \pm \sqrt{(a^2-2c^2)}\}/2$, $\{a \mp \sqrt{(a^2-2c^2)}\}/2$; and $\{a \pm \sqrt{(2c^2-a^2)}\}/2$, $\{a \mp \sqrt{(2c^2-a^2)}\}/2$; these two solutions are obviously not both real in any given case. **19.** The sides of the excised rectangle are $\frac{1}{2}\{d+\sqrt{(4a^2-4ab+d^2)}\}$, $\frac{1}{2}\{d-\sqrt{(4a^2-4ab+d^2)}\}$. In order that the problem may be possible, these values must be both real and both positive, and less than $a$ and $b$ respectively. **20.** £200, £300. **21.** The sides of the corner are $3a/5$, $4a/5$. **22.** $\sqrt{(ace/bd)}$, $\sqrt{(bda/ce)}$, $\sqrt{(ceb/da)}$, $\sqrt{(dac/eb)}$, $\sqrt{(ebd/ac)}$. **23.** $d(\sqrt{b}+\sqrt{a})/(\sqrt{b}-\sqrt{a})$ miles. **24.** Distance $46\frac{2}{3}$ miles, and rates of A, B, C 4, 7, and 10 miles an hour respectively; or distance 30 miles, and rates 3, 6, 10 miles an hour. **25.** 13, 5.

## LXXIII.

**1.** 30,240; 6048. **2.** 1,814,400; 3360; 201,801,600. **3.** 30,240. **4.** $n=5$. **5.** $r(n-1)(n-2)\ldots(n-r+1)$. **6.** 1 : 11. **7.** $n=2r$. **8.** 210. **9.** 300. **10.** $n=2r$. **11.** $(2n)!$; $(n!)^2$. **12.** $n=s^2+s$. **13.** $(n-1)!/2$. **14.** 210. **15.** ${}_5P_3 - {}_4P_2 + 4 = 52$. **16.** ${}_7P_7 - 2{}_6P_6 + {}_5P_5$. **17.** $3^n$. **18.** $2^n n!$; $2^n(n-1)!$. **19.** $m!$. **20.** ${}_9P_1 \times {}_8P_8 \times {}_5P_5 = 362,880$. **21.** $21^5.5^4$. **22.** $4^5$. **23.** $(4!)^2=576$, if the vowels and consonants be given, otherwise ${}_{21}C_4 \times {}_5C_4 \times (4!)^2 = 17,236,800$. **24.** $4!3!=144$. **25.** 780. **26.** $(m+1)^n$. **27.** $5!6!=86,400$. **28.** $15^3$; $15!$. **29.** $(2n)!$. **30.** ${}_8C_2=28$; ${}_8C_3 3!/2=56$; ${}_8C_4 3!/2=210$; ${}_8C_5 4!/2=672$; ${}_8C_6 5!/2=1680$; ${}_8C_7 6!/2=2880$. **31.** ${}_8P_4 \times {}_6P_3 \times 4! \times 3! = 29,030,400$. **32.** $(2m+n)!/m!(m+n)!$. **33.** $3p^3$; $3!(p!)^3$. **34.** $2^5=32$; $5!/2!3!=10$. **35.** 25,401,600.

## LXXIV

**1.** $n=6$. **2.** ${}_{100}C_3=161,700$. **3.** ${}_{10}C_3 : 2{}_{10}C_4 = 2:7$. **4.** $n=12$, $r=4$. **5.** ${}_{n-s}C_{r-s}r!$. **6.** ${}_8C_4+{}_9C_6=154$. **8.** $n=2r$. **10.** ${}_4C_3 \times 8 = 32$. **11.** $n^2(2n-1)$. **12.** ${}_7C_4=35$. **13.** ${}_{10}C_2 \times {}_5C_1 \times 3! = 1350$. **14.** ${}_{11}C_2 \times 9! = 19,958,400$. **15.** ${}_{21}P_4 \times {}_5C_3 \times {}_5P_3 = 86,184,000$. **16.** ${}_nC_2 - n = n(n-3)/2$.

**18.** ${}_{20}C_{10}=184{,}756$ days, *i.e.* over 505 years. **19.** $3r-4$. **20.** ${}_{10}C_5=252$. **21.** $l!m!n!/p!q!r!(l-p)!(m-q)!(n-r)!$. **22.** ${}_4C_2\times{}_{10}C_9+{}_4C_3\times{}_{10}C_8+{}_4C_4\times{}_{10}C_7=360$. **23.** 18,480. **24.** $rs+2$. **25.** $lmn+\frac{1}{2}\Sigma l^2m-\Sigma lm$. **26.** ${}_4C_2 3!3!+{}_4C_1 3!3!/2!=288$. **27.** ${}_3C_1\times{}_6C_2\times{}_9C_3=3780$; ${}_3C_1\times{}_6C_2\times{}_9C_3\times{}_2C_1\times{}_4C_2\times{}_6C_3\times{}_1C_1\times{}_2C_2\times{}_3C_3=907{,}200$. **29.** $12!/(4!)^3=34{,}650$. **30.** ${}_nC_m m!$; ${}_nC_m$. **31.** $(nr)!/(r!)^n$. **33.** $1+{}_4C_1 4!/3!+{}_4C_2 4!/2!+{}_4C_3 4!/1!+{}_4C_4 4!=209$. **34.** $2^6-1=63$. **36.** $12!/(2!)^6=7{,}484{,}400$; $12!/4!(2!)^4=1{,}247{,}400$; $12!/(4!)^2(2!)^2!=207{,}900$; $12!/4!(2!)^2=4{,}989{,}600$. **38.** If $u_n$ denote the number of selections $+1$ (corresponding to a selection of 0 things), then it is easily shown that $u_n=2u_{n-1}$; whence both results readily follow. **39.** 28, if he must give all his six votes; 84, if he need not give them all. **40.** $\frac{1}{8}n(n-1)(n-2)(n-3)(n-4)$. **41.** $\Sigma n_1 n_2 \ldots n_s$.

## LXXV.

**1.** $-684x^2y^{17}$. **2.** $10{,}560y^3x^8$. **3.** $-2268$; 1. **4.** 326,592. **5.** $(-1)^r n!\{n^2-(4r-3)n+4r^2-8r+2\}/r!(n-r+2)!$. **6.** Follows from the identity $(1+x)^{m+n}\equiv(1+x^m)(1+x^n)$. **9.** 130. **10.** 360; 120; 360; 720; 1260; 5; 7. **11.** 7. **13.** Follows from the identity $(1-x^2)^n\equiv(1-x)^n(1-x)^n$. **14.** $\Sigma a^7+7\Sigma a^6b+21\Sigma a^5b^2+42\Sigma a^5bc+35\Sigma a^4b^3+105\Sigma a^4b^2c+140\Sigma a^3b^3c+210\Sigma a^3b^2c^2$. **15.** 252. **16.** $\Sigma a^6+6\Sigma a^5b+15\Sigma a^4b^2+30\Sigma a^4bc+20\Sigma a^3b^3+60\Sigma a^3b^2c+120\Sigma a^3bcd+90\Sigma a^2b^2c^2+180\Sigma a^2b^2cd$. **17.** $-480$. **18.** 25,920. **19.** 627. **20.** $-291{,}368$. **21.** $-565{,}180$. **22.** 44,803.

## LXXVI.

**4.** $\sum_{abc} a^3+3(b+c)(c+a)(a+b)\equiv\Sigma a^3$; each of the four $=-(b+c)(c+a)(a+b)$. **5.** $9a+3b+c=d$. **7.** $\{(a^3-b^3)^2/b^3\}^2=(b^3+c^3-2a^3)^2$. **8.** 1. **9.** $\Sigma x^2=\Sigma xy$. **10.** $ab+3\lambda(a+b)=0$. **12.** The given relation may be written $(x-y)^2+(z-u)^2+(y-z)^2=0$. **14.** The converse is $a:b=c:d$, or $a:b=d:c$. **16.** The converse is $a:b=c:d$, or $a:b=-c:d$. **17.** The given relation is $\frac{1}{2}(1+x-y)\{(x-1)^2+(x+y)^2+(y+1)^2\}=0$. **22.** The given condition gives $y^2+z^2-x^2=-2yz$, etc. **23.** $\Sigma A^4-2\Sigma A^2B^2$ contains the factor $\Sigma A$. **24.** $\Sigma a/(1+a)=1$. $A=l(amn-hnl+fl^2-glm)$, etc.; $D=l(bn^2+cm^2-2fmn)$, etc. **29.** The given condition leads to $y^3+z^3-3xyz=-x^3$, etc. **30.** $y+z=0$. **31.** $(a^2+b^2)^2d^4+2ab(c^2a^2+c^2b^2-2a^2b^2)d^2+a^2b^2(a^2-c^2)(b^2-c^2)=0$. **34.** $a^3-2ab^2+c^3=0$. **35.** $3\Sigma a^2b-2\Sigma a^3-12abc\equiv\Pi(b+c-2a)$. **37.** If $u\equiv\Sigma x$, $v=\Sigma xy$, $w=xyz$, the system becomes $u=0$, $uv-w=a$, $u^3+3uv+3w=b$, *i.e.* $u=0$, $w=-a$, $3w=b$, etc. **38.** On eliminating $x$, we get $\{(a-b)y-(c-a)z\}^2=0$, etc. **41.** The given equations lead to $x=a$, $y=-z$; or $y=b$, $z=-x$, or $z=c$, $x=-y$, etc. **42.** The given condition leads to $\Sigma x^2+\Sigma xy=0$, *i.e.* $\frac{1}{2}\Sigma(y+z)^2=0$, etc. **43.** $\Sigma xy=0$. **45.** Put $\Sigma a^2l^2=s$, then $l^2=-(bc+s)/(c-a)(a-b)$, etc. Then calculate $\Sigma m^2n^2bc(b-c)^2$ in terms of $s$, $a$, $b$, $c$. **47.** $1:1:1$ and all the permutations of $-2\pm\sqrt{2}:1:1$.

## LXXVIII.

**22.** $y=9-\frac{8}{9}(x-3)^2$. **23.** $y=3(x+1)(2x-1)$; turning value $-27/8$, a minimum.

## LXXIX.

**2.** $1-2x+2x^2$; $<3/10^6$. **3.** $1-x-6x^2$; $<2/10^6$. **4.** $1-2x+4x^2$; $<8/10^6$. **5.** $1-x-x^2$; $<4/10^6$. **6.** $\frac{1}{2}-\frac{1}{4}x+\frac{3}{8}x^2$; $<4/10^7$. **7.** $1+2x+2x^2$; $<3/10^6$. **8.** $1+3x+6x^2$; $<1/10^5$. **9.** $1+6x+18x^2$; $<4/10^5$.

## LXXXI.

**1.** 1·47. **2.** −2·89, −1·25, 2·06. **3.** −1·77. **4.** 1·13. **7.** 1·42. **8.** One between −2 and −1; another between 0 and 1. **9.** One between −3 and −2; another between 0 and 1. **10.** Minimum when $x=y=2$; maximum when $x=y=-2$. **11.** The minimum value is 120, corresponding to $x=4$, $y=6$. **12.** $AB^2$ and $\frac{3}{4}AB^2$. **14.** The vertices of the minimum square bisect the sides of the given square. **15.** When the trapezium is a parallelogram. **16.** When P bisects AB. **17.** When PAQ is isosceles. **18.** Maximum when $x=y=\pm 5$; minimum when $x=-y=\pm 5$. **19.** Strictly speaking, there are no turning values. **20.** Maximum when $\sqrt{2}x=\sqrt{5}y=\pm 5$; minimum when $\sqrt{2}x=-\sqrt{5}y=\pm 5$. **21.** When the triangle is isosceles. **22.** When the triangle is isosceles. **23.** The height of the maximum rectangle is half its base. **24.** When the edges of the corner are parallel to the sides of the page, the ratio is a maximum; the distance is then $\sqrt{2}d$, where $d$ is the distance of the line of print from the top of the page. **25.** $AD=(2-\sqrt{2})c/4$. **26.** When $CL=ab^2/(a^2+b^2)$, $CM=a^2b/(a^2+b^2)$.

THE END

*Printed by* R. & R. CLARK, LIMITED, *Edinburgh*.

# Black's Literary Epoch Series

# NINETEENTH-CENTURY PROSE

BY J. H. FOWLER, M.A.

ASSISTANT MASTER AT CLIFTON COLLEGE.

*Small Crown* 8*vo*, 135 *pp*. *Price* 1*s*. *net*. *Bound in Cloth.*

CONTENTS.—General Introduction.—COLERIDGE, First Literary Impressions.—DE QUINCEY, The Vision of Sudden Death.—MACAULAY, The Siege of Namur.—CARLYLE, The Election of Abbot Samson.—THACKERAY, The Last Years of George III.—RUSKIN, The Lamp of Memory.

## OPINIONS OF THE PRESS.

"Mr. Fowler is thoroughly pertinent in his remarks, and expert in seeing what is wanted by the precise stage of mind that has to be dealt with."—*Manchester Guardian.*

"The Editors . . . have given us worthy work."—*Methodist Times.*

"The typical writers dealt with in the volume of prose are Coleridge, De Quincey, Macaulay, Carlyle, Thackeray, and Ruskin. The author has gone to work in a business-like manner, and has succeeded in producing a book of inestimable value to the student, and one which will be fully appreciated by the teacher."—*Secondary Education.*

"The delicacy of literary touch shown in Mr. Fowler's selections, his criticisms, and his notes, make an excellent book."—*Education.*

"In the hands of an experienced teacher each extract might form the basis of an interesting hour's talk, which would send his pupils to the authors themselves."—*Guardian.*

"The end sought to be attained is a very high one, and it may be said unhesitatingly that the class which studies Mr. Fowler's volume will derive therefrom not merely general literary principles, but also the power of reading critically some of our chief prose writers. . . . The book is distinctly good."—*Yorkshire Post.*

"The selections are good and the criticism far beyond what would be expected in so modest a manual."—*Educational Review.*

"The method adopted is to print selected passages of six poets and prose writers with brief notes on biography and general and technical criticism, and seems sensible and well carried out on the whole."—*Athenæum.*

"The pieces for analysis have been judiciously chosen, and the judgments expressed imply wide catholicity of taste, and are remarkably free from bias. These books are calculated to induce in those who use them an intelligent appreciation and a genuine love of what is best in our literature."—*Publishers' Circular.*

"We welcome this aid towards the proper teaching of English literature."—*Preparatory Schools' Review.*

A. & C. BLACK, SOHO SQUARE, LONDON.

www.ingramcontent.com/pod-product-compliance
Lightning Source LLC
Chambersburg PA
CBHW020905210326
41598CB00018B/1777
*9783744645805*